AF811356

Electro-Fermentation for Synthesis of Chemicals, Biochemicals and Biofuels

Surajbhan Sevda
Editor

Electro-Fermentation for Synthesis of Chemicals, Biochemicals and Biofuels

Springer

Editor
Surajbhan Sevda
Environmental Bioprocess Laboratory
Department of Biotechnology
National Institute of Technology Warangal
Warangal, Telangana, India

ISBN 978-981-96-6357-6 ISBN 978-981-96-6358-3 (eBook)
https://doi.org/10.1007/978-981-96-6358-3

© The Editor(s) (if applicable) and The Author(s), under exclusive license to Springer Nature Singapore Pte Ltd. 2025

This work is subject to copyright. All rights are solely and exclusively licensed by the Publisher, whether the whole or part of the material is concerned, specifically the rights of translation, reprinting, reuse of illustrations, recitation, broadcasting, reproduction on microfilms or in any other physical way, and transmission or information storage and retrieval, electronic adaptation, computer software, or by similar or dissimilar methodology now known or hereafter developed.
The use of general descriptive names, registered names, trademarks, service marks, etc. in this publication does not imply, even in the absence of a specific statement, that such names are exempt from the relevant protective laws and regulations and therefore free for general use.
The publisher, the authors and the editors are safe to assume that the advice and information in this book are believed to be true and accurate at the date of publication. Neither the publisher nor the authors or the editors give a warranty, expressed or implied, with respect to the material contained herein or for any errors or omissions that may have been made. The publisher remains neutral with regard to jurisdictional claims in published maps and institutional affiliations.

This Springer imprint is published by the registered company Springer Nature Singapore Pte Ltd.
The registered company address is: 152 Beach Road, #21-01/04 Gateway East, Singapore 189721, Singapore

If disposing of this product, please recycle the paper.

Preface

The quest for sustainable technologies that address the dual challenges of environmental degradation and resource depletion has driven the global scientific community to explore innovative solutions. Among these, electro-fermentation has emerged as a promising interdisciplinary approach that bridges microbiology, electrochemistry, and bioengineering. By leveraging the interplay between microorganisms and electrochemical systems, electro-fermentation holds tremendous potential for producing valuable chemicals, biochemicals, and biofuels while simultaneously contributing to waste valorization and carbon sequestration.

This book, *Electro-Fermentation for Synthesis of Chemicals, Biochemicals and Biofuels*, provides a comprehensive overview of the principles, applications, and industrial implications of electro-fermentation. The chapters collectively illuminate how this cutting-edge technology is redefining the paradigms of biochemical production, carbon management, and sustainable resource utilization.

The book begins with an introduction to the basic principles of electro-fermentation (Chap. 1), offering readers foundational knowledge to appreciate the intricate science underpinning this transformative process. Building on this, subsequent chapters delve into the myriad applications of electro-fermentation. These include its role in biochemical production (Chap. 2), biofuel synthesis (Chap. 4), and its capacity to drive microbial electrosynthesis systems for CO_2 sequestration and biochemical generation (Chap. 3).

The focus then shifts to the industrial and environmental impact of electro-fermentation. Chapters explore industrial applications (Chap. 5), the utilization of food and solid waste for biofuel and high-value product generation (Chaps. 7 and 8), and advanced strategies for improving bioproduct synthesis in anaerobic environments (Chap. 9). These discussions emphasize electro-fermentation's ability to integrate seamlessly into a circular carbon economy (Chaps. 10 and 16).

The transformative potential of EF is further highlighted through case studies on dark fermentation and electro-fermentation for CO_2 conversion (Chap. 11) and microbial electrosynthesis systems (Chaps. 12 and 15). Applications in waste treatment and syngas utilization (Chaps. 13 and 14) demonstrate EF's pivotal role in addressing pressing environmental concerns.

Finally, the book underscores the importance of Shewanella as a model organism (Chap. 15), emphasizing its exoelectrogenic pathways as a cornerstone for advancing electro-fermentation strategies. By integrating these diverse insights, the text presents electro-fermentation not merely as a scientific advancement but as a cornerstone for fostering a sustainable and regenerative global economy.

This book is intended to serve as a vital resource for researchers, practitioners, and industry professionals seeking to understand and harness the potential of electro-fermentation. It is our hope that the knowledge presented here will inspire further innovation and collaboration, accelerating the adoption of electro-fermentation as a key enabler in achieving global sustainability goals.

I extend our gratitude to the contributors who have shared their expertise and vision to make this volume a reality. I believe their collective efforts will serve as a beacon for those venturing into this exciting field of research and application.

Warangal, India Surajbhan Sevda

Contents

About the Editor

Dr. Surajbhan Sevda completed his Ph.D. in 2013 from the Indian Institute of Technology Delhi, New Delhi, India. He is currently an assistant professor at the Department of Biotechnology, National Institute of Technology Warangal, India. Prior to this, he was a scientific officer (research scientist) at IIT Guwahati, India. Dr. Sevda has authored over 52 research articles in scientific journals and contributed 44 book chapters. He has also edited eight books covering topics such as green energy, solid waste management, mathematical and statistical applications in food engineering, electrochemistry, and bioreactor design for vaccine production. He obtained his Bachelor of Engineering in Biotechnology from the University of Rajasthan in 2006 and his Master of Technology in Fermentation Technology from the Institute of Chemical Technology (formerly UDCT), University of Mumbai, India, in 2008. In 2018, he was a visiting scientist at the University of Calgary, Canada.

Chapter 1
Basic Principles of Electro-Fermentation Process

Soubhagya Nayak and Surajbhan Sevda

1 Introduction to Electro-fermentation

Fermentation can be defined as a phenomenon that does not require oxygen for the degradation of glucose and is utilized by several distinct types of microbes as well as cells. Energy from fermentation can only be produced by glycolysis when coupled with one or two other processes. Fermentation promotes the accumulation of volatile fatty acids or alcohol by inhibiting the phenomenon of anaerobic degradation, i.e., methanogenesis (Chandrasekhar et al., 2021).

Conventional fermentation processes frequently involve redox aberrations in their metabolic pathways, which lead to the concurrent processing of a series of oxidation and reduction reactions in order to acquire equilibrium. Redox imbalance may affect the production and purity of conventional fermentations as the average state of oxidation of the end products and substrates must be the same (Chu et al., 2020). Traditional fermentation typically yields compounds along with a multitude of byproducts, to maintain intracellular redox equilibrium. The aggregation or scarcity of the equivalent of intracellular reduction creates hindrances in the metabolic process leading to the product or leads to thermodynamic constraints. The aggregation or absence of intracellular reducing equivalents could result in thermodynamic constraints or obstruct metabolic flux toward the end product (Gong et al., 2020). Various kinds of methods can be adopted to establish redox equilibrium, like sparging of gases, regulating pH and also adding co-substrates. Oxygen can be purged into the growth medium in the form of an electron acceptor. It helps to create a balance between the redox equivalents produced during the microbial metabolic processes (Rago et al., 2019). In the anaerobic fermentation process, N_2 gas has been used to restrict the aggregation of gaseous byproducts (Angenent et al., 2004). H_2 is sparged when

S. Nayak · S. Sevda (✉)
Environmental Bioprocess Laboratory, Department of Biotechnology, National Institute of Technology Warangal, Warangal, Telangana State, India
e-mail: sevdasuraj@nitw.ac.in; sevdasuraj@gmail.com

© The Author(s), under exclusive license to Springer Nature Singapore Pte Ltd. 2025

S. Sevda (ed.), *Electro-Fermentation for Synthesis of Chemicals, Biochemicals and Biofuels*, https://doi.org/10.1007/978-981-96-6358-3_1

more donors of electrons are required, or other low-molecular-soluble compounds (formate, acetate, etc.) are utilized to improve reductive process (Lee et al., 1999).

Electro-fermentation (EF) can be defined as an advanced domain of biotechnology that utilizes electrochemical techniques to direct and regulate fermentative phenomena.

It incorporates the fundamentals of conventional industrial fermentation along with applied electrochemistry (Virdis et al., 2022). EF is capable of getting over a few of the drawbacks of traditional fermentation. This method has the potential to overcome thermodynamic constraints and enhance the particularity of metabolic pathways, thus improving and regulating microbial fermentation (Moscoviz et al., 2016). This technique is capable of processing many kinds of wastewater streams and generating a diverse array of chemicals. It regulates the metabolic processes of electrogens in the presence of electrodes, in which the anodic terminal acts as an electron acceptor and the cathodic terminal acts as an electron donor (Velvizhi et al., 2022).

The two primary categories of EF are cathodic electro-fermentation (CEF) and anodic electro-fermentation (AEF) (Sriram et al., 2022). As an electron donor during the CEF, the cathode forms more reduced products, while the anode forms oxidized products during the AEF (Rago et al., 2015). AEF is capable of transforming aerobic processes into anoxic processes by substituting an anode in place of oxygen, which acts as an electron acceptor. Thus, by preventing substrate degradation in the form of undesired biomass production, which reduces the amount of energy needed for the growth of cells, as well as by reducing full substrate oxidation into CO_2, high productivity can be maintained or even increased (Vassilev et al., 2021). In the case of CEF, the electrode contributes electrons that are obtained from the anode in order to assist in the production of a reservoir of redox cofactors, which include NADH, with increased production of ATP (Kracke & Krömer, 2014).

Acidogenic fermentation is performed by a group of bacteria, which is a crucial step involved in anaerobic digestion. It integrates the phenomena of hydrolysis as well as methanogenesis. When acidogenic compounds are produced, syntrophic interaction takes place between anaerobic bacteria. Acetogenic and glucose-degrading bacteria can syntrophically break down glucose and yield acetate, hydrogen, and carbon dioxide (Liu et al., 2021a, 2021b).

2 Microbial Metabolism in Electro-fermentation

Being the only electron recipient, the anode gathers free electrons from the cells of microbes directly or through electron mediators. In general, there are two kinds of electron transfer from microorganisms toward the electrode: direct and indirect. This is due to the fact the type of transfer relies on how the cell and anode come into contact. When extrinsic elements like redox compounds assist in the transport of electrons to the electrode, electricity is generated through indirect transport systems (Thapa et al., 2022). Bacterial species like *Shewanella oneidensis* and *Geobacter sulfurreducens* exhibit direct electron transfer because they possess electrically conducting proteins (Reguera et al., 2006). Through cytochrome-rich conducting pili or direct physical connection, specific electroactive proteins in the cell membrane, such as Omc for Geobacter sp. and Mtr for Shewanella sp., facilitate the transfer of electrons between the membrane of the cell and electrode (Kracke et al., 2015).

Some bacteria that are electrically active are lacking in the cellular apparatus and proteins necessary to conduct exo-electron transport. Since these microorganisms are unable to communicate directly with the electrode's surface and extract electrons from organic compounds, they are usually considered to be poor or moderate electrogens. The incorporation of electron mediators enhances electron recovery (Roller, 1984). Electrons from bacteria are regularly transported to the electrode when mediators of electron are introduced to the medium. These chemical molecules possess greater redox potentials and can participate in redox chemical kinetics. They can be cyclic or heterocyclic. In order to enhance the current density, the mediators transfer electrons between anode and the bacterium. There was a surge in current density by 10 times when neutral red was introduced into the MFC seeded with E. coli, because neutral red has a greater negative value of redox potential than compared to E. coli (Park, 2000). Several additional chemicals, including flavins, thionine, ADQS, methylene blue, and methylene red, are frequently utilized for the purpose of electron shuttling in MFCs (Christwardana et al., 2018). It has been discovered that certain bacteria may generate chemicals which aid the electrons flow from the electrode to the source. Rabaey and colleagues were the first to describe the presence of this sort of self-synthesized electron mediator in the case of Pseudomonas aeruginosa. When *Enterococcus faecium* developed its own electron mediator, it enhanced the power density 13 times more than when other types of bacteria were contained (Rabaey & Boon, 2005).

Pilli are present in both gram negative and gram positive bacteria. Pili are capable of conducting electricity in certain electroactive bacteria, especially those of the genera Shewanella and Geobacter (Gorby et al., 2006). Nanowires, also referred to as electro-conductive pili, are extremely conductive but just a couple of micrometers in length (El-Naggar et al., 2010).

3 Electrochemical Principles

Electrochemistry is the investigation regarding the way electrical and chemical effects interact. The field of electrochemistry emerged in 1800 after Alessandro Volta challenged Luigi Galvani's theory that animals could generate electricity. Volta declared that the voltaic battery was the first contemporary electric battery to be invented. As a result, new research areas pertaining to the chemical generation of energy and the chemical reactions with electricity have evolved (Sun, 2023). The analysis of the electrical energy generated by chemical interactions and the chemical reactions caused by an electric current shift is at the core of electrochemistry (Hamnett, 2023). Faraday was the first person to coin the terms ion, anode, and cathode. The Greek term of wanderer, "ion," was commonly used to designate to the groups that were active in Faraday's electrolysis processes. Anodic currents are created by species that disperse in the direction of the anode and cathodic currents are created by species that disperse in the direction of the cathode. Generally, the cathode undergoes reduction, while the anode undergoes oxidation. The two types of cells are called electrolytic and galvanic (Purkait et al., 2024). Electrochemical cells can be classified into two subtypes: galvanic and electrolytic cells. When a conducting material is utilized to link the electrodes, galvanic cells show impulsive reactions, in comparison with electrolytic cells, which require a higher voltage to perform the electrochemical process (Bruckenstein, 1977). In the case of a galvanic cell, the two electrodes are designated by the anode and cathode, but in an electrolytic cell, the electrodes must perform specific roles (Changmai et al., 2022).

It is initially required to comprehend the processes of oxidation and reduction in electrochemistry (Petrucci et al., 2000). In a redox (oxidation and reduction) reaction, electrons move between particles or alter the degree of atom oxidation. Moreover, there should be an equivalent number of electrons acquired during reduction and oxidation overall. One of the two components of a reaction, one involving gaining electrons and the other involving loss of electrons, is called a half-reaction. The activities in electrochemical systems that impact the transfer of charges along the membrane system, the outer circuit, and the junction between the ionic conductor and an electronic conductor occur when they are referred to as "electrochemical cells." These operations involve the flow of ions and electrons. These systems are defined by the presence of two electrodes spaced not less than one electrolyte step apart. A charge imbalance area which develops in the innermost monomolecular layer of adsorbed water molecules and ions makes up the phase shift region among the two distinct phases (Hamnett, 2023). When a metal is submerged in liquid, it forms two layers that encircle the component and reflect the scattering of uneven charges (ions) across the liquid–solid boundary. The initial layer being the surface charge layer, which retains to the interface of the metal's electron liquid molecules that are polar to produce 2 narrow positive and negative charge planes. The "compact Helmholtz layer" and outer Helmholtz layer is the portion of the inner layer that is close to both electrodes. The free ions in the subsequent layer, designated by the term "diffuse layer," are drawn to the outermost layer by heat movement as well as electrical attraction. A negatively charged inner layer that matches the surplus cations in the outermost layer is formed in this second layer by weakly binding some of the bigger anions onto the metal, if water-soluble ions were prevalent (Jurado & Espinosa-Marzal, 2017). By performing as a type of capacitor, this equilibrium dual layer generates half-cell potential, also known as "the electrode solution potential (Pavlov, 2017)."

MES may serve as a suitable substitute technology, in comparison with traditional anaerobic treatment of wastewater and bioenergy harvesting technologies. Microorganisms are the main driving force in MES, as they may degrade organic waste via metabolic processes that occur on the anode and generate biofuel, or recover beneficial chemicals on the cathode through reduction (Wang & Ren, 2013). The electrochemistry and physiology of bacteria together form the foundation of the

MES operating principle. Extensive study on MFCs was carried out prior to the start of MES research. In general, the MES consists of an electrolyte(s), a cathode, and an anode. It is possible to use an ion exchange membrane to divide the anodic and cathodic chambers. Microbes in the anodic chamber decompose substrates by forming biofilms or remaining planktonic, producing protons, electrons, and other compounds as byproducts. The anode receives the liberated electrons and transfers them to the cathode via an external load. The circuit is completed when the protons permeate, dissipate via the exchange membrane, or just diffuse across the cathode where they are reduced through the incoming electrons. Electric current created by the external load is carried by the electrons. In MFCs, the movement of electrons is often used instantly to generate power, but in the case of MECs, it can be utilized to evolve hydrogen and other beneficial chemicals (Roy & Pandit, 2018).

Materials made of carbon and graphene are frequently employed as electrodes because of their remarkable chemical stability, high conductivity, and accessibility. Electron transport and conductivity of electricity could be enhanced by altering the electrode's characteristics. The extent of permeability, surface area, as well as biocompatibility of the substance determine whether biofilm adherence is promoted or inhibited. O_2 and N_2, two functional groups on the surface, could boost electrochemical processes at the electrode surface (Roy et al., 2015). The type of electrode material influences the development of biofilm. Carbon cloth has a greater roughness on the surface, which makes it a suitable medium for enabling growth of biofilm. There have also been reports of biofilm growth on metal electrodes made of copper, silver, gold, or steel. Metal electrodes are not preferred choices for the anode material in MFC because they are known to be hazardous to bacteria, in contrast to carbon-based materials. According to Baudler et al., electroactive bacteria are impervious to the toxicity of metals. It has been shown that the ability of electroactive bacteria to synthesize biofilms and produce significant quantities of electricity is not hindered by the antibacterial characteristics of these metals (Schmidt et al., 2015).

Material composed of carbon produced by biomass can be utilized as an electrode in many different systems, including supercapacitors. The exceptional characteristics of biomass-derived carbon substances, which include their significant surface area, chemical resilience, better conductivity for electricity, and biological compatibility, enable them to be often used as electrodes in BESs. BES electrodes have been constructed using a variety of biomass materials, including corn straw, sewage effluent, food waste, wood, and forest residual (Baek & Lee, 2024). Since algae derived carbon substances match numerous properties with traditional carbon-derived electrode materials like felt, brushes, and carbon cloth, they have emerged as potential alternatives for electrodes in BES. These characteristics comprise their usefulness as a medium for support and their broad surface area, which promotes the establishment of biofilms (Chakraborty et al., 2020). The distinct tubular structure of microtubular electrodes allows them to be a potential substitute for traditional flat electrodes. The benefits of microtubular shape over traditional planar structures were illustrated by research on flow-based batteries and solid-oxide fuel cell designs. Microtubular electrodes can be utilized in the form of gas diffusion electrodes in the phenomenon of gas-phase electrocatalysis. GDEs can reduce the path length of gas diffusion as well as enhance the concentration of gas around the electrocatalyst, which is crucial in order to obtain higher current generation throughout the electrochemical gas-phase kinetics (Rabiee et al., 2022).

4 Synergistic Effects in Electro-fermentation

Electroactive bacteria can be beneficial for extracellular degradation of contaminants, but their enrichment is mainly limited to the surface of the electrode of the bio-electrochemical system and is difficult for the purpose of scaling up. In order to overcome this, an innovative method has been implemented, involving the need for the anaerobic granular sludge to be electrochemically stimulated and the resultant electroactive granular sludge to be used in an electrode-free bioreactor to eliminate pollutants. After seven days of electrochemical stimulation, EGS developed rapidly, attaining its highest possible current generation of 112.5 A/m^3, and the number of genera resembling electroactive bacteria grew from 5% to approximately 20%. The conductivity increased onefold. The polysaccharide composition of extracellular polymeric substances remains unaltered which indicates a stronger redox system. There has been no reduction in performance or electroactivity throughout the batch process for a period of 60 days. The efficiency and microbial population of granular sludge were found to exhibit an array of long-lasting improvements after a basic seven day electrochemical stimulation (Cheng et al., 2024).

Microbial metabolic processes can produce reactive oxygen species (ROS), a phenomenon known as microbial advanced oxidation. The regular transition between anaerobic and aerobic respiration is necessary for this process to succeed. Aerobic facultative bacteria generate ROS like H_2O_2, when their intercellular electrons are absorbed via oxygen. Microbes that are able to carry out extracellular respiration in anaerobic conditions transmit intercellular electrons to extracellular undissolved acceptors of electrons, like Fe^{3+} elements. This reduces Fe^{3+} to Fe^{2+}, which then reacts with hydrogen peroxide to produce radicals of hydroxyl (Song et al., 2024). The formation of microbial advanced oxidation depends substantially on extracellular electron transport. The absence of appropriate electron acceptors and the poor conductivity of sediments enable extracellular electron transfer-based metabolic processes of bacteria unusual in these environments. This is demonstrated by the low concentrations and poor respiration of extracellular respiratory bacteria in wetlands, which restricts the application of microorganisms able to respire extracellularly for the treatment of pollutants, which includes microbial advanced oxidation (Lovley, 2017). Microbes that can perform extracellular respiration can be enhanced and their extracellular electron transfer may be facilitated by making use of electric stimulation (Wang et al., 2020). The polarization of protein amide groups by an electric field is being demonstrated to enhance the polarity of amide group C–O linkages across the outermost layer of biofilm (Liu et al., 2021a, 2021b). This can raise the relative capacitance of protein amide groups, which boosts the transfer of electrons in extracellular as well as intracellular proteins through proton-coupled transfer of electrons.

The enhanced conductivity and capacitance of anaerobic biofilms could facilitate the exchange of electrons in microbial accumulation even more. Wang and coworkers reported that the richness of Geobacter anodireducens which is an extracellular respiratory bacterium, multiplied from 24 to 50% through electrical stimulation (Du et al., 2018).

Extreme low-frequency electromagnetic fields have a surprising potential impact upon microbial fermentation mechanisms. The products of fermentation may be affected by the development and metabolic processes of microorganisms, which is influenced by electromagnetic fields. Addressing the impact of electromagnetic fields is especially important when it comes to the fermentation process of cocoa beans, since it enhances the nutritional value and sensual reach of the beans through careful manipulation of the fermentation environment. The purpose of this work is to utilize widely recognized models and controlled experiments to deliver empirical data on the impact of electromagnetic fields upon microbial development. The objective of the study is to increase the knowledge about the way electromagnetic fields and microbial activity interact during the fermentation process of cocoa beans (Guzmán-Armenteros et al., 2024). The models of Gompertz, Baranyi, and logistic are often utilized in the study of microbial kinetics. The proliferation of the population of microbes is characterized by the logistic model, which incorporates factors such as the density at the initial phase, time, rate of growth, and ultimate density. The Baranyi model includes the lag, exponential, and stationary stages of growth of microbes and accounts for dynamic circumstances. Since the Gompertz model suggests a gradual reduction in growth rate over time, it is appropriate for microbes with prolonged lag phases and lower development rates (Nguimkeu, 2014).

5 Factors Affecting Electro-fermentation Efficiency

Operating circumstances, design features, cathodic factors, and anodic parameters all affected MES efficiency. The preferred anode material should, taking into account the anodic parameters, exhibit the necessary biocompatibility, have a substantial specific surface area, be non-corrosive as well as non-hazardous, significantly conduct, remain stable chemically in wastewater, and be locally and economically accessible. Considering the low reduction dynamics and cathodic restrictions, the primary constraining considerations are the cathode's slow rate of accepting electrons and reduction processes, which limit the MES's generation of products. Additionally, microbial inoculum, wastewater utilized as anolyte, catalyst used, and catholyte, all of which influence MES performance are necessary components for its development (Gil et al., 2003). The overall design of MES, such as its form, material used, spacing between the electrodes, characteristics, and specific surface area, as well as the distribution of wastewater flow, all affect its efficiency. The material preferred for the electrodes should be conducting, biocompatible, and have a large specific surface area, low resistance, and be stable chemically in solutions of electrolytes. A number of structural layouts, including single-chambered MFC, were suggested as means to lower the internal resistance of MFC. In this type of design, the cathode electrode is directly exposed to air (Pandit et al., 2012).

The production is influenced by the development and metabolism of microbes. In order to promote the kinetics of microbial development, the system needs to run under ideal circumstances. The efficiency of MES is restricted by the supply of protons at the cathode. The delayed movement of protons across the membrane is compensated for by the inclusion of buffers or NaCl. The progress of redox reactions and microbial growth is also affected by temperature (Li et al., 2010). The microbial part, which includes characteristics of the inoculum, its potential to form biofilms, its ability to transmit electrons, and the kinetics of substrate utilization, is essential for MES. When handling diverse wastewater, biological optimization requires the choice of appropriate bacterial consortium as well as methods for adjusting bacterial populations to reactor-optimized circumstances (Roy & Pandit, 2018).

Carbon-based substances, nutritive media, wastewater discharges, etc., are some examples of feedstock that are necessary to produce the end product. They are efficient for generating hydrogen, as demonstrated by MFC and MEC. dos Passos et al. have used a variety of substrates, such as glucose, lactic acid, acetic acid, butyric acid, acetate, and fermentation byproduct; but acetate exhibited the highest power density when assessed. In spite of their small size and dependence on photosynthesis, microalgae offer a novel substrate toward the production of hydrogen. They require light and water, and their production of ethanol, CO_2, and biohydrogen varies according to the culture circumstances. Therefore, although these kinds of substrates have certain disadvantages, with modification they can generate high yields and could turn out to be affordable in the long term (Huang et al., 2020). Many complicated substrates, such as milk, glycol, starch, and hybrid substrates, may be utilized by the MFC and MEC to recover biohydrogen (Montpart et al., 2015). In the case of MEC, milk and hybrid feedstock produced notable hydrogen generation outcomes, but starch and glycol were not used to achieve results that were optimal (Krishnan et al., 2019). In comparison with anaerobic digestion (AD), MEC with food waste (FW) generated greater amounts of hydrogen. Subsequently, organic materials decomposed apart much faster in AD than it did in MEC (Huang et al., 2020).

The pH and conductivity of the electrolyte have significant effects on the total efficiency of bio-electrochemical systems. The efficiency of the systems (MFC and MEC) is significantly influenced by the electrolyte's ion concentration. Protons are stimulated to travel across electrodes by increasing conductivity, which is produced by lowering their internal resistance and raising buffer concentration (Yang et al., 2017). A buffer with an elevated concentration, such as NaCl, can be introduced into the solution to enhance its conductivity (Gadkari et al., 2020). The highest level of hydrogen reclamation was identified at a pH of 6.5. Consequently, since the catholyte's pH regulates and limits the amount of hydrogen that can be produced, it must be maintained (Yossan et al., 2013). Since microorganisms are susceptible to pH changes, these variations may have an impact on microbial respiration, which in turn affects the flow of electrons outside of cells. Although a lot of the parameters are related to pH and require more comprehensive studies to fully understand the way in which MEC components respond to varying pH levels, the majority of studies have documented using neutral pH (Sharma et al., 2024).

One of the most significant parameters influencing microbial activity is temperature. The primary design element of MEC is the anaerobic environment, and temperature has a considerable impact on the activity of microbes. Several electrochemical studies have shown that temperature greatly influences the synthesis of biohydrogen. In contrast to methanogenic MEC, a low range of temperature favors synthesis of biohydrogen (Ki et al., 2017). A mid-range temperature is essential for the proliferation of microbes, to generate biohydrogen (Zhang et al., 2019). In addition to hydrodynamic force, organic load rate, and amount of oxygen in dissolved state, HRT is reported as one of the major operating factors in the case of MEC. The generation of hydrogen can be remarkably enhanced through the loading of catalyst across the surface of the electrodes (Liu et al., 2014). The generation of hydrogen is influenced by various HRTs when using a MEC. The generation of hydrogen surged substantially when the HRT was reduced from 10 to 4 h (Gil-Carrera et al., 2013).

6 Electro-fermentation Reactor Design

One of the most important factors influencing the efficiency of microbial electrosynthesis and electro-fermentation is the design of the reactor. A simpler membrane-less reactor design has been developed for the phenomenon of EF (Roy et al., 2022). No such considerable amount of reactor transformation has been made in the context of EF over the last 2 years. The existing reactor layouts and older developments have been thoroughly explained in prior publications (Bhagchandanii et al., 2020). In recent assessments, the material used for electrodes has been thoroughly examined as an essential attribute (Zhang et al., 2021). Basic elements such as the mechanism of electron transport, the impact of pH on product the extraction process, and the influence of salinity on the entire MES system have previously been discussed upon in prior publications (Roy et al., 2022). Various reactor designs have been investigated for MES. MES reactors with an H-shaped double compartment configuration are frequently used (Mohanakrishna et al., 2020). A number of aspects, such as significant ohmic losses, high energy consumption, poor mass transfer rate, operational complexity in continuous mode, and the need for a substantial amount of space, limit the implications of this particular design for scale-up purposes. In certain investigations, parallel-plate double chamber reactors have been utilized to get over these constraints by reducing the distance between the cathode and anode chambers (Tahir et al., 2020). Certain groups have attempted single-chamber, membrane-less reactors in an effort to reduce the cost related to the usage of membranes (Li et al., 2019). There are also reports of a multicompartment module reactor comprising three separate chambers, including one chamber used for product recovery (Das et al., 2021). Three-chamber reactors coupled with double cathode have also been listed (Vassilev et al., 2019). The operational and handling challenges related to these type of reactors outweigh the benefit, even if these modifications will enable the desired product to be synthesized and extracted simultaneously. High energy consumption is necessary for this kind of reactor to operate since the ohmic losses expand concurrently with the growing distance between the electrodes. There was a demonstration of a rotating bed biofilm reactor model for the MES utilization. In order to adequately create hydrogen bubbles, the cathode electrode was positioned toward the bottom of the single-chambered reactor (Cai et al., 2022). A more effective and advantageous BES will arise from a combination of anodic and cathodic processes. Whenever it comes to operating mode, continuous mode will maximize the yield of products because it supplies the reactors with greater amounts of carbon than batch mode. A few investigations have demonstrated that running MES continually can generate methane as well as acetic acid over an extended period in a steady manner (Cai et al., 2022). The generation of products with additional value from the generated waste stream using energy from renewable sources as a source of electrons is one of the major objective associated with these bio-electrochemical systems (del Pilar Anzola Rojas et al., 2018). Both self-generated reactions as well as fuel cells that produce hydrogen were used as energy sources in EF (Lai & Lan, 2021).

In order to perform both basic and applied investigations on aspects like testing materials for electrodes, reactor architecture, microbial catalysts, as well as electron transport phenomena, the majority of MES research were carried out using lab-scale reactors. A pure strain of *Methanococcus maripaludis* was used within a 50 L reactor to exhibit the generation of methane from carbon dioxide in a recent study on MES (Enzmann & Holtmann, 2019). Electro-fermentation in bio-electrochemical systems (BESs) coupled with the process of anaerobic digestion (AD) is an interesting approach, since it effectively converts carbon dioxide to methane at the biocathode, utilizing microorganisms in the form of catalysts (Noori et al., 2020). In order to improve BES-AD efficiency in respect to methane production, different designs were proposed recently. Both the anode and the cathode in a single-chamber BES-AD arrangement utilize a common electrolyte, whereas separators—such as exchange of cations, anion exchange, even both of them at once—are required in double chamber as well as multiple-chamber reactors to keep the anode and cathode electrodes apart. It also suggests redox reactions take place in different chambers in multiple-chamber BES-AD designs, but in the case of single-chamber arrangements, they proceed in the same chamber (Flores-Rodriguez et al., 2019). When designing a dual-chamber BES-AD, an ion exchange membrane divides the chambers, which contain the anode and cathode electrodes (Song et al., 2016). The electrolysis of water or the breakdown of organic compounds can take place at the anode electrode, generating ions of hydrogen. These ions then move across a proton-exchange membrane into the cathode chamber, where they can generate hydrogen gas at the cathode electrode. Hydrogen-utilizing microbes could use the produced hydrogen gas for generating methane from carbon dioxide as a substrate. In regard to functionality, production, and expenses, many designs possess certain benefits and drawbacks. A single-chamber BES-AD layout appears to be more feasible and cost-effective, as it lacks a separating membrane. When these systems are run at greater potential, they become susceptible to oxygen contamination from electrolysis of water across the anode. In terms of efficiency, a dual-chamber BES–AD system has a substantially higher cathodic current generation than a single-chamber version, that led to a higher methane concentration (Liu et al., 2017). It has been proposed to utilize a three-chamber BES-AD setup to enhance methane while reclaiming nutrients at the same time (Zeppilli et al., 2017). In this configuration, an anion-exchange membrane (AEM) and a proton-exchange membrane (PEM) divided the three chambers. The accumulation chamber, which the AEM used to isolate from the cathode chamber, was positioned between the anode and the PEM. Municipal wastewater that was laden with NH_3 was processed in the anode chamber, and external carbon dioxide was transformed into CH_4 in the cathode chamber. It was recently proposed to combine carbon dioxide recovery, biogas upgrading by isolating carbon dioxide from methane, and treatment of wastewater into a three-chamber BES configuration known as a microbial electrolytic capturing, segregation, and regeneration cell (MESC) (Jin et al., 2017). This layout segregates the electrolytes and electrodes using a bipolar membrane (BPM) across the anode and the regeneration cell and an AEM separating the regeneration cell and the cathode. This system, while running in batch mode, could eliminate 80% of the COD from influent wastewater and elevate methane in biogas to as much as 98% with 20% efficacy. The system could attain an optimal methane concentration of 97.5% in continuous mode. An extra chamber, a type of adsorption chamber, was added to the MESC in order to further alter it. Therefore, the whole BES consisted of 4 chambers: anode, regeneration, adsorption, and cathode chambers (Kokkoli et al., 2018). This configuration is quite effective in segregating carbon dioxide from raw gas and potentially reaches a maximum saturation of 99–100% of methane, with batch mode operation and an external supply voltage of 1.2 V. The system could extract over 88% of COD that was soluble in the form of acetate from wastewater.

7 Monitoring and Control in Electro-fermentation

An approach that utilizes electromagnetic radiation with frequency that vary between 300 and 300,000 MHz is called microwave heating (MWH). Many studies have investigated into the viability of a number of possible applications for microwaves for fermentation processes, including the synthesis of hydrolases, the formation of biohydrogen, lowering of accumulative production of gases, the reduction of fermentation times, and the safe storage of fermentation products. It significantly impacts the dipole rotation as well as the conduction of ions in macromolecules (Wang et al., 2021). The application of MWH promotes the process of fermentation by allowing the dough to absorb energy. Juice, yogurt, and other liquid-based products are hence typically fermented using MWH (Gholamhosseinpour et al., 2020). Kapcsándi and coworkers assessed the effect of implementing this technique on fermentation duration and discovered that the duration of fermentation was decreased by 40% as compared to that of the control sample (Wang et al., 2021). The impact of microwave prior to treatment on enzymes throughout solid-state fermentation was examined in another investigation. The findings suggested that MWH may improve hydrolase yield, and after microwave pretreatment, there was a substantial rise in glucose amylase as well as protease activity when in contrast to the samples that were untreated (Melikoglu, 2012).

The traditional method of heat treating food items depends on the conduction and convection of heat. Wheat flour and dough are poor thermal conductors, it is challenging to heat them evenly using conventional heating techniques. An efficient substitute for traditional heating includes radiofrequency heating (RFH). The material which is heated receives a direct transfer of electromagnetic radiation from RFH, and because of its extended wavelength, which allows it to permeate further into the substance, the heating becomes more uniform (Altemimi et al., 2019). In order to make hazardous microbes ineffective prior to fermentation, wheat flour can be pretreated by RFH (Villa-Rojas et al., 2017). In contrast to traditional heating, the RFH technique provides a more even distribution of temperatures. Additionally, since the method of heat transfer doesn't involve contact with food items, pollutants are efficiently evaded in the finished product (Siefarth et al., 2014). The heat produced from the heating coil in RFH apparatus allows it to maintain the fermentation environment at roughly 35 °C, mimicking the conditions of a conventional fermentation tank. When it involves the productivity of fermentation and the management of fermentation circumstances, RFH offers additional advantages. Since it has a significant amount of water in it, it can be ideal for making fermented dough (Bajd & Serša, 2011).

The technique for heating batter by circulating an electric current across it is referred to as ohmic heating (OH). Energy passes straight inside the batter in the case of OH to prevent interior solid parts from being destroyed (Knirsch et al., 2010). OH is being noted for use in a number of food processing procedures, such as pasteurization, blanching, extraction, fermentation, dehydration, and sterilizing (Ferreira et al., 2019). In particular, OH along with a sufficiently strong electric field can support the process of fermentation via both thermal and non-thermal actions. Electricity has the ability to alter the kind and amount of products of fermentation and shorten the lag phase of fermentation of different microbes. The main processes are the rapid temperature adjustment, the improvement of the permeability of cells, and the dispersion of micronutrients from substrates (Gavahian & Tiwari, 2020). Its ability to preserve the superior sensory and nutritional qualities of items while being environmentally conscious over conventional methods is one of its key benefits. In contrast to traditional heating, the effect of excess heating and unwanted surface peeling on the final products can be minimized through the OH technique (Gally et al., 2017).

The fields of cell biology, bioanalysis, and medicine have made extensive use of electrochemical impedance spectroscopy (EIS). EIS could offer important insights into the condition of bacterial cells since it can assess even minute variations in the electrical characteristics of electrode contacts in real time while avoiding labeling. It provides insight into the workings of various biological processes, including microbial proliferation, metabolism, as well as biofilm development (Saulnier et al., 2024). Voltammetric as well as amperometric technologies comprise some of other types of electrochemical methods utilized for investigating resistance to antibiotics and microbial activity. They can reveal details regarding the ability of microbes or their metabolites to transmit electrons across electrodes. Microbes produce electrons by oxidatively converting organic compounds, but only a small percentage of species of bacteria have been proven to be capable of transmitting electrons to an inert electrode (Sundaresan et al., 2022).

Liquid chromatography (LC) is a crucial tool for the monitoring of fermentation. Pyruvate is an indirect indicator of NADH, the reducing cofactor which promotes bacterial development, and in the wine-making process, measuring the pyruvic acid level during malo-lactic fermentation provides insight into the condition of the bacteria. Acids such as lactic, acetic, pyruvic, formic, propionic, and butyric that are indicative about the quality as well as stability of dairy-based goods can be produced via lactose fermentation. In a single run of LC assessment, all of these compounds can be readily identified. It is extensively utilized in the food sector to assess the final product's quality. For instance, research has demonstrated a direct correlation between the quantity and variety of organic acids contained in grape juice and its quality. The organic acid spectrum has a direct impact on the quality of the final product when producing sour cream through fermentation (Freitag, 2005).

In order to produce recombinant proteins, researchers suggest applying a reliable ^{1}H NMR approach enabling high-throughput monitoring as well as absolute estimation of extracellular metabolites in fermentations of E. coli. The quantitative approach was created using the concentration estimation method centered around pulse length. The artificially created electronic signal was calibrated using an external sample with a concentration that was known. This signal was then utilized in the form of a reference to determine the absolute measurement of nutrients as well as metabolites produced during fermentation. As representations of significant nitrogen as well as carbon metabolism, alanine, betaine, phenylalanine, glucose, and acetate are the 5 major metabolites detected in the E. coli fermentation samples for which absolute quantitative results has been provided (Forsberg et al., 2024).

In another study, near infrared spectroscopy in combination with a nano-composite colorimetric sensory array was used to detect the active components in the fermentation of kombucha. Headspace analysis revealed that the kombucha broth comprised 67 volatile chemicals belonging to 7 distinct chemical families. The GC-MS (gas chromatography-mass spectrometer) was utilized for this purpose (Kang et al., 2023).

8 Applications of Electro-fermentation

In the course of acidogenic fermentation, short-chain fatty acid (SCFA) molecules are generated in the form of intermediary products (Rai & Singh, 2016). Wastewater streams with high carbon content as well as carbon dioxide can act as potent substrates for the generation of these intermediate products during electro-fermentation (Chandrasekhar et al., 2021). Integration of the MEC technique with anaerobic fermentation proved to be quite beneficial in improving the rate of conversion of fatty acid molecules into hydrogen (Marone et al., 2017). The phenomenon of electro-fermentation could be employed for the production of biopolymers through the anoxygenesis process, the generation of biodiesel through a photosynthetic system, and the synthesis of bioelectricity (Saratale et al., 2017). Synthesis of SCFAs is achieved by a sequence of metabolic processes that are mediated by acetogens and acidogenic bacteria. The processes of hydrolysis and solubilization are considered to be the ones that limit the rate at which SCFAs generate waste activated sludge (WAS) (Gao et al., 2024).

Carboxylic acids with carbon chain lengths ranging from C6 to C10 are referred to as medium chain fatty acids (MCFAs). These acids include capric acid, heptanoic acid, caproic acid, caprylic acid, and pelargonic acid. They have been utilized as flexible precursors to acquire biofuels and raw materials in a variety of industries (Gu et al., 2024). The most widely accessible compounds in fermentation discharges are acetate, propionate, and butyrate, which are capable of being used as precursor substances in BES with applied energy conditions to produce MCFAs (Tharak et al., 2021). MCFAs exhibit a low level of solubility due to the presence of large chains of hydrophobic carbon. They can be easily segregated from fermentation broth at low expense (Agler et al., 2012). The fundamental fermentation method that leads to SCFA production and the chain extension process for upgrading SCFAs to MCFAs are the two stages in the WAS fermentation process that result in the production of MCFAs (Gu et al., 2024). The byproducts formed during anaerobic fermentation are known as bio-solvents. Splitting off the oxidation and reduction reaction depends on the synthesis of the final product through electro-fermentation (Zhao et al., 2015). Electrons that are generated inside anodic chambers can be utilized to limit the production of VFA molecules and boost the generation of alcohol compounds during electro-fermentation (Chandrasekhar et al., 2021).

First-generation biomass is usually represented by sugar- and starch-based products, oil crops, seeds such as palm and sunflower. Biomasses that are starchy and sugar-based can be fermented readily in anaerobic environments (Ananthi et al., 2024). Subsequently, using first-generation biomasses has significant disadvantages, such as using up a lot of agricultural land and interfering with the production of food grains. This impact makes severe shortages of foods with overuse of fertilizers (Dragone & Fernandes, 2010). The second generation of biomasses for the generation of biological hydrogen is composed of inedible residues from crops, such as lignin-based biomasses from agricultural and forest residues, food processing wastes, and waste from industries (Singh et al., 2011). Agro-industrial, agronomic, waste papers, carrot pulp, and other high-protein and carbohydrate wastes are the solid organic matter used as feedstock for the synthesis of biohydrogen (De Vrije et al., 2010; Guo et al., 2014). In order to produce biohydrogen, liquid organic wastes such as distillery waste waters have been utilized as second-generation biomasses (Sridevi et al., 2014). Algae possess a high carbohydrate content and thus can be used as an effective third-generation biomass for producing biohydrogen (Ananthi et al., 2024). Third-generation biomass has several benefits, such as being intensely CO_2 absorbent, easy to cultivate, and quickly producing biomass with a high percentage of carbohydrates that is without lignin and hemicellulose (Ananthi et al., 2024).

Sugar-based resources like sugarcane or starch-rich feedstock like corn, potatoes, cereals, and grains are the primary sources of first-generation bioethanol (Pavão et al.,

2023). Monomeric and dimeric sugars that exist in sugarcane crops are fermented to produce bioethanol with the aid of yeast. It is possible to recycle the yeast utilized to ferment sugarcane and molasses, which will save money and time (Aro, 2016). Agricultural and industrial lignocellulosic products that are not edible are utilized to produce second-generation bioethanol. These wastes include rice husk, wheat straw, maize stalks, olive pomace, bagasse, coconut husk, paper pulp industry waste, and fruit peels (Branco et al., 2019). Chitin and cellulose are structurally similar as they are both made of *N*-acetyl-D-glucosamine units. Chitosan is made from chitin with the help of an alkaline or acidic deacetylation process. Chitosan and chitin are further viable options for the synthesis of bioethanol (Aranda-Martinez et al., 2017; Govindaraj et al., 2023). The process of generating bioethanol from second-generation biomass is an intricate task in comparison with that of first-generation biomass. It involves a number of steps, including physical pretreatment, chemical pretreatment, detoxification of the pretreated material, the breakdown of the pretreated plant matter rich in cellulosic content (producing monomers and oligomers of simple sugars), the hydrolysis product fermentation and finally condensing it using distillation and molecular filtering to the required levels of purity. Bioethanol production from second-generation biomass is quite less than that from first-generation biomass (Jain & Kumar, 2024). Macro and microalgae serve as the sources for the production of third-generation bioethanol. These are mainly categorized as photoautotrophic lifeforms, which efficiently absorb atmospheric carbon during cultivation by converting carbon dioxide into useful components like lipids and carbohydrates (Choudhary et al., 2020).

Algae store carbohydrates such as cellulose, starch, pentose, and hexose, which are fermented to produce bioethanol. Rapid growth, high lipid and carbohydrate content, and lack of effect on food supply enable algae to be a potent source for bioethanol production (Jain & Kumar, 2024). Algae have the ability to withstand a variety of wastewater streams, such as municipal wastewater and industrial wastewater. They assist in the treatment of wastewater as well as in nutrient cycling. Since they don't contain lignin, have a short life cycle, can be cultivated economically, and can be used for biological remediation, microalgae provide useful substrate for bioethanol generation (Choudhary et al., 2020). Some microalgae store essential lipids that can be used to produce biodiesel (Hena et al., 2018). Genetically modified feedstock and genomically produced microbes are used for developing fourth generation bioethanol. Genetic engineering makes it easier for organisms to acquire desired characteristics. These characteristics include enhanced lipid synthesis, better photosynthesis and carbon fixation, and the use of different sugar types including pentoses and hexoses. Compared to their wild counterparts, genetically engineered algae are more environmentally friendly, have a higher quantity of energy, and emit less pollutants, making them an attractive source of biomass for bioethanol (Cavelius et al., 2023). Deoxygenation and compression are applied to all feed gases, waste from industry or byproduct gases (which mostly contain carbon monoxide), and syngas (which mostly contain carbon monoxide and hydrogen) that are produced during the process of gasification of feedstock. The resultant compressed gas, which is high in carbon monoxide and hydrogen is pumped into the bioreactor, that is filled with specially developed fermenting microorganisms suspended in a nutrient-rich liquid. These microorganisms produce bioethanol and other byproducts and utilize carbon dioxide and hydrogen for growth and metabolism (Handler et al., 2016). Escherichia coli carry out anaerobic fermentation of glycerol to yield ethanol (Adnan et al., 2014).

Implementing mevalonate as a model target, electro-fermentation was enhanced through the use of a strain that was developed. Mevalonate is a crucial precursor. Mevalonate synthesis in genetically modified Escherichia coli has been depicted. The mevalonate synthesis route in E. coli needs 2 NADPH molecules to produce mevalonate. According to Matsumoto et al.'s hypothesis, mevalonate production is enhanced by reductive electro-fermentation because it facilitates NADPH replenishment. Since electro-fermentation markedly improved the synthesis of ethanol, they developed an adhE gene-deficient C43 (DE3) to enhance the production of acetyl-CoA, the precursor of mevalonate. Elimination of adhE gene is quite helpful in boosting mevalonate synthesis, most probably in regards to cofactor renewal and actyl-CoA availability. Electro-fermentation combined with metabolic engineering increased mevalonate synthesis (Matsumoto et al., 2023).

The process of electro-fermentation offers supplementary electrons, which in turn elevates the production of reductive biochemicals. In an endeavor to raise succinate generation, E. coli Suc260 was electro-fermented using neutral red as an electron shuttle. The synthesis of succinate was raised to 9.59% through electro-fermentation, relative to the process of anaerobic fermentation, with an equivalent amount of cell growth and use of glucose. The actual succinate production (0.8 g/g) was less compared to the projected optimal yield of succinate (1.12 g/g). It could be due to the fact that utilizing neutral red as an electron shuttle during the electro-fermentation of E. coli suc260, there was minimal current and electricity (Ye et al., 2023). Microbial electrosynthesis (MES) can be used to synthesize mono carbon compounds like formate and methanol, which can then be used as substrates or platform chemicals for an independent bioeconomy (Müller, 2019). The primary byproduct of the reduction of carbon dioxide process utilizing MES is acetate (Roy et al., 2021). The maximum yield of acetic acid, approximately 12 g/L, has been observed in dual cathode reactors utilizing enriched microbial cultures and *Sporomusa ovata* (Vassilev et al., 2019).

In an effort to increase polyhydroxybutyrate (PHB) deposition in R. eutropha H16, electro-fermentation was carried out with a hydrogen fuel cell providing the electric current. The electro-fermentation system combined with environmentally friendly energy technology is implemented in this investigation to improve the precision of target products, boost growth of bacteria, decrease the need for additives for redox balancing or pH regulation, and increase the rate of carbon conversion by affecting the conditions of fermentation and the metabolism of bacteria with green energy. R. eutropha H16 was electro-fermented using electric current of varying strengths in order to examine the impact of electric current on development of cells, generation of biomass, glycerol utilization, and PHB deposition. The electric current provided during the process of electro-fermentation has a prominent impact on the growth of R. eutropha H16, the generation of biomass, the utilization of substrate, and the deposition of PHB. Electro-fermentation is capable of deflecting the flow of carbon into the pathways involved in metabolism for the purpose of synthesizing more metabolites apart from feedstock and the production of PHB (Lai & Lan, 2021).

The higher temperatures have been reported to support thermophilic bacteria and promote faster reaction rates, both of which increase the generation of volatile organic compounds. Previous studies found that mesophilic fermentation produced 0.298 g COD/g of volatile suspended solids, whereas thermophilic fermentation produced 0.368 g COD/g of short-chain fatty acids (Lu et al., 2024). Investigations were conducted regarding the effects of electrical voltage and operating temperatures on the generation of volatile fatty acids in mixed-culture electro-fermentation reactors. When identical reactors were run at various voltages and temperatures, the variations in VFA components and other factors were observed.

Corynebacterium species are the primary producers of lysine and are also able to transmit electrons outside of cells. They are excellent options for electro-fermentations (Kracke et al., 2015). Through the application of polarized electrodes, Xafenias et al. attempted to determine and comprehend the circumstances under which C. glutamicum's lysine production could be electrochemically sustained. They exposed this particular strain toward 12 distinct circumstances related to the gas environment (nitrogen and carbon dioxide), the potentials of the electrodes (reductive, open circuit, oxidative), the presence or the absence of a readily soluble electron accepting compound (with and without nitrate), and the presence or absence of an easily soluble redox mediator in order to investigate lysine synthesis in anaerobic electro-fermentation circumstances. The significance and various functions of carbon dioxide, reductive potentials, as well as redox mediator circumstances for lysine electro-fermentations have been demonstrated for the very first time (Xafenias et al., 2017). Although lysine yields along with concentrations seemed similar to those of earlier investigations using the same strain and oxygen as the acceptor of electrons, a notable impact was observed on the consumption of glucose by using the combination of reducing power, AQ2S as a redox mediator, and carbon dioxide as an appropriate gas environment. This promotes the idea that bio-electrochemical systems can improve traditional fermentation processes.

Triacylglycerides are the primary type of fats and lipids, and they are excellent, substantial precursors that are readily obtained from feedstock. Over 20% of the dry matter of the majority of oleaginous microalgae can be retained in lipids. Additionally, in adverse circumstances, several species may acquire over 50% lipid oil (Chisti, 2007). Despite huge amount of oil and fat, algae may also generate glycerol, proteins and amino acid, coupled with cellulose and various carbohydrates (Guo et al., 2015). It is possible to ferment such cellulose and other carbohydrates to produce bioethanol or biogas. Algae can be processed via gasification, hydrothermal liquefaction, pyrolysis, direct combustion, and anaerobic digestion to produce biofuels, just as other types of biomass (Borowitzka & Moheimani, 2013). Governments and businesses have supported the advancement of research and commercialization of algal biofuels over the past few decades and have funded several pilot programs involving the fuel (Singh & Gu, 2010). Although it is technically feasible to produce ethanol and diesel from algal cells, the financial implication of the process lies behind the ultimate commercialization of algae biofuels. The algal-based biofuel industry has moved to different fields for the production of highly valuable algae products like docosahexaenoic acid and eicosapentaenoic acid or to generate energy from other feedstock like energy crops or sugar because the pilot projects were not profitable (Su et al., 2017).

9 Challenges and Future Directions

The efficiency of BES is primarily influenced by a variety of parameters, including biological interaction, electrochemical processes, design and engineering elements, ambient conditions of operation, and several more, all of which require ongoing evaluation of multidisciplinary systems. It is crucial to consider the consistency as well as accuracy of the outcomes gained throughout the testing process prior to drawing any inferences about the investigation. During an extended period of operation, an adequate set of experimental duplicates may deal with such difficulties in order to verify the durability, operational stability, as well as variability of the electrochemical reactor (Jadhav et al., 2022).

Although industrial fermentation methods such as electro-fermentation are in high demand after, achieving industrial or pilot fermentation is still a challenging task. Electro-fermentation is quite a recent term and was introduced only a decade ago. Large reactor layouts may be studied and their potential optimizations are yet to be accomplished because the size is still not enhanced (Moscoviz et al., 2016). There are not many pilot-scale research on this electrochemical fermentation method that are currently available. System performance is the key concern for the scale-up, precisely as it is for other types of bio-electrochemical approaches. Increased reactor capacity leads to a large amount of energy expenditure as well as a reduced actual output volume. The major objective is to keep the electron transfer efficacy of the exoelectrogen toward the anode steady. The relation among the actual density and the surface of anode was not meticulously quantified (Redwood et al., 2012). The development of an electrochemically active system that is quite capable in maintaining the necessary electron translocation within the biofilm is crucial. A significant factor in ultimate electro-fermentation efficacy is the device's overall loss of energy. The pH gradient, which is a prevalent concern in the majority of bio-electrochemical systems irrespective of membrane development, is the primary reason for these losses of energy (Chandrasekhar et al., 2021).

Acknowledgements This research work was supported by the Scheme for Promotion of Academic and Research Collaboration (SPARC), India, Grant no SPARC/2024-2025/ENSU/P3267.

References

Adnan, N. A. A., Suhaimi, S. N., Abd-Aziz, S., Hassan, M. A., & Phang, L. Y. (2014). Optimization of bioethanol production from glycerol by Escherichia coli SS1. *Renewable Energy*, *66*, 625–633. https://doi.org/10.1016/j.renene.2013.12.032

Agler, M. T., Spirito, C. M., Usack, J. G., Werner, J. J., & Angenent, L. T. (2012). Chain elongation with reactor microbiomes: Upgrading dilute ethanol to medium-chain carboxylates. *Energy and Environmental Science*, *5*(8), 8189–8192. https://doi.org/10.1039/c2ee22101b

Altemimi, A., Aziz, S. N., Al-Hilphy, A. R., Lakhssassi, N., Watson, D. G., & Ibrahim, S. A. (2019). Critical review of radio-frequency (RF) heating applications in food processing. *Food Quality and Safety*, *3*(2), 81–91. https://doi.org/10.1093/fqsafe/fyz002

Ananthi, V., Bora, A., Ramesh, U., Yuvakkumar, R., Raja, K., Ponnuchamy, K., Muthusamy, G., & Arun, A. (2024). A review on the technologies for sustainable biohydrogen production. *Process Safety and Environmental Protection*, *186*(April), 944–956. https://doi.org/10.1016/j.psep.2024.04.034

Angenent, L. T., Karim, K., Al-Dahhan, M. H., Wrenn, B. A., & Domíguez-Espinosa, R. (2004). Production of bioenergy and biochemicals from industrial and agricultural wastewater. *Trends in Biotechnology*, *22*(9), 477–485. https://doi.org/10.1016/j.tibtech.2004.07.001

Aranda-Martinez, A., Ortiz, M. Á. N., García, I. S. A., Zavala-Gonzalez, E. A., & Lopez-Llorca, L. V. (2017). Ethanol production from chitosan by the nematophagous fungus Pochonia chlamydosporia and the entomopathogenic fungi Metarhizium anisopliae and Beauveria bassiana. *Microbiological Research*, *204*(June), 30–39. https://doi.org/10.1016/j.micres.2017.07.009

Aro, E. M. (2016). From first generation biofuels to advanced solar biofuels. *Ambio*, *45*(1), 24–31. https://doi.org/10.1007/s13280-015-0730-0

Baek, G., & Lee, J. (2024). Algae-derived electrodes in bioelectrochemical systems. *Journal of Environmental Chemical Engineering*, *12*(3), 112535. https://doi.org/10.1016/j.jece.2024.112535

Bajd, F., & Serša, I. (2011). Continuous monitoring of dough fermentation and bread baking by magnetic resonance microscopy. *Magnetic Resonance Imaging*, *29*(3), 434–442. https://doi.org/10.1016/j.mri.2010.10.010

Bhagchandanii, D. D., Babu, R. P., Sonawane, J. M., Khanna, N., Pandit, S., Jadhav, D. A., Khilari, S., & Prasad, R. (2020). A comprehensive understanding of electro-fermentation. *Fermentation*, *6*(3), 1–31. https://doi.org/10.3390/FERMENTATION6030092

Borowitzka, M. A., & Moheimani, N. R. (2013). Sustainable biofuels from algae. *Mitigation and Adaptation Strategies for Global Change*, *18*(1), 13–25. https://doi.org/10.1007/s11027-010-9271-9

Branco, R. H. R., Serafim, L. S., & Xavier, A. M. R. B. (2019). Second generation bioethanol production: On the use of pulp and paper industry wastes as feedstock. *Fermentation*, *5*(1), 1–30. https://doi.org/10.3390/fermentation5010004

Bruckenstein, S. (1977). I. Introduction, pp. 223–254.

Cai, W., Cui, K., Liu, Z., Jin, X., Chen, Q., Guo, K., & Wang, Y. (2022). An electrolytic-hydrogen-fed moving bed biofilm reactor for efficient microbial electrosynthesis of methane from CO_2. *Chemical Engineering Journal*, *428*(August 2021), 132093. https://doi.org/10.1016/j.cej.2021.132093

Cavelius, P., Engelhart-Straub, S., Mehlmer, N., Lercher, J., Awad, D., & Brück, T. (2023). The potential of biofuels from first to fourth generation. *PLoS Biology*, *21*(3), 1–20. https://doi.org/10.1371/JOURNAL.PBIO.3002063

Chakraborty, I., Sathe, S. M., Dubey, B. K., & Ghangrekar, M. M. (2020). Bioresource technology waste-derived biochar: Applications and future perspective in microbial fuel cells. *Bioresource Technology*, *312*(April), 123587. https://doi.org/10.1016/j.biortech.2020.123587

Chandrasekhar, K., Kumar, A. N., Kumar, G., Kim, D. H., Song, Y. C., & Kim, S. H. (2021). Electro-fermentation for biofuels and biochemicals production: Current status and future directions. *Bioresource Technology, 323*(December 2020), 124598. https://doi.org/10.1016/j.biortech.2020.124598

Changmai, M., Das, P. P., Mondal, P., Pasawan, M., Sinha, A., Biswas, P., Sarkar, S., & Purkait, M. K. (2022). Hybrid electrocoagulation-microfiltration technique for treatment of nanofiltration rejected steel industry effluent. *International Journal of Environmental Analytical Chemistry, 102*(1), 62–83. https://doi.org/10.1080/03067319.2020.1715381

Cheng, L., Li, S. N., Zhou, X. T., Xiao, X., & Yu, Y. Y. (2024). Short-term electrochemical stimulation induces sustained enhancement of extracellular pollutant degradation in anaerobic granular sludge. *Journal of Water Process Engineering, 63*(May), 105555. https://doi.org/10.1016/j.jwpe.2024.105555

Chisti, Y. (2007). Biodiesel from microalgae. *Biotechnology Advances, 25*(3), 294–306. https://doi.org/10.1016/j.biotechadv.2007.02.001

Choudhary, P., Assemany, P. P., Naaz, F., Bhattacharya, A., de Siqueira Castro, J., do Couto Couto, E. D. A., Calijuri, M. L., Pant, K. K., & Malik, A. (2020). A review of biochemical and thermochemical energy conversion routes of wastewater grown algal biomass. *Science of the Total Environment, 726*(271), 137961. https://doi.org/10.1016/j.scitotenv.2020.137961

Christwardana, M., Frattini, D., Accardo, G., Yoon, S. P., & Kwon, Y. (2018). Effects of methylene blue and methyl red mediators on performance of yeast based microbial fuel cells adopting polyethylenimine coated carbon felt as anode. *Journal of Power Sources, 396*(June), 1–11. https://doi.org/10.1016/j.jpowsour.2018.06.005

Chu, N., Liang, Q., Jiang, Y., & Zeng, R. J. (2020). Microbial electrochemical platform for the production of renewable fuels and chemicals. *Biosensors and Bioelectronics, 150*(September 2019), 111922. https://doi.org/10.1016/j.bios.2019.111922

Das, S., Chakraborty, I., Das, S., & Ghangrekar, M. M. (2021). Application of novel modular reactor for microbial electrosynthesis employing imposed potential with concomitant separation of acetic acid. *Sustainable Energy Technologies and Assessments, 43*(July 2020), 100902. https://doi.org/10.1016/j.seta.2020.100902

Dragone, G., & Fernandes, B. D. (2010, June) Third generation biofuels from microalgae third generation biofuels from microalgae.

Du, Q., Mu, Q., Cheng, T., Li, N., & Wang, X. (2018). Real-time imaging revealed that exoelectrogens from wastewater are selected at the center of a gradient electric field. https://doi.org/10.1021/acs.est.8b01468

Enzmann, F., & Holtmann, D. (2019). Rational scale-up of a methane producing bioelectrochemical reactor to 50 L pilot scale. *Chemical Engineering Science, 207*, 1148–1158. https://doi.org/10.1016/j.ces.2019.07.051

Ferreira, M. V. S., Cappato, L. P., Silva, R., Rocha, R. S., Guimarães, J. T., Balthazar, C. F., Esmerino, E. A., Freitas, M. Q., Rodrigues, F. N., Granato, D., & Neto, R. P. (2019) Ohmic heating for processing of whey-raspberry flavored beverage. *Food Chemistry, 297*(June), 125018. https://doi.org/10.1016/j.foodchem.2019.125018

Flores-Rodriguez, C., Nagendranatha Reddy, C., & Min, B. (2019). Enhanced methane production from acetate intermediate by bioelectrochemical anaerobic digestion at optimal applied voltages. *Biomass and Bioenergy, 127*(May), 105261. https://doi.org/10.1016/j.biombioe.2019.105261

Forsberg, J., Rasmussen, C. T., van den Berg, F. W., Engelsen, S. B., & Aru, V. (2024). Fermentation analytical technology (FAT): monitoring industrial E. coli fermentations using absolute quantitative 1H NMR spectroscopy. *Analytica Chimica Acta, 1311*(March). https://doi.org/10.1016/j.aca.2024.342722

Freitag, R. (2005). Liquid chromatography | Biotechnology applications. In: *Encyclopedia of analytical science*, pp. V5-258–V5-267. https://doi.org/10.1016/B0-12-369397-7/00330-7

Gadkari, S., Fontmorin, J. M., Yu, E., & Sadhukhan, J. (2020). Influence of temperature and other system parameters on microbial fuel cell performance: Numerical and experimental investigation. *Chemical Engineering Journal, 388*(January), 124176. https://doi.org/10.1016/j.cej.2020.124176

Gally, T., Rouaud, O., Jury, V., Havet, M., Ogé, A., & Le-Bail, A. (2017). Proofing of bread dough assisted by ohmic heating. *Innovative Food Science and Emerging Technologies, 39*, 55–62. https://doi.org/10.1016/j.ifset.2016.11.008

Gao, Q., Li, L., Zhang, Y., Zhou, H., Jiang, J., Wei, L., Wang, G., Ding, J., & Zhao, Q. (2024). Advanced oxidation processes (AOPs)-based sludge pretreatment techniques for enhanced short-chain fatty acids production: A critical review. *Chemical Engineering Journal, 489*(April), 151496. https://doi.org/10.1016/j.cej.2024.151496

Gavahian, M., & Tiwari, B. K. (2020). Moderate electric fields and ohmic heating as promising fermentation tools. *Innovative Food Science and Emerging Technologies, 64*(June), 102422. https://doi.org/10.1016/j.ifset.2020.102422

Gholamhosseinpour, A., Hashemi, S. M. B., Jahromi, L. R., & Sourki, A. H. (2020). Conventional heating, ultrasound and microwave treatments of milk: Fermentation efficiency and biological activities. *International Dairy Journal, 110*. https://doi.org/10.1016/j.idairyj.2020.104809

Gil-Carrera, L., Escapa, A., Carracedo, B., Morán, A., & Gómez, X. (2013). Performance of a semi-pilot tubular microbial electrolysis cell (MEC) under several hydraulic retention times and applied voltages. *Bioresource Technology, 146*, 63–69. https://doi.org/10.1016/j.biortech.2013.07.020

Gil, G., Chang, I. S., Kim, B. H., Kim, M., Jang, J. K., Park, H. S., & Kim, H. J. (2003). Operational parameters affecting the performance of a mediator-less microbial fuel cell. *Biosensors and Bioelectronics, 18*, 327–334.

Gong, Z., Yu, H., Zhang, J., Li, F., & Song, H. (2020). Microbial electro-fermentation for synthesis of chemicals and biofuels driven by bi-directional extracellular electron transfer. *Synthetic and Systems Biotechnology, 5*(4), 304–313. https://doi.org/10.1016/j.synbio.2020.08.004

Gorby, Y. A., Yanina, S., McLean, J. S., Rosso, K. M., Moyles, D., Dohnalkova, A., Beveridge, T. J., Chang, I. S., Kim, B. H., Kim, K. S., & Culley, D. E. (2006). Electrically conductive bacterial nanowires produced by Shewanella oneidensis strain MR-1 and other microorganisms. *Proceedings of the National Academy of Sciences, 103*(30). https://doi.org/10.1073/pnas.0604517103

Govindaraj, V., Subramani, A. K., Gopalakrishnan, R., Kim, S. K., Raval, R., & Raval, K. (2023). Bioethanol: a new synergy between marine chitinases from Bacillus haynesii and ethanol production by Mucor circinelloides. *Fermentation, 9*(1). https://doi.org/10.3390/fermentation9010040

Gu, X., Sun, J., Wang, T., Li, J., Wang, H., Wang, J., & Wang, Y. (2024) Comprehensive review of microbial production of medium-chain fatty acids from waste activated sludge and enhancement strategy. *Bioresource Technology, 402*(April), 130782. https://doi.org/10.1016/j.biortech.2024.130782

Guo, X. M., Trably, E., Latrille, E., Carrere, H., & Steyer, J. P. (2014). Predictive and explicative models of fermentative hydrogen production from solid organic waste: Role of butyrate and lactate pathways. *International Journal of Hydrogen Energy, 39*(14), 7476–7485. https://doi.org/10.1016/j.ijhydene.2013.08.079

Guo, Y., Yeh, T., Song, W., Xu, D., & Wang, S. (2015). A review of bio-oil production from hydrothermal liquefaction of algae. *Renewable and Sustainable Energy Reviews, 48*, 776–790. https://doi.org/10.1016/j.rser.2015.04.049

Guzmán-Armenteros, T. M., Villacís-Chiriboga, J., Guerra, L. S., & Ruales, J. (2024). Electromagnetic fields effects on microbial growth in cocoa fermentation: A controlled experimental approach using established growth models. *Heliyon, 10*(3). https://doi.org/10.1016/j.heliyon.2024.e24927

Hamnett, A. (2023). Fundamentals of electrochemistry. In: *Encyclopedia of chemical physics and physical chemistry: Volume 1: Fundamentals*, pp. 477–521. https://doi.org/10.1524/zpch.1995.189.part_1.147a

Handler, R. M., Shonnard, D. R., Griffing, E. M., Lai, A., & Palou-Rivera, I. (2016). Life cycle assessments of ethanol production via gas fermentation: Anticipated greenhouse gas emissions for cellulosic and waste gas feedstocks. *Industrial and Engineering Chemistry Research, 55*(12), 3253–3261. https://doi.org/10.1021/acs.iecr.5b03215

Hena, S., Znad, H., Heong, K. T., & Judd, S. (2018). Dairy farm wastewater treatment and lipid accumulation by Arthrospira platensis. *Water Research, 128*, 267–277. https://doi.org/10.1016/j.watres.2017.10.057

Huang, J., Feng, H., Huang, L., Ying, X., Shen, D., Chen, T., Shen, X., Zhou, Y., & Xu, Y. (2020). Continuous hydrogen production from food waste by anaerobic digestion (AD) coupled single-chamber microbial electrolysis cell (MEC) under negative pressure. *Waste Management, 103*, 61–66. https://doi.org/10.1016/j.wasman.2019.12.015

Jadhav, D. A., Chendake, A. D., Vinayak, V., Atabani, A., Abdelkareem, M. A., & Chae, K. J. (2022). Scale-up of the bioelectrochemical system: Strategic perspectives and normalization of performance indices. *Bioresource Technology, 363*(September), 127935. https://doi.org/10.1016/j.biortech.2022.127935

Jain, S., & Kumar, S. (2024). A comprehensive review of bioethanol production from diverse feedstocks: Current advancements and economic perspectives. *Energy, 296*(February), 131130. https://doi.org/10.1016/j.energy.2024.131130

Jin, X., Zhang, Y., Li, X., Zhao, N., & Angelidaki, I. (2017). Microbial electrolytic capture, separation and regeneration of CO_2 for biogas upgrading. *Environmental Science and Technology, 51*(16), 9371–9378. https://doi.org/10.1021/acs.est.7b01574

Jurado, L. A., & Espinosa-Marzal, R. M. (2017). Insight into the electrical double layer of an ionic liquid on graphene. *Scientific Reports, 7*(1), 1–12. https://doi.org/10.1038/s41598-017-04576-x

Kang, W., Lin, H., Ahmad, W., Li, H., & Chen, Q. (2023). Determination of active constituents in Kombucha fermentation broth using nano-composite colorimetric sensor based on selected volatile markers determined by GC-MS. *Microchemical Journal, 195*(September), 109493. https://doi.org/10.1016/j.microc.2023.109493

Ki, D., Parameswaran, P., Popat, S. C., Rittmann, B. E., & Torres, C. I. (2017). Maximizing coulombic recovery and solids reduction from primary sludge by controlling retention time and pH in a flat-plate microbial electrolysis cell. *Environmental Science: Water Research and Technology, 3*(2), 333–339. https://doi.org/10.1039/c6ew00305b

Knirsch, M. C., Dos Santos, C. A., de Oliveira Soares, A. A. M., & Penna, T. C. V. (2010). Ohmic heating—A review. *Trends in Food Science and Technology, 21*(9), 436–441. https://doi.org/10.1016/j.tifs.2010.06.003

Kokkoli, A., Zhang, Y., & Angelidaki, I. (2018) Microbial electrochemical separation of CO_2 for biogas upgrading. *Bioresource Technology, 247*(September 2017), 380–386. https://doi.org/10.1016/j.biortech.2017.09.097

Kracke, F., & Krömer, J. O. (2014). Identifying target processes for microbial electrosynthesis by elementary mode analysis. *BMC Bioinformatics, 15*(1), 1–14. https://doi.org/10.1186/s12859-014-0410-2

Kracke, F., Vassilev, I., & Krömer, J. O. (2015). Microbial electron transport and energy conservation—The foundation for optimizing bioelectrochemical systems. *Frontiers in Microbiology, 6*(JUN), 1–18. https://doi.org/10.3389/fmicb.2015.00575

Krishnan, S., Din, M. F. M., Taib, S. M., Nasrullah, M., Sakinah, M., Wahid, Z. A., Kamyab, H., Chelliapan, S., Rezania, S., & Singh, L. (2019). Accelerated two-stage bioprocess for hydrogen and methane production from palm oil mill effluent using continuous stirred tank reactor and microbial electrolysis cell. *Journal of Cleaner Production, 229*, 84–93. https://doi.org/10.1016/j.jclepro.2019.04.365

Lai, Y. H., & Lan, J. C. W. (2021). Enhanced polyhydroxybutyrate production through incorporation of a hydrogen fuel cell and electro-fermentation system. *International Journal of Hydrogen Energy*, *46*(31), 16787–16800. https://doi.org/10.1016/j.ijhydene.2020.08.117

Lee, P. C., Lee, W. G., Kwon, S., Lee, S. Y., & Chang, H. N. (1999). Succinic acid production by Anaerobiospirillum succiniciproducens: Effects of the H_2/CO_2 supply and glucose concentration. *Enzyme and Microbial Technology*, *24*(8–9), 549–554. https://doi.org/10.1016/S0141-0229(98)00156-2

Li, F., Sharma, Y., Lei, Y., Li, B., & Zhou, Q. (2010). Microbial fuel cells: The effects of configurations, electrolyte solutions, and electrode materials on power generation. *Applied Biochemistry and Biotechnology*, *160*(1), 168–181. https://doi.org/10.1007/s12010-008-8516-5

Li, X., Zeng, C., Lu, Y., Liu, G., Luo, H., & Zhang, R. (2019). Development of methanogens within cathodic biofilm in the single-chamber microbial electrolysis cell. *Bioresource Technology*, *274*(September 2018), 403–409. https://doi.org/10.1016/j.biortech.2018.12.002

Liu, C., Ren, L., Yan, B., Luo, L., Zhang, J., & Awasthi, M. K. (2021a). Electron transfer and mechanism of energy production among syntrophic bacteria during acidogenic fermentation: A review. *Bioresource Technology*, *323*(November 2020), 124637. https://doi.org/10.1016/j.biortech.2020.124637

Liu, S. Y., Charles, W., Ho, G., Cord-Ruwisch, R., & Cheng, K. Y. (2017). Bioelectrochemical enhancement of anaerobic digestion: Comparing single- and two-chamber reactor configurations at thermophilic conditions. *Bioresource Technology*, *245*(August), 1168–1175. https://doi.org/10.1016/j.biortech.2017.08.095

Liu, X., Zhan, J., Liu, L., Gan, F., Ye, J., Nealson, K. H., Rensing, C., & Zhou, S. (2021b). In situ spectroelectrochemical characterization reveals cytochrome-mediated electric syntrophy in Geobacter coculture. https://doi.org/10.1021/acs.est.1c00356

Liu, Y. P., Wang, Y. H., Wang, B. S., & Chen, Q. Y. (2014). Effect of anolyte pH and cathode Pt loading on electricity and hydrogen co-production performance of the bio-electrochemical system. *International Journal of Hydrogen Energy*, *39*(26), 14191–14195. https://doi.org/10.1016/j.ijhydene.2014.02.127

Lovley, D. R. (2017). Syntrophy goes electric: Direct interspecies electron transfer. *Annual Review of Microbiology*, *71*, 643–664. https://doi.org/10.1146/annurev-micro-030117-020420

Lu, S. et al. (2024) Temperature dependence of acids recovery from wastewater. *Resources, Conservation & Recycling*, *203*(January), 107407. https://doi.org/10.1016/j.resconrec.2024.107407

Marone, A., Ayala-Campos, O. R., Trably, E., Carmona-Martínez, A. A., Moscoviz, R., Latrille, E., Steyer, J. P., Alcaraz-Gonzalez, V., & Bernet, N. (2017). Coupling dark fermentation and microbial electrolysis to enhance bio-hydrogen production from agro-industrial wastewaters and by-products in a bio-refinery framework. *International Journal of Hydrogen Energy*, *42*(3), 1609–1621. https://doi.org/10.1016/j.ijhydene.2016.09.166

Matsumoto, T., Higuma, K., Yamada, R., & Ogino, H. (2023). Mevalonate production by electro-fermentation in Escherichia coli via Mtr-based electron transfer system. *Biochemical Engineering Journal*, *191*(December 2022), 108772. https://doi.org/10.1016/j.bej.2022.108772

Melikoglu, M. (2012). Solid-state fermentation of wheat pieces by aspergillus oryzae: Effects of microwave pretreatment on enzyme production in a biorefinery. *International Journal of Green Energy*, *9*(6), 529–539. https://doi.org/10.1080/15435075.2011.622026

Mohanakrishna, G., Reesh, I. M. A., Vanbroekhoven, K., & Pant, D. (2020). Microbial electrosynthesis feasibility evaluation at high bicarbonate concentrations with enriched homoacetogenic biocathode. *Science of the Total Environment*, *715*, 137003. https://doi.org/10.1016/j.scitotenv.2020.137003

Montpart, N., Rago, L., Baeza, J. A., & Guisasola, A. (2015). Hydrogen production in single chamber microbial electrolysis cells with different complex substrates. *Water Research*, *68*, 601–615. https://doi.org/10.1016/j.watres.2014.10.026

Moscoviz, R., Toledo-Alarcón, J., Trably, E., & Bernet, N. (2016). Electro-fermentation: How to drive fermentation using electrochemical systems. *Trends in Biotechnology*, *34*(11), 856–865. https://doi.org/10.1016/j.tibtech.2016.04.009

El-Naggar, M. Y., Wanger, G., Leung, K. M., Yuzvinsky, T. D., Southam, G., Yang, J., Lau, W. M., Nealson, K. H., & Gorby, Y. A. (2010). Electrical transport along bacterial nanowires from Shewanella oneidensis MR-1. *Proceedings of the National Academy of Sciences*, *107*(42), 18127–18131. https://doi.org/10.1073/pnas.1004880107

Müller, V. (2019). New horizons in acetogenic conversion of one-carbon substrates and biological hydrogen storage. *Trends in Biotechnology*, *37*(12), 1344–1354. https://doi.org/10.1016/j.tibtech.2019.05.008

Nguimkeu, P. (2014). A simple selection test between the Gompertz and Logistic growth models. *Technological Forecasting and Social Change*, *88*, 98–105. https://doi.org/10.1016/j.techfore.2014.06.017

Noori, M. T., Vu, M. T., Ali, R. B., & Min, B. (2020). Recent advances in cathode materials and configurations for upgrading methane in bioelectrochemical systems integrated with anaerobic digestion. *Chemical Engineering Journal*, *392*(October 2019), 123689. https://doi.org/10.1016/j.cej.2019.123689

Pandit, S., Ghosh, S., Ghangrekar, M. M., & Das, D. (2012). Performance of an anion exchange membrane in association with cathodic parameters in a dual chamber microbial fuel cell. *International Journal of Hydrogen Energy*, *37*(11), 9383–9392. https://doi.org/10.1016/j.ijhydene.2012.03.011

Park, D. O. O. H. (2000). Electricity generation in microbial fuel cells using neutral red as an electronophore. *66*(4), 1292–1297.

Pavão, L. V., Santos, L. F., Oliveira, C. M., Cruz, A. J., Ravagnani, M. A., & Costa, C. B. (2023). Flexible heat integration system in first-/second-generation ethanol production via screening pinch-based method and multiperiod model. *Energy*, *271*(January 2022). https://doi.org/10.1016/j.energy.2023.127017

Pavlov, D. (2017) *Lead-carbon electrodes, lead-acid batteries: Science and technology*. https://doi.org/10.1016/b978-0-444-59552-2.00015-8

Petrucci, R. H. et al. (2000). Book & media reviews. *74*(5), 491–492.

del Pilar Anzola Rojas, M., Zaiat, M., Gonzalez, E. R., De Wever, H., & Pant, D. (2018). Effect of the electric supply interruption on a microbial electrosynthesis system converting inorganic carbon into acetate. *Bioresource Technology*, *266*(May), 203–210. https://doi.org/10.1016/j.biortech.2018.06.074

Purkait, M. K., Das, P. P., & Sharma, M. (2024). Fundamentals of electrochemistry in water treatment. In *Electrocoagulation based treatment of water and wastewater* (pp. 47–67). https://doi.org/10.1016/b978-0-443-13892-8.00004-8

Rabaey, K., & Boon, N. (2005). Microbial phenazine production enhances electron transfer in biofuel cells. *Environmental Science & Technology*, *39*(9), 3401–3408.

Rabiee, H., Ge, L., Hu, S., Wang, H., & Yuan, Z. (2022). Microtubular electrodes: An emerging electrode configuration for electrocatalysis, bioelectrochemical and water treatment applications. *Chemical Engineering Journal*, *450*(P1), 138476. https://doi.org/10.1016/j.cej.2022.138476

Rago, L., Ruiz, Y., Baeza, J. A., Guisasola, A., & Cortés, P. (2015). Microbial community analysis in a long-term membrane-less microbial electrolysis cell with hydrogen and methane production. *Bioelectrochemistry*, *106*, 359–368. https://doi.org/10.1016/j.bioelechem.2015.06.003

Rago, L., Pant, D., & Schievano, A. (2019). Electro-fermentation-microbial electrochemistry as new frontier in biomass refineries and industrial fermentations. In *Advanced bioprocessing for alternative fuels, biobased chemicals, and bioproducts: Technologies and approaches for scale-up and commercialization*. Elsevier Inc. https://doi.org/10.1016/B978-0-12-817941-3.00014-0

Rai, P. K., & Singh, S. P. (2016). Integrated dark- and photo-fermentation: Recent advances and provisions for improvement. *International Journal of Hydrogen Energy*, *41*(44), 19957–19971. https://doi.org/10.1016/j.ijhydene.2016.08.084

Redwood, M. D., Orozco, R. L., Majewski, A. J., & Macaskie, L. E. (2012). Electro-extractive fermentation for efficient biohydrogen production. *Bioresource Technology*, *107*, 166–174. https://doi.org/10.1016/j.biortech.2011.11.026

Reguera, G. *et al.* (2006) 'Biofilm and nanowire production leads to increased current in Geobacter sulfurreducens fuel cells', *Applied and Environmental Microbiology*, *72*(11), pp. 7345–7348. Available at: https://doi.org/10.1128/AEM.01444-06.

Roller, S. D., Bennetto, H. P., Delaney, G. M., Mason, J. R., Stirling, J. L., & Thurston, C. F. (1984). Electron-transfer coupling in microbial fuel cells: 1. Comparison of redox-mediator reduction rates and respiratory rates of bacteria. *Journal of Chemical Technology and Biotechnology. Biotechnology*, *34*(1), 3–12.

Roy, M., Yadav, R., Chiranjeevi, P., & Patil, S. A. (2021). Direct utilization of industrial carbon dioxide with low impurities for acetate production via microbial electrosynthesis. *Bioresource Technology*, *320*(PA), 124289. https://doi.org/10.1016/j.biortech.2020.124289

Roy, M., Aryal, N., Zhang, Y., Patil, S. A., & Pant, D. (2022). Technological progress and readiness level of microbial electrosynthesis and electrofermentation for carbon dioxide and organic wastes valorization. *Current Opinion in Green and Sustainable Chemistry*, *35*, 100605. https://doi.org/10.1016/j.cogsc.2022.100605

Roy, S., & Pandit, S. (2018). Microbial electrochemical system: Principles and application. In *Biomass, biofuels, biochemicals: Microbial electrochemical technology: Sustainable platform for fuels, chemicals and remediation*. Elsevier B.V. https://doi.org/10.1016/B978-0-444-64052-9.00002-9

Roy, S., Schievano, A., & Pant, D. (2015). Electro-stimulated microbial factory for value added product synthesis. *Bioresource Technology*, *213*, 129–139. https://doi.org/10.1016/j.biortech.2016.03.052

Saratale, R. G., Kuppam, C., Mudhoo, A., Saratale, G. D., Periyasamy, S., Zhen, G., Koók, L., Bakonyi, P., Nemestóthy, N., & Kumar, G. (2017). Bioelectrochemical systems using microalgae—A concise research update. *Chemosphere*, *177*, 35–43. https://doi.org/10.1016/j.chemosphere.2017.02.132

Saulnier, J., Jose, C., & Lagarde, F. (2024). Electrochemical techniques for label-free and early detection of growing microbial cells and biofilms. *Bioelectrochemistry*, *155*(July 2023), 108587. https://doi.org/10.1016/j.bioelechem.2023.108587

Schmidt, I., Langner, M., & Greiner, A. (2015). Does it have to be carbon? Metal anodes in microbial fuel cells and related bioelectrochemical systems. *Energy & Environmental Science*, 2048–2055. https://doi.org/10.1039/c5ee00866b

Sharma, A., Mehdi, S. E. H., Pandit, S., Eun-Oh, S., & Natarajan, V. (2024). Factors affecting hydrogen production in microbial electrolysis cell (MEC): A review. *International Journal of Hydrogen Energy*, *61*(January), 1473–1484. https://doi.org/10.1016/j.ijhydene.2024.02.193

Siefarth, C., Tran, T. B. T., Mittermaier, P., Pfeiffer, T., & Buettner, A. (2014). Effect of radio frequency heating on yoghurt, I: Technological applicability, shelf-life and sensorial quality. *Foods*, *3*(2), 318–335. https://doi.org/10.3390/foods3020318

Singh, A., Olsen, S. I., & Nigam, P. S. (2011). A viable technology to generate third-generation biofuel. *Journal of Chemical Technology and Biotechnology*, *86*(11), 1349–1353. https://doi.org/10.1002/jctb.2666

Singh, J., & Gu, S. (2010). Commercialization potential of microalgae for biofuels production. *Renewable and Sustainable Energy Reviews*, *14*(9), 2596–2610. https://doi.org/10.1016/j.rser.2010.06.014

Song, X., Yu, Q., Zhang, Z., & Zhang, Y. (2024). Enhancing the microbial advanced oxidation of P-nitrophenol in sediment through accelerating extracellular respiration with electrical stimulation. *Journal of Hazardous Materials*, *475*(June), 134898. https://doi.org/10.1016/j.jhazmat.2024.134898

Song, Y. C., Feng, Q., & Ahn, Y. (2016). Performance of the bio-electrochemical anaerobic digestion of sewage sludge at different hydraulic retention times. *Energy and Fuels*, *30*(1), 352–359. https://doi.org/10.1021/acs.energyfuels.5b02003

Sridevi, K., Sivaraman, E., & Mullai, P. (2014). Back propagation neural network modelling of biodegradation and fermentative biohydrogen production using distillery wastewater in a hybrid upflow anaerobic sludge blanket reactor. *Bioresource Technology, 165*(C), 233–240. https://doi.org/10.1016/j.biortech.2014.03.074

Sriram, S., Wong, J. W. C., & Pradhan, N. (2022). Recent advances in electro-fermentation technology: A novel approach towards balanced fermentation. *Bioresource Technology, 360*(July), 127637. https://doi.org/10.1016/j.biortech.2022.127637

Su, Y., Song, K., Zhang, P., Su, Y., Cheng, J., & Chen, X. (2017). Progress of microalgae biofuel's commercialization. *Renewable and Sustainable Energy Reviews, 74*(February), 402–411. https://doi.org/10.1016/j.rser.2016.12.078

Sun, P. (2023). Fundamentals and applications of nanoelectrochemistry. *Encyclopedia of Nanomaterials, 01018*, V1-479–V1-495. https://doi.org/10.1016/B978-0-12-822425-0.00107-X

Sundaresan, V., Do, H., Shrout, J. D., & Bohn, P. W. (2022). Electrochemical and spectroelectrochemical characterization of bacteria and bacterial systems. *Analyst, 147*(1), 22–34. https://doi.org/10.1039/d1an01954f

Tahir, K., Miran, W., Jang, J., Shahzad, A., Moztahida, M., Kim, B., & Lee, D. S. (2020). A novel MXene-coated biocathode for enhanced microbial electrosynthesis performance. *Chemical Engineering Journal, 381*(July 2019), 122687. https://doi.org/10.1016/j.cej.2019.122687

Thapa, B. S., Kim, T., Pandit, S., Song, Y. E., Afsharian, Y. P., Rahimnejad, M., Kim, J. R., & Oh, S. E. (2022). Overview of electroactive microorganisms and electron transfer mechanisms in microbial electrochemistry. *Bioresource Technology, 347*(December 2021), 126579. https://doi.org/10.1016/j.biortech.2021.126579

Tharak, A., & Venkata Mohan, S. (2021). Electrotrophy of biocathodes regulates microbial-electro-catalyzation of CO_2 to fatty acids in single chambered system. *Bioresource Technology*. https://doi.org/10.1016/j.biortech.2020.124272

Vassilev, I., Kracke, F., Freguia, S., Keller, J., Krömer, J. O., Ledezma, P., & Virdis, B. (2019). Microbial electrosynthesis system with dual biocathode arrangement for simultaneous acetogenesis, solventogenesis and carbon chain elongation. *Chemical Communications, 55*(30), 4351–4354. https://doi.org/10.1039/c9cc00208a

Vassilev, I., Averesch, N. J., Ledezma, P., & Kokko, M. (2021). Anodic electro-fermentation: Empowering anaerobic production processes via anodic respiration. *Biotechnology Advances, 48*, 107728. https://doi.org/10.1016/j.biotechadv.2021.107728

Velvizhi, G., Sarkar, O., Rovira-Alsina, L., Puig, S., & Mohan, S. V. (2022). Conversion of carbon dioxide to value added products through anaerobic fermentation and electro fermentation: A comparative approach. *International Journal of Hydrogen Energy, 47*(34), 15442–15455. https://doi.org/10.1016/j.ijhydene.2021.12.205

Villa-Rojas, R., Zhu, M. J., Marks, B. P., & Tang, J. (2017). Radiofrequency inactivation of Salmonella Enteritidis PT 30 and Enterococcus faecium in wheat flour at different water activities. *Biosystems Engineering, 156*, 7–16. https://doi.org/10.1016/j.biosystemseng.2017.01.001

Virdis, B., Hoelzle, R. D., Marchetti, A., Boto, S. T., Rosenbaum, M. A., Blasco-Gómez, R., Puig, S., Freguia, S., & Villano, M. (2022). Electro-fermentation: Sustainable bioproductions steered by electricity. *Biotechnology Advances, 59*(February). https://doi.org/10.1016/j.biotechadv.2022.107950

De Vrije, T., Budde, M. A., Lips, S. J., Bakker, R. R., Mars, A. E., & Claassen, P. A. (2010). Hydrogen production from carrot pulp by the extreme thermophiles Caldicellulosiruptor saccharolyticus and Thermotoga neapolitana. *International Journal of Hydrogen Energy, 35*(24), 13206–13213. https://doi.org/10.1016/j.ijhydene.2010.09.014

Wang, B., Liu, W., Zhang, Y., & Wang, A. (2020). Intermittent electro field regulated mutualistic interspecies electron transfer away from the electrodes for bioenergy recovery from wastewater. *Water Research, 185*, 1–13. https://doi.org/10.1016/j.watres.2020.116238

Wang, H., & Ren, Z. J. (2013). A comprehensive review of microbial electrochemical systems as a platform technology. *Biotechnology Advances*, *31*(8), 1796–1807. https://doi.org/10.1016/j.biotechadv.2013.10.001

Wang, Z., Huang, J., Ma, S., Wang, X., Sun, B., Wang, F., Li, L., & Bao, Q. (2021). Novel heating technologies to improve fermentation efficiency and quality in wheat products: A short review. *Grain and Oil Science and Technology*, *4*(2), 81–87. https://doi.org/10.1016/j.gaost.2021.01.001

Xafenias, N., Kmezik, C., & Mapelli, V. (2017). Enhancement of anaerobic lysine production in Corynebacterium glutamicum electrofermentations. *Bioelectrochemistry*, *117*, 40–47. https://doi.org/10.1016/j.bioelechem.2017.06.001

Yang, Y., Qin, M., Yang, X., & He, Z. (2017). Enhancing hydrogen production in microbial electrolysis cells by in situ hydrogen oxidation for self-buffering pH through periodic polarity reversal. *Journal of Power Sources*, *347*, 21–28. https://doi.org/10.1016/j.jpowsour.2017.02.046

Ye, X., Zheng, T., Zhang, W., Dong, W., Wu, H., Ma, J., & Jiang, M. (2023). Membrane engineering of Escherichia coli based on "Building bridges" and "Digging tunnels" to improve electrofermentation of succinate. *Bioelectrochemistry*, *152*(December 2022), 108455. https://doi.org/10.1016/j.bioelechem.2023.108455

Yossan, S., Xiao, L., Prasertsan, P., & He, Z. (2013). Hydrogen production in microbial electrolysis cells: Choice of catholyte. *International Journal of Hydrogen Energy*, *38*(23), 9619–9624. https://doi.org/10.1016/j.ijhydene.2013.05.094

Zeppilli, M., Mattia, A., Villano, M., & Majone, M. (2017). Three-chamber bioelectrochemical system for biogas upgrading and nutrient recovery. *Fuel Cells*, *17*(5), 593–600. https://doi.org/10.1002/fuce.201700048

Zhang, L., Wang, Y. Z., Zhao, T., & Xu, T. (2019). Hydrogen production from simultaneous saccharification and fermentation of lignocellulosic materials in a dual-chamber microbial electrolysis cell. *International Journal of Hydrogen Energy*, *44*(57), 30024–30030. https://doi.org/10.1016/j.ijhydene.2019.09.191

Zhang, S., Jiang, J., Wang, H., Li, F., Hua, T., & Wang, W. (2021). A review of microbial electrosynthesis applied to carbon dioxide capture and conversion: The basic principles, electrode materials, and bioproducts. *Journal of CO_2 Utilization*, *51*(July), 101640. https://doi.org/10.1016/j.jcou.2021.101640

Zhao, Z., Zhang, Y., Wang, L., & Quan, X. (2015). Potential for direct interspecies electron transfer in an electric-anaerobic system to increase methane production from sludge digestion. *Nature Publishing Group*, (April), 1–12. https://doi.org/10.1038/srep11094

Chapter 2
Revolutionizing Biochemical Synthesis with Electro-Fermentation: Principles, Applications, and Sustainability

Jayesh Manohar Sonawane, Ankisha Vijay, Vikas Sharma, and Jesse Greener

1 Introduction

Fermentation is a pillar of biochemical production of goods ranging from biofuel to drugs and food. It takes advantage of the metabolic potential of microbes to convert organic matter (often carbohydrates) into valuable compounds such as alcohols, organic acids, and gases. Traditionally, fermentation fueled breakthroughs in brewing, biofuel production, and food preservation (Fernández-Domínguez et al., 2020). However, traditional fermentation has drawbacks, ranging from inefficient resource use and energy-consuming processes, which limit its economic and environmental viability (Karimi et al., 2021). Electro-fermentation is a hybrid approach that merges fermentation with applied electrochemistry to overcome such limitations. By driving an external electrical current into the fermenter, electro-fermentation provides enhanced control over metabolic cycles and other redox processes. The result is a process that maximizes yield efficiency and product selectivity and contributes to sustainable production objectives (Virdis et al., 2022) and can significantly advance industrial outcomes involving biochemical processes (Chandrasekhar et al., 2021). As such, electro-fermentation is an important sub-domain within the emerging movement towards bioelectrochemical systems (BES) and shares basic advantages of more environmentally sound biochemical processes (Horváth-Gönczi et al., 2023).

J. M. Sonawane (✉)
Department of Microbiology, University of Massachusetts Amherst, Amherst, MA, USA
e-mail: jsonawane@umass.edu

A. Vijay · V. Sharma
Department of Biotechnology, Jaypee Institute of Information Technology Noida, Noida, India

J. Greener
Département de Chimie, Faculté des Sciences et de Génie, Université Laval, Québec, QC, Canada

CHU de Québec, Centre de Recherche, Université Laval, Québec, QC, Canada

© The Author(s), under exclusive license to Springer Nature Singapore Pte Ltd. 2025

S. Sevda (ed.), *Electro-Fermentation for Synthesis of Chemicals, Biochemicals and Biofuels*, https://doi.org/10.1007/978-981-96-6358-3_2

Electro-fermentation is one promising advance in this paradigm, expanding the scope of microbial electrosynthesis and allowing a more rapid translation of substrates into valuable bioproducts (Bhagchandanii et al., 2020). However, unlike most BES that control electron transfer to electrogenic cells, electro-fermentation modifies the redox environment around the cells, thus controlling reaction details by changing concentrations of molecular electron mediators.

This chapter covers the basics of electro-fermentation, including the microorganisms used, techniques for boosting production, the existing industrial use, and the long-term future of this potentially disruptive technology in creating a sustainable biochemical industry.

2 The Basics of Fermentation

2.1 Fermentation as a Metabolic Process

Fermentation is a necessary process for creating energy in an oxygen-poor process environment, in which bacteria, fungi, and yeasts break down. The central biological purpose of fermentation is to enzymatically convert sugars (carbohydrates) into simpler molecules, like ethanol and lactic acid, or gases, like carbon dioxide and is often a necessary process for creating energy in low-oxygen conditions (Sharma et al., 2020).

The metabolism of fermentation centers around glycolysis, in which glucose is oxidized into pyruvate and electrons reduce NAD+, regenerating NADH. The specific chemical by-products depend largely on the microorganism and its phenotype. For example, the redox and physicochemical conditions such as pH, temperature, substrates, and electron acceptor concentrations play an important role in determining the microorganism phenotype and thc relative microbial populations.

Generally, standard fermentation, redox balance is controlled by cycling electron carriers, including NAD+/NADH, and this constrains the flexibility to optimize biochemical production above the metabolic limitations of the microorganisms themselves (Wang et al., 2020). Various chemical products are obtained through the derivatization of pyruvate. For example, this can include lactate in parallel with NADH oxidization to NAD+. In contrast, ethanol is produced after pyruvate is first decarboxylated to acetaldehyde, which is then reduced to ethanol in parallel with NADH oxidization (Sharma et al., 2020; Wang et al., 2020).

2.2 Traditional and Modern Fermentation Techniques

Traditional fermenting techniques have advanced from manually intensive to more sophisticated and automated to meet industrialized applications. Alcohol, biofuels, organic acids, pharmaceuticals, enzymes, and vitamins are fermentation products. However, despite the improvements in genetic engineering, process control, and bioreactor design, conventional fermentation is largely limited by natural microbial mechanisms, which negatively impact product control, yields, and purity (Sharma et al., 2020; Siddiqui et al., 2023).

The goal of optimizing metabolic flux is one of the main obstacles in scaling up traditional fermentation due to uncontrollable side-reactions (leading to waste by-products), which is most notably in multi-step processes. When microbial activity outgrows nutrient resources, fermentation can reach stationary or death phases, challenging scale up to industrial levels, especially in closed bulk reactors (Du et al., 2022). Moreover, traditional fermentation processes are highly resource-intensive, requiring energy and molecular inputs to maintain ideal growth conditions (largely temperature and buffers).

These issues have inspired other fermentation methods, such as electro-fermentation, which attempts to circumvent the limitations of classical fermentation by providing external electrochemical stimuli to manipulate microbial metabolism (Chandrasekhar et al., 2021).

3 Principles of Electro-fermentation

3.1 Defining Electro-fermentation

Redox control is the key to electro-fermentation as the external electric current regulates microbial metabolism by adjusting electron flows in the fermentation vessel (Kumar et al., 2018). While traditional fermentation relies on microbes' endogenous metabolism to maintain redox equilibrium, in electro-fermentation, electrodes are external electron donors or acceptors that can tune the redox condition of the microbial substrate more precisely.

Typical electro-fermentation systems include electrodes welded onto the bioreactor and connected to an external power source. Such electrodes can serve as biocathodes (electric donors) or bioanodes (electric acceptors), depending on the metabolic pathway under study. For instance, electro-fermentation can promote the synthesis of alcohols or organic acids by selectively decreasing certain intermediates (for instance, pyruvate or acetyl-CoA) through selective electron transfers (Bhagchandanii et al., 2020; Salar-García et al., 2024). Figure 1 demonstrates the basic fermentation process and some different products and applications.

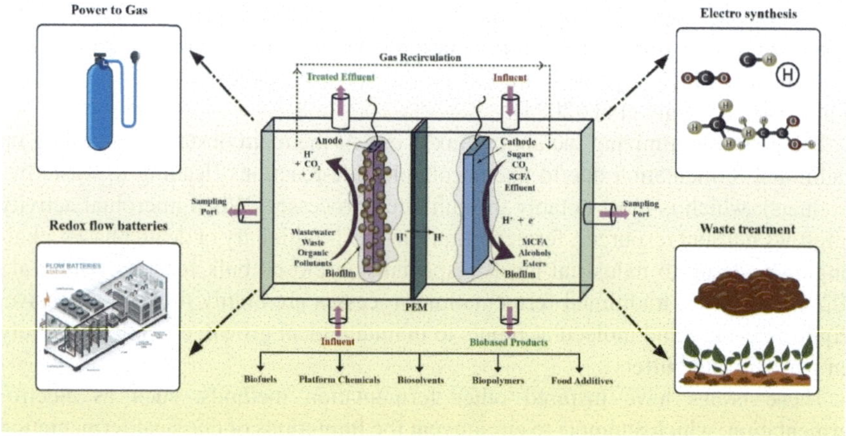

Fig. 1 Schematic of electro-fermentation and its various applications

3.2 *Mechanisms of Electro-fermentation*

The electro-fermentation process is the electron exchange between microbial cells and outside electrodes. Fermenting microorganisms typically produce electrons as the substrate is oxidized and pass them to internal electron acceptors or waste them. However, in electro-fermentation, outside electrodes serve as electron donors or acceptors and, therefore, directly modify microbial redox states to allow more precise, precise biochemical production.

- **Electron Flow**: Electrons can indirectly flow into the microbial cell through mediated electron transport, usually by mediators NAD+ and NADH. At a biocathode the electrode acts as the electron donor, usually passing negative charge to the microbes via NADH after reduction of NAD+ or via H_2 after reduction of protons. Conversely, at a bioanode, electrons flow out of the microbe after reduction of NAD+ and protons. This gives us more freedom to modulate microbial metabolism, releasing or supplying electrons at will (Bhagchandanii et al., 2020; Gong et al., 2020).
- **Redox Balancing**: The flexibility to control the redox conditions within the fermentation chamber is essential in optimizing metabolism. Specific metabolic routes can be selected by tuning the electrode potential to produce more biochemicals—organic acids, alcohols, or fatty acids. Research has found that redox regulation by electro-fermentation yields far better product yields and better use of resources than traditional processes (Bhagchandanii et al., 2020; Chandrasekhar et al., 2021; Virdis et al., 2022). Table 1 summarizes a comparison of traditional fermentation and electro-fermentation.

Table 1 Comparison of traditional fermentation and electro-fermentation

Aspect	Traditional fermentation	Electro-fermentation
Energy source	Chemical substrates (e.g., glucose)	External electrical inputs along with substrates
Control over metabolism	Limited, often constrained by redox balance	High control via modulation of redox conditions through electrodes
Product yield	Lower yield due to metabolic constraints	Higher yield with targeted redox adjustments
By-product generation	Significant by-product formation	Reduced by-products with selective pathway regulation
Sustainability	Poor due to low resource consumption efficiencies	High due to improved resource consumption and energy efficiencies
Applications	Biofuels, food, pharmaceuticals	Biofuels, bioplastics, fine chemicals, environmental applications

4 Biochemical Production Using Electro-fermentation

4.1 Biochemical Products from Electro-fermentation

Electro-fermentation has shown great potential in producing various biochemicals, many of which are difficult to obtain in high yields through conventional fermentation processes (Kumar et al., 2018). Figure 2 shows different values added products from glucose using an electrode as an electron donor in electro-fermentation. Some of the key products include:

- **Short-chain fatty acids (SCFAs)**: The most common short-chain fatty acids that can be made via electro-fermentation are acetate, butyrate, and propionate. They are the precursors of industrial chemicals such as biofuels and plastics. Electro-fermentation, by elevating the redox state and offering more electron acceptors, has significantly increased SCFA production (Bajracharya et al., 2022; Virdis et al., 2022).
- **Alcohols**: Ethanol and butanol are some of the most frequent electro-fermentation products. By manipulating electron transfer in the microbial cell, scientists can boost alcohol yields and minimize unwanted by-products. Electro-fermentation of ethanol proved especially efficient when mixed with some strains of microbial organisms, including *Clostridium acetobutylicum* (Awate et al., 2017).
- **Organic Acids**: Other organic acids generated by electro-fermentation include lactic acid and succinic acid. They are essential for the production of medicines, food components and bioplastics. By manipulating electrode potentials, electro-fermentation can be modulated to yield organic acid yields that are purified as much as possible (Bhagchandanii et al., 2020; Di Lorenzo et al., 2022).

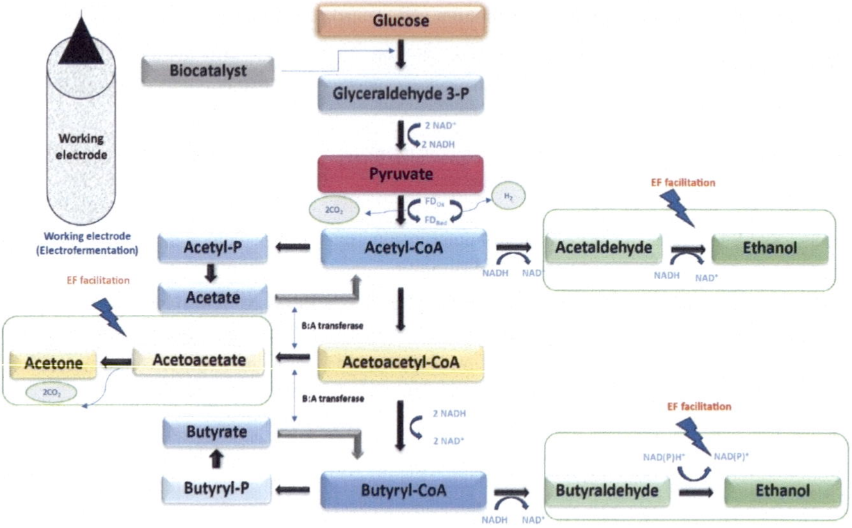

Fig. 2 Metabolism of glucose into different value-added products using electro-fermentation

4.2 Microorganisms Used in Electro-fermentation

The selection of microbes determines the efficiency and effectiveness of electro-fermentation. Microbes that simultaneously transfer electrons and ferment are needed to enable biochemical production and redox control. In electro-fermentation systems, these species are selected based on many factors, such as the substrate type, the product, or the electrode potential that regulates microbial metabolism. In most cases, a wide range of bacteria, yeasts, and archaea are used because they are metabolically diverse and highly electrochemically resistant (Bhagchandanii et al., 2020; Chandrasekhar et al., 2021; Harnisch et al., 2024).

- **Bacteria**: Electro-fermentation relies heavily on bacteria that can eat EET. *Geobacter sulfurreducens* and *Shewanella oneidensis* are electroactive bacteria using electrodes to direct electrons. These bacteria are frequently deployed in systems where the redox state is modulated, like those producing organic acids and biofuels. Furthermore, fermenting bacteria such as *Clostridium acetobutylicum* and *Escherichia coli* generate alcohols and solvents via electro-fermentation (Gong et al., 2020; Shi et al., 2024).

For acetone–butanol–ethanol (ABE) fermentation, *Clostridium acetobutylicum* is best known for butanol, acetone, and ethanol. In electro-fermentation, strict redox control can control the yields of these products. In the same way, *Escherichia coli* can be genetically engineered to make many kinds of biofuels and chemicals, and electro-fermentation can boost production by providing a second electron donor or acceptor (Alberto García Mogollón et al., 2023; Li et al., 2022).

- **Yeasts**: Yeast species like *Saccharomyces cerevisiae* are also involved in electro-fermentation, specifically alcohol production. *S. cerevisiae* is widely used and adapted for ethanol fermentation, and electrochemical control can further improve ethanol yields by eliminating the requirement for external electron acceptors. By varying the redox state, electro-fermentation allows for more precise control over by-product production, improving ethanol purity and processing efficiency. Yeasts are particularly useful for high-end biochemical manufacturing where metabolic regulation is critical (Harnisch et al., 2024; Joshi et al., 2019).
- **Archaea**: Archaea, although less common than bacteria and yeasts, are used in electro-fermentation (such as *Methanococcus maripaludis*) because of their unique metabolic abilities. Archaea are particularly well-adapted to environments with extreme conditions, like warm water or salty soils. Because they can employ a variety of electron donors and acceptors, they are good candidates for electro-fermenters that produce methane or hydrogen. New experiments have shown that archaea can be used in bioelectrochemical platforms, where electro-fermentation provides bioenergy or specialty chemicals (Harnisch et al., 2024; Pfeifer et al., 2021) (Table 2).

Table 2 Microorganisms used in electro-fermentation and their key products

Microorganism	Type	Common products	Characteristics and applications
Geobacter sulfurreducens	Bacteria	Organic acids	Electroactive, ideal for SCFA production
Shewanella oneidensis	Bacteria	Alcohols, organic acids	Electroactive, adaptable to biofuel applications
Clostridium acetobutylicum	Bacteria	Butanol, acetone, ethanol	Enhances ABE fermentation with redox control
Escherichia coli	Bacteria	Alcohols, solvents	Genetically adaptable for varied bioproducts
Saccharomyces cerevisiae	Yeast	Ethanol	Improves ethanol yield with electrical modulation
Methanococcus maripaludis	Archaea	Methane, hydrogen	Suited for bioenergy production in harsh conditions

4.3 *Optimization of Electro-fermentation*

Optimizing electro-fermentation systems involves turning into several broad operational variables to maximize biochemical production. Such parameters include electrode potential, microbial strain choice, pH, temperature, and substrate concentration. Optimization demands a deep understanding of microbial metabolism and electrochemical engineering to achieve the desired production result (Du et al., 2022; Salar-García et al., 2024).

- **Electrode Potential**: Controlling electrode potential directly determines the redox state of the fermentation conditions. As an electron donor's electrode potential is higher, it favors reductive metabolic routes, while an electron acceptor's electrode potential is higher and favors oxidative metabolism. The process can be biased locally to make specific compounds, like organic acids or alcohols (Adilkhanova et al., 2024; Chandrasekhar et al., 2021).
- **Electrode Material and Design**: The choice of electrode material impacts electro-fermentation efficiency. Carbon fibers, like graphite or carbon felt, are used extensively because of their surface area and biocompatibility. These layers are sometimes chemically or biologically functionalized by monolayers or biofilms, respectively, to improve increase electron transfer rates, and minimize electrochemical losses. Additionally, the placement of the electrodes in the bioreactor can also influence the flow of electrons, which impacts microbial growth and product yields (Lekshmi et al., 2023).
- **Temperature and pH**: Fermentation metabolisms and their biochemical products are highly sensitive to pH settings, which directly influence electron movement and proton concentration. The proton concentration plays a critical role in facilitating reduction reactions, including the production of H_2, which can function as an electron shuttle. For example, a pH range between 5.5 and 6.5 is typically optimal for the synthesis of organic acids such as lactic acid, whereas a different pH window is required for alcohol production. Likewise, temperature regulation is essential for achieving optimal microbial growth and fermentation efficiency, with each microbial strain having specific temperature thresholds to maximize productivity (Bhagchandanii et al., 2020; Gong et al., 2020).
- **Substrate Concentration and Feed Methods**: Substrate concentration is critical to achieve maximum biochemical yield in electro-fermentation plants. Substrate shortages restrict microbial activity, while excesses result in substrate inhibition and the creation of unwanted by-products. Continuous or fed-batch fermentation is a feed strategy that can keep substrate levels high at any given moment. Continuous feeding combined with live monitoring of electrode potential and metabolite levels makes it possible to dynamically modulate the system to optimize yields and productivity (Bhagchandanii et al., 2020; Chandrasekhar et al., 2021).

5 Advantages and Limitations of Electro-fermentation

5.1 Advantages of Electro-fermentation

Electro-fermentation offers several key advantages over conventional fermentation processes, particularly in terms of energy efficiency, metabolic control, and sustainability.

- **Enhanced Metabolic Control**: Using electrochemical modifications, electro-fermentation offers unprecedented control over microbial metabolism. This allows for fine-tuning metabolic pathways, yielding more products we want and fewer by-products. The control of redox balance via electrode potential is particularly useful in producing certain biochemicals, like organic acids, alcohols, and biofuels (Barbosa et al., 2021; Bhagchandanii et al., 2020).
- **Energy Efficiency**: One of the primary advantages of electro-fermentation is that it can be utilized to conserve energy in biochemical production. Electro-fermentation can eliminate the need for expensive and energy-consuming substrates when electrodes transfer electrons to microbial cells. This applies especially in the biofuel sector, where feedstocks and energy inputs are the largest scaling impediments (Bhagchandanii et al., 2020; Virdis et al., 2022).
- **Sustainability and Waste Valorization**: Electro-fermentation is intrinsically more sustainable than traditional fermentation, as it incorporates renewable electricity as an energy source. It also enables the valorization of waste streams by turning waste or commodity substrates into valuable biochemicals. For instance, organic or industrial waste could potentially be used as feedstock in electro-fermentation plants as part of the circular economy (Salar-García et al., 2024; Virdis et al., 2022).

5.2 Limitations of Electro-fermentation

Despite its many advantages, electro-fermentation is not without its challenges.

- **Electrode Biofouling**: Electrode biofouling is a common issue in bioelectrochemical devices since later microorganisms settle on the electrode surface, which increases mass transfer resistance and percent electron flow. This leads to lower process efficiency and higher operating expenses due to the constant cleaning or replacement of electrodes (Bhagchandanii et al., 2020; Salar-García et al., 2024).
- **Economic Viability**: While electro-fermentation could significantly contribute to biochemical production, it remains an economic enigma, at least at the industrial scale. Electrochemical infrastructure—such as high-performance electrodes and power supplies—can be prohibitively expensive. Besides, the scale of electro-fermentation technologies is an open question, with further studies required to

establish whether the large-scale deployment is possible (Bhagchandanii et al., 2020; Chandrasekhar et al., 2021; Kumar et al., 2018; Salar-García et al., 2024).

- **Microbial Suitability**: Not all strains of bacteria are suited to electro-fermentation, and finding suitable strains capable of surviving under the influence of external electrodes is a topic of ongoing study. Despite some bacteria and yeasts successfully adapting for electro-fermentation, strain diversity and robustness remain constrained (Gong et al., 2020).

6 Applications and Future Directions

6.1 Industrial Applications of Electro-fermentation

The industrial applications of electro-fermentation range far beyond fuels, bioplastics, medicines, and food. By giving microbes precise regulation of their metabolism and accelerating biochemical reactions, electro-fermentation promises new opportunities for low-cost, high-volume production of valuable chemicals.

- **Biofuels Synthesis**: Perhaps the most attractive application of electro-fermentation is the production of biofuels (bioethanol, biobutanol, methane). Traditional biofuels are reliant on substrates and by-products. Electro-fermentation solves these problems by regulating electron flow and increasing yield and purity. Electrochemical regulation can, e.g., dramatically improve ethanol production from *Saccharomyces cerevisiae* and allow ethanol to yield more and be free from contaminants (Alberto García Mogollón et al., 2023). Also used in butanol production is *Clostridium acetobutylicum*, where electro-fermentation allows better control of the metabolic flux and, hence, the overall yield of ABE fermentation (Alberto García Mogollón et al., 2023; Harnisch et al., 2024). Figure 3 demonstrates the different biofuel production using the electro-fermentation process.

Another important industrial electro-fermentation product is methane, which is obtained through microbial electrosynthesis using archaea like *Methanococcus maripaludis* (Wang & Liu, 2024). Electro-fermentation systems have already been demonstrated to convert CO_2 and H_2O into methane, providing a sustainable way of generating biogas at low environmental costs (Virdis et al., 2022).

- **Bioplastics and Biochemicals**: Electro-fermentation can produce several bioplastics, including polyhydroxyalkanoates (PHAs) and bacteria-derived biodegradable polymers. Electro-fermentation alters the redox balance of the fermenter, delivering PHA from bacteria such as *Ralstonia eutropha* for increased yield and minimizing conventional bioplastic manufacturing at a reduced cost. It would reduce the plastic pollution problem globally by providing an environmentally friendly alternative to petrochemical plastics (Dalton et al., 2022).

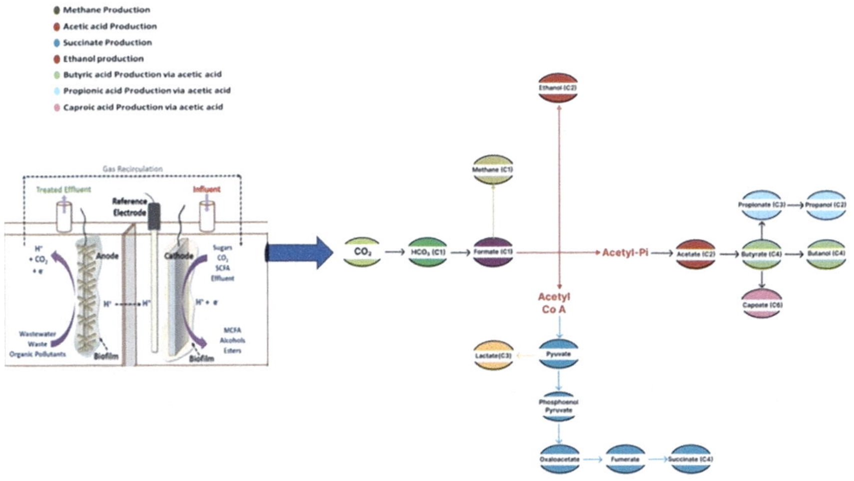

Fig. 3 Production of different biofuels using the electro-fermentation process

Besides bioplastics, electro-fermentation produces value-added biochemicals such as organic acids (succinic acid, lactic acid), alcohols, and bioactive compounds. Electro-fermentation, for instance, could save money on the production of succinic acid, one of the steps in the manufacture of other industrial chemicals, by tracking electron flows and finding the best pathways for microbial carbon fixation (Alberto García Mogollón et al., 2023; Bhagchandanii et al., 2020; Joshi et al., 2019; Salar-García et al., 2024) (Table 3).

- **Medicines and Fine Chemicals**: In the pharmaceutical sector, electro-fermentation can transform antibiotics, amino acids, and secondary metabolites in drug manufacturing by enhancing microbial metabolism, thereby enhancing biosynthesis. Redox state fine-tuning makes selective biosynthesis of extremely complex molecules that would otherwise be too difficult or expensive to make by chemical synthesis or fermentation. The latter is most relevant when producing high-quality natural products with profound therapeutic implications, such as

Table 3 Key products and applications of electro-fermentation

Product type	Example compounds	Industrial applications
Short-chain fatty acids (SCFAs)	Acetate, butyrate, propionate	Bioplastics, biofuels, industrial chemicals
Alcohols	Ethanol, butanol	Biofuels, solvents, pharmaceuticals
Organic acids	Lactic acid, succinic acid	Food preservatives, pharmaceuticals, bioplastics
Biogas	Methane, hydrogen	Renewable energy, waste valorization

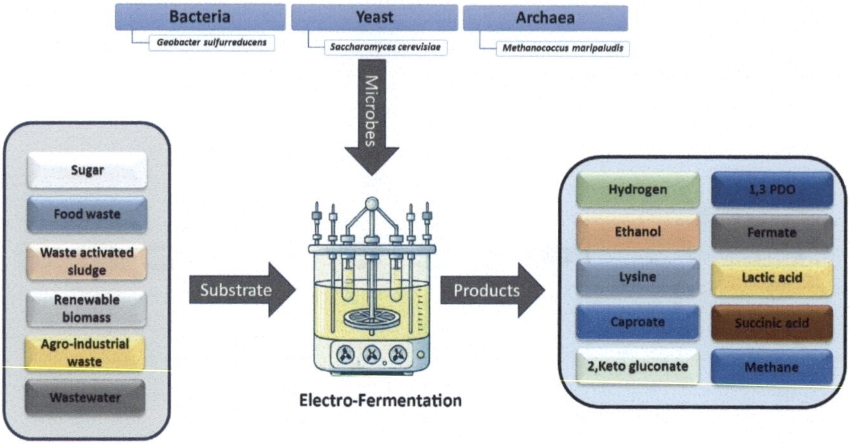

Fig. 4 Various applications and microbes involved in electro-fermentation

polyketides and non-ribosomal peptides. Furthermore, electro-fermentation can also help offset the environmental footprint of pharmaceutical production by reducing energy and waste (Bhagchandanii et al., 2020; Ferraiuolo et al., n.d.).

- **Food Processing**: Electro-fermentation can also assist food processing by increasing the yield of safe food acids (lactic and citric acid) commonly used as preservatives, flavorings, and acidulants. In addition, electro-fermentation can also refine the microbial fermentation pathways we employ for food preservation, cheesemaking, and creating flavor compounds to make our products more sustainable and efficient (Aguirre-Garcia et al., 2024; Bhagchandanii et al., 2020). Figure 4 represents the various industrial applications and microbes involved in electro-fermentation.

6.2 Environmental and Sustainability Benefits

- **Emission Minimization**: In combination with electrochemical inputs and renewable electricity, electro-fermentation would dramatically reduce the carbon emissions from biochemical production. In biofuels and bioplastics, e.g., electro-fermentation allows direct use of CO_2 as a feedstock to create useful chemicals like methane, ethanol, and organic acids. This fuels carbon capture and utilization efforts and mitigates climate change through carbon-neutral supply chains (Bhagchandanii et al., 2020; Dessì et al., 2021).
- **Energy Efficiency**: Electro-fermentation maximizes energy efficiency by eliminating the requirement for chemical energy (glucose or other carbohydrates) as electron donors. Giving electrons directly through electrodes can help reduce the overall energy consumed by microbial fermentations. This is especially true for

biofuels, where high-energy substrates can be replaced by low-cost, renewable electricity sources (Bhagchandanii et al., 2020; Jia et al., 2020).

- **Valorization of Waste and Towards Circular Economies**: Electro-fermentation can be a valuable tool in the growing waste-to-wealth movement via transform waste streams and low-value substrates into high-value chemicals. So organic residues from agriculture, food processing or sewerage treatment plants can, e.g., serve as feedstock for electro-fermentation. These generate valuable biochemicals and help create circular economies through waste reuse (Alvarez-Pugliese et al., 2024; Bhagchandanii et al., 2020).

6.3 Future of Electro-fermentation

Electro-fermentation has the potential to grow, particularly when coupled to new technologies in microbial electrosynthesis, synthetic biology, and bioelectrochemical systems, which can enable scale -up to industrial levels.

- **Microbial Electrosynthesis and Bioelectrochemical Systems**: Microbial electrosynthesis (MES) is an electro-fermentation method that uses electricity to turn CO_2 and waste into valuable compounds for microbes. Future work in MES aims to produce carbon-free biofuels, biochemicals, and food additives, which shows the possibility of powering green, sustainable production (Harnisch et al., 2024) (Table 4).

- **Compatibility with Synthetic Biology**: Synthetic biology combined with electro-fermentation offers immense promise for broadening the type of products that can be produced. By genetically engineering microbes with increased electroactivity and metabolic elasticity, scientists hope to create customized strains that can yield new compounds with high specificity and efficiency. Combining CRISPR gene editing and metabolic pathway engineering with electro-fermentation promises to

Table 4 Optimization parameters for electro-fermentation

Parameter	Description	Typical adjustments
Electrode potential	Controls redox balance in fermentation	Fine-tuning based on desired product (e.g., acids vs. alcohols)
Electrode material	Impacts electron transfer efficiency	Carbon-based materials like graphite or carbon felt
pH and temperature	Affects microbial metabolism and yield	Adjustments based on microbial strain (e.g., pH 5.5–6.5 for acid production)
Substrate concentration	Impacts microbial growth and product yield	Continuous or fed-batch feed methods

transform the manufacturing of medicines, specialty chemicals, and biomaterials (Bhagchandanii et al., 2020; Ramírez Rojas et al., 2022).

- **Industrial Scaling Electro-Fermentation**: Although much of the existing work in electro-fermentation is done in the lab, industrial scaling remains a daunting challenge. The success of designing bioreactors, developing electrodes, and optimizing the process will play a key role in making electro-fermentation commercially feasible. Attempts to build inexpensive, durable electrodes and optimize electron transfers will drive industrial-scale electro-fermentation (Bhagchandanii et al., 2020; Chandrasekhar et al., 2021).

7 Conclusion

Electro-fermentation is a revolution in biochemical production, breaking the limit of microbial technology by integrating electrochemical modulation into conventional fermentation. This combination improves metabolic regulation, energy efficiency, and sustainable production. By leveraging external electrochemical stimuli, electro-fermentation optimizes microbial metabolic pathways so that a diverse range of biochemicals, biofuels, and bio-based materials can be produced at high selectivity, yield, and efficiency levels. This capability brings multiple benefits in diverse fields, ranging from biofuels and drugs to food, bioplastics, and even management of the environment.

One of the most compelling aspects of electro-fermentation is that it has the potential to solve global problems like climate change, waste management, and sustainable use of resources. Driving microbial synthesis with external electricity (often from renewable sources), electro-fermentation removes the need for chemical inputs and high-energy reactions and reduces the carbon emissions of industrial biochemical production. Additionally, finely tuned microbial pathways improve product specificity, decrease by-product waste, and enable easier downstream processing, a significant cost and environmental benefit.

As microbial electrosynthesis, synthetic biology, and bioelectrochemical systems are being explored, tools and techniques are emerging that further expand the boundaries of electro-fermentation. Synthetic biology, for example, allows strains of microbes to be modified for higher electroactivity or specific pathways, making electro-fermentation more adaptable to generate specific molecules. Bioelectrochemical developments—including new electrode materials and reactors—also increase efficiency and scalability, making electro-fermentation an attractive option for industrial-scale use.

At the crossroads of biotechnology and electrochemistry, electro-fermentation promises a world of ever-new advances that will fuel the sustainable manufacture of chemicals, fuels, and materials necessary for the circular economy. Such a combination reimagines the mechanisms for managing biochemical production and is closely aligned with global sustainability objectives by lowering environmental impact and encouraging cleaner, greener manufacturing. As electro-fermentation technologies advance, they could open up new sectors and help build a more sustainable industrial base.

References

Adilkhanova, A., Ormantayeva, A., Kaziullayeva, A., Olaifa, K., Eghtesadi, N., Abbas, A. H., Calvio, C., Pham, T. T., Ajunwa, O. M., & Marsili, E. (2024). Electrofermentation increases concentration of poly γ-glutamic acid in Bacillus subtilis biofilms. *Microbial Biotechnology, 17*(3). https://doi.org/10.1111/1751-7915.14426

Aguirre-Garcia, Y. L., Nery-Flores, S. D., Campos-Muzquiz, L. G., Flores-Gallegos, A. C., Palomo-Ligas, L., Ascacio-Valdés, J. A., Sepúlveda-Torres, L., & Rodríguez-Herrera, R. (2024). Lactic acid fermentation in the food industry and bio-preservation of food. *Fermentation, 10*(3). https://doi.org/10.3390/fermentation10030168

Alberto García Mogollón, C., Carlos Quintero Díaz, J., & Omar Gil Posada, J. (2023). Production of acetone, butanol, and ethanol by electro-fermentation with Clostridium saccharoperbutylacetonicum N1-4. *Bioelectrochemistry, 152*, 108414. https://doi.org/10.1016/j.bioelechem.2023.108414

Alvarez-Pugliese, C. E., Donneys-Victoria, D., Cardona-Velez, W. J., & Botte, G. G. (2024). Perspectives on electrochemical valorization of organic waste. *Current Opinion in Electrochemistry, 46*, 101508. https://doi.org/10.1016/j.coelec.2024.101508

Awate, B., Steidl, R. J., Hamlischer, T., & Reguera, G. (2017). Stimulation of electro-fermentation in single-chamber microbial electrolysis cells driven by genetically engineered anode biofilms. *Journal of Power Sources, 356*, 510–518. https://doi.org/10.1016/j.jpowsour.2017.02.053

Bajracharya, S., Krige, A., Matsakas, L., Rova, U., & Christakopoulos, P. (2022). Advances in cathode designs and reactor configurations of microbial electrosynthesis systems to facilitate gas electro-fermentation. *Bioresource Technology, 354*, 127178. https://doi.org/10.1016/j.biortech.2022.127178

Barbosa, S. G., Peixoto, L., Alves, J. I., & Alves, M. M. (2021). Bioelectrochemical systems (BESs) towards conversion of carbon monoxide/syngas: A mini-review. *Renewable and Sustainable Energy Reviews, 135*, 110358. https://doi.org/10.1016/j.rser.2020.110358

Bhagchandanii, D. D., Babu, R. P., Sonawane, J. M., Khanna, N., Pandit, S., Jadhav, D. A., Khilari, S., & Prasad, R. (2020). A comprehensive understanding of electro-fermentation. *Fermentation, 6*(3). https://doi.org/10.3390/FERMENTATION6030092

Chandrasekhar, K., Naresh Kumar, A., Kumar, G., Kim, D. H., Song, Y. C., & Kim, S. H. (2021). Electro-fermentation for biofuels and biochemicals production: Current status and future directions. *Bioresource Technology, 323*. https://doi.org/10.1016/j.biortech.2020.124598

Dalton, B., Bhagabati, P., De Micco, J., Padamati, R. B., & O'Connor, K. (2022). A review on biological synthesis of the biodegradable polymers polyhydroxyalkanoates and the development of multiple applications. *Catalysts, 12*(3). https://doi.org/10.3390/catal12030319

Dessì, P., Rovira-Alsina, L., Sánchez, C., Dinesh, G. K., Tong, W., Chatterjee, P., Tedesco, M., Farràs, P., Hamelers, H. M. V., & Puig, S. (2021). Microbial electrosynthesis: Towards sustainable biorefineries for production of green chemicals from CO_2 emissions. *Biotechnology Advances, 46*, 107675. https://doi.org/10.1016/j.biotechadv.2020.107675

Di Lorenzo, R. D., Serra, I., Porro, D., & Branduardi, P. (2022). State of the art on the microbial production of industrially relevant organic acids. *Catalysts, 12*(2). https://doi.org/10.3390/catal12020234

Du, Y. H., Wang, M. Y., Yang, L. H., Tong, L. L., Guo, D. S., & Ji, X. J. (2022). Optimization and scale-up of fermentation processes driven by models. *Bioengineering, 9*(9). https://doi.org/10.3390/bioengineering9090473

Fernández-Domínguez, D., Astals, S., Peces, M., Frison, N., Bolzonella, D., Mata-Alvarez, J., & Dosta, J. (2020). Volatile fatty acids production from biowaste at mechanical-biological treatment plants: Focusing on fermentation temperature. *Bioresource Technology, 314*, 123729. https://doi.org/10.1016/j.biortech.2020.123729

Ferraiuolo, S. B., Cammarota, M., Schiraldi, C., Odile, &, & Restaino, F. (n.d.). *Streptomycetes as platform for biotechnological production processes of drugs*. https://doi.org/10.1007/s00253-020-11064-2/Published

Gong, Z., Yu, H., Zhang, J., Li, F., & Song, H. (2020). Microbial electro-fermentation for synthesis of chemicals and biofuels driven by bi-directional extracellular electron transfer. *Synthetic and Systems Biotechnology, 5*(4), 304–313. https://doi.org/10.1016/j.synbio.2020.08.004

Harnisch, F., Deutzmann, J. S., Boto, S. T., & Rosenbaum, M. A. (2024). Microbial electrosynthesis: Opportunities for microbial pure cultures. *Trends in Biotechnology, 42*(8), 1035–1047. https://doi.org/10.1016/j.tibtech.2024.02.004

Horváth-Gönczi, N. N., Bagi, Z., Szuhaj, M., Rákhely, G., & Kovács, K. L. (2023). Bioelectrochemical systems (BES) for biomethane production—Review. *Fermentation, 9*(7). https://doi.org/10.3390/fermentation9070610

Jia, X., Li, M., Wang, Y., Wu, Y., Zhu, L., Wang, X., & Zhao, Y. (2020). Enhancement of hydrogen production and energy recovery through electro-fermentation from the dark fermentation effluent of food waste. *Environmental Science and Ecotechnology, 1*, 100006. https://doi.org/10.1016/j.ese.2019.100006

Joshi, J., Dhungana, P., Prajapati, B., Maharjan, R., Poudyal, P., Yadav, M., Mainali, M., Yadav, A. P., Bhattarai, T., & Sreerama, L. (2019). Enhancement of ethanol production in electrochemical cell by Saccharomyces cerevisiae (CDBT2) and Wickerhamomyces anomalus (CDBT7). *Frontiers in Energy Research, 7*. https://doi.org/10.3389/fenrg.2019.00070

Karimi, F., Mazaheri, D., Saei Moghaddam, M., Mataei Moghaddam, A., Sanati, A. L., & Orooji, Y. (2021). Solid-state fermentation as an alternative technology for cost-effective production of bioethanol as useful renewable energy: A review. *Biomass Conversion and Biorefinery*. https://doi.org/10.1007/s13399-021-01875-2

Kumar, P., Chandrasekhar, K., Kumari, A., Sathiyamoorthi, E., & Kim, B. S. (2018). Electro-fermentation in aid of bioenergy and biopolymers. *Energies, 11*(2). https://doi.org/10.3390/en11020343

Lekshmi, G. S., Bazaka, K., Ramakrishna, S., & Kumaravel, V. (2023). Microbial electrosynthesis: Carbonaceous electrode materials for CO_2 conversion. *Materials Horizons, 10*(2), 292–312. https://doi.org/10.1039/D2MH01178F

Li, J., Zhang, Y., Sun, K., Liu, W., Yan, H., & Meng, J. (2022). Optimization of a cathodic electro-fermentation process for enhancing co-production of butanol and hydrogen via acetone-butanol-ethanol fermentation of Clostridium beijerinckii. *Energy Conversion and Management, 251*, 114987. https://doi.org/10.1016/j.enconman.2021.114987

Pfeifer, K., Ergal, İ, Koller, M., Basen, M., Schuster, B., & Rittmann, S.K.-M.R. (2021). Archaea biotechnology. *Biotechnology Advances, 47*, 107668. https://doi.org/10.1016/j.biotechadv.2020.107668

Ramírez Rojas, A. A., Swidah, R., & Schindler, D. (2022). Microbes of traditional fermentation processes as synthetic biology chassis to tackle future food challenges. *Frontiers in Bioengineering and Biotechnology, 10*. https://doi.org/10.3389/fbioe.2022.982975

Salar-García, M. J., Ortiz-Martínez, V. M., Sánchez-Segado, S., Valero Sánchez, R., Sáez López, A., Lozano Blanco, L. J., & Godínez-Seoane, C. (2024). Sustainable production of biofuels and biochemicals via electro-fermentation technology. *Molecules, 29*(4). https://doi.org/10.3390/molecules29040834

Sharma, R., Garg, P., Kumar, P., Bhatia, S. K., & Kulshrestha, S. (2020). Microbial fermentation and its role in quality improvement of fermented foods. *Fermentation, 6*(4). https://doi.org/10.3390/fermentation6040106

Shi, X., Liang, Y., Wen, G., Evlashin, S. A., Fedorov, F. S., Ma, X., Feng, Y., Zheng, J., Wang, Y., Shi, J., Liu, Y., Zhu, W., Guo, P., & Kim, B. H. (2024). Review of cathodic electroactive bacteria: Species, properties, applications and electron transfer mechanisms. *Science of the Total Environment, 946*, 174332. https://doi.org/10.1016/j.scitotenv.2024.174332

Siddiqui, S. A., Erol, Z., Rugji, J., Taşçı, F., Kahraman, H. A., Toppi, V., Musa, L., Di Giacinto, G., Bahmid, N. A., Mehdizadeh, M., & Castro-Muñoz, R. (2023). An overview of fermentation in the food industry—Looking back from a new perspective. *Bioresources and Bioprocessing, 10*(1). https://doi.org/10.1186/s40643-023-00702-y

Virdis, B., Hoelzle, R. D., Marchetti, A., Boto, S. T., Rosenbaum, M. A., Blasco-Gómez, R., Puig, S., Freguia, S., & Villano, M. (2022). Electro-fermentation: Sustainable bioproductions steered by electricity. *Biotechnology Advances, 59*, 107950. https://doi.org/10.1016/j.biotechadv.2022.107950

Wang, Q., Li, H., Feng, K., & Liu, J. (2020). Oriented fermentation of food waste towards high-value products: A review. *Energies, 13*(21). https://doi.org/10.3390/en13215638

Wang, L., & Liu, H. (2024). Microbial electrosynthesis of single cell protein and methane by coupling fast-growing Methanococcus maripaludis with microbial electrolysis cells. *Bioresource Technology, 393*, 130025. https://doi.org/10.1016/j.biortech.2023.130025

Chapter 3
Microbial Electrosynthesis Systems for CO_2 Sequestration and Biochemical Production

Sneha Prakash Naik, Isha Bodhe, C. Reshma, G. Velvizhi, Devu Pengadeth, and Gunda Mohanakrishna

1 Introduction

The global carbon footprint has increased significantly over the last century due to the increased greenhouse gas (GHG) emissions causing an everlasting environmental impact. The rise in GHG emissions, waste generation, and the energy crisis are aftermath of rapid urbanization, industrialization, and population growth. GHGs comprise 63% of CO_2, 24% of CH_4, and 3% of N_2O (ElMekawy et al., 2016). These GHGs cause critical damage to the earth's atmosphere and will lead to severe weather conditions which is evident from global annual emissions peaking at 35.8 Gt CO_2 in 2023 (Mohanakrishna et al., 2018; Liu et al., 2024). To resolve this crisis, several attempts at technological advancements have been made. The focal point of research is the development of CO_2 capture, utilization, and storage systems (CCUS), along with the generation of clean fuel and sustainable production of valuable biochemicals (Bui et al., 2018; Granovskii et al., 2007; Liu et al., 2015). Lowering the atmospheric CO_2 concentration and achieving 'negative CO_2 emissions' or 'carbon neutrality' would always remain a formidable task; hence, it is essential to create innovative technologies to lower the emissions to an acceptable limit. Over the years, numerous methods have surfaced for carbon capture and utilization (CCU) to convert waste CO_2

S. P. Naik · D. Pengadeth · G. Mohanakrishna (✉)
Center for Energy and Environment, School of Advanced Sciences, KLE Technological University, Hubli, Karnataka, India
e-mail: gmohanak@yahoo.com

I. Bodhe · C. Reshma · G. Velvizhi
CO2 Research & Green Technologies Centre, Vellore Institute of Technology, Vellore, India

I. Bodhe
School of Biosciences and Technology, Vellore Institute of Technology, Vellore, India

I. Bodhe · C. Reshma · G. Velvizhi
School of Advanced Sciences, Vellore Institute of Technology, Vellore, India

© The Author(s), under exclusive license to Springer Nature Singapore Pte Ltd. 2025
S. Sevda (ed.), *Electro-Fermentation for Synthesis of Chemicals, Biochemicals and Biofuels*, https://doi.org/10.1007/978-981-96-6358-3_3

streams into resources with added value. This approach has significantly changed the perspective where CO_2 is considered as a resource rather than a waste by-product (Qasem et al., 2018). In general, the CCU technology would reduce GHGs (i) directly by absorbing CO_2 and (ii) indirectly by substituting green fuels and chemicals as alternatives to fossil fuels. The global adoption of these strategies might offset more than 10% of GHGs while meeting 6–9% of the world's primary energy demands. Currently, the most common techniques employed for the conversion of CO_2 are chemical and thermal reduction, photocatalysis, electrocatalysis, biological and bioelectrochemical reduction (ElMekawy et al., 2016; Mohanakrishna et al., 2016); however, these processes face several shortcomings.

Microbial electrosynthesis (MES) is a process where the biocathode is involved in the reduction of carbon compounds for the production of useful chemicals and energy (Bian et al., 2020; Mohanakrishna et al., 2020). In MES, the electron transfer and microbes must work together synergistically to use CO_2 as the carbon source. The electrochemically active bacteria (EAB) produces organic matter from inorganic CO_2, which is difficult to store and can be converted into high-value compounds (Chang et al., 2006; Tian et al., 2019). MES is an effective process for CO_2 reduction and the generation of useful biochemicals (Fig. 1). The recent literature reviews provided an overview of the main features of the technology, including electron transport and basic microbiology regarding the state-of-the-art and its prospects in addition to techno-economic viability (Bian et al., 2020; Kondaveeti et al., 2020; Logan et al., 2019).

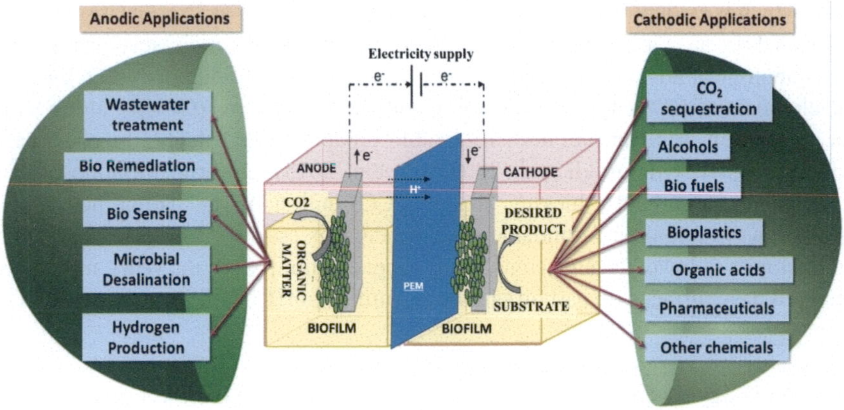

Fig. 1 Schematic representation of the multifaceted application of microbial electrosynthesis (MES)

2 Fundamentals of Microbial Electrosynthesis

MES is an emerging bioremediation and bioelectrogenesis technology, in which microbes growing on the conductive electrode surface act as a biocatalyst to facilitate the anodic oxidation or cathodic reduction reaction (Fig. 2). The performance of MES is dependent on several components, including the biocatalyst, electrode materials, reactor design, and membrane. The microbial population among all these elements could be regarded as the key factor of MES since they facilitate the biochemical reactions that use CO_2 to produce multicarbon organic compounds (Lodge & Ueki, 2016). MES can be carried out using single or mixed bacterial cultures to produce value-added products from CO_2. A unique class of bacteria such as homoacetogens and acetoclastic methanogens takes up CO_2 anaerobically and chemolithoautotrophically, allowing their conversion to valuable compounds using hydrogen as an electron donor.

2.1 *Principle of MES*

The discovery of microbial electroactivity occurred in 1911, and it has been long known that microbes can interact with electrodes. This groundbreaking discovery demonstrated that microorganisms could generate electricity and an electromotive force (EMF) comparable to a galvanic cell during the breakdown of organic matter. The experiment employed a combination of early signal processing techniques to quantify this novel phenomenon. Microbial electrochemical technologies (METs) which are also termed bioelectrochemical systems (BES) are often viewed as solutions for wastewater management problems by converting wastewater contaminants into biofuels or electricity (Logan et al., 2010, Venkata Mohan et al., 2014; Velvizhi & Venkata Mohan, 2015). In BES, at the anode, microbial activity leads to the oxidation

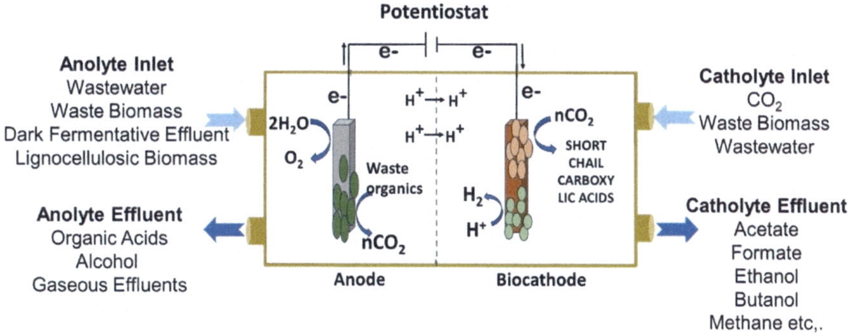

Fig. 2 Schematic representation of a microbial electrosynthesis system. The system is divided into two compartments, namely oxidation and reduction chambers. An external potential is applied to the system, and inorganic or organic matter converts into simpler products via anaerobes

of biodegradable substrates, such as waste products, resulting in the generation of electrical current. This current can be directly harvested using microbial fuel cells (MFCs) to produce energy and for the bioelectrochemical treatment of complex wastewater (Mohanakrishna et al., 2010, Velvizhi & Venkata Mohan, 2015, Fornero et al., 2010). Alternatively, bioelectrochemical energy equivalents can be channeled through microbial electrolysis cells (MECs) to generate hydrogen (Kracke et al., 2019; Venkata Mohan et al., 2014) and other valuable products (Wang & Ren, 2013). Furthermore, the cathode chamber can be employed for the production of organic molecules in the MES or the elimination of contaminants through microbial remediation cells (MRCs) (Wang & Ren, 2013).

2.2 Microbiology of MES

Enrichment of a particular set of bacteria to use CO_2 as a substrate for a high yield of desired chemicals is crucial for MES. The noteworthy microorganisms found in several MES studies are homoacetogens, owing to their ability to fix CO_2 and produce acetic acid as the end product (Annie Modestra et al., 2015; Mohanakrishna et al., 2015; Tremblay et al., 2015). In MES, chemoautotrophs (such as *Cupriavidus necator, Rhodococcus opacus, Synechococcus sp., and Nostoc punctiforme)* synthesize organic molecules and energy through the fixation of CO_2. Microbes involved in the MES processes are referred as exoelectrogens, which transfer the electrons directly to an electrode (Logan & Rabaey, 2012). Apart from direct electron transfer, other mechanisms such as mediated and mediator-less electron transfer mechanisms are involved in the anodic biofilm vicinity for electrochemical potential generation. The cathode, which is the counterpart of the anode, functions for the reduction reactions in the cathodic compartment where oxygen reduces to water molecules. In the case of MES, microorganisms dwelling on the cathode participate in the electron transfer in the presence of CO_2 (reduction reaction) which leads to the synthesis of organic molecules. Acetogens are preferred microbial catalysts for MES and other autotrophic bioprocesses like gas fermentation because CO_2 serves as their final electron acceptor, setting them apart from other autotrophic microorganisms. This indicates that a high coulombic efficiency is achieved since the majority of the electrons from a cathode or other electron donors end up in the reduced products. Through the Wood–Ljungdahl pathway, acetogens convert CO_2 mostly into acetate, although they can also produce ethanol, butyrate, 2,3-butanediol, and 1-butanol (Aryal et al., 2017).

Acetogens, such as *Acetobacterium sp.*, typically proliferate in methanogenesis-inhibiting MES reactors, those inoculated with environmental samples and treated to prevent methanogenesis. These organisms predominantly produce acetate from CO_2 and by applying potential. Additionally, a variety of acetogenic species, including *Moorella thermoacetica, Clostridium aceticum, Clostridium ljungdahlii, Acetobacterium woodii, Sporomusa ovata, sphaeroides,* and *silvacetica,* have been used to illustrate the effectiveness of a pure culture-driven MES system (Mohanakrishna et al., 2015, Ragsdale & Pierce, 2008). Exoelectrogens primarily control the anodes, although depending on the kind of reaction taking place, the communities that form in cathodic biofilms can be more diverse and incomprehensible (Siegert et al., 2015). The colony of bacteria, typically Methanobacterium among other participants at the cathode (biocathode) in single-chamber microbial electrolysis cells intended for methane production by consuming hydrogen. However, the type of methanogen that predominates could change according to the cathode material and voltage applied. According to research, these hydrogenotrophic methanogens may even directly transport electrons to the cathode in order to profit from the hydrogen created by the system's microorganisms. *Methanosarcina sp., Methanosaeta sp., Methanococcus maripaludis,* and *Methanobacterium strains* have all been shown to exhibit direct electron transfer in certain studies (Feng et al., 2015). In an investigation involving a biocathode, Mani et al. (2020) developed a power source from bacteria found in textile wastewater that was enhanced at the biocathode. The microbial consortia and their growth characteristics were interesting as they grow on an electrode without an additional substrate. The combination of bacteria that was effective at absorbing CO_2 from the atmosphere can also help to assimilate the nitrates. The nitrate reducers and CO_2 fixers collaborated to help the bacteria form clumps or biofilms and produce power. The work discovered a novel class of bacteria that may generate energy from wastewater without the use of complex sugars (Mani et al., 2020).

2.3 Electron Transfer Mechanisms

Extracellular electron transfer (EET) is a widespread microbial process crucial for biogeochemical cycling, which allows certain microorganisms to utilize insoluble extracellular compounds as an electron acceptors or donors. In MES systems, microbes can achieve direct electron transfer (DET) from electrodes (cathodes or conductive materials). This eliminates the need for mediators or self-produced chemicals, making it a cost-effective strategy for bioelectro commodity production using external electrical input (Karthikeyan et al., 2019). Several electrotrophic bacteria, including acetogens, methanogens, and photosynthetic iron oxidizers, have been shown to perform direct electron uptake (EU) from cathodes. Acetogens like *Sporomusa ovata* can convert CO_2 into acetate via this process, while methanogens like *Methanobacterium palustre* can directly convert CO_2 to methane (Park et al., 1999). Phototrophs like *Rhodopseudomonas palustris* TIE-1 can also utilize cathodes as electron donors for their photoautotrophic growth. These examples highlight the diverse applications of direct EU in MES for biochemical production, including fuels and chemicals, from various substrates like CO_2 and Fe (III) (Bose et al., 2014). The anode in MES acts as the sole electron acceptor, receiving electrons from microbes either directly or through the mediation of external electron shuttles. These two electron transfer mechanisms are categorized as direct and indirect, respectively, based on the type of interaction between the bacterial cell and the anode electrode. In direct electron transfer, electrons are channelled from bacteria to the electrode surface without the involvement of intermediary molecules. On the contrary, indirect electron transfer (IET) relies on external redox compounds to facilitate electron flow. These mediators function by accepting electrons from bacterial cells and then shuttling them to the anode for current generation (Christwardana et al., 2018).

Electroactive bacteria utilize specialized proteins for electron transfer to the extracellular space for reduction of electrodes. *Shewanella oneidensis* employs multiheme cytochromes like MtrA and OmcA, while *Geobacter sulfurreducens* utilizes the outer membrane porin-cytochrome complex (Pcc) for exoelectrogenic activity (Liu et al., 2014; Richardson et al., 2012). Cytochromes, with their heme groups, facilitate electron transfer as evident from the correlation between higher cytochrome content and increased current in biofilms. *Shewanella* and *Geobacter* also produce conductive pili, particularly type IV pili, which contribute to electron transfer and biofilm formation. In *G. sulfurreducens*, these pili, also known as nanowires, have been directly linked to electron transfer through atomic force microscopy (AFM) analysis (Okamoto et al., 2012; Vellingiri et al., 2019). Various ways to enhance the electroactivity of microorganisms include immobilization of biofilms, entrapment of bacteria via nanoparticles or chemical entrapment, or application of magnetic fields. Emerging data suggest that incorporating conductive nanomaterials can significantly enhance microbial electroactivity. Studies have demonstrated that carbon nanotubes (CNTs) and graphene oxide (GO) promote electron transfer efficiency (EET) in exoelectrogenic bacteria.

3 Factors Influencing MES Performance

The operational parameters play a crucial role in enhancing the overall performance of MES systems (Fig. 3). Optimizing process parameters can be done based on the end product, microbial community dynamics, and power generation. The performance of MES systems can be regulated by various physical, chemical, and biological parameters, namely temperature, pH, hydraulic retention time (HRT), applied potential, electrode material, membrane/separator, substrate concentration, reactor configuration, and biocatalyst (Kong et al., 2020).

3.1 Temperature

Temperature is a vital factor that influences the MES performance. Typically, MES systems have been studied under mesophilic conditions (20–37 °C), and it is found to be more suitable as well. The operational temperature influences the biofilm stability, which in turn influences the bacterial growth and product generation. Also, it has been observed that the metabolic characteristics of each exoelectrogenic species depends on temperature, and they are found to be species specific (Al-Mamun et al.,

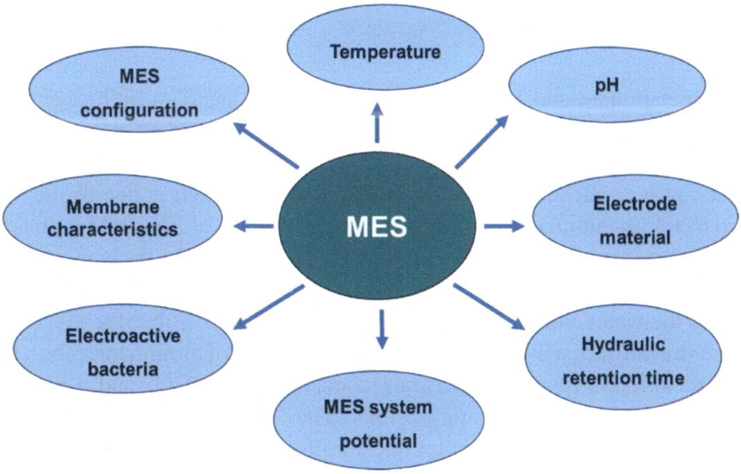

Fig. 3 Factors influencing MES performance

2023). Although the mesophilic range has its merits, operation at elevated temperatures (thermophilic conditions) enables better activity and offers more durability to the MES systems, as it enhances the substrate diffusion and minimizes the activation overpotential (Kong et al., 2020). The psychrophilic temperature (10 °C) has a negative impact on microbial biofilm proliferation, and transformation to such a temperature range causes the microbes to end up in a lethargic state with reduced performance and lower by-product recovery (Ivase et al., 2020). Therefore, it is advisable to maintain the operation temperature within the mesophilic range to achieve appreciable results.

The influence of temperature on the bacterial community was demonstrated by Zhang et al., (2018) in a study utilizing CO_2 as the carbon source. In the MES operation, a higher yield of methane (0.8–2.06 mM) with a production rate of 0.06–0.157 mM L^{-1} h^{-1} was observed by altering the operating temperature from the mesophilic range to the thermophilic range (15 to 50 °C). However, a drop in methane production was observed when the temperature reached to 70 °C. Acetate producers including *M. thermoacetica*, *M. thermoautotrophica*, and *Thermoanaerobacter kivui* showed a significant improvement in the acetate production rate under the temperature range of 25–60 °C, and it decreased as the temperature approached 70 °C. This is due to the inactivation of bacterial activity as the acetate producer approaches thermophilic temperature causing a decrease in proliferation, suggesting the mesophilic temperature range is suitable for acetate production (Yang et al., 2021).

3.2 pH

Microbial metabolism and bioelectrochemical performance are influenced by pH, and it also determines the end product formation and process efficiency. Typically, neutral pH is favourable for MES and for cathodic reduction reactions. Changes in the pH can be noticed during the process progress which is due to the volatile fatty acid (VFA) accumulation within the system (Kong et al., 2020; Mohanakrishna et al., 2016). Apart from this, pH also influences the soluble CO_2, the inorganic carbon source. A study demonstrates that with an increased CO_2 concentration, the CO_2 gets dissolved in the water and releases protons in carbonic acid. Subsequently, catholyte pH declines toward 4.5, which inactivates the metabolism of biofilm microbes. Hence, it is wise to utilize a regulated substrate concentration of CO_2 and a pH ranging between 6.5 and 8.0 for MES applications. The fluctuation in pH influences the physiological conditions such as ionic concentration, proton-motive force, membrane potential, and biofilm development that in turn regulate the process performance and chain elongation (Mohanakrishna et al., 2020). Cathodic pH is also manifested by several other parameters, such as CO_2 dissolution process and the hydrogen evolution reaction (HER) (Chen et al., 2023; Li et al., 2022). Several studies were performed with and without pH adjustment to generate targeted products from the MES process. The availability of substrate for microbial conversion of CO_2 to acetate and the overall yield can be enhanced by regulating the process pH under optimal conditions. In the case of acetate production from CO_2 with the mixed microbial consortium, it is found in the range of 5.8 and 6.0 (Batlle-Vilanova et al., 2016). Alkaline conditions in cathodic compartment exhibit deleterious effects on the growth and development of biofilm microbes and hinder biocathodic functions for acetate production (Kong et al., 2020; Mohanakrishna et al., 2016).

3.3 Hydraulic Retention Time (HRT)

The selection of an appropriate HRT is an important parameter in MES systems. HRT directs the stability of the MES process and determines the chain elongation, when the low molecular weight organic compounds are considered as substrates. From CO_2 reduction, the MES system generates an intermediate product such as acetate, which in turn gets converted to high molecular organics such as butyrate and caproate (Chen et al., 2023). Chain elongation is a long-term process, and it requires a high retention time to accumulate the prerequisite molecule (intermediate compound) to yield the final product. One of the earliest such research investigations found that the chain elongation process took 100 days to detect a C6 molecule (Vassilev et al., 2018). On the contrary, integrating MES with a membrane bioreactor operated at short HRT showed that lower HRT has a favorable effect on the process as it lowered the membrane fouling complications (Neoh et al., 2016). Adaptation of a short HRT period enables lesser energy consumption and a smaller reactor size for operation, thereby providing long-term stability and expediency to MESs (Kong et al., 2023). However, operating MES at a short HRT could cause a reduced substrate utilization that could affect the system's substrate removal efficiency.

3.4 Electrode Materials

Electrode material plays a key role in the productivity and efficacy of the MES process. An ideal electrode material should support the MES performance by having properties such as higher conductivity (for electron flow), high surface area, and hydrophilicity to provide adhesion for biofilm and to aid biofilm establishment and an efficient rate of electron transfer between microbes and electrode. Additionally, low cost, high mass transfer rate, adequate mechanical strength, chemical stability, corrosion resistance, and biocompatibility also need to be considered for the selection of suitable electrode materials (Chen et al., 2023). The most commonly explored cathodic electrode materials are carbon-based, modified carbon, metal-based, and hybrid electrodes (Zhang et al., 2021). Recently, researchers have focused more on electrode surface modification with a suitable metal-based catalyst through several surface treatment processes to help the cathodic reduction reaction. This approach reduces the activation energy and facilitates the electron transfer mechanism in the system. NanoWeb-RVC is a conductive and stable scaffold electrode, having porous and rough surfaces, and has also been proposed as a suitable cathodic electrode for MES as it enables biofilm development (Jourdin et al., 2014).

Exposing electrode material to various pretreatment methods like high-temperature, ammonia gas, or simple heat treatment, chemical pretreatment methods can enhance the activity of microbial niche embedded on the electrode surface with improved power densities (Al-Mamun et al., 2023). To boost the efficiency of electrode material, the interaction with microbes (*Sporomusa ovata*) and to increase the electron transfer mechanism, a modified carbon cloth electrode with polymer and nanoparticles was applied for acetate production which resulted in a sixfold to sevenfold increment in acetate production (Zhang et al., 2013). In another study by Mohanakrishna et al., (2016), an MES system was operated with a VITO-CoRE™ electrode under two different poised potentials (−800 mV and −600 mV) that showed biocompatibility to develop homoacetogenic biofilm on cathode and resulted in maximum acetate production of 5.45 g/L at −800 mV.

3.5 Electrochemical Potential

In MES systems, the reaction kinetics and thermodynamic reactions are increased by the application of an external potential. As per theory, in the bioelectrochemical process, microbial reactions can be carried out with less energy input. However, the end product efficiency is found to be less, which can be attributed to the energy losses by several factors associated with the MES. It was found that the applied potential on MES for biochemical production is equal to theoretical potentials and overpotential/energy losses prevailed in the system (Kong et al., 2020). Cathode potential influences the coulombic efficiency and thereby the reduction reaction as well. Each biochemical reduction reaction proceeds at a particular electrochemical potential in microbial metabolism. It is important to identify the appropriate electrochemical potential. The electrochemical potential also depends on the reactants or substrates involved in the reduction reaction. Apart from the above, cathodic biofilm/bacteria should host the appropriate biochemistry to proceed with the target biochemical reactions. It was observed that moving toward the negative side for cathodic reaction (about −0.6 V to −1.1) improves electron transfer and reduction reaction (Srikanth et al., 2014). For acetate production, the study with −0.8 V and −0.6 V revealed that −0.8 V of reducing cathode potential has a beneficial effect on acetic acid production (Mohanakrishna et al., 2016). The cathodic potential also depends on the type of electrode and materials used.

3.6 Electroactive Bacteria

The operation of MES relies on the catalytic metabolic conversion process driven by electroactive microbes embedded inside the biofilm and adhering to the cathode electrode. The microbial communities or bacterial cells that are present on the biocathode reduce the terminal electron acceptor (TEA) for CO_2 fixation and promote the end product synthesis. MES employing pure and mixed microbial cultures has been explored, and each inoculum type demonstrated its advantages and limitations in diverse studies. The end product quality and its production rate mainly depend on the selected microbial inoculum; therefore, the choice of inoculum should be on the basis of the targeted end product (Mohanakrishna et al., 2015). CO_2 reduction into organic acid in MES uses acetogenic bacteria, namely *Sporomusa silvacetica, Sporomusa sphaeroides, Clostridium ljungdahlii, Clostridium aceticum, Moorella thermoacetica,* and *Acetobacterium woodii.* Electron acceptors like nitrate, fumarate, and chlorinated solvents present in the MESs can be reduced by the electrons released by the electrode assisted with pure culture microorganisms such as *Geobacter, Sporomusa Ovata, and Clostridium Ljungdahlii* (Al-Mamun et al., 2023). In MES system, mixed microbial culture acetate production from CO_2 reduction demonstrated an increased production rate, compared to pure culture (Batlle-Vilanova et al., 2016). Compared to pure culture, an electroacetogenic microbiome has more ability to produce higher acetate yield under applied potentials (Marshall et al., 2013). This improvement in yield reported by multiple studies is due to the versatility of mixed microbial cultures that also enable working with harsh environmental conditions.

3.7 Membrane Characteristics

Dual-chambered MES systems have anodic and cathodic compartments separated by an ion exchange membrane. The application of a separator across the chambers helps to avoid mixing of the electrolyte solution (anolyte and catholyte) and enable better oxygen diffusion, thereby achieving a favourable environment in the MES system while maintaining the product quality without contamination (Al-Mamun et al., 2023). Ideally, the membrane inside MES should possess features like high ion selectivity, proton exchange, chemical stability, mechanical strength, low electrical resistance, minimal oxygen and fuel crossover, porosity, and compatibility and should be cost-effective. Generally, the membrane materials used in the MES operations are polymer electrolyte membranes, ultrafiltration membranes, forward osmosis (FO) membranes, ceramic membranes, and composite membranes (Grossi et al., 2014). Typically, polymer electrolyte-based membranes like proton exchange membrane (PEM) or cation exchange membrane (CEM) were used in most of the research studies. The commercial names for such CEM membranes can be Hyflons, Zirfons, and Ultrexs CMI-7000, and membranes are also considered better alternatives to Nafion (PEM) (Leong et al., 2013).

The selectively permeable nature of membranes allows anions and cations to move across the compartments creating a pH imbalance/difference in the system's microenvironment. This leads to the accumulation of H^+ ions in the anodic compartment causing a negative effect on MES function (Zhang et al., 2010). Additionally, membranes also create ohmic resistances leading to an increase in the energy consumption inside the system (Logan & Rabaey, 2012). The pH gradient as specified earlier when generated by the membrane across the chamber can also cause a variation in the system's redox reactions that also leads to potential losses. The improper placement of the membrane between the anode and cathode electrodes could also affect the process efficiency and lead to poor system performance (Zhang et al., 2010). From an economic point of view, the usage of membrane is not feasible when upscaling the MES system as it would require a high capital cost for installation and maintenance (Thulluru et al., 2023). Hence, membrane-less MES configurations have been explored for treating wastewater/biomass to recover high value-added products and energy as they also contribute to minimize the energy losses and to lower the internal resistance.

3.8 MES Configuration

Constructing an MES with a definite configuration greatly influences the reactor performance and efficacy. The major characteristics influencing MES design and the product formation can be listed as anode/cathode electrode spacing, distance from the electrode (anode/cathode) to membrane, the total reactor volume, electrolyte solution (anolyte and catholyte), reactor headspace gas, surface area of electrodes and membrane, source of substrates (CO_2 gas or any other substrate), operation mode (batch, fed-batch, and continuous mode), and removal of intermediate/by-product from the system (Thulluru et al., 2023).

The H-shaped reactor is the most studied configuration that comprises anode and cathode compartments separated by a membrane The limitation of H-shaped reactor design is the need for additional downstream processing steps that increase the cost for by-product and biocatalyst removal from the system (Gildemyn et al., 2015). The overpotential/ohmic loss created inside the reactor during unfavorable operation conditions can be minimized by reducing the distance between the electrodes which was proved when the electrode spacing from the membrane was reduced to 8 cm from 13 cm. Optimum electrode and membrane distance resulted in increased acetate production rate accompanied by a high current generation and high CO_2 reduction (Song et al., 2020). Introducing an extraction chamber and cathode compartment of the MES reactor is found to be advantageous to collect the products in a continuous process to eliminate feedback regulation by the products in the cathode chamber (Li et al., 2022). The MES reactor comprises three chambers, among which two of them are cathode, and are used to evaluate the process efficiency of the cathodic chamber operated at different pH conditions. One of the cathode chambers operated at pH 6.9 exhibited greater conversion of acetate and proved to be optimum for the acetogenesis and chain elongation process. Later, ethanol was produced through solventogenesis in the second cathode chamber at pH 4.9 showing that the application of multichambered MES helps to improve production efficiency with low-cost configurations (Vassilev et al., 2018).

3.9 *Nature of Substrate*

Most of the MES were operated for CO_2 reduction for acetate generation. Acetate is the dominant product that can be formed through homoacetogenic metabolism. Inorganic carbon sources like gaseous CO_2 and liquid bicarbonate have an impact on homoacetogenic bacterial growth and the chain elongation step (Christodoulou et al., 2017; Mohanakrishna et al., 2016). The application of alkali-based CO_2 capture helps in generating liquid bicarbonate, cutting down the operating cost (Li et al., 2022). It is important to note that a lower concentration of carbon source can retard the acetate formation and consecutively influence the chain elongation. Sparse CO_2 solubility and limited mass transfer of CO_2 are the major limitations associated with MES. Escape of CO_2 from the MES reactor before being reducing under cathodic electroactive bacteria is another challenge. This limitation can be overcome by implementing some of the additional approaches like CO_2 solubility improvement by stirring the catholyte and assisting CO_2 diffusion by introducing hollow fiber membranes (HFM) (Wu et al., 2022). Recirculation of headspace gas through the catholyte is another approach that helps in maximum CO_2 reduction (Bajracharya et al., 2022). The liquid phase CO_2 has four forms of ions such as CO_2, H_2CO_3, HCO_3^-, and CO_3^{2-}. The form of ions depends on pH and solubility in the catholyte. A study evaluated the ability of inorganic carbon sources, i.e., gaseous CO_2 and bicarbonate at two different applied cathode potentials of –800 mV and –600 mV for acetate production. They have observed that in MES replacing the bicarbonate with gaseous CO_2 as a carbon source improves the substrate and biofilm interaction and reduction results into increased acetate production (Mohanakrishna et al., 2016). Apart from the inorganic carbon substrates, several other organic molecules also act as substrates for chain elongation to produce ethanol, butyrate, polyhydroxy butyrate, caproate, etc. (Mohanakrishna et al., 2020).

4 Application of MES Systems

4.1 MES and CO_2 Sequestration

Through the closed-loop circulation of resources and wastes, the development of bio-based fuels using atmospheric CO_2 as a feedstock would reduce our reliance on fossil fuels (D'Amato et al., 2017). A model MES designed for wastewater treatment and CO_2 sequestration consists of an anode fed with organic substrates in which the oxidation reaction occurs in the presence of microbes releasing protons and electrons. Through an external circuit, electrons move to the cathode chamber and the protons move through the PEM membrane toward the cathode that bubbled with CO_2. The reduction reaction occurs in the cathode in the presence of CO_2 by utilizing the electrons from the circuit, and another group of microorganisms known as chemolithoautotrophs transform CO_2 into a variety of products, including fuels, biochemicals, alcohols (Annie Modestra et al., 2022; Izadi et al., 2021), biogas (He et al., 2023), organic acids (Im et al., 2022), and organic precursors for plastics (Ibrahim et al., 2023).

4.2 CO_2 Bioconversion into Useful Products

Various pathways for CO_2 fixation to value-added products along with electron exchange mechanisms are explained in Fig. 4. A variety of inorganic substances are being oxidized by anoxygenic phototrophs, which are metabolically adaptable microorganisms such as the green and purple sulfur bacteria. These include solid-phase conductive substances, as well as H_2S, H_2, and Fe^{2+}, through an intriguing process known as extracellular electron uptake (EEU) (Guzman, 2019). Guzman, (2019) demonstrated that EEU is associated with the photosynthetic electron transport chain using *Rhodopseudomonas palustris* as a model system and also indicated that this process causes cells to become highly reduced in relation to the intracellular pools of NADH and NADPH. Their findings demonstrate that EEU causes the Calvin cycle to fix CO_2 in biomass. Additionally, the Calvin cycle also serves as

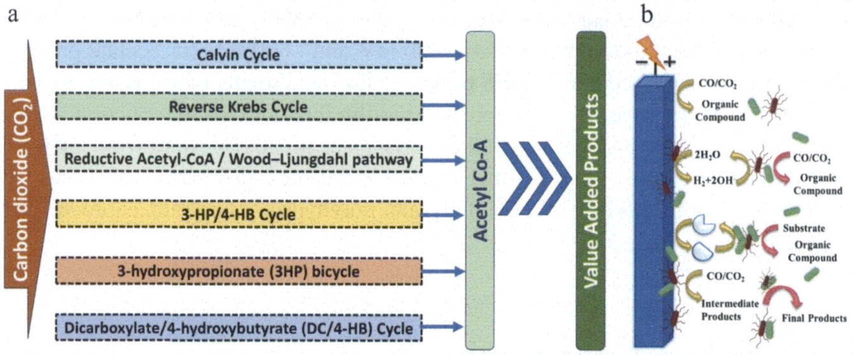

Fig. 4 **a** Overview of the 6 CO_2 fixation pathways to value-added products, **b** Electron transfer between bacterial cells and electrodes

the main electron sink through mutation analysis (Guzman, 2019). The bioconversion of CO_2 using the MES technique as a strategy creates valuable resources while lowering atmospheric CO_2. MES primarily produces C1 to C4 bioproducts through cathodic CO_2 bioelectrochemical conversion. Polyhydroxy butyrate (PHB), which is a bioplastic, can also be produced by MEC through the integration of fermentative products in the cathodic reduction reactions. On the other hand, the precursors for bioplastics can be produced from CO_2 by acetogenic or methanogenic bacteria, which can be further reduced to PHB under cathodic reaction in MES. Compared to conventional polymers made from petroleum, PHBs provide a more environmentally friendly option. Through utilizing CO_2 as a substrate, certain microorganisms can create useful biochemicals including organic acids, alcohols, and even medicines (Reddy et al., 2020). As an alternative to goods derived from fossil fuels, CO_2-based products can be utilized as a building blocks for more sophisticated chemical components.

The two electron transfer pathways such as direct electron transfer (DET) and mediated electron transfer (MET) assist in microbiological CO_2 reduction in MES. Under the DET mechanism, some chemicals transport electrons straight from the outer membrane of the bacteria to the electrode surface. Electron mediators transport electrons between the microorganisms and the electrode in the MET mechanism. Usually, these mediators are tiny organic molecules that can cross through the electrode and the microbial cell wall. The bacterial conversion of CO_2 property is used as a catalyst to produce value-added compounds. Syngas fermentation is a process where microorganisms consume syngas (CO_2 and CO) to produce fuels and chemicals. MES can also be defined as a combination of syngas fermentation in the presence of biocathode microenvironment.

4.2.1 CO_2 Conversion to Alcohols

MES can be used to produce various alcohols, including ethanol, butanol, and propanol. Microbes within the MES cathodic biofilm utilize electrons provided by the cathode to reduce CO_2 to intermediates, namely formate and acetate. Specific bacteria within the mixed culture of cathodic biofilms further convert these intermediates into alcohols. The key pathway for MES-based alcohol production involves the Wood–Ljungdahl pathway (acetyl-CoA pathway) or the reductive glycine pathway. The Wood–Ljungdahl pathway is the dominant pathway that uses CO_2 as the sole carbon source. Microbes like acetogens reduce CO_2 to formate and then to acetyl-CoA, a central intermediate. Specific bacteria like *Clostridium ljungdahlii* or *Clostridium autoethanogenum* can further convert acetyl-CoA to ethanol through acetaldehyde (Christodoulou et al., 2017) while the reductive glycine pathway also utilizes CO_2 but requires additional one-carbon compounds like methanol or formate. It involves the formation of glycine and its subsequent conversion to acetate, a precursor for ethanol production (Vassilev et al., 2018). Several bacteria have been identified as a promising candidates for MES-based alcohol production, acetogenic bacteria such as *C. ljungdahlii, C. autoethanogenum and Moorella thermoacetica* (Quraishi et al., 2021), and other ethanol producers like *Escherichia coli* (engineered strains) (Jiang et al., 2020). This process conversion could be hindered by challenges such as selectivity, controlling the metabolic pathways of the microbial community to favour alcohol production over other products (e.g., organic acids) remains a challenge and low product concentration. Current MES designs typically produce low alcohol concentrations, requiring further downstream processing for economic viability (Vassilev et al., 2018).

4.2.2 CO_2 Conversion to Fatty Acids

MES can be used to produce short-chain fatty acids (SCFAs) such as acetate, butyrate, and caproate. The process is similar to alcohol production, with microbes utilizing electrons to convert CO_2 to intermediates and further conversion to fatty acids. The primary pathway for MES-based fatty acid production involves the same Wood–Ljungdahl pathway used for alcohol production. Microbes convert CO_2 to acetyl-CoA, which can then be channelled into fatty acid synthesis through several enzymes like fatty acid synthase (FAS). Several bacteria are promising candidates for MES-based fatty acid production. Acetogenic bacteria are similar to alcohol production, *Clostridium* species and *Moorella thermoacetica* are potential candidates due to their efficient CO_2 fixation and acetyl-CoA production (Song et al., 2023), and other fatty acid producers such as *E. coli* (engineered strains) and *Synechococcus elongatus* (cyanobacteria) can also be explored for MES-based fatty acid production (Zhao et al., 2023). This complex metabolite conversion could face hurdles, such as chain elongation, downstream processing, and product separation (Christodoulou et al., 2017; Song et al., 2023). The most common bioproduct generated after CO_2 conversion in MES is often acetate. As minor products, other products like butyrate and oxo-butyrate have been developed. Reducing the accumulated acetate in MES further produces more reduced products (Bian et al., 2020; Ganigué et al., 2015).

4.2.3 CO_2 Conversion to Other Value-Added Products

MES also holds potential for producing other valuable products from CO_2 including lactate, volatile organic compounds (VOCs), and methane. Lactate is a precursor for biodegradable plastics such as polylactic acid which can be produced by the bioelectrochemical action of hydrogen and lactic acid in MES. VOCs can be produced by some of the microbes which can convert CO_2 to specific VOCs with industrial applications. Value-added CO_2 products can be obtained by integrating MES with chain elongation and bioconversion systems. This method states that the MES generates precursory molecules including H_2, CH_4, and acetate/SCFA, which can then undergo multistep bioconversions to become longer chain carboxylates, biofuels, bioplastics, polysaccharides, and proteins. For instance, the syngas platform can use the CH_4 produced in MES as a feedstock to produce ethylene, formaldehyde, and methanol (Anburajan et al., 2019; Chu et al., 2020). The challenges associated with these processes could overcome by understanding the pathways and metabolic engineering under a bioelectrochemical environment (Quraishi et al., 2021).

5 Challenges and Future Perspectives

The primary obstacles to the practical use of MES technology in the real world have several challenges which can be listed as high internal resistance, limited energy generation, current instability, by-products, and expensive setup facilities also, the problem of weak cathode kinetics is also an important obstacle. However, some research was pertinently made to address this. Selecting low-cost electrode materials made from waste products through green synthesis (such as activated carbon and biochar) can be a good way to scale-up issues from an economical perspective. Moreover, the development of chemical and biological fouling on the membrane surface affects the function of the membrane/separator and necessitates regular cleaning and membrane replacement. Membrane biofouling can be managed by physical, chemical, and biological means to preserve the long-term function of MES. The latest advancement of the MES platform is for CO_2 by-product synthesis. Comprehensive plans for effective anodic electron generation and by-product extraction were put out in an effort to reduce energy usage.

Carbonaceous materials can also be combined with other materials to improve the specific area of the electrode, which in turn increases current density. However, research into engineering carbonaceous electrodes with the appropriate physicochemical properties will help to characterize both cell–surface interactions and electron transfer efficiencies. Dip-coating was used to create unique MXene-coated electrodes with improved charge transfer efficiency, electrical conductivity, and selective microbial enrichment, increasing cathodic current output (Idumah et al., 2021). MES systems have been researched extensively in the recent decade for diverse applications, including clean energy development, environment and health monitoring, biosensors, and sustainable chemicals production. The operation of MES depends upon many factors, and these factors that govern the operational outcome. Apart from this, enrichment of biocatalysts and identification of biocompatible electrodes require intense research.

Understanding of exoelectrogenic microorganisms (EEMs) and their electron transfer mechanisms is still in its infancy; hence, intense research should be performed by developing effective screening techniques based on novel EET mechanisms. Synthetic biology emerges as a powerful tool for augmenting BES performance. This approach is being leveraged not only to enhance the performance of exoelectrogenic organisms but also to enable electrochemical activity in non-producers (Choi et al., 2014). Electrode materials play a crucial role in MES performance, and studies have reported that in comparison with metal electrode, carbonaceous materials are more biocompatible for MES applications as they offer better chemical stability and are suitable for effective microbial growth and biofilm development. Carbonaceous materials such as graphene, graphene oxide, graphite, graphite felt, graphitic carbon nitride ($g\text{-}C_3N_4$), activated carbon, carbon felt, carbon dots, and carbon fibers are effective for MES. Biochar that is being produced from several other biomass vaporization methods can be developed for biocompatible electrodes in MES. The application of 3D manufacturing techniques helps to develop novel reactor designs and 3D electrodes with high efficiency. 3D bioprinting is an innovative approach to engineering the biofilm-coated electrodes and to develop reliable theoretical models at the molecule/system level to realize the significant rate-limiting steps, reactions at the electrode–microbe interface, and extracellular electron transfer mechanisms. The MES platform offers a great opportunity for energy-driven bioproduction of various products. Addressing the identified challenges for scale-up helps with the commercialization of MES technology for bioremediation and sustainable value addition simultaneously.

6 Conclusions

MES technology offers an environmentally benign approach to waste remediation and biochemical production. However, MES technology at present is at its nascent stage and needs to overcome the significant microbial and technical hurdles. The cost associated with MES technology is high due to the need for electrode materials, membranes, and their maintenance. Despite these challenges, the MES approach to biochemical production is relevant from the sustainable development perspective, and it needs further exploration. The economic and cost-benefit analysis concerning the technology is required to identify key factors that influence the production cost and scale-up. The integration of MES with other processes can help to improve the carboxylate utilization for energy and secondary by-product generation. With the current advances in the field of bioelectrochemical technology, more robust MES systems with better productivity and efficiency can be developed.

7 Questions

1. What are the practical complications that could arise from implementing MES systems?
2. The cost breakdown of MES systems is a hurdle that needs to be overcome, the use of expensive electrode materials and membranes would only increase the cost of the system. How can this be overcome with the current technological advancements?

References

Al-Mamun, A., Ahmed, W., Jafary, T., Nayak, J. K., Al-Nuaimi, A., & Sana, A. (2023). Recent advances in microbial electrosynthesis system: Metabolic investigation and process optimization. *Biochemical Engineering Journal, 196*, Article 108928.

Amulya, K., & Mohan, S. V. (2019). Fixation of CO2, electron donor and redox microenvironment regulate succinic acid production in Citrobacter amalonaticus. *Science of the Total Environment, 695*, Article 133838.

Anburajan, P., Naresh Kumar, A., Sabapathy, P. C., Kim, G.-B., Cayetano, R. D., Yoon, J.-J., Kumar, G., & Kim, S. H. (2019). Polyhydroxy butyrate production by *Acinetobacter junii* BP25, *Aeromonas hydrophila* ATCC 7966, and their co-culture using a feast and famine strategy. *Bioresource Technology, 293*, Article 122062.

Angermayr, S. A., & Hellingwerf, K. J. (2013). On the use of metabolic control analysis in the optimization of cyanobacterial biosolar cell factories. *Journal of Physical Chemistry, 117*(38), 11169–11175.

Annie Modestra, J., Matsakas, L., Rova, U., & Christakopoulos, P. (2022). Prospects and trends in bioelectrochemical systems: Transitioning from CO_2 towards a low-carbon circular bioeconomy. *Bioresource Technology, 364*, Article 128040.

Annie Modestra, J., Navaneeth, B., & Venkata Mohan, S. (2015). Bio-electrocatalytic reduction of CO_2: Enrichment of homoacetogens and pH optimization towards enhancement of carboxylic acids biosynthesis. *Journal of CO2 Utilization, 10*, 78–87.

Aryal, N., Tremblay, P. L., Lizak, D. M., & Zhang, T. (2017). Performance of different *Sporomusa* species for the microbial electrosynthesis of acetate from carbon dioxide. *Bioresource Technology, 233*, 184–190.

Asimakopoulos, K., Grimalt-Alemany, A., Lundholm-Høffner, C., Gavala, H. N., & Skiadas, I. V. (2021). Carbon sequestration through syngas biomethanation coupled with H2 supply for a clean production of natural gas grade biomethane. *Waste Biomass Valori, 12*(11), 6005–6019.

Bajracharya, S., Krige, A., Matsakas, L., Rova, U., & Christakopoulos, P. (2022). Dual cathode configuration and headspace gas recirculation for enhancing microbial electrosynthesis using *Sporomusa ovata. Chemosphere, 287*, Article 132188.

Batlle-Vilanova, P., Puig, S., Gonzalez-Olmos, R., Balaguer, M. D., & Colprim, J. (2016). Continuous acetate production through microbial electrosynthesis from CO2 with microbial mixed culture. *Journal of Chemical Technology and Biotechnology, 91*(4), 921–927.

Bian, B., Bajracharya, S., Xu, J., Pant, D., & Saikaly, P. E. (2020). Microbial electrosynthesis from CO2: Challenges, opportunities and perspectives in the context of circular bioeconomy. *Bioresource Technology, 302*, Article 122863.

Blasco-Gómez, R., Ramió-Pujol, S., Bañeras, L., Colprim, J., Balaguer, M. D., & Puig, S. (2019). Unravelling the factors that influence the bio-electrorecycling of carbon dioxide towards biofuels. *Green Chemistry, 21*(3), 684–691.

Bose, A., Gardel, E. J., Vidoudez, C., Parra, E. A., & Girguis, P. R. (2014). Electron uptake by iron-oxidizing phototrophic bacteria. *Nature Communications, 5*(1), 3391.

Bui, M., Adjiman, C. S., Bardow, A., Anthony, E. J., Boston, A., Brown, S., Fennell, P. S., Fuss, S., Galindo, A., Hackett, L. A., Hallett, J. P., Herzog, H. J., Jackson, G., Kemper, J., Krevor, S., Maitland, G. C., Matuszewski, M., Metcalfe, I. S., Petit, C., Reimer, J., ... Mac Dowell, N. (2018). Carbon capture and storage (CCS): the way forward. *Energy & Environmental Science, 11*(5), 1062–1176.

Chang, I. S., Moon, H., Bretschger, O., Jang, J. K., Park, H. I., & Nealson, K. H. (2006). Electrochemically active bacteria (EAB) and mediator-less microbial fuel cells. *Journal of Microbiology and Biotechnology, 16*(2), 163–177.

Chen, G., Wang, R., Sun, M., Chen, J., Iyobosa, E., & Zhao, J. (2023). Carbon dioxide reduction to high–value chemicals in microbial electrosynthesis system: Biological conversion and regulation strategies. *Chemosphere*, 140251.

Choi, D., Lee, S. B., Kim, S., Min, B., Choi, I.-G., & Chang, I. S. (2014). Metabolically engineered glucose-utilizing Shewanella strains under anaerobic conditions. *Bioresource Technology, 154*, 59–66.

Christodoulou, X., Okoroafor, T., Parry, S., & Velasquez-Orta, S. B. (2017). The use of carbon dioxide in microbial electrosynthesis: Advancements, sustainability and economic feasibility. *Journal of CO2 Utilization, 18*, 390–399.

Christwardana, M., Frattini, D., Accardo, G., Yoon, S. P., & Kwon, Y. (2018). Effects of methylene blue and methyl red mediators on performance of yeast based microbial fuel cells adopting polyethylenimine coated carbon felt as anode. *Journal of Power Sources, 396*, 1–11.

Chu, N., Liang, Q., Jiang, Y., & Zeng, R. J. (2020). Microbial electrochemical platform for the production of renewable fuels and chemicals. *Biosensors & Bioelectronics, 150*, Article 111922.

D'Amato, D., Droste, N., Allen, B., Kettunen, M., Lähtinen, K., Korhonen, J., Leskinen, P., Matthies, B. D., & Toppinen, A. (2017). Green, circular, bio economy: A comparative analysis of sustainability avenues. *Journal of Cleaner Production, 168*, 716–734. https://doi.org/10.1016/j.jclepro.2017.09.053

Dessì, P., Rovira-Alsina, L., Sánchez, C., Dinesh, G. K., Tong, W., Chatterjee, P., Tedesco, M., Farràs, P., Hamelers, H. M. V., & Puig, S. (2021). Microbial electrosynthesis: Towards sustainable biorefineries for production of green chemicals from CO2 emissions. *Biotechnology Advances, 46*, Article 107675.

ElMekawy, A., Hegab, H. M., Mohanakrishna, G., Elbaz, A. F., Bulut, M., & Pant, D. (2016). Technological advances in CO2 conversion electro-biorefinery: A step toward commercialization. *Bioresource Technology, 215*, 357–370.

Feng, J., Qian, Y., Wang, Z., Wang, X., Xu, S., Chen, K., & Ouyang, P. (2018). Enhancing the performance of *Escherichia coli*-inoculated microbial fuel cells by introduction of the phenazine-1-carboxylic acid pathway. *Journal of Biotechnology, 275*, 1–6.

Feng, Y., Zhang, Y., Chen, S., & Quan, X. (2015). Enhanced production of methane from waste activated sludge by the combination of high-solid anaerobic digestion and microbial electrolysis cell with iron–graphite electrode. *Chemical Engineering Journal, 259*, 787–794. https://doi.org/10.1016/j.cej.2014.08.048

Fornero, J. J., Rosenbaum, M., & Angenent, L. T. (2010). Electric power generation from municipal, food, and animal wastewaters using microbial fuel cells. *Electroanalysis, 22*(7–8), 832–843.

Ganigué, R., Puig, S., Batlle-Vilanova, P., Balaguer, M. D., & Colprim, J. (2015). Microbial electrosynthesis of butyrate from carbon dioxide. *Chemical Communications, 51*(15), 3235–3238.

Geppert, F., Liu, D., Weidner, E., & ter Heijne, A. (2019). Redox-flow battery design for a methane-producing bioelectrochemical system. *International Journal of Hydrogen Energy, 44*(39), 21464–21469.

Gildemyn, S., Verbeeck, K., Slabbinck, R., Andersen, S. J., Prévoteau, A., & Rabaey, K. (2015). Integrated production, extraction, and concentration of acetic acid from CO_2 through microbial electrosynthesis. *Environmental Science & Technology Letters, 2*(11), 325–328.

Granovskii, M., Dincer, I., & Rosen, M. A. (2007). Greenhouse gas emissions reduction by use of wind and solar energies for hydrogen and electricity production: Economic factors. *International Journal of Hydrogen Energy, 32*(8), 927–931.

Grossi, E. N., Chao, L. Y., Flynn, M., & Hogan, J. A. (2014). The effects of membrane selection on bioelectrochemical system performance. In *44th International Conference on Environmental Systems.*

Guzman, M. S. (2019). *Understanding the Physiology of Extracellular Electron Uptake in Purple Nonsulfur Bacteria.* Washington University in St.

He, Y., Li, J., Zhang, L., Zhu, X., Pang, Y., Fu, Q., & Liao, Q. (2023). 3D-printed GA/PPy aerogel biocathode enables efficient methane production in microbial electrosynthesis. *Chemical Engineering Journal, 459*, Article 141523.

Ibrahim, I., Ikhmal, M., Balachandran, K., Me, M., Loh, K. S., Bakar, M., Jong, B., & Lim, S. S. (2023). Role of microbial electrosynthesis system in CO_2 capture and conversion: A recent advancement toward cathode development. *Frontiers in Microbiology, 14*, 1192187.

Idumah, C. I., Ezeani, E. O., Ezika, A. C., & Timothy, U. J. (2021). Recent advancements in flame retardancy of MXene polymer nanoarchitectures. *Safety in Extreme Environments, 3*(3), 253–273.

Im, C., Valgepea, K., Modin, O., & Nygård, Y. (2022). *Clostridium ljungdahlii* as a biocatalyst in microbial electrosynthesis–effect of culture conditions on product formation. *Bioresource Technology Reports, 19*, Article 101156.

Ivase, T. J. P., Nyakuma, B. B., Oladokun, O., Abu, P. T., & Hassan, M. N. (2020). Review of the principal mechanisms, prospects, and challenges of bioelectrochemical systems. *Environmental Progress & Sustainable Energy, 39*(1), 13298.

Izadi, P., Fontmorin, J. M., Virdis, B., Head, I. M., & Yu, E. H. (2021). The effect of the polarised cathode, formate and ethanol on chain elongation of acetate in microbial electrosynthesis. *Applied Energy, 283*, Article 116310.

Jiang, Y., Chu, N., Qian, D. K., & Zeng, R. J. (2020). Microbial electrochemical stimulation of caproate production from ethanol and carbon dioxide. *Bioresource Technology, 295*, Article 122266.

Jourdin, L., Freguia, S., Donose, B. C., Chen, J., Wallace, G. G., Keller, J., & Flexer, V. (2014). A novel carbon nanotube modified scaffold as an efficient biocathode material for improved microbial electrosynthesis. *Journal of Materials Chemistry A, 2*(32), 13093–13102.

Karthikeyan, R., Singh, R., & Bose, A. (2019). Microbial electron uptake in microbial electrosynthesis: A mini-review. *Journal of Industrial Microbiology and Biotechnology, 46*(9–10), 1419–1426.

Kondaveeti, S., Abu-Reesh, I. M., Mohanakrishna, G., Bulut, M., & Pant, D. (2020). Advanced routes of biological and bio-electrocatalytic carbon dioxide (CO_2) mitigation toward carbon neutrality. *Frontiers in Energy Research, 8.*

Kong, F., Ren, H. Y., Pavlostathis, S. G., Nan, J., Ren, N. Q., & Wang, A. (2020). Overview of value-added products bioelectrosynthesized from waste materials in microbial electrosynthesis systems. *Renewable and Sustainable Energy Reviews, 125*, Article 109816.

Kong, W., Wang, Q., Huang, L., Quan, X., & Li Puma, G. (2023). Highly efficient and photostable conversion of bicarbonate to acetate by a pyrrolic nitrogen enriched Ag_3PO_4/g-C_3N_4 photocathode with an intermediate band level in photo-assisted microbial electrosynthesis systems. *Chemical Engineering Journal, 471*, 144673. https://doi.org/10.1016/j.cej.2023.144673

Kracke, F., Wong, A. B., Maegaard, K., Deutzmann, J. S., Hubert, M. A., Hahn, C., Jaramillo, T. F., & Spormann, A. M. (2019). Robust and biocompatible catalysts for efficient hydrogen-driven microbial electrosynthesis. *Communications Chemistry, 2*(1), 45.
Leong, J. X., Daud, W. R. W., Ghasemi, M., Liew, K. B., & Ismail, M. (2013). Ion exchange membranes as separators in microbial fuel cells for bioenergy conversion: A comprehensive review. *Renewable and Sustainable Energy Reviews, 28*, 575–587.
Li, W., Jiafang, Z., Menggen, L., Aixin, Z., & Ning, H. (2022). Simulation of cathode for synthesizing organic acids by MES reduction of CO2. *Bioelectrochemistry, 143*, Article 107984.
Liew, F. E., Nogle, R., Abdalla, T., Rasor, B. J., Canter, C., Jensen, R. O., Wang, L., Strutz, J., Chirania, P., De Tissera, S., Mueller, A. P., Ruan, Z., Gao, A., Tran, L., Engle, N. L., Bromley, J. C., Daniell, J., Conrado, R., Tschaplinski, T. J., Giannone, R. J., … Köpke, M. (2022). Carbon-negative production of acetone and isopropanol by gas fermentation at industrial pilot scale. *Nature Biotechnology, 40*(3), 335–344.
Liu, Y., Wang, Z., Liu, J., Levar, C., Edwards, M. J., Babauta, J. T., Kennedy, D. W., Shi, Z., Beyenal, H., Bond, D. R., Clarke, T. A., Butt, J. N., Richardson, D. J., Rosso, K. M., Zachara, J. M., Fredrickson, J. K., & Shi, L. (2014). A trans-outer membrane porin-cytochrome protein complex for extracellular electron transfer by *Geobacter sulfurreducens* PCA. *Environmental Microbiology Reports, 6*(6), 776–785.
Liu, Z., Deng, Z., Davis, S. J., & Ciais, P. (2024). Global carbon emissions in 2023. *Nature Reviews Earth & Environment, 5*(4), 253–254.
Liu, Z., Guan, D., Wei, W., Davis, S. J., Ciais, P., Bai, J., Peng, S., Zhang, Q., Hubacek, K., Marland, G., Andres, R. J., Crawford-Brown, D., Lin, J., Zhao, H., Hong, C., Boden, T. A., Feng, K., Peters, G. P., Xi, F., … He, K. (2015). Reduced carbon emission estimates from fossil fuel combustion and cement production in China. *Nature, 524*(7565), 335–338.
Lodge, T. P., & Ueki, T. (2016). Mechanically tunable, readily processable ion gels by self-assembly of block copolymers in ionic liquids. *Accounts of Chemical Research, 49*(10), 2107–2114.
Logan, B. E. (2010). Scaling up microbial fuel cells and other bioelectrochemical systems. *Applied Microbiology and Biotechnology, 85*(6), 1665–1671. https://doi.org/10.1007/s00253-009-2378-9
Logan, B. E., & Elimelech, M. (2012). Membrane-based processes for sustainable power generation using water. *Nature, 488*(7411), 313–319.
Logan, B. E., & Rabaey, K. (2012). Conversion of wastes into bioelectricity and chemicals by using microbial electrochemical technologies. *Science, 337*(6095), 686–690.
Logan, B. E., Rossi, R., Ragab, A., & Saikaly, P. E. (2019). Electroactive microorganisms in bioelectrochemical systems. *Nature Reviews Microbiology, 17*(5), 307–319.
Lu, L., Guest, J. S., Peters, C. A., Zhu, X., Rau, G. H., & Ren, Z. J. (2018). Wastewater treatment for carbon capture and utilization. *Nature Sustainability, 1*(12), 750–758.
Lu, Z., Chang, D., Ma, J., Huang, G., Cai, L., & Zhang, L. (2015). Behaviour of metal ions in bioelectrochemical systems: A review. *Journal of Power Sources, 275*, 243–260.
Mani, P., VT, F., Bowman, K., TS, C., Keshavarz, T., & Kyazze, G. (2020). Development of an electroactive aerobic biocathode for microbial fuel cell applications. *Environmental Microbiology Reports, 12*(5), 607–612.
Marshall, C. W., Ross, D. E., Fichot, E. B., Norman, R. S., & May, H. D. (2013). Long-term operation of microbial electrosynthesis systems improves acetate production by autotrophic microbiomes. *Environmental Science & Technology, 47*(11), 6023–6029. https://doi.org/10.1021/es400341b
Masnadi, M. S., El-Houjeiri, H. M., Schunack, D., Li, Y., Englander, J. G., Badahdah, A., Monfort, J.-C., Anderson, J. E., Wallington, T. J., Bergerson, J. A., Gordon, D., Koomey, J., Przesmitzki, S., Azevedo, I. L., Bi, X. T., Duffy, J. E., Heath, G. A., Keoleian, G. A., McGlade, C., … Brandt, A. R. 2018. Global carbon intensity of crude oil production. *Science, 361*(6405), 851–853.
Mohanakrishna, G., Seelam, J. S., Vanbroekhoven, K., & Pant, D. (2015). An enriched electroactive homoacetogenic biocathode for the microbial electrosynthesis of acetate through carbon dioxide reduction. *Faraday Discussions, 183*, 445–462.

Mohanakrishna, G., Vanbroekhoven, K., & Pant, D. 2016. Imperative role of applied potential and inorganic carbon source on acetate production through microbial electrosynthesis. *Journal of CO2 Utilization 15*, 57–64.

Mohanakrishna, G., Venkata Mohan, S., & Sarma, P. N. (2010). Bio-electrochemical treatment of distillery wastewater in microbial fuel cell facilitating decolorization and desalination along with power generation. *Journal of Hazardous Materials, 177*(1), 487–494.

Mohanakrishna, G., Reesh, I. M. A., Vanbroekhoven, K., & Pant, D. (2020). Microbial electrosynthesis feasibility evaluation at high bicarbonate concentrations with enriched homoacetogenic biocathode. *Science of the Total Environment, 715*, Article 137003.

Mohanakrishna, G., Vanbroekhoven, K., & Pant, D. (2018). Impact of dissolved carbon dioxide concentration on the process parameters during its conversion to acetate through microbial electrosynthesis. *Reaction Chemistry & Engineering, 3*(3), 371–378.

Neoh, C. H., Noor, Z. Z., Mutamim, N. S. A., & Lim, C. K. (2016). Green technology in wastewater treatment technologies: Integration of membrane bioreactor with various wastewater treatment systems. *Chemical Engineering Journal, 283*, 582–594.

Neoh, C. H., Yung, P. Y., Noor, Z. Z., Razak, M. H., Aris, A., Din, M. F. M., & Ibrahim, Z. (2017). Correlation between microbial community structure and performances of membrane bioreactor for treatment of palm oil mill effluent. *Chemical Engineering Journal, 308*, 656–663.

Okamoto, A., Hashimoto, K., & Nakamura, R. (2012). Long-range electron conduction of Shewanella biofilms mediated by outer membrane C-type cytochromes. *Bioelectrochemistry, 85*, 61–65.

Park, D. H., Laivenieks, M., Guettler, M. V., Jain, M. K., & Zeikus, J. G. (1999). Microbial utilization of electrically reduced neutral red as the sole electron donor for growth and metabolite production. *Applied and Environmental Microbiology, 65*(7), 2912–2917.

Patil, S. A., Arends, J. B. A., Vanwonterghem, I., van Meerbergen, J., Guo, K., Tyson, G. W., & Rabaey, K. (2015). Selective enrichment establishes a stable performing community for microbial electrosynthesis of acetate from CO2. *Environmental Science & Technology, 49*(14), 8833–8843.

Qasem, N. A. A., Ben-Mansour, R., & Habib, M. A. (2018). An efficient CO2 adsorptive storage using MOF-5 and MOF-177. *Applied Energy, 210*, 317–326.

Quraishi, M., Wani, K., Pandit, S., Gupta, P. K., Rai, A. K., Lahiri, D., Jadhav, D. A., Ray, R. R., Jung, S. P., Thakur, V. K., & Prasad, R. (2021). Valorisation of CO2 into value-added products via Microbial Electrosynthesis (MES) and electro-fermentation technology. *Fermentation, 7*(4), 291.

Ragsdale, S. W., & Pierce, E. (2008). Acetogenesis and the Wood–Ljungdahl pathway of CO2 fixation. *Biochimica et Biophysica Acta (BBA)-Proteins and Proteomics, 1784*(12), 1873–1898.

Reddy, M. V., Kumar, G., Mohanakrishna, G., Shobana, S., & Al-Raoush, R. I. (2020). Review on the production of medium and small chain fatty acids through waste valorization and CO_2 fixation. *Bioresource Technology, 309*, 123400.

Richardson, D. J., Butt, J. N., Fredrickson, J. K., Zachara, J. M., Shi, L., Edwards, M. J., White, G., Baiden, N., Gates, A. J., Marritt, S. J., & Clarke, T. A. (2012). The 'porin–cytochrome' model for microbe-to-mineral electron transfer. *Molecular Microbiology, 85*(2), 201–212.

Romans-Casas, M., Feliu-Paradeda, L., Tedesco, M., Hamelers, H. V. M., Bañeras, L., Balaguer, M. D., Puig, S., & Dessì, P. (2024). Selective butyric acid production from CO2 and its upgrade to butanol in microbial electrosynthesis cells. *Environmental Science and Ecotechnology, 17*, Article 100303.

Rozendal, R. A., Jeremiasse, A. W., Hamelers, H. V. M., & Buisman, C. J. N. (2008). Hydrogen production with a microbial biocathode. *Environmental Science & Technology, 42*(2), 629–634.

Sharma, M., Aryal, N., Sarma, P. M., Vanbroekhoven, K., Lal, B., Benetton, X. D., & Pant, D. (2013). Bioelectrocatalyzed reduction of acetic and butyric acids via direct electron transfer using a mixed culture of sulfate-reducers drives electrosynthesis of alcohols and acetone. *Chemical Communications, 49*(58), 6495–6497.

Slate, A. J., Whitehead, K. A., Brownson, D. A. C., & Banks, C. E. (2019). Microbial fuel cells: An overview of current technology. *Renewable and Sustainable Energy Reviews, 101*, 60–81.

Song, X., Wang, W., Cao, X., Wang, Y., Zou, L., Ge, X., Zhao, Y., Si, Z., & Wang, Y. (2020). *Chlorella vulgaris* on the cathode promoted the performance of sediment microbial fuel cells for electrogenesis and pollutant removal. *Science of the Total Environment, 728*, Article 138011.

Song, X., Zhang, H., Zhang, Y., Goh, B., Bao, B., Mello, S. S., Sun, X., Zheng, W., Gazzaniga, F. S., Wu, M., Qu, F., Yin, Q., Gilmore, M. S., Oh, S. F., & Kasper, D. L. (2023). Gut microbial fatty acid isomerization modulates intraepithelial T cells. *Nature, 619*(7971), 837–843.

Siegert, M., Yates, M. D., Spormann, A. M., & Logan, B. E. (2015). Methanobacterium dominates biocathodic archaeal communities in methanogenic microbial electrolysis cells. *ACS Sustainable Chemistry & Engineering, 3*(7), 1668–1676. https://doi.org/10.1021/acssuschemeng.5b00367

Srikanth, S., Singh, D., Vanbroekhoven, K., Pant, D., Kumar, M., Puri, S. K., & Ramakumar, S. S. V. (2018). Electro-biocatalytic conversion of carbon dioxide to alcohols using gas diffusion electrode. *Bioresource Technology, 265*, 45–51.

Straub, M., Demler, M., Weuster-Botz, D., & Dürre, P. (2014). Selective enhancement of autotrophic acetate production with genetically modified *Acetobacterium woodii*. *Journal of Biotechnology, 178*, 67–72.

Thulluru, L. P., Ghangrekar, M. M., & Chowdhury, S. (2023). Progress and perspectives on microbial electrosynthesis for valorisation of CO2 into value-added products. *Journal of Environmental Management, 332*, 117323.

Tian, S., Wang, H., Dong, Z., Yang, Y., Yuan, H., Huang, Q., Song, T., & Xie, J. (2019). Mo2C-induced hydrogen production enhances microbial electrosynthesis of acetate from CO2 reduction. *Biotechnology for Biofuels, 12*(1), 71.

Tremblay, P.-L., Höglund, D., Koza, A., Bonde, I., & Zhang, T. (2015). Adaptation of the autotrophic acetogen *Sporomusa ovata* to methanol accelerates the conversion of CO_2 to organic products. *Scientific Reports, 5*(1), 16168.

Vassilev, I., Hernandez, P. A., Batlle-Vilanova, P., Freguia, S., Krömer, J. O., Keller, J., Ledezma, P., & Virdis, B. (2018). Microbial electrosynthesis of isobutyric, butyric, caproic acids, and corresponding alcohols from carbon dioxide. *ACS Sustainable Chemistry & Engineering, 6*(7), 8485–8493.

Vellingiri, A., Song, Y. E., Munussami, G., Kim, C., Park, C., Jeon, B.-H., Lee, S.-G., & Kim, J. R. (2019). Overexpression of c-type cytochrome, CymA in *Shewanella oneidensis* MR-1 for enhanced bioelectricity generation and cell growth in a microbial fuel cell. *Journal of Chemical Technology & Biotechnology, 94*(7), 2115–2122.

Velvizhi, G., & Mohan, S. V. (2015). Bioelectrogenic role of anoxic microbial anode in the treatment of chemical wastewater: Microbial dynamics with bioelectro-characterization. *Water Research, 70*, 52–63.

Venkata Mohan, S., Velvizhi, G., Annie Modestra, J., & Srikanth, S. (2014). Microbial fuel cell: Critical factors regulating bio-catalyzed electrochemical process and recent advancements. *Renewable and Sustainable Energy Reviews, 40*, 779–797.

Wang, H., & Ren, Z. J. (2013). A comprehensive review of microbial electrochemical systems as a platform technology. *Biotechnology Advances, 31*(8), 1796–1807.

Wu, S., Wang, H., Yuan, G., Hu, B., Sun, Z., Yan, S., & Li, Y. (2022). Carbon dioxide flow behavior through nanopores: Implication for CO_2 sequestration in unconventional gas reservoirs. *Industrial & Engineering Chemistry Research, 61*(45), 16869–16882. https://doi.org/10.1021/acs.iecr.2c03123

Yang, X., Wang, Q., Qing, S., Gao, Z., Tong, X., & Yang, N. (2021). Modulating electronic structure of an Au-Nanorod-Core–PdPt-Alloy-Shell catalyst for efficient alcohol electro-oxidation. *Advanced Energy Materials, 11*(26). https://doi.org/10.1002/aenm.202100812

Zhang, S., Jiang, J., Wang, H., Li, F., Hua, T., & Wang, W. (2021). A review of microbial electrosynthesis applied to carbon dioxide capture and conversion: The basic principles, electrode materials, and bioproducts. *Journal of CO2 Utilization, 51*, 101640.

Zhang, S., Lu, H., & Lu, Y. (2013). Enhanced stability and chemical resistance of a new nanoscale biocatalyst for accelerating CO2 absorption into a carbonate solution. *Environmental Science and Technology, 47*(23), 13882–13888.
Zhang, X., Cheng, S., Huang, X., & Logan, B. E. (2010). Improved performance of single-chamber microbial fuel cells through control of membrane deformation. *Biosensors and Bioelectronics, 25*(7), 1825–1828.
Zhang, P., Liu, J., Qu, Y., Li, D., He, W., & Feng, Y. (2018). Nanomaterials for facilitating microbial extracellular electron transfer: Recent progress and challenges. *Bioelectrochemistry, 123*, 190–200. https://doi.org/10.1016/j.bioelechem.2018.05.005
Zhao, J., Ma, H., Wu, W., Bacar, M. A., Wang, Q., Gao, M., Wu, C., Xia, C., & Qian, D. (2023). Conversion of liquor brewing wastewater into medium chain fatty acids by microbial electrosynthesis: Effect of cathode potential and CO_2 supply. *Fuel, 332*, Article 126046.

Chapter 4
Biofuel Production by Electro-Fermentation

Sujeeth Nachimuthu Krishnan, Aaron Mani Annadurai, Harithaa Sathyanarayanan, Brindha Matharasi Murugan, and Murugesan Gnanadesigan

1 Introduction

1.1 Need for Biofuels

The intensifying global demand for finite resources like gas, oil, and charcoal has led to scarcity, exacerbated by industrialization and population growth, crucial for socio-economic development of developing countries. These glitches can be rectified by the generation of alternative biofuels derived from organic sources like plants, animals, biomass, or microorganisms (Aravindh et al., 2023). Although alternative fuels such as natural gas, liquefied petroleum gas, and electricity exist, they necessitate adaptations to conventional internal combustion (IC) engines powered by fossil fuels. Biofuels, being similar to fossil fuels, can be accommodated by existing vehicles without significant modifications. This gives biofuels an added advantage over other alternative fuels (Moshood et al., 2021).

Using biofuels instead of conventional fuels can reduce carbon, sulphur, and nitrogen emissions, contributing to overcome global warming (Mahapatra et al., 2021). However, their primary drawback is the high initial investment required and due to its lower yields. As a result, extensive research is being conducted to improve biofuel yields. While biofuels aid in environmental sustainability, achieving economic sustainability is also crucial for their widespread adoption (Rial, 2024). To overcome this, various solutions are put forward by researchers to address this issue,

S. Nachimuthu Krishnan · A. Mani Annadurai · H. Sathyanarayanan · B. M. Murugan · G. Murugesan (✉)
Natural Products Research Laboratory, Department of Microbial Biotechnology, School of Biotechnology & Genetic Engineering, Bharathiar University, Coimbatore, Tamil Nadu, India
e-mail: contactdrgd@buc.edu.in

S. Nachimuthu Krishnan
e-mail: sujeeth.mbt@buc.edu.in

© The Author(s), under exclusive license to Springer Nature Singapore Pte Ltd. 2025
S. Sevda (ed.), *Electro-Fermentation for Synthesis of Chemicals, Biochemicals and Biofuels*, https://doi.org/10.1007/978-981-96-6358-3_4

such as developing new generations of biofuels (1G, 2G, 3G & 4G) that improve upon the flaws of previous ones, converting waste organic biomass into fuels, and utilizing electro-fermentation (Pikula et al., 2020; Salar-García et al., 2024).

1.2 Need for Electro-Fermentation

Electro-fermentation represents an innovative approach to traditional fermentation processes, incorporating electricity/electrons to enhance microbial productivity. This technique involves the direct supply or removal of electrons via electrodes, which serve as a reducing source for the microorganisms. By modulating the fermentation environment and metabolic pathways, electro-fermentation can effectively eliminate undesirable byproducts, thereby increasing both the yield and purity of the target metabolite. This interdisciplinary concept seamlessly integrates principles from electrochemistry and industrial fermentation, offering a promising avenue for improved bioproduction strategies (Rago et al., 2019).

Electro-fermentation offers several significant advantages over conventional fermentation processes, including (1) enhanced biomass yield through increased ATP production, (2) improved control over redox and pH imbalances, and (3) facilitation of single-target product synthesis (Rago et al., 2019). Despite these benefits, the technique is not without drawbacks. It requires higher energy input and substantial initial investments in specialized equipment such as electrodes and membranes (Salar-García et al., 2024). Nevertheless, ongoing research in this field continues to yield promising results. Some studies have reported up to a threefold increase in product yield with minimal energy input, highlighting the potential of this technology to revolutionize fermentation processes in various industrial applications (Virdis et al., 2022).

The substantial initial investments required for electro-fermentation are projected to yield significant long-term returns through improved product yields. Consequently, this technology has emerged as a promising trend in bioprocessing, with the potential to enhance the production of both biofuels and high-value chemicals derived from fermentation processes. Notably, electro-fermentation shows particular promise in augmenting the yields of various biofuels, including ethanol, butanol, methane, and even electricity (Chandrasekhar et al., 2021).

2 Mechanism of Electro-Fermentation

Electron transfer serves as the primary driver of metabolic flux, which is defined as the rate of metabolite passage through a biochemical pathway over time. This electron flow is facilitated by various donors, including NADH, $FADH_2$, and H_2O, as well as acceptors such as NAD+, FAD, and CO_2, depending on cellular requirements (Gong et al., 2020). Among these redox couples, the NAD+/NADH ratio plays a particularly crucial role in maintaining efficient metabolite production, regulating intracellular redox state, as measured by the Oxidation-Reduction Potential (ORP), and ensuring overall cellular health.

The NAD+/NADH ratio serves as a critical metabolic regulator in cellular processes. A high NAD+/NADH ratio promotes anabolic pathways, leading to energy storage, while a low ratio favours catabolic processes, resulting in increased energy harvesting (Katsyuba et al., 2020). Under normal conditions, cells maintain this ratio through various enzymatic mechanisms to regulate metabolism. In electro-fermentation, however, direct electron transfer to microorganisms via electrodes allows for the manipulation of the NAD+/NADH ratio, consequently altering metabolic pathways (Gong et al., 2020). This controlled alteration can be tailored to optimize the production of desired metabolites. The process effectively converts extracellular electrons into intracellular reducing power, providing a novel means of directing cellular metabolism.

In electro-fermentation, the direction of electron flow between microorganisms and electrodes is determined by the desired end-product. When an oxidized product is targeted, microorganisms are induced to donate electrons, with the electrode serving as the electron acceptor. This process is termed anodic electro-fermentation, characterized by the working electrode's role in accepting electrons. Conversely, for the production of reduced compounds, microorganisms accept electrons provided by the electrode, which acts as an electron donor. This latter scenario is referred to as cathodic electro-fermentation, distinguished by the working electrode's function in donating electrons. These modes allow for precise control over redox conditions, enabling the selective production of either oxidized or reduced metabolites (Moscoviz et al., 2016).

Electrodes (e^- acceptor) + **Microbes** (e^-donor) → **Oxidized Products**

(Anodic electrofermentation)

Electrodes (e^- donor) + **Microbes** (e^-acceptor) → **Reduced products**

(Cathodic electrofermentation)

3 Electrofermenting Microorganisms

All living organisms require at least one source each of carbon, nitrogen, and electrons (reduction) for survival. Organisms are classified based on their electron source: Those utilizing organic molecules are termed organotrophs, while those employing inorganic molecules are known as lithotrophs. In electro-fermentation, a unique class of microorganisms can utilize solid electroconductive materials, specifically electrodes, as electron donors or acceptors. These specialized organisms are designated as electro-active bacteria (EAB), exoelectrogens, or electrotrophs. Unlike conventional electron sources, electrodes cannot be directly assimilated by most living organisms, making EAB distinct in their ability to engage in extracellular electron transfer (Rago et al., 2019).

3.1 Direct Extracellular Electron Transfer

Among the most well-characterized electro-active bacteria (EAB) are *Shewanella oneidensis* and *Geobacter sulfurreducens*. These organisms possess a sophisticated transmembrane c-type cytochrome system that facilitates the transport of electrons across the cell membrane. This mechanism, known as direct extracellular electron transfer (Direct EET), enables the passage of electrons from the extracellular environment to the intracellular space through direct contact with the electrode. There are three types of Direct EETs (Fig. 1).

3.1.1 Direct Contact Via Cytochromes

Cytochromes are integral membrane proteins containing one or more heme groups that facilitate electron-transfer processes (Deutzmann et al., 2015). Various electron-transfer mechanisms have been identified, with direct electron transfer mediated by c-type cytochromes being particularly notable in genera such as Geobacter and Shewanella. The electrode-microbe interactions are largely governed by key components including quinones, flavins, ferredoxin, and cytochromes, which are believed to play crucial roles in catalytic processes (Fig. 2).

Direct electron transfer is considered one of the most efficient methods for electron transport, particularly in cathodic systems. This mechanism, involving the transfer of electrons from the cathode to corresponding microorganisms, has been extensively demonstrated in biocathodes for beta proteobacteria and diverse microbial communities (Bhagchandanii et al., 2020).

In microbe–electrode interactions, electron flow occurs bidirectionally: from microbe to electrode (anodic process) and from electrode to microbe (cathodic process). The broad range of redox potentials observed in cytochrome C-based electron-transfer channels suggests a flexible directionality of electron flow, accommodating both cathodic and anodic processes. Fundamentally, electrons migrate from electron donors with lower redox potentials to electron acceptors with higher oxidation potentials. During electro-fermentation, the application of electrical stimulation, whether through positive or negative potentials, can induce reorientation of reducing

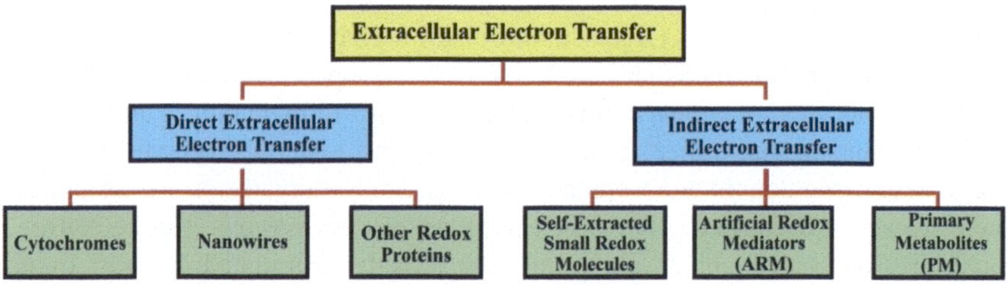

Fig. 1 Schematic representation of extracellular electron transfer

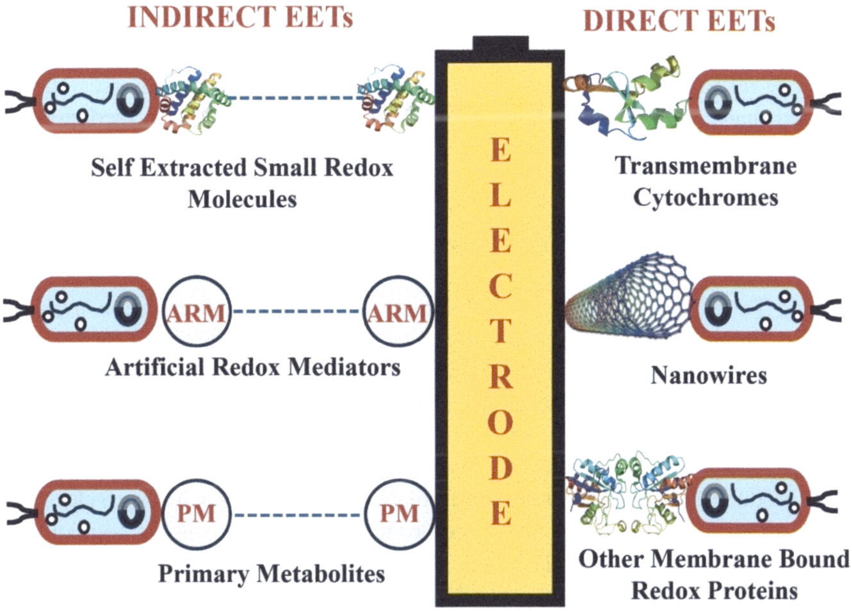

Fig. 2 Schematic representation of extracellular electron-transfer mechanism

equivalents (electrons and protons) within the microbial electron transport chain (Sravan et al., 2019).

3.1.2 Direct Contact and Transfer Via Nanowires

In conduction-based extracellular electron transfer (EET), microorganisms generate high current densities while facilitating electron transmission through specialized biological structures. This mechanism, termed "transfer via biological projections," utilizes pili or nanowires to conduct electrons. The uniqueness of this approach lies in its use of distinct projections for electron transfer. These biological conduits transport electrons from the microorganisms to electron carrier proteins, with nanowires supporting high current densities that corroborate this mechanism.

The pili involved in this process often promote cell accumulation within biofilms. Notably, these structures exhibit metal-like properties, characterized by increased temperature and decreased conductivity with rising current. While these properties facilitate electron transfer, they may potentially compromise the efficiency of microbial electrolytic systems (MESs). Multi-Heme Cytochromes (MHCs) play a crucial role in facilitating electron transport in electro-active microorganisms. These proteins form a polypeptide chain through covalent linkages, with adjacent MHCs typically separated by a distance of 1-2 nm. The unique structural configuration of MHCs significantly enhances extracellular electron transfer (EET) efficiency (Bhagchandanii et al., 2020). While initial hypotheses suggested that PilA protein in pili was primarily responsible for long-distance electron transfer, subsequent research has revealed that polymerized hexa-heme cytochromes are the principal mediators of this process.

3.1.3 Direct Contact and Transfer Via Other Redox Proteins

The iron [Fe(0)] corroding and electro-methanogenic archaeon *Methanococcus maripaludis* exhibit apparent direct electron uptake capabilities, mediated by surface-associated redox enzymes such as hydrogenases and presumably, formate dehydrogenases. These enzymes, when released from cells and adsorbed onto suitable redox-active surfaces, catalyse the formation of H_2 or formate. *M. maripaludis* cells rapidly consume these low molecular weight compounds, preventing accumulation of the products. The hydrogenases and formate dehydrogenase can absorb electrons from cathodic surfaces and catalyse the production of small, soluble compounds. These compounds subsequently serve as catabolic substrates for hydrogenotrophic microorganisms (Deutzmann et al., 2015).

3.2 Indirect Extracellular Electron Transfer

Indirect extracellular electron transfer (Indirect EET) refers to the process by which bacteria transfer electrons without direct contact with an electrode, instead utilizing mediators to facilitate the electron transfer. This mechanism can be further categorized into three distinct types (Fig. 1).

3.2.1 Indirect Transfer Via Self-Extracted Small Redox Molecules

In numerous Iron-Reducing Bacteria, electron transfer is facilitated by small, self-secreted redox molecules, primarily flavins which can bind to cytochromes and get reduced to transfer electrons from bacteria to extracellular metal oxides. Other redox-active compounds, including quinones, phenazines, and pyocyanin, also participate in this process. Then, researchers studied this mechanism and incorporated artificially synthesized redox molecules resembling them (Liu et al., 2018). In the case of Shewanella species, Flavin Mononucleotide (FMN) or Riboflavin serves as the primary mediators in this indirect electron-transfer mechanism (Gong et al., 2020).

3.2.2 Indirect Transfer Via Artificial Redox Mediators

Artificial redox mediators (ARMs) such as methyl viologen, neutral red, anthraquinone-2,6-disulphonate, potassium ferricyanide, and phospholipid polymers play a role in Indirect EET (Gong et al., 2020). Initially, ARMs undergo oxidation and subsequently transfer their electrons to Flavin Mononucleotide (FMN). This interaction facilitates EET through two distinct pathways. Firstly, the reduced form of FMN ($FMNH_2$) activates cytochromes, enabling electron transfer from ARMs to metal oxides. Secondly, $FMNH_2$ directly transfers its excess electrons to the metal oxides (Liu et al., 2018).

3.2.3 Indirect Transfer Via Primary Metabolites (PMs)

In acetogens and methanogens, hydrogen is the primary electron donor synthesized during primary metabolism. Microorganisms electrochemically consume or produce hydrogen and formate to balance their redox states (Gong et al., 2020). Additionally, organisms such as *Candida melibiosica* can transfer electrons without artificial mediators, indicating electron transfer through primary metabolites (Babanova et al., 2011).

4 Biofuel Production Using Electro-Fermentation

Biofuels production such as biohydrogen, biomethane, bioethanol, and biobutanol can be enhanced using electro-fermentation (Fig. 3). This can be achieved due to direct or indirect contact of microorganisms with the electrodes.

Various factors can affect the yield such as organism used, chamber setup, type of fermentation, voltage applied, and electrodes (Salar-García et al., 2024) (Table 1).

4.1 Biomethane

Biomethane can be transported through pipelines for any distance quickly, combined with natural gas, or used alone. It can also be compressed into cylinders and transported by road or rail for short- to medium-distance travel or liquefied for longer-distance transport. After refining, the biomethane should meet local standards so that it can be blended with natural gas and used as vehicle fuel (Leahy et al., 2010). The choice of this business model is more dependent on supportive public policies and regulatory changes than earlier models because biomethane is more expensive to produce than biogas which requires less processing (Moraes et al., 2022).

Presently, the most significant renewable gas supply alternative is biomethane, which blends seamlessly with natural gas. Reductions of more than 80% of green

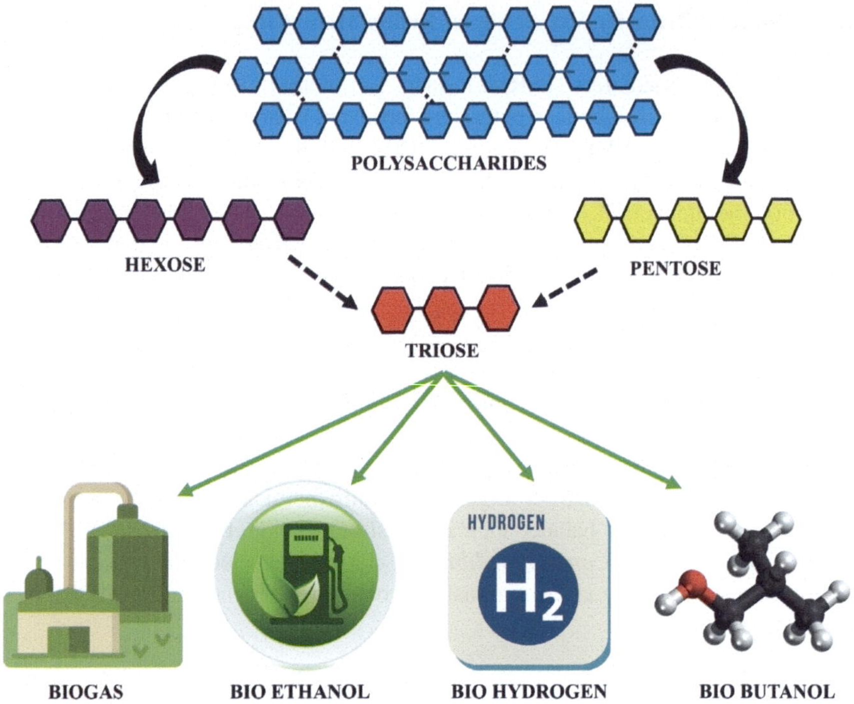

Fig. 3 Schematic representation of bioproducts production from polysaccharides

house gas emissions as compared to natural gas seem possible with the installation of sophisticated facilities and ongoing advancements in crop farming and plant operation (Adams, 2018). Less than 3% of the need for bioenergy is currently satisfied by biogas and biomethane, which also make up only 0.3% of primary energy. Without emitting any net carbon emissions, they can offer the system benefits of natural gas (storage and flexibility). This becomes an essential characteristic as economies move toward decarbonization. Biomethane being a vital cooking fuel for poor nations, it also offers a sustainable supply of power that can benefit people looking for local, decentralized energy sources. The concepts of the circular economy are well-suited to biogas and biomethane, which convert various organic wastes into higher-value products (Horváth-Gönczi et al., 2023).

The biomethane/biogas is produced using anaerobic chambers contained with a substrate that supports microbial growth and the electro-fermentation is done by connecting two electrodes either as single or double chambers containing a headspace where the biomethane in gaseous form can be collected. The electron transfer helps in various metabolic changes and improved yield. The headspace must be sparged with inert gases or nitrogen before fermentation, and after fermentation, the biomethane is along with some impurities such as hydrogen and carbon dioxide which need to be purified before further processing (Horváth-Gönczi et al., 2023; Zhen et al., 2016).

Table 1 Biofuel production in microbial electrochemical systems using various organisms and systems

S. no	Biofuel type	Organism used	Setup	Fermentation type	Maximum voltage	Electrodes	Yield	References
1	Methane	Anaerobic sludge containing *Geobacter metallireducens* and *Methanosaeta harundinacea in* glucose containing medium	500 ml single-chamber MES	Batch	+ 0.6	Graphite brush – anode	+ 350% in 33 days	Zhao et al., (2015)
						Graphite rod – cathode		
2	Methane	*Egeria densa* in domestic waste water mixed with lake water	0.8 L single-chamber membrane free MES	Semi-continuous	+ 1.0	Ti/RuO_2 mesh plates – anode and cathode	+ 110% in 40 days	Zhen et al., (2016)
3	Methane	Anaerobic sludge	150 mL single-chamber three-electrode MES	Batch	– 0.6 V	Graphite plates – cathode and anode	+ 72.2% in 18 days	Ren et al., (2018)
						Saturated calomel – reference electrode		
4	Biohythane (hydrogen + methane)	Mixed consortium from STP containing *Clostridiaceae*, *Bacteroides*, and Proteobacteria	1 L anaerobic MES	Fed-Batch	100 Ω	Graphite – both anode and cathode	+ 131.25% in 48 h	Shanthi Sravan et al., (2018)
					– 0.6 V		+ 87.5% in 48 h	
5	Hydrogen	*Thermotoga neapolitana* in modified ATCC 1977 culture medium	200 ml single-chamber MES	Fed-Batch	± 1.2 V	Carbon cloth – both anode and cathode	+ 193.5% in 72 h	d'Ippolito et al., (2021)

(continued)

Table 1 (continued)

S. no	Biofuel type	Organism used	Setup	Fermentation type	Maximum voltage	Electrodes	Yield	References
6	Butanol	*Clostridium saccharoperbutylacetonicum* N1-4 strains in fermentation medium	200 ml anaerobic dual-chamber bioreactor	Batch	– 600 mV	Carbon brush – both anode and cathode	+ 150% in 18 h	Alberto García Mogollón et al., (2023)
7	Butanol	*Clostridium pasteurianum* in MP2 medium	300 ml two-chamber H-type reactor with MP2 media in CC and 0.1 K buffer in AC	Batch	+ 1.5 V	Graphite felt – both anode and cathode Ag\AgCl$_3$ – reference electrode	+ 260% in 72 h	Khosravanipour Mostafazadeh et al., (2016)
8	Ethanol	*Ethanoligenens harbinense* YUAN-3 strain in modified PYG medium	290 ml single-chamber anode EFS	Batch	– 0.4 V	Carbon paper – anode Stainless-steel mesh – cathode	+ 130% in 65 h	Gu et al., (2022)
9	Ethanol	*Clostridium autoethanogenum* DSM 10061 in modified DSM 640 medium, excluding trypticase peptone	Two-chamber H-type reactor 1L in cathode and 0.3 L in anode	Batch	– 600 mV	Graphite rod – both anode and cathode Ag\AgCl$_3$ -- reference electrode	– 22% in 3–4 days	Martínez-Ruano et al., (2023)
10	Ethanol	*Saccharomyces cerevisiae* CDBT2 in PYN media supplemented with glucose or lignocellulosic biomass hydrolysate	300 ml anodic chamber MES	Batch	4 V	Graphite felt – cathode Platinum wire – anode	+ 19.8% in 4 days	Joshi et al., (2019)

(continued)

Table 1 (continued)

S. no	Biofuel type	Organism used	Setup	Fermentation type	Maximum voltage	Electrodes	Yield	References
11	Ethanol	*Saccharomyces cerevisiae* CDBT7 in PYN media supplemented with glucose or lignocellulosic biomass hydrolysate	300 ml cathodic chamber MES	Batch	4 V	Graphite felt – cathode Platinum wire – anode	+ 23.7% in 4 days	Joshi et al., (2019)
12	Ethanol	*Saccharomyces cerevisiae* CDBT2 in PYN media supplemented with glucose or lignocellulosic biomass hydrolysate	300 ml cathodic chamber MES	Batch	4 V	Graphite felt – cathode Platinum wire – anode	+ 7.1% in 4 days	Joshi et al., (2019)
13	Ethanol	*Saccharomyces cerevisiae* CDBT7 in PYN media supplemented with glucose or lignocellulosic biomass hydrolysate	300 ml anodic chamber MES	Batch	4 V	Graphite felt – cathode Platinum wire – anode	+ 1% in 4 days	Joshi et al., (2019)
14	Hydrogen	Clostridiaceae-dominant anaerobic sludge in fermentation medium	0.5 L two-chamber MES	Batch	– 0.4 V and + 0.9 V	90% platinum with 10% iridium grids Both anode and cathode – saturated calomel reference electrode	+ 244.6% in 20 h	Toledo-Alarcón et al., (2019)

4.2 Biohydrogen

It is regarded as a biofuel that is renewable and a perfect energy source to fight the pressing need to reduce emissions from fossil fuels. Although several biohydrogen technologies have been created, the process of converting via microbial processes combined with electro-fermentation is still in the stages of development. It is a good fuel since burning it will not have negative effects on the atmosphere compared to burning fossil fuels. The greatest specific energy value of molecular hydrogen is 143 GJ/tonne, and when it burns, energy is released explosively. Through the conversion of chemical energy, fuel cells quietly generate electricity. Automobile manufacturers in the USA, Japan, and the EU have been creating hydrogen-fuelled vehicles and hydrogen fuelling stations, demonstrating the significance of hydrogen as an energy source for the transportation sector. The primary concern with hydrogen as an energy source is its production process (Show et al., 2019). Benefits from combining the production of biohydrogen with waste treatment into a single reaction include the simultaneous synthesis of fuel that is carbon neutral and waste treatment (Mogili et al., 2022).

Biohydrogen is produced in MECs where electrolytic microbes colonize around anode electrode and oxidize the substrate by undergoing fermentation getting necessary nutrients along with electrons and protons. The electrons are attracted towards the anode and are accelerated towards cathode by applying external voltage. In the meantime, protons reach the cathode via a semi permeable proton exchanging membrane through the substrate, where the reached protons (H^+) combine with the electrons (e^-) which are reached through applied external voltage to give hydrogen (H_2) (Kadier et al., 2018). The chemical reaction can be described as,

Anode: $CH_3COO^- + 4H_2O \rightarrow 2HCO_3^- + 9H^+ + 8e^-$

Cathode: $8H^+ + 8e^- \rightarrow 4H_2$

4.3 Bioethanol

Bioethanol is a clean-burning, renewable fuel produced by fermenting carbohydrates that significantly contributes to environmental conservation by reducing global warming and conserving fossil fuels. Producing bioethanol from waste or biomass is a strategy to reduce crude oil consumption and environmental pollution. Lignocellulosic biomasses, such as corn, sugar, molasses, and non-food sources like grasses and trees, are used as feedstocks for ethanol synthesis. Bioethanol shares many physical and chemical properties with ethanol but requires different resources for its production. In its purest form, bioethanol is a clear, colourless liquid with a faint, distinctive smell, freezing at − 112 °C and boiling at 78 °C (Hahn-Hägerdal et al., 2001; Mogili et al., 2022). With roughly 85.1% of global biofuel output in 2017, bioethanol is the most prevalent type of biofuel. The main sources of bioethanol are sugarcane in Brazil (20 billion litres yearly) and maize in the USA (60 billion litres). In 2017, the combined contribution of these two nations to the global bioethanol output was close to 87%. The bioethanol's high-octane number (89 MON and 107 RON) makes it a desirable biofuel. When compared to gasoline, using bioethanol in an internal combustion engine has additional benefits. Bioethanol's increased heat of vaporization, improved flame speed, and wider flammability limit result in a shorter burn time and an increased compression ratio. Bioethanol's high oxygen content, combined with very little sulphur, lowers emissions, promotes clean burning, and increases combustion efficiency (Kunamalla et al., 2022). To produce bioethanol, the inoculum (majorly yeasts) is allowed to metabolize the substrate in MES which will give ethanol as an end-product, and when a constant voltage is applied across electrodes, the yield is enhanced (Joshi et al., 2019).

4.4 Biobutanol

Compared to bioethanol, biobutanol is a superior biofuel produced through the fermentation of acetone, butanol, and ethanol (ABE). Research on the kinetic and physiological models for fermentation using lignocellulosic biomass (LCB) provides valuable insights for process optimization, which is crucial for the sustainable biobutanol production (Bhandiwad et al., 2014). The two primary stages of the ABE fermentation process are acid fermentation and carbon dioxide, hydrogen, and acetic and butyric acid production from sugars by the clostridia. Since it promotes redox balance, more butyrate than acetate is produced during this process. Acetone, butanol, and ethanol are formed from the generated acids during the second phase of solvent fermentation. When used as fuel, biobutanol offers 30% greater energy value than ethanol. It can be blended at high percentages with any fuel, including diesel and gasoline (Eloka-Eboka & Maroa, 2023). For Clostridium species, the oxidative and reductive pathways were followed in the metabolism of glucose, the formation of butyrate and butanol. NAD+ is converted to NADH in the oxidative route, where glucose enters glycolysis to create pyruvate. Depending on the microbe, pyruvate is broken down into a variety of compounds, including butyrate, ethanol, lactate, acetate, and butanol, along with a tiny amount of hydrogen generation. The reductive route converts glucose into butanol when NADH is oxidized to NAD+.

5 Challenges in Electro-Fermentation

Although electro-fermentation increases yield compared to normal fermentation, there are significant challenges in scaling it up for large-scale production. Electro-fermentation is a relatively new and underutilized fermentation strategy involves numerous variables, including electrode material, size, distance, electrolyte type, reactor design, voltage requirements for different culture volumes, operation setup, fermentation type, organisms used (pure or mixed cultures), ion exchange membrane type, and fermentation time. Despite the recent growth of this field over the past decade, there is insufficient experimental data to draw definitive conclusions. The biochemistry underlying the enhanced rate, yield, and titre of the product remains poorly understood, necessitating further studies on metabolic pathways. Without this understanding, predicting the yield of desired end-products remains challenging, complicating large-scale production (Virdis et al., 2022).

During large-scale production, there is a significant possibility of reduced efficiency due to high input energy and costs, which may not directly correlate with the yield of end-products. Increased resistance of internal materials can affect electron flow and lead to energy loss, compromising the primary goals of the system, such as energy and economic sustainability. The lack of experimental studies further fails to disprove this prediction (Chandrasekhar et al., 2021).

The interdisciplinary nature of electro-fermentation, encompassing biochemistry, electrochemistry, and material science, presents a significant challenge as it requires comprehensive expertise across these fields to address existing issues. Recently, metabolic engineering of electro-active bacteria (EAB) has been developed to enhance yield and potentially improve electron transfer. However, other challenges also need to be addressed (Chandrasekhar et al., 2021; Gong et al., 2020).

6 Conclusion

Thus, the electro-fermentation technique in biofuel production demonstrates significant promise for enhancing sustainability by improving yields in lab-scale experiments. However, this technique is still in its early stages of development and requires extensive experimental data to address the numerous variables and challenges it presents. To transition from lab-scale to industrial-scale production, further research is needed to optimize factors such as electrode material, reactor design, and metabolic pathways.

Appendices

AC	Anode Chamber
CC	Cathode Chamber
EFS	Electro-fermentation System
MES	Microbial Electrolysis System
MON	Motor Octane Number
RON	Research Octane Number
Yield increase % =	Yield in electro-fermentation/Yield in control × 100

References

Adams, P. (2018). GHG emissions from biomethane gas-to-grid injection via anaerobic digestion. In *Greenhouse Gases Balances of Bioenergy Systems* (pp. 141–158). Elsevier. https://doi.org/10.1016/B978-0-08-101036-5.00009-4

Alberto García Mogollón, C., Carlos Quintero Díaz, J., & Omar Gil Posada, J. (2023). Production of acetone, butanol, and ethanol by electro-fermentation with Clostridium saccharoperbutylacetonicum N1-4. *Bioelectrochemistry, 152*, 108414. https://doi.org/10.1016/j.bioelechem.2023.108414

Aravindh, S., Ruban, P., Sujeeth, N. K., & Gnanadesigan, M. (2023). Exploring the biodiesel production potential of marine microalgae Nodosilinea nodulosa SNMVBTAR001 from Kerala coastal region, India. *Biocatalysis and Agricultural Biotechnology, 53*, Article 102882. https://doi.org/10.1016/j.bcab.2023.102882

Babanova, S., Hubenova, Y., & Mitov, M. (2011). Influence of artificial mediators on yeast-based fuel cell performance. *Journal of Bioscience and Bioengineering, 112*(4), 379–387. https://doi.org/10.1016/j.jbiosc.2011.06.008

Bhagchandanii, D. D., Babu, R. P., Sonawane, J. M., Khanna, N., Pandit, S., Jadhav, D. A., Khilari, S., & Prasad, R. (2020). A comprehensive understanding of electro-fermentation. *Fermentation, 6*(3), 92. https://doi.org/10.3390/fermentation6030092

Bhandiwad, A., Shaw, A. J., Guss, A., Guseva, A., Bahl, H., & Lynd, L. R. (2014). Metabolic engineering of Thermoanaerobacterium saccharolyticum for n-butanol production. *Metabolic Engineering, 21*, 17–25. https://doi.org/10.1016/j.ymben.2013.10.012

Chandrasekhar, K., Naresh Kumar, A., Kumar, G., Kim, D.-H., Song, Y.-C., & Kim, S.-H. (2021). Electro-fermentation for biofuels and biochemicals production: Current status and future directions. *Bioresource Technology, 323*, Article 124598. https://doi.org/10.1016/j.biortech.2020.124598

Deutzmann, J. S., Sahin, M., & Spormann, A. M. (2015). Extracellular enzymes facilitate electron uptake in biocorrosion and bioelectrosynthesis. *MBio, 6*(2). https://doi.org/10.1128/mBio.00496-15

d'Ippolito, G., Squadrito, G., Tucci, M., Esercizio, N., Sardo, A., Vastano, M., Lanzilli, M., Fontana, A., & Cristiani, P. (2021). Electrostimulation of hyperthermophile Thermotoga neapolitana cultures. *Bioresource Technology, 319*, Article 124078. https://doi.org/10.1016/j.biortech.2020.124078

Eloka-Eboka, A. C., & Maroa, S. (2023). Biobutanol fermentation research and development: feedstock, process and biofuel production. In *Advances and Developments in Biobutanol Production* (pp. 79–103). Elsevier. https://doi.org/10.1016/B978-0-323-91178-8.00007-2

Gong, Z., Yu, H., Zhang, J., Li, F., & Song, H. (2020). Microbial electro-fermentation for synthesis of chemicals and biofuels driven by bi-directional extracellular electron transfer. *Synthetic and Systems Biotechnology, 5*(4), 304–313. https://doi.org/10.1016/j.synbio.2020.08.004

Gu, J., Li, Z., Xie, G., Yang, Y., Liu, B., Ren, Z. J., & Xing, D. (2022). Electro-fermentation enhances H2 and ethanol co-production by regulating electron transfer and substrate transmembrane transport. *Chemical Engineering Journal, 429*, Article 132223. https://doi.org/10.1016/j.cej.2021.132223

Hahn-Hägerdal, B., Wahlbom, C. F., Gárdonyi, M., van Zyl, W. H., Otero, R. R. C., & Jönsson, L. J. (2001). Metabolic Engineering of Saccharomyces cerevisiae for Xylose Utilization. In *Metabolic Engineering* (pp. 53–84). https://doi.org/10.1007/3-540-45300-8_4

Horváth-Gönczi, N. N., Bagi, Z., Szuhaj, M., Rákhely, G., & Kovács, K. L. (2023). Bioelectrochemical systems (BES) for biomethane production—review. *Fermentation, 9*(7), 610. https://doi.org/10.3390/fermentation9070610

Joshi, J., Dhungana, P., Prajapati, B., Maharjan, R., Poudyal, P., Yadav, M., Mainali, M., Yadav, A. P., Bhattarai, T., & Sreerama, L. (2019). Enhancement of ethanol production in electrochemical cell by saccharomyces cerevisiae (CDBT2) and wickerhamomyces anomalus (CDBT7). *Frontiers in Energy Research, 7*. https://doi.org/10.3389/fenrg.2019.00070

Kadier, A., Kalil, M. S., Chandrasekhar, K., Mohanakrishna, G., Saratale, G. D., Saratale, R. G., Kumar, G., Pugazhendhi, A., & Sivagurunathan, P. (2018). Surpassing the current limitations of high purity H2 production in microbial electrolysis cell (MECs): Strategies for inhibiting growth of methanogens. *Bioelectrochemistry, 119*, 211–219. https://doi.org/10.1016/j.bioelechem.2017.09.014

Katsyuba, E., Romani, M., Hofer, D., & Auwerx, J. (2020). NAD+ homeostasis in health and disease. *Nature Metabolism, 2*(1), 9–31. https://doi.org/10.1038/s42255-019-0161-5

Khosravanipour Mostafazadeh, A., Drogui, P., Brar, S. K., Tyagi, R. D., Le Bihan, Y., Buelna, G., & Rasolomanana, S.-D. (2016). Enhancement of biobutanol production by electromicrobial glucose conversion in a dual chamber fermentation cell using C. pasteurianum. *Energy Conversion and Management, 130*, 165–175. https://doi.org/10.1016/j.enconman.2016.10.050

Kunamalla, A., Mailaram, S., Shrirame, B. S., Kumar, P., & Maity, S. K. (2022). Hydrocarbon biorefinery: A sustainable approach. In *Hydrocarbon Biorefinery* (pp. 1–44). Elsevier. https://doi.org/10.1016/B978-0-12-823306-1.00004-2

Leahy, S. C., Kelly, W. J., Altermann, E., Ronimus, R. S., Yeoman, C. J., Pacheco, D. M., Li, D., Kong, Z., McTavish, S., Sang, C., Lambie, S. C., Janssen, P. H., Dey, D., & Attwood, G. T. (2010). The genome sequence of the rumen methanogen methanobrevibacter ruminantium reveals new possibilities for controlling ruminant methane emissions. *PLoS ONE, 5*(1), Article e8926. https://doi.org/10.1371/journal.pone.0008926

Liu, X., Shi, L., & Gu, J.-D. (2018). Microbial electrocatalysis: Redox mediators responsible for extracellular electron transfer. *Biotechnology Advances, 36*(7), 1815–1827. https://doi.org/10.1016/j.biotechadv.2018.07.001

Mahapatra, S., Kumar, D., Singh, B., & Sachan, P. K. (2021). Biofuels and their sources of production: A review on cleaner sustainable alternative against conventional fuel, in the framework of the food and energy nexus. In *Energy Nexus* (Vol. 4). Elsevier Ltd. https://doi.org/10.1016/j.nexus.2021.100036

Martínez-Ruano, J., Suazo, A., Véliz, F., Otálora, F., Conejeros, R., González, E., & Aroca, G. (2023). Electro-fermentation with clostridium autoethanogenum: Effect of pH and neutral red addition. *Environmental Technology & Innovation, 31*, Article 103183. https://doi.org/10.1016/j.eti.2023.103183

Mogili, N. V., Murugesan, N., Ayothiraman, S., Gautam, R., Deshavath, N. N., & Reddy Erva, R. (2022). Biohydrogen production from wastewater and organic solid wastes. In *Waste-to-Energy Approaches Towards Zero Waste* (pp. 165–195). Elsevier. https://doi.org/10.1016/B978-0-323-85387-3.00009-4

Moraes, B. de S., Palacios-Bereche, R., Martins, G., Nebra, S. A., Fuess, L. T., Silva, A. J., Clementino, W. da S., Bajay, S. V., Manduca, P. C., Lamparelli, R. A., & Berni, M. D. (2022). Biogas production: Technologies and applications. In *Biofuels and Biorefining* (pp. 215–282). Elsevier. https://doi.org/10.1016/B978-0-12-824116-5.00009-X

Moscoviz, R., Toledo-Alarcón, J., Trably, E., & Bernet, N. (2016). Electro-fermentation: How to drive fermentation using electrochemical systems. *Trends in Biotechnology, 34*(11), 856–865. https://doi.org/10.1016/j.tibtech.2016.04.009

Moshood, T. D., Nawanir, G., & Mahmud, F. (2021). Microalgae biofuels production: A systematic review on socioeconomic prospects of microalgae biofuels and policy implications. *Environmental Challenges, 5*, Article 100207. https://doi.org/10.1016/j.envc.2021.100207

Pikula, K., Zakharenko, A., Stratidakis, A., Razgonova, M., Nosyrev, A., Mezhuev, Y., Tsatsakis, A., & Golokhvast, K. (2020). The advances and limitations in biodiesel production: Feedstocks, oil extraction methods, production, and environmental life cycle assessment. *Green Chemistry Letters and Reviews, 13*(4), 275–294. https://doi.org/10.1080/17518253.2020.1829099

Rago, L., Pant, D., & Schievano, A. (2019). Electro-fermentation—microbial electrochemistry as new frontier in biomass refineries and industrial fermentations. In *Advanced Bioprocessing for Alternative Fuels, Biobased Chemicals, and Bioproducts* (pp. 265–287). Elsevier. https://doi.org/10.1016/B978-0-12-817941-3.00014-0

Ren, G., Hu, A., Huang, S., Ye, J., Tang, J., & Zhou, S. (2018). Graphite-assisted electro-fermentation methanogenesis: Spectroelectrochemical and microbial community analyses of cathode biofilms. *Bioresource Technology, 269*, 74–80. https://doi.org/10.1016/j.biortech.2018.08.078

Rial, R. C. (2024). Biofuels versus climate change: Exploring potentials and challenges in the energy transition. *Renewable and Sustainable Energy Reviews, 196*, Article 114369. https://doi.org/10.1016/j.rser.2024.114369

Salar-García, M. J., Ortiz-Martínez, V. M., Sánchez-Segado, S., Valero Sánchez, R., Sáez López, A., Lozano Blanco, L. J., & Godínez-Seoane, C. (2024). Sustainable production of biofuels and biochemicals via electro-fermentation technology. *Molecules, 29*(4), 834. https://doi.org/10.3390/molecules29040834

Shanthi Sravan, J., Butti, S. K., Sarkar, O., Vamshi Krishna, K., & Venkata Mohan, S. (2018). Electrofermentation of food waste—regulating acidogenesis towards enhanced volatile fatty acids production. *Chemical Engineering Journal, 334*, 1709–1718. https://doi.org/10.1016/j.cej.2017.11.005

Show, K.-Y., Yan, Y., & Lee, D.-J. (2019). Bioreactor and bioprocess design for biohydrogen production. In *Biohydrogen* (pp. 391–411). Elsevier. https://doi.org/10.1016/B978-0-444-64203-5.00016-2

Sravan, J. S., Butti, S. K., Sarkar, O., & Mohan, S. V. (2019). Electrofermentation. In *Microbial Electrochemical Technology* (pp. 723–737). Elsevier. https://doi.org/10.1016/B978-0-444-64052-9.00029-7

Toledo-Alarcón, J., Moscoviz, R., Trably, E., & Bernet, N. (2019). Glucose electro-fermentation as main driver for efficient H2-producing bacteria selection in mixed cultures. *International Journal of Hydrogen Energy, 44*(4), 2230–2238. https://doi.org/10.1016/j.ijhydene.2018.07.091

Virdis, B., D. Hoelzle, R., Marchetti, A., Boto, S. T., Rosenbaum, M. A., Blasco-Gómez, R., Puig, S., Freguia, S., & Villano, M. (2022). Electro-fermentation: Sustainable bioproductions steered by electricity. *Biotechnology Advances, 59*, 107950. https://doi.org/10.1016/j.biotechadv.2022.107950

Zhao, Z., Zhang, Y., Wang, L., & Quan, X. (2015). Potential for direct interspecies electron transfer in an electric-anaerobic system to increase methane production from sludge digestion. *Scientific Reports, 5*(1), 11094. https://doi.org/10.1038/srep11094

Zhen, G., Kobayashi, T., Lu, X., Kumar, G., & Xu, K. (2016). Biomethane recovery from Egeria densa in a microbial electrolysis cell-assisted anaerobic system: Performance and stability assessment. *Chemosphere, 149*, 121–129. https://doi.org/10.1016/j.chemosphere.2016.01.101

Chapter 5
Industrial Application of Electro-Fermentation for Biochemical Synthesis

Vashisht Vishakha and Meenu Chhabra

1 Introduction

Electro-fermentation (EF) is an electrochemically navigated procedure that prompts biochemical reactions. It utilizes a synergistic procedure of amalgamating microorganisms with electrochemistry to bio-catalyze the redox reactions. The relationship between the biochemical and electrochemical reactions assists in overcoming the thermodynamic restriction of conventional fermentation processes, intending toward redox-balancing and simultaneously synchronizing the electro-metabolic pathways of unique biobased products (Sravan et al., 2018). Electrodes function as sources or sinks of electrons in a microbiological environment, allowing them to control the kinetics of electron transfer and establish electrochemically driven fermentation (Moscoviz et al., 2016). For ages, fermentation has been rudimentary in food and beverages (ethanol, wine, beer, fermented milk, etc.), antibiotic generation (penicillin, streptomycin, etc.), whole microbial cells (probiotics), solvent generation (ethanol, butanol, acetone, etc.), wastewater treatment, biopolymer synthesis [polyhydroxybutyrate (PHB), polyhydroxyalkanoate (PHA)], flavor and perfume synthesis, vitamins (B3, B12, C etc.), enzyme production, and cosmetics formation (skincare serum) (Albuquerque et al., 2007; Doran, 2013; Humphrey & Lee, 1992). The main goal of the food processing industry is to increase the preservation period and productivity of the product, limiting the texture and organoleptic characteristics of foodstuff for a prolonged period. Fermentation become the predominant biotechnological evolutionary platform for many products (Dahiya et al., 2015). However, it is a labor-exhaustive and economically inimical job. The conventional fermentation method melds upstream and downstream processes that require procuring purified

V. Vishakha · M. Chhabra (✉)
Department of Bioscience and Bioengineering, Indian Institute of Technology Jodhpur, Jodhpur, Rajasthan, India
e-mail: meenuchhabra@iitj.ac.in

© The Author(s), under exclusive license to Springer Nature Singapore Pte Ltd. 2025
S. Sevda (ed.), *Electro-Fermentation for Synthesis of Chemicals, Biochemicals and Biofuels*, https://doi.org/10.1007/978-981-96-6358-3_5

substances, optimizing energy inputs, ionic utilization, and maintaining sterile conditions, which is time-consuming and capital-intensive. A highly optimized strain is used in the culture media, which is extremely selective for its metabolic pathway. Conventional fermentation pathways are often redox imbalanced, with challenges in product selection and conversion efficiency from substrate to product (Bhagchandanii et al., 2020). To counteract the redox equivalents of the microbe's metabolism, nitrogen is sparged inside the reactor during anaerobic fermentation which serves as an electron acceptor, in contrast, air serves as an electron acceptor during aerobic fermentation. Hydrogen (H_2) gas is bubbled when more electron donors are needed to boost reductive metabolism which is called micro-bubbling. One of the main disadvantages of micro-bubbling is that its solubility in the gas phase limits the effectiveness of mass transfer (Angenent et al., 2004; Bajracharya et al., 2016). The fermenters are pressed to high pressure to make the gas phase more soluble, which raises the overall cost. The idea of utilizing acids, bases, chelating agents, buffers, and antifoam agents to change the medium conditions in addition to these limitations raises the overall operational costs (Liu et al., 2008). The monitoring of the osmotic conditions and salinity of the solutions is a worry resulting from these adjustments. It might be challenging to separate soluble metabolites, contaminants, and other secondary chemicals from the culture medium in a traditional fermentation process. The product itself may prevent bacterial growth and metabolism by feedback inhibition (Roy et al., 2016). To surpass all traditional methods in the optimization of pure cultures for industrial and biorefinery operations, EF will be effective in product specification by electrostimulation.

2 Electro-Fermentation and Microbe-electrode Interaction

Microbial electrochemical technologies include microbiological surfaces that transmit electrons to electrodes and vice versa during electro-fermentation (Thrash & Coates, 2008). The electro-fermentation process is carried out in two chambers: anodic and cathodic. Microbes during the substrate expedition produce electrons along with adenosine triphosphate (ATP) which is further transferred to the anodic electrode through biofilm formation via an external circuit, the anodic electrode circulates it to the cathodic electrode. Using an electrode as an electron source, microorganisms in the cathodic chamber make the desired bio-product (Choi & Sang, 2016). Microbes obtain their energy in the form of ATP from different metabolic pathways which is directly proportional to the potential energy difference, ΔE^o (in volts), across the electron donor and acceptor (Thrash & Coates, 2008). ΔE^o can be calculated from Gibbs free energy in Joule/mol:

$\Delta E^o = -\Delta G/nF$

ΔG = Gibbs free energy

n = Number of electrons involved in a reaction

F = Faraday's constant (96485.3 C/mol)

The maximum amount of energy that a reaction can release (exergonic) or demand (endergonic) is represented by the Gibbs free energy of that reaction. The reaction will produce electricity or need electricity is governed by ΔE°. Based on Gibbs free energy, electrochemical cells are divided into microbial fuel cells (MFCs) and microbial electrolytic cells (MECs) (Bajracharya et al., 2016). On transfer of electrons, bioelectricity is produced in microbial fuel cells and electricity needs to be provided in microbial electrolytic cells for H_2 gas production. The applied/in situ potential, supplied to the electrodes, plays a vital role in regulating the transfer of redox species (H^+ and e^-) between the microbe and electrode contact (Liu et al., 2008). Extracellular electron transfer (EET) rates are influenced by both external stimulation and the presence of a conductive material in a microbiological condition which offers process flexibility and specificity (Lovley, 2017). Electrodes are fabricated of graphite, graphite flake, carbon cloth, and carbon mesh that provide a microporous structure and large surface area for bacterial attachment and allow electrostatic interactions between microbes and the electrode surface. By interacting with electrodes, electroactive bacteria drive their metabolism through extracellular electron transfer (EET). Most EABs can transfer electrons between the interior and outside surroundings either directly or indirectly. Direct EET involves direct contact between the electrode surface and EAB unescorted by any dissociative redox compounds (Bajracharya et al., 2016; Liu et al., 2008). Indirect methods include multi-heme cytochromes, nanowires, redox proteins (NADH and FADH), shuttle-mediated EET [flavin mononucleotide (FMN), riboflavin, phenazine-1-carboxamide (PCN), 1-hydro-xyphenazine (1-OHPHZ), pyocyanin (PYO), and phenazine-1-carboxylate (PCA)], and artificial redox mediator (methyl viologen, potassium ferricyanide, neutral red, anthraquinone-2, 6-disulfonate (AQDS), and artificial-synthesized phospholipid polymers) (Fig. 1) (Albuquerque et al., 2007; Binod & Pandey, 2009; Doran, 2013; Humphrey & Lee, 1992; Schievano et al., 2016).

Reaction kinetics and product synthesis are impacted by anode/cathode reactions involving differences in oxidation and reduction potentials. During BEF functioning, the specific product synthesis is determined by the input electricity/potential. An overall positive potential of 1.10 V is produced as a byproduct of the reduction of the electron acceptor (cathode) (reduction of O_2 to H_2O vs. $E^{\circ} = 0.82$ V) and oxidation of the electron donor (anode) (oxidation of acetate to HCO^{3-} vs. SHE (E°) = −0.28 V) with a higher cathode electrode potential (Schievano et al., 2016). However, an input of potential is necessary since applying a larger anode potential (acetate to HCO^{3-} vs. SHE (E0) = −0.28 V) than cathode reduction potential (H^+/H_2 vs. SHE (E°) = −0.41 V) resulting in an overall negative cell potential of ~0.13 V (Kotloski & Gralnick, 2013). Compared to other types of reactions, the existence of H_2O as an electron donor (H_2O/O_2; E°=0.82V vs. SHE) necessitates larger energy inputs. During microbe–electrode interactions, redox suppressors including riboflavin, methyl viologen, thionin, neutral red, etc., can also speed up electron transfer rates (Kotloski & Gralnick, 2013).

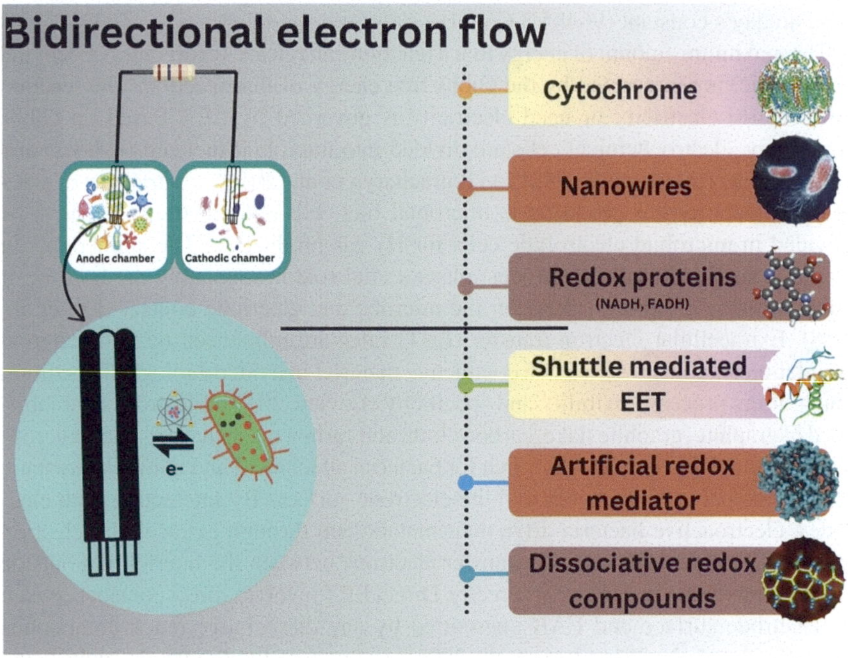

Fig. 1 Direct and indirect flow of electrons through different mediators

3 Implementation of Electro-Fermentation in Industries

3.1 Bioalcohal Formation

The demand for alcohol, particularly ethanol, as a powerful energy substitute is still rising globally (Mathew et al., 2015). Historically, alcohol has been made using a variety of fermentation techniques, which typically take 48–90 h to reach their maximum concentration. Bioethanol is a sustainable, ecological, high-octane, and low-carbon vehicle fuel, fundamentally extracted from the monosaccharides (sugars), and polysaccharides (starches, and cellulosic matter) found in plants (Abel et al., 2021). Being a carbon-neutral fuel, it emits 32–62% lower greenhouse gases than gasoline (Abel et al., 2021; An et al., 2015). Bioethanol is derived from food crops such as sugarcane juice, corn, maize, wheat, beet molasses, and cane molasses. Due to ethical concerns, researchers have shifted their interest toward lignocellulosic biomasses such as corn straw, rice straw, and wheat straw for ethanol production. Lignocellulosic biomasses are made up of cellulose, hemicellulose, and lignin. Lignin being recalcitrant, required pretreatment (separate hydrolysis and fermentation [SHF]) which escalates overall production costs (Vega et al., 2024). Several amalgamated production systems have been created, such as consolidated bioprocessing (CBP), advanced simultaneous saccharification and fermentation (SSF), and

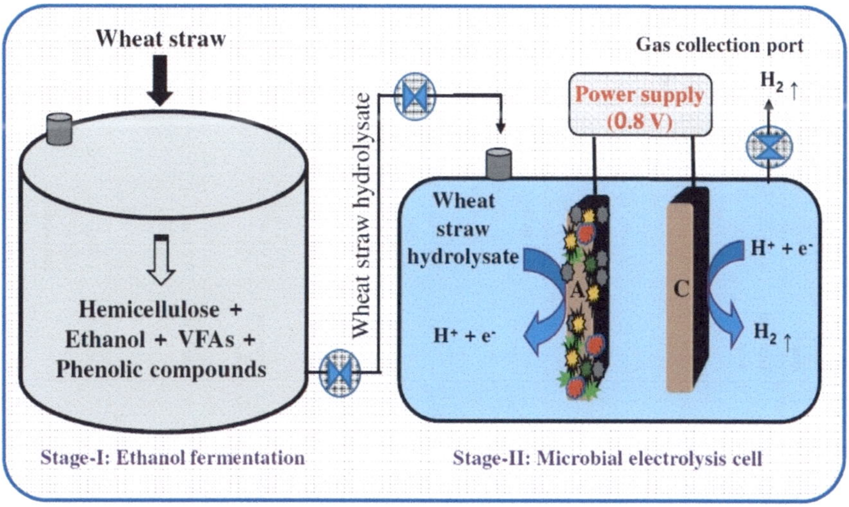

Fig. 2 Schematic presentation of the manufacture of MEC to produce bioethanol and hydrogen (Kumar et al., 2018)

simultaneous saccharification and co-fermentation (SSCF) for bioethanol production (Dhungana et al., 2022). These systems decrease overall production time, operating costs, and inhibitors, and surge the rate of hydrolysis abreast of integrating enzymatic hydrolysis and fermentation in a single reactor (Foust et al., 2009). It is believed that employing lignocellulolytic (or ligninolytic and cellulolytic) and fermenting microbial consortium in a single reactor to perform synchronized saccharification and electro-fermentation (SSEF) will abolish the demand for chemicals and a rigorous pretreatment approach (Joshi et al., 2018). Additionally, an integrative method was created that combines the manufacture of MEC to produce bioethanol and hydrogen while simultaneously degrading phenolic chemicals and their derivatives (furfurals) from the bioethanol byproducts (Kumar et al., 2018) (Fig. 2).

Today, there is a growing demand for biofuels, especially butanol. Butanol is regarded as a second-generation biofuel with a higher energy density and less volatility than ethanol. The bio-based 1-butanol market was expanded at an annual rate of 9.3% between 2017 and 2021 (Engel et al., 2019). The energy density of butanol is 29.2 MJ/L, comparable to gasoline's (32.5 MJ/L) and almost 38% higher than that of the fuel additive ethanol. However, low productivity is a significant limiting issue that can be addressed by BES (Khosravanipour Mostafazadeh et al., 2016). Researchers have used various BES methods to produce alcohol which is given in Table 1.

Table 1 Production of different biobased products through EF

S.no	Electro-fermentation	Electrolytic cell	Electrode material	Microorganism	Substrate	Voltage applied	Reference electrode	Synthesized product	Conducting material	Production rate	References
1	Anodic EF	Single-chambered MEC	Rayon felt graphite	*Cellulomonas uda*	Cellobiose	+0.24 V versus Ag/AgCl	Ag/AgCl	Bioethanol	Titanium wire	$0.056 \pm 0.003 g^{-1} l^{-1} h^{-1}$	Awate et al., (2017)
2	Cathodic EF	Two-chambered MFC	Graphite	Acetate-reducing inoculum used from an upflow anaerobic sludge blanket reactor treating distillery wastewater (Royal Nedalco)	Acetate	−550 mV versus NHE	Normal hydrogen electrode (NHE)	Bioethanol	Gold wire	74.6 ± 6.3%	Steinbusch et al., (2010)
3	Cathodic EF	Two-chambered MEC	Graphite felt electrode	Clostridium pasteurianum DSM 525	Glucose	+0.045 V versus SHE	Standard hydrogen electrode (SHE)	Biobutanol	–	12.2 mM	Choi et al., (2014)
4	Cathodic EF	Two-chambered MFC	Graphite	Inoculum obtained from an anaerobic glucose-fed pH-stat reactor	Glycerol	−0.9 V versus SHE	Standard hydrogen electrode (SHE)	1,3-propanediol (PDO)	–	0.50 mol of PDO/mol glycerol	(Zhou et al., 2013)
5	Cathodic EF	Two-chambered MEC	Carbon cloth	*Clostridium kluyveri*	Carbon dioxide (CO_2)	+0.197 V versus SHE	SHE	Butyrate, ethanol, butanol	Titanium rod	1.82 mMC/d	Ganigué et al., (2015)
6	Cathodic EF	Two-chambered MEC	Carbon cloth	*Methanobacterium palustre*	CO_2	−0.7 V (versus Ag/AgCl)	Ag/AgCl	Methane (CH_4)	Titanium wire	$4.5 L m^{-2} d^{-1}$	Cheng et al., (2009)
7	Cathodic EF	Two-chambered MEC	Carbon paper/ carbon rod	*Consortia*	CO_2	+199 mV versus SHE	SHE	CH_4	Metal wire	$9.2 L m^{-2} d^{-1}$	Villano et al., (2010)
8	Cathodic MEC	Two-chambered MEC	Graphite	*Acetobacterium* spp.	CO_2	+210 mV versus SHE	SHE	Acetate, Hydrogen(H_2)	Titanium wire	$1.04 gL^{-1} d^{-1}$ and $0.2 g L^{-1} d^{-1}$ Respectively	Marshall et al., (2013)

(continued)

Table 1 (continued)

S.no	Electro-fermentation	Electrolytic cell	Electrode material	Microorganism	Substrate	Voltage applied	Reference electrode	Synthesized product	Conducting material	Production rate	References
9	Cathodic MEC	Two-chambered MEC	Graphite	Consortia	Wastewater	+0.41 V	–	H_2	Copper wire	0.0125mg H_2/mg Chemical oxygen demand (COD)	DITZIG et al., (2007)
10	Cathodic MEC	Two-chambered MEC	Carbon felt/ Platinum loaded	Consortia	Acetate	−0.42 V versus NHE	NHE	H_2	Metal wire	627 ± 39 μmol	Chae et al., (2009)
11	Anodic EF	Two-chambered MEC	Carbon fibre	*Shewanella oneidensis*	Glycerol	+0.4 V versus SHE	SHE	Ethanol, Acetate	Metal wire	Ethanol yield: 52 ± 4 %; Acetate yield: 13 ± 6 %	Flynn et al., (2010)
12	Anodic EF	Two-chambered MEC	–	*S. oneidensis*	Lactate	0 mV versus NHE	Ag/AgCl	Acetoin	Metal wire	0.24 g L^{-1}	Bursac et al., (2017)
13	Anodic EF	Two-chambered MEC	Carbon cloth	*Pseudomonas putida*	Glucose	+0.197 V versus SHE	Ag/AgCl	2-keto gluconic acid	Titanium wire	0.25 ± 0.02 $mmolg_{CDW}^{-1}$ h^{-1}	Yu et al., (2018)
14	Anodic EF	Two-chambered MFC	Carbon paper	*Klebsiella pneumoniae*	Glycerol	–	–	Acetate	Metal wire	21.7 mM	Kim et al., (2019)
15	Cathodic EF	Two-chambered MEC	Carbon fiber	Wastewater	Glucose	+196 mV versus SHE	Ag/AgCl	CH_4	Copper wire	408.3 mL CH_4/g COD	Choi et al., (2017)
16	Anodic EF	Two-chambered MFC	Graphite granule	Consortia	Acetate	–	–	CH_4	Graphite rod	0.41 mol/mol of acetate	Clauwaert et al., (2008)
17	Cathodic EF	Two-chambered bioreactors	Graphite plate, Carbon cloth	Wastewater+ cow dung	CO_2	−0.8 V versus Ag/AgCl	Ag/AgCl	Short-chain fatty acid (SCFA), CH_4	Copper wire	SCFA:2050 mg/L CH_4:41.2 mL/day	Velvizhi et al., (2022)
18	EF	Bench-scale anaerobic EF reactors	Graphite	Mixed consortium from sewage treatment plant	Food waste	−0.6 V	–	Acetic acid, Butyric acid, Propanoic acid	Copper wire	Acetic acid: 3221 mg/L Butyric acid: 900 mg/L Propanoic acid: 474 mg/L	Shanthi Sravan et al., (2018)

(continued)

Table 1 (continued)

S.no	Electro-fermentation	Electrolytic cell	Electrode material	Microorganism	Substrate	Voltage applied	Reference electrode	Synthesized product	Conducting material	Production rate	References
19	EF + Dark fermentation	Cathodic MEC	Graphite	*Clostridium thermocellum*	Lignocellulosic biomass	0.100 V	-	H_2	Metal wire	1.11±0.13 L H_2/L	Lalaurette et al., (2009)
20	EF	MEC	Carbon cloth	*Acetobacterium, Geobacter, Desulfovibrio* and *Methanobrevibacter*	Anaerobic digested sludge	0.8 V	–	CH_4, H_2	Metal wire	0.83 L/L d and 92.58%	Jia et al., (2020)
21	Cathodic EF	BES reactor	Graphite	Mixed culture	Glycerol	1.44 V versus SHE	SHE	1,3-propanediol	Metal wire	0.57 ± 0.09 mol/mol glycerol	Zhou et al., (2015)
22	Cathodic EF	MFC	Graphite	*Clostridium pasteurianum*	Glucose	0.045 V versus SHE	SHE	Butanol	Metal wire	Increased from 78% to 96%	Choi et al., (2014)
23	Anodic EF	MEC	Graphite	*Clostridium cellobioparum Geobacter sulfurreducens*	Glycerol	0.46 V versus SHE	SHE	Ethanol	Metal wire	10 g/L	Speers et al., (2014)

4 Biogas Production

In an anaerobic reactor, methanogens typically use hydrogen, formate, and acetate from pure chemicals or different biowastes as substrates to create biomethane (Kumar et al., 2018; Wang et al., 2009). Acetoclastic methanogens run the first pathway (acetate substrate), whereas hydrogenotrophic methanogenesis runs the second pathway (using H_2 substrate) (Wang et al., 2009). Factors such as pH, hydraulic retention time, substrate load, temperature, C/N ratio, acclimation duration, and substrate composition impact the production of biogas. On a lab scale, it is simple to manage two or more of these factors for any given experimental setting. However, on a big scale, it is challenging to maintain control over an extended length of time, especially when it comes to the feed composition. It was underlined, although, that while optimizing such feedstock shows promise, it is constrained by the large organic contents provided by grass silage and maize, which account for most of the biogas generated (Mézes et al., 2017). Any brief drop in feedstock could increase the percentage of energy crops in the substrates, which would accelerate the evolution of biogas. A rigorous economic evaluation of any agricultural biogas plant is desirable because energy crops require less land than food crops (Kumar et al., 2018; Mézes et al., 2017). In addition to the primary parameters listed above, research has been done on how microbial populations evolve during an operation. In addition to previous co-feeding techniques, acclimatization of the mixed cultures (defined or undefined) has demonstrated promise in improving performance and consistency (Kumar et al., 2018; Wang et al., 2009). On the other hand, MEC-based EF may offer several benefits over traditional biogas generation. Methanogens are said to produce CH_4 mostly from H_2 gas produced at the cathode and partially by acetate conversion during this process, which involves direct electron acceptance (from the cathode) (Logan & Rabaey, 2012). Thus, in addition to the CO_2 consumption throughout the process, around 86% enriched CH_4 generation was noted as compared to traditional anaerobic digestion (Villano et al., 2010). Various examples have been mentioned in Table 1. When the feed contains a very high-strength wastewater or organic load, anaerobic digesters are required to support EF, which performs effectively on low-strength wastewater. It is important to realize that EF-based technology (as it has been created thus far) may not be able to function as a stand-alone process and, as a result, may not have the capacity to completely replace the current methodology; rather, it will need the assistance of other techniques to reach greater efficiency.

Soon, integrated bioprocesses might be useful for treating wastewater and waste materials while sustainably generating bioenergy.

5 Short-Chain Fatty Acids Production

In EF, inserting electrodes into a microbial fermentative environment has a catalytic impact that increases electron kinetics and reaction rates (Moscoviz et al., 2016; Nikhil et al., 2015; Redwood et al., 2012). BES is made possible by introducing electrodes into the microbial environment, which regulates the rate kinetics and benefits the synthesis of products from a particular substrate (Appleby, 1971; Stamenkovic et al., 2017). By lowering the activation energy (E_a) and heat of activation (H_a), the differences in conventional processes with the electrode character can get around the rate constraints (Villano et al., 2017). The heat of adsorption of the reaction (H_{ad}) has increased due to the lowered E_a and H_a, leading to improved platform chemical recovery. The impact of electrocatalysis on the microbial environment can drastically alter the rate of biological reaction constants. The fermentation rate using waste material as feedstock to produce several metabolic end products, including short-chain fatty acids (SCFA) or volatile fatty acids (VFA) is described in this communication using EF. SCFAs have great commercial value and are created during the acidogenesis of organic substrates. By utilizing the metabolism of anaerobic microbiomes, chain elongation is currently showing promise as a bioprocess to produce medium-chain fatty acids (MCFAs) and SCFAs. SCFA are straight-chain mono-carboxylates with six to twelve carbon atoms in length that can be produced from organic waste streams like food waste, organic wastewater, and waste-activated sludge (Stamenkovic et al., 2017; Villano et al., 2017). Because of their limited solubility in acidic environments, and their high carbon/oxygen ratio makes their separation easier after production. SCFAs are a potential and cost-effective intermediate for the synthesis of biochemicals and biofuels. For improved substrate conversion and the conversion of VFA-rich effluents into H_2, anaerobic/dark fermentation in conjunction with MEC technology is a very providential option. The bio-electrochemical stimulation of monosaccharide acidification with anaerobic bacteria enhances the effectiveness of SCFA synthesis. Additionally, produced SCFA can be a valuable electron donor in some different bioprocesses.

For electron donor applications, electro-fermentation can be utilized to create biopolymers by anoxygenesis, biodiesel production using photosynthetic equipment, and bioelectricity generation, etc. (Saratale et al., 2017; Sivagurunathan et al., 2018). Since SCFA is a very promising VFA that is lightweight, pollution-free, and easily biodegradable, it is a valuable commodity for the apparel, fruit, beverage, polymer, and plastics industries. Through an acetogenic process, feedstock or CO_2 is converted to SCFA via a series of microbially driven metabolic processes. Table 1 consists of various examples of SCFA formation from different substrates including waste material.

6 Acid Production

EF demands methods that achieve optimum productivity to lower the cost of upstream and downstream processing for technological commercialization. EF may be able to modify the fermentation process to produce organic acid selectively (Lovley, 2017). Under non-sterile conditions, EF maximizes biomass conversion toward the buildup of high-titer selective organic acids. The applied potential could increase metabolic activity and prevent an excessive buildup of feedback inhibition during fermentation. Numerous organic acids, including citric acid (CA), acetic acid (AA), lactic acid (LA), succinic acid (SA), and muconic acid (MA), can be produced from biomass. Organic acid is used in a variety of industries, including the leather, automobile, textile, and pharmaceutical sectors (Uçkun Kiran et al., 2014). Carboxylic and hydroxyl groups are present in acids that can be further transformed into different biochemicals (Uçkun Kiran et al., 2014). One such well-established method for extracting a variety of biobased products from various organic feedstock is anaerobic digestion (AD). Several strategies are being investigated to maximize energy conversion efficiency and operational stability in conventional AD, and bio-electro-fermentation (BEF) has worked well (Schievano et al., 2016). The electrons from the working electrode in BEF are used to reduce CO_2 into a wide range of marketable organic compounds that serve as building blocks for other products and reduce carbon footprints by capturing CO_2.

Acetate is produced by pyruvate, which is generated with the help of the glycolysis pathway. A class of facultative autotrophic bacteria known as acetogens produces acetate by using carbon monoxide (CO) or H_2 and reducing CO_2. Acetate is produced using a variety of substrates in addition to CO_2, including pentoses, methyl compounds, hexoses, etc. Only acetogens are found in the genera *Sporomusa* and *Acetobacterium,* but both acetogenic and non-acetogenic species are found in the *Clostridium* genera. *Clostridium*, *Acetobacterium woodii*, and *Moorella thermoacetica* are the acetogens that are most frequently researched. These bacteria were selected for biocathode-driven CO_2 fixation in BEF due to their metabolic versatility.

For SA production in both aerobic and anaerobic settings, the glycolysis pathway converts glucose to phosphoenolpyruvate (PEP) and then to pyruvate. PEP functions as a substrate for the carboxylase-catalyzed anaplerotic process in anaerobic circumstances and is transformed into oxaloacetate with the aid of the enzymes PEP carboxylase or PEP carboxykinase. By combining CO_2 and the enzymes pyruvate carboxylase, pyruvate can also be transformed into oxaloacetate or malate. Malate dehydrogenase helps transform the oxaloacetate from PEP into malate. With the aid of fumarase and fumarate reductase, the malate produced can be converted to fumarate and then to succinate (Fig. 4) (Park & Zeikus, 1999).

MA, also known as 2,4-hexadienedioic acid, is primarily utilized as a key and dominating precursor molecule in the production of polyamides and polyesters. It is utilized as a starting material for the synthesis of a variety of high-value chemicals, as bulk material for muconic homo and copolymer polymer processes, and to produce its derivatives, such as adipic acid, caprolactam, and terephthalic acid.

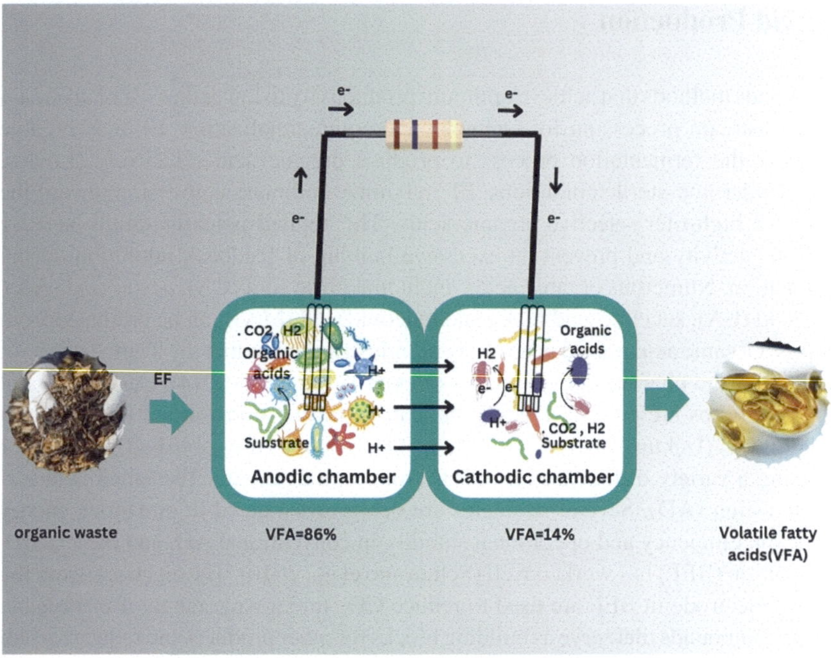

Fig. 3 EF process flow diagram for organic acid recovery from biomass

There are three isomeric forms of muconic acid: cis,cis-muconic acid, cis,trans-muconic acid, and trans,trans-muconic acid. The aromatic compounds of a petroleum base, including toluene, phenol, benzene, guaiacol, and benzoate, are oxidized by a specific microorganism to produce muconic acid (Schievanoet al., 2016a). Aromatic compounds will be obtained from the biomass feedstocks through the utilization of conversion technologies. Aromatic compounds are converted to catechol by phenol 2-monooxygenase which is an intermediary molecule for their conversion to muconic acid. Further, catechol 1,2-dioxygenase will facilitate the ortho-cleavage of catechol to produce muconic acid (Fig. 3). Table 1 gives various examples of acid production from EF using feedstock as substrate.

7 Cosmetic Production

The cosmetics industry caters to the needs of consumers who are affected by a variety of dermatitis. Establishing the quantity of synthetic ingredients in cosmetics by substituting them with natural components aligns with contemporary pro-ecological trends. Alcohols are crucial in the production of cosmetics because of their utilization in cosmetics, resins, solvents, and polymer preservation. 1,3-Propanediol (PDO) is significant in cosmetics, and glycerol acts as the substrate for its biological production

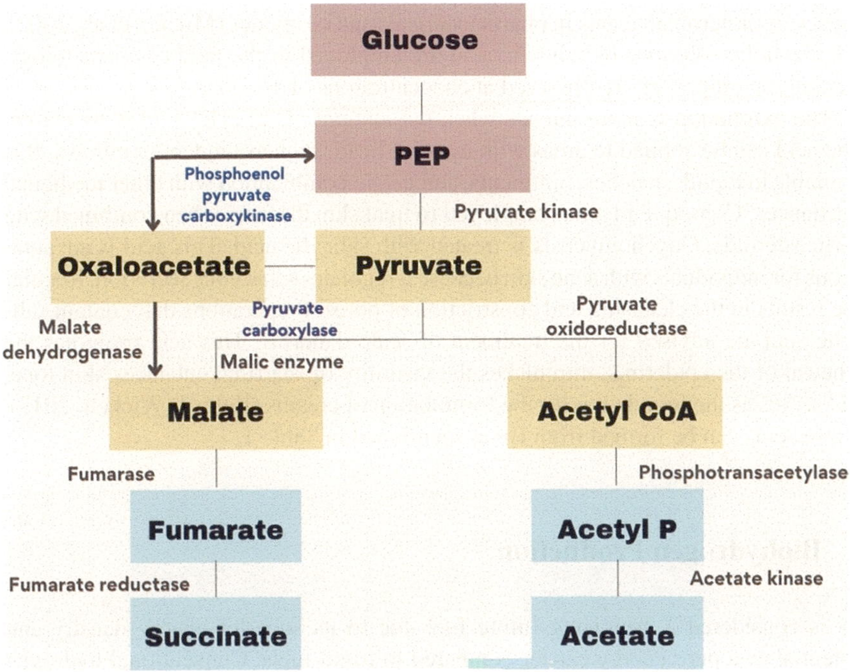

Fig. 4 Flow chart of metabolic pathway of acetate and succinate production

(Kumar et al., 2018; Schievano et al., 2016). n-Butyl alcohol is a primary aliphatic alcohol that has been historically employed as a solvent in nail care cosmetic products. However, a new concentration of use data suggests that it is also being used at low concentrations in eye makeup, personal hygiene, and grooming cosmetic products ("Final Report of the Addendum to the Safety Assessment of N-Butyl Alcohol as Used in Cosmetics," 2008). Hair growth serums and body moisturizers contain alcohols, which are intended to moisturize and stimulate the scalp, promote hair growth, eliminate dandruff, and prevent hair loss (Martins Ferreira et al., 2012). However, the carcinogenic effects of topical ethanol are the primary concern for public health, as there is unambiguous evidence of the carcinogenicity of ethanol when applied on the skin and consumed orally in the form of alcoholic beverages. So far, there is insufficient evidence to link the use of topical ethanol with an elevated risk of skin cancer. The epidemiological evidence regarding the association between oral cancer and the use of ethanol in the oral cavity in the form of mouthwashes or mouth rinses is limited and conflicting. Ethanol may facilitate the transdermal absorption of xenobiotics and function as a skin penetration enhancer (Poddębniak & Kalinowska-Lis, 2024). Acids are widely used in the cosmetic industry due to their broad antimicrobial spectrum, excellent stability, and non-volatility. The types of acids used in cosmetics include benzoic acid (BA) dehydroacetic acid (DHA), benzoic acid (BA), sorbic acid (SOA), salicylic acid (SA), and their sodium salts have made them widely used as antimicrobial agents in pharmaceuticals and cosmetics (Mikami et al., 2002). 2% alcoholic solutions of salicylic acid are employed in the field of cosmetology. Keratolytic properties are observed at concentrations of 10–20%.

The exfoliation is more intense when the amount of salicylic acid is increased. The acid can be applied to areas with a strong bead, such as the feet or elbows. It is available in liquids, patches, ointments, and gel, in combination with other medicinal substances. Consequently, it is employed to treat skin diseases when combined with corticosteroids. Onychomycosis is treated with salicylic acid. This acid is advantageous for individuals with acne skin because it regulates sebaceous secretion. Regular use results in the cleansing and constriction of pores. Preparations that contain salicylic acid are advised for the treatment of scalp dandruff. This acid promotes the renewal of the epidermis, diminishes the visibility of scarring, enhances skin tone, and serves as the foundation for the formulation of creams (Kelm & Wickett, 2017). These acids can be formed from EF as mentioned in Table 1.

8 Biohydrogen Production

H_2 is considered a promising future fuel due to its superior energy density and potential as a pure energy carrier compared to fossil fuels. Conventional hydrogen production methods, such as fossil fuel gasification and water electrolysis, are cost-intensive and unsustainable due to their high energy requirements (Møller et al., 2017). Hydrogen fuels can contribute significantly to net zero emissions, decarbonizing industries with high emissions, and long-distance transportation (Rahman et al., 2016). When H_2 fuel burns it produces water as burnt product, because of this it is considered the cleanest fuel. The Indian Ministry of Finance announced Rs 19,700 crore for green hydrogen production and aims to produce five million tons of green hydrogen with a 45% reduction in carbon emissions by 2030 (Budget, 2023).

The conversion of organics into H_2 has been investigated using a variety of biomass and inoculants, with MEC serving as a typical microbial BES. In MEC, an external power source is required, which is supplied by connecting a nonrenewable source of energy, i.e., a battery. Several researchers have reported increased H_2 yield in a combined treatment process of dark fermentation and MEC using fermentation effluent (Oncel et al., 2014). This is because the effluent from a dark fermentation process is abundant in organics that cannot be further fermented to produce hydrogen (Table 1, Fig. 5). EAB was also subjected to functional regulation and adaptive evolution approaches to improve commercial efficiencies and hydrogen recovery through EF. For biohydrogen production, water need to be split into H_2 and O_2. A huge amount of energy is required to split water molecules (Oncel et al., 2014). The Gibs free energy value of water is $\Delta G^{\circ} = 237$ kJ/mol. Hydrogen production through MEC comes under the category of grey/ blue hydrogen. Alternatively, MFC can serve as a medium to produce green hydrogen effectively without using any use of fossil fuel. Figure 6 illustrates the proposed pathway of biohydrogen production using microalgae in MFC. Microalgae contains hydrogenase enzymes, which are essential

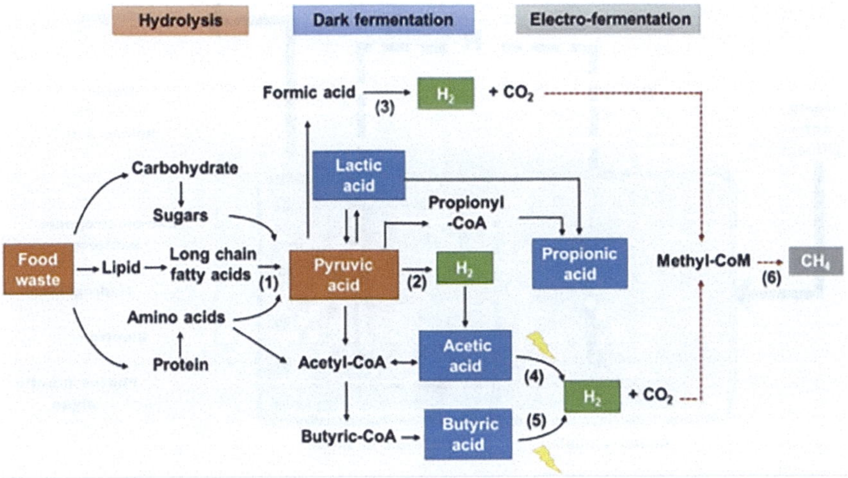

Fig. 5 Illustration of the sequential system's metabolic pathways and metabolites, in conjunction with hydrogen production through dark fermentation and EF (Jia et al., 2020)

for the generation of H_2. Hydrogenases are of two types: Fe-Fe and Ni-Fe. Fe-Fe has the highest rate of H_2 production, but it is extremely vulnerable to O_2 with concentrations as low as .001% (King et al., 2022; Oncel et al., 2014). Fe-Fe hydrogenases consist of two endogenous hydrogenases HYD1 and HYD2. HYD1 produces 80% of H_2 while Hyd2 produces 20% of H_2 in the cell (Xu et al., 2019). Oxygen inactivates these enzymes irreversibly (Yarkent et al., 2024). The genes synthesizing HDY1 can be altered genetically to increase biohydrogen production (Masukawa et al., 2012). The algal bacterial consortium can enhance hydrogen production up to many folds under nutrient-deprived conditions (Fakhimi et al., 2020).

9 Future Perspective

Both MFC and MEC can potentially enhance microbial processes and could significantly impact the development of biomass refinery chains. The way the electrical potential and current affect the metabolism from EET requires a more detailed understanding. Another significant gap that must be addressed is the absence of fermenter designs that incorporate electrodes to facilitate the efficient steering of the EF. Furthermore, it will be crucial to determine the feasibility and stability of a surface-based technology in long-term operations when it is in contact with intricate culture broths. Additionally, a techno-economical assessment must identify the potential additional expenses associated with the EF to realize these improvements in process efficiency. EF also provides a novel framework for the investigation of interactions between electroactive bacteria and fermenters in defined co-cultures

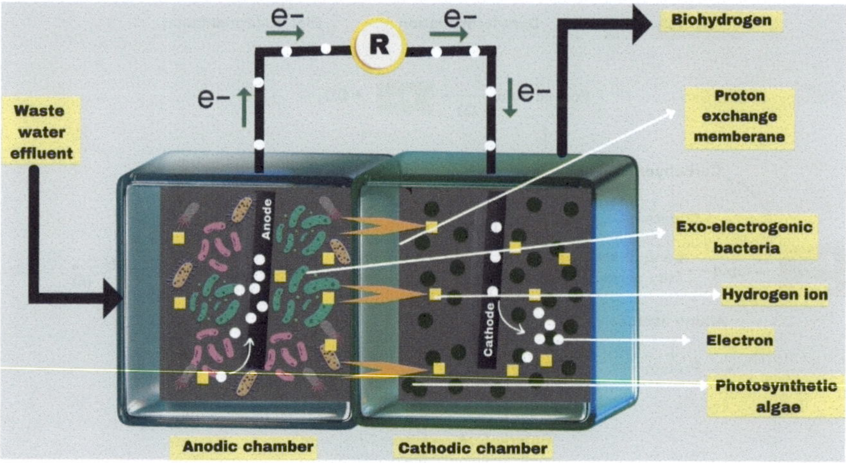

Fig. 6 Proposed schematic of biohydrogen production through microalgae in MFC

and mixed cultures. The external control of electrodes can have a significant and immediate impact on electroactive bacteria, making it an excellent instrument for observing these interactions under well-controlled conditions. Interspecies interactions can also be influenced by the electrode's material, such as direct electron transfer, which necessitates a conductive surface that is suitable for the attachment of electroactive bacteria. Novel opportunities to observe specific interspecies interactions are emerging, with novel metabolic functions to be explored, because of the increasing number of electroactive bacteria that have been discovered in recent years (Lovley, 2017). This creates a vast and thrilling research area for the development of new, enhanced EF processes that utilize a diverse array of substrates, microbial catalyzers, and targeted products.

As the world is more concerned about drastically drifting climate impacts, we need to focus on alternative sources of energy that can tackle the global challenge of carbon emission. Biohydrogen is the cleanest fuel and has the potency to replace fossil fuel, but it is necessary to investigate a more continuous process mode of biohydrogen production. Consequently, additional research and development are necessary to enhance the rates and yields of H_2 production and optimize the dimensions of bioreactors. This will enable H_2 to replace non-renewable energy sources, such as fossil fuels, and will maintain the world's energy supplies. Furthermore, it is imperative to conduct comprehensive research and analysis of gas membrane technology for H_2 separation in conjunction with the H_2 fermentation process to ascertain the feasibility of an integrated application. Thus, it would be advantageous to establish a sustainable energy supply by minimizing waste and contamination through its utilization as a resource and utilizing the hydrogen produced as a fuel in fuel cell technology to generate electricity and run automobiles in the future. Microalgae has the highest potency to produce green hydrogen in MFC but its sensitivity to oxygen hampers its performance. Genetic engineering approaches are required to make genes resistant that are responsible for sensitivity against oxygen to escalate the rate of biohydrogen production.

References

Abel, R., Coney, K., Johnson, C., Thornton, M., Zigler, B., & McCormick, R. (2021). *Global ethanol-blended-fuel vehicle compatibility study*. https://doi.org/10.2172/1832216

Albuquerque, M. G. E., Eiroa, M., Torres, C., Nunes, B. R., & Reis, M. A. M. (2007). Strategies for the development of a side stream process for polyhydroxyalkanoate (PHA) production from sugar cane molasses. *Journal of Biotechnology, 130*(4), 411–421. https://doi.org/10.1016/j.jbiotec.2007.05.011

An, L., Zhao, T. S., & Li, Y. S. (2015). Carbon-neutral sustainable energy technology: Direct ethanol fuel cells. *Renewable and Sustainable Energy Reviews, 50*, 1462–1468. https://doi.org/10.1016/j.rser.2015.05.074

Angenent, L. T., Karim, K., Al-Dahhan, M. H., Wrenn, B. A., & Domíguez-Espinosa, R. (2004). Production of bioenergy and biochemicals from industrial and agricultural wastewater. *TRENDS in Biotechnology, 22*(9), 477–485.

Appleby, A. J. (1971). Electrocatalysis and fuel cells. *Catalysis Reviews, 4*(1), 221–244.

Awate, B., Steidl, R. J., Hamlischer, T., & Reguera, G. (2017). Stimulation of electro-fermentation in single-chamber microbial electrolysis cells driven by genetically engineered anode biofilms. *Journal of Power Sources, 356*, 510–518. https://doi.org/10.1016/j.jpowsour.2017.02.053

Bajracharya, S., Vanbroekhoven, K., Buisman, C. J. N., Pant, D., & Strik, D. P. (2016). Application of gas diffusion biocathode in microbial electrosynthesis from carbon dioxide. *Environmental Science and Pollution Research, 23*, 22292–22308.

Bhagchandanii, D. D., Babu, R. P., Sonawane, J. M., Khanna, N., Pandit, S., Jadhav, D. A., Khilari, S., & Prasad, R. (2020). A comprehensive understanding of electro-fermentation. *Fermentation, 6*(3), 92. https://doi.org/10.3390/fermentation6030092

Binod, P., & Pandey, A. (2009). Bioprocess technology and product development. *In Bioprocess and bioproducts–technology, trends and opportunities* (pp. 10–27).

Bursac, T., Gralnick, J. A., & Gescher, J. (2017). Acetoin production via unbalanced fermentation in *Shewanella oneidensis*. *Biotechnology and Bioengineering, 114*(6), 1283–1289. https://doi.org/10.1002/bit.26243

Chae, K.-J., Choi, M.-J., Kim, K.-Y., Ajayi, F. F., Chang, I.-S., & Kim, I. S. (2009). A solar-powered microbial electrolysis cell with a platinum catalyst-free cathode to produce hydrogen. *Environmental Science & Technology, 43*(24), 9525–9530.

Cheng, S., Xing, D., Call, D. F., & Logan, B. E. (2009). Direct biological conversion of electrical current into methane by electromethanogenesis. *Environmental Science & Technology, 43*(10), 3953–3958.

Choi, K.-S., Kondaveeti, S., & Min, B. (2017). Bioelectrochemical methane (CH4) production in anaerobic digestion at different supplemental voltages. *Bioresource Technology, 245*, 826–832. https://doi.org/10.1016/j.biortech.2017.09.057

Choi, O., Kim, T., Woo, H. M., & Um, Y. (2014). Electricity-driven metabolic shift through direct electron uptake by electroactive heterotroph Clostridium pasteurianum. *Scientific Reports, 4*(1), 6961.

Choi, O., & Sang, B.-I. (2016). Extracellular electron transfer from cathode to microbes: application for biofuel production. *Biotechnology for Biofuels, 9*, 1–14.

Clauwaert, P., Tolêdo, R., van der Ha, D., Crab, R., Verstraete, W., Hu, H., Udert, K. M., & Rabaey, K. (2008). Combining biocatalyzed electrolysis with anaerobic digestion. *Water Science and Technology, 57*(4), 575–579. https://doi.org/10.2166/wst.2008.084

Dahiya, S., Sarkar, O., Swamy, Y. V., & Venkata Mohan, S. (2015). Acidogenic fermentation of food waste for volatile fatty acid production with co-generation of biohydrogen. *Bioresource Technology, 182*, 103–113. https://doi.org/10.1016/j.biortech.2015.01.007

Dhungana, P., Prajapati, B., Maharjan, S., & Joshi, J. (2022). *Current trends in lignocellulosic bioethanol production.*

Ditzig, J., Liu, H., & Logan, B. (2007). Production of hydrogen from domestic wastewater using a bioelectrochemically assisted microbial reactor (BEAMR). *International Journal of Hydrogen Energy, 32*(13), 2296–2304. https://doi.org/10.1016/j.ijhydene.2007.02.035

Doran, P. M. (2013). Reactor engineering. In *Bioprocess engineering principles* (pp. 761–852). Elsevier. https://doi.org/10.1016/B978-0-12-220851-5.00014-9

Engel, M., Holtmann, D., Ulber, R., & Tippkötter, N. (2019). Increased biobutanol production by mediator-less electro-fermentation. *Biotechnology Journal, 14*(4). https://doi.org/10.1002/biot.201800514

Fakhimi, N., Gonzalez-Ballester, D., Fernández, E., Galván, A., & Dubini, A. (2020). Algae-bacteria consortia as a strategy to enhance H2 production. *Cells, 9*(6), 1353. https://doi.org/10.3390/cells9061353

Final Report of the Addendum to the Safety Assessment of n-Butyl Alcohol as Used in Cosmetics. (2008). *International Journal of Toxicology, 27*(2_suppl), 53–69. https://doi.org/10.1080/10915810802244504

Flynn, J. M., Ross, D. E., Hunt, K. A., Bond, D. R., & Gralnick, J. A. (2010). Enabling unbalanced fermentations by using engineered electrode-interfaced bacteria. *MBio, 1*(5). https://doi.org/10.1128/mBio.00190-10

Foust, T. D., Aden, A., Dutta, A., & Phillips, S. (2009). An economic and environmental comparison of a biochemical and a thermochemical lignocellulosic ethanol conversion processes. *Cellulose, 16*, 547–565.

Ganigué, R., Puig, S., Batlle-Vilanova, P., Balaguer, M. D., & Colprim, J. (2015). Microbial electrosynthesis of butyrate from carbon dioxide. *Chemical Communications, 51*(15), 3235–3238.

Humphrey, A. E., & Lee, S. E. (1992). Industrial fermentation: Principles, processes, and products. In *Riegel's handbook of industrial chemistry* (pp. 916–986). Springer Netherlands. https://doi.org/10.1007/978-94-011-7691-0_24

Jia, X., Li, M., Wang, Y., Wu, Y., Zhu, L., Wang, X., & Zhao, Y. (2020). Enhancement of hydrogen production and energy recovery through electro-fermentation from the dark fermentation effluent of food waste. *Environmental Science and Ecotechnology, 1*, Article 100006. https://doi.org/10.1016/j.ese.2019.100006

Joshi, J., Bhattarai, T., & Sreerama, L. (2018). Efficient methods of pretreatment for the release of reducing sugars from lignocellulosic biomass native to Nepal and characterization of pretreated lignocellulosic biomass. *International Journal of Advanced Biotechnology and Research, 9*, 9–23.

Kelm, G. R., & Wickett, R. R. (2017). The role of fatty acids in cosmetic technology. In *Fatty acids* (pp. 385–404). Elsevier. https://doi.org/10.1016/B978-0-12-809521-8.00012-X

Khosravanipour Mostafazadeh, A., Drogui, P., Brar, S. K., Tyagi, R. D., Le Bihan, Y., Buelna, G., & Rasolomanana, S.-D. (2016). Enhancement of biobutanol production by electromicrobial glucose conversion in a dual chamber fermentation cell using C. pasteurianum. *Energy Conversion and Management, 130*, 165–175. https://doi.org/10.1016/j.enconman.2016.10.050

Kim, M. Y., Kim, C., Ainala, S. K., Bae, H., Jeon, B.-H., Park, S., & Kim, J. R. (2019). Metabolic shift of Klebsiella pneumoniae L17 by electrode-based electron transfer using glycerol in a microbial fuel cell. *Bioelectrochemistry, 125*, 1–7. https://doi.org/10.1016/j.bioelechem.2018.08.002

King, S. J., Jerkovic, A., Brown, L. J., Petroll, K., & Willows, R. D. (2022). Synthetic biology for improved hydrogen production in *Chlamydomonas reinhardtii. Microbial Biotechnology, 15*(7), 1946–1965. https://doi.org/10.1111/1751-7915.14024

Kotloski, N. J., & Gralnick, J. A. (2013). Flavin electron shuttles dominate extracellular electron transfer by Shewanella oneidensis. *MBio, 4*(1). https://doi.org/10.1128/mBio.00553-12

Kumar, P., Chandrasekhar, K., Kumari, A., Sathiyamoorthi, E., & Kim, B. (2018). Electro-fermentation in aid of bioenergy and biopolymers. *Energies, 11*(2), 343. https://doi.org/10.3390/en11020343

Lalaurette, E., Thammannagowda, S., Mohagheghi, A., Maness, P.-C., & Logan, B. E. (2009). Hydrogen production from cellulose in a two-stage process combining fermentation and electrohydrogenesis. *International Journal of Hydrogen Energy, 34*(15), 6201–6210. https://doi.org/10.1016/j.ijhydene.2009.05.112

Liu, Y., Zheng, P., Sun, Z., Ni, Y., Dong, J., & Wei, P. (2008). Strategies of pH control and glucose-fed batch fermentation for production of succinic acid by *Actinobacillus succinogenes* CGMCC1593. *Journal of Chemical Technology & Biotechnology, 83*(5), 722–729. https://doi.org/10.1002/jctb.1862

Logan, B. E., & Rabaey, K. (2012). Conversion of wastes into bioelectricity and chemicals by using microbial electrochemical technologies. *Science, 337*(6095), 686–690.

Lovley, D. R. (2017). Syntrophy goes electric: Direct interspecies electron transfer. *Annual Review of Microbiology, 71*(1), 643–664. https://doi.org/10.1146/annurev-micro-030117-020420

Marshall, C. W., Ross, D. E., Fichot, E. B., Norman, R. S., & May, H. D. (2013). Long-term operation of microbial electrosynthesis systems improves acetate production by autotrophic microbiomes. *Environmental Science & Technology, 47*(11), 6023–6029.

Martins Ferreira, L., Binz, T., & Yegles, M. (2012). The influence of ethanol containing cosmetics on ethyl glucuronide concentration in hair. *Forensic Science International, 218*(1–3), 123–125. https://doi.org/10.1016/j.forsciint.2011.10.015

Masukawa, H., Kitashima, M., Inoue, K., Sakurai, H., & Hausinger, R. P. (2012). Genetic engineering of cyanobacteria to enhance biohydrogen production from sunlight and water. *Ambio, 41*, 169–173.

Mathew, A. S., Wang, J., Luo, J., & Yau, S.-T. (2015). Enhanced ethanol production via electrostatically accelerated fermentation of glucose using Saccharomyces cerevisiae. *Scientific Reports, 5*(1), 15713.

Mézes, L., Bai, A., Nagy, D., Cinka, I., & Gabnai, Z. (2017). *Optimization of raw material composition in an agricultural biogas plant*.

Mikami, E., Goto, T., Ohno, T., Matsumoto, H., & Nishida, M. (2002). Simultaneous analysis of dehydroacetic acid, benzoic acid, sorbic acid and salicylic acid in cosmetic products by solid-phase extraction and high-performance liquid chromatography. *Journal of Pharmaceutical and Biomedical Analysis, 28*(2), 261–267. https://doi.org/10.1016/S0731-7085(01)00564-7

Møller, K., Sheppard, D., Ravnsbæk, D., Buckley, C., Akiba, E., Li, H.-W., & Jensen, T. (2017). Complex metal hydrides for hydrogen, thermal and electrochemical energy storage. *Energies, 10*(10), 1645. https://doi.org/10.3390/en10101645

Moscoviz, R., Toledo-Alarcón, J., Trably, E., & Bernet, N. (2016). Electro-fermentation: how to drive fermentation using electrochemical systems. *Trends in Biotechnology, 34*(11), 856–865.

Nikhil, G. N., Venkata Subhash, G., Yeruva, D. K., & Venkata Mohan, S. (2015). Synergistic yield of dual energy forms through biocatalyzed electrofermentation of waste: Stoichiometric analysis of electron and carbon distribution. *Energy, 88*, 281–291. https://doi.org/10.1016/j.energy.2015.05.043

Oncel, S. S., Kose, A., Faraloni, C., Imamoglu, E., Elibol, M., Torzillo, G., & Sukan, F. V. (2014). Biohydrogen production using mutant strains of Chlamydomonas reinhardtii: The effects of light intensity and illumination patterns. *Biochemical Engineering Journal, 92*, 47–52. https://doi.org/10.1016/j.bej.2014.06.019

Park, D. H., & Zeikus, J. G. (1999). Utilization of electrically reduced neutral red by *Actinobacillus succinogenes* : physiological function of neutral red in membrane-driven fumarate reduction and energy conservation. *Journal of Bacteriology, 181*(8), 2403–2410. https://doi.org/10.1128/JB.181.8.2403-2410.1999

Poddębniak, P., & Kalinowska-Lis, U. (2024). A survey of preservatives used in cosmetic products. *Applied Sciences, 14*(4), 1581. https://doi.org/10.3390/app14041581

Rahman, S. N. A., Masdar, M. S., Rosli, M. I., Majlan, E. H., Husaini, T., Kamarudin, S. K., & Daud, W. R. W. (2016). Overview biohydrogen technologies and application in fuel cell technology. *Renewable and Sustainable Energy Reviews, 66*, 137–162. https://doi.org/10.1016/j.rser.2016.07.047

Redwood, M. D., Orozco, R. L., Majewski, A. J., & Macaskie, L. E. (2012). Electro-extractive fermentation for efficient biohydrogen production. *Bioresource Technology, 107*, 166–174. https://doi.org/10.1016/j.biortech.2011.11.026

Roy, S., Schievano, A., & Pant, D. (2016). Electro-stimulated microbial factory for value added product synthesis. *Bioresource Technology, 213*, 129–139. https://doi.org/10.1016/j.biortech.2016.03.052

Saratale, R. G., Kuppam, C., Mudhoo, A., Saratale, G. D., Periyasamy, S., Zhen, G., Koók, L., Bakonyi, P., Nemestóthy, N., & Kumar, G. (2017). Bioelectrochemical systems using microalgae—a concise research update. *Chemosphere, 177*, 35–43. https://doi.org/10.1016/j.chemosphere.2017.02.132

Schievano, A., Pepé Sciarria, T., Vanbroekhoven, K., De Wever, H., Puig, S., Andersen, S. J., Rabaey, K., & Pant, D. (2016a). Electro-fermentation—merging electrochemistry with fermentation in industrial applications. *Trends in Biotechnology, 34*(11), 866–878. https://doi.org/10.1016/j.tibtech.2016.04.007

Shanthi Sravan, J., Butti, S. K., Sarkar, O., Vamshi Krishna, K., & Venkata Mohan, S. (2018). Electrofermentation of food waste—regulating acidogenesis towards enhanced volatile fatty acids production. *Chemical Engineering Journal, 334*, 1709–1718. https://doi.org/10.1016/j.cej.2017.11.005

Sivagurunathan, P., Kuppam, C., Mudhoo, A., Saratale, G. D., Kadier, A., Zhen, G., Chatellard, L., Trably, E., & Kumar, G. (2018). A comprehensive review on two-stage integrative schemes for the valorization of dark fermentative effluents. *Critical Reviews in Biotechnology, 38*(6), 868–882. https://doi.org/10.1080/07388551.2017.1416578

Speers, A. M., Young, J. M., & Reguera, G. (2014). Fermentation of glycerol into ethanol in a microbial electrolysis cell driven by a customized consortium. *Environmental Science & Technology, 48*(11), 6350–6358.

Sravan, J. S., Butti, S. K., Sarkar, O., Krishna, K. V., & Mohan, S. V. (2018). Electrofermentation of food waste–regulating acidogenesis towards enhanced volatile fatty acids production. *Chemical Engineering Journal, 334*, 1709–1718.

Stamenkovic, V. R., Strmcnik, D., Lopes, P. P., & Markovic, N. M. (2017). Energy and fuels from electrochemical interfaces. *Nature Materials, 16*(1), 57–69.

Steinbusch, K. J. J., Hamelers, H. V. M., Schaap, J. D., Kampman, C., & Buisman, C. J. N. (2010). Bioelectrochemical ethanol production through mediated acetate reduction by mixed cultures. *Environmental Science & Technology, 44*(1), 513–517.

Thrash, J. C., & Coates, J. D. (2008). Direct and indirect electrical stimulation of microbial metabolism. *Environmental Science & Technology, 42*(11), 3921–3931.

Uçkun Kiran, E., Trzcinski, A. P., Ng, W. J., & Liu, Y. (2014). Enzyme production from food wastes using a biorefinery concept. *Waste and Biomass Valorization, 5*, 903–917.

Vega, L. P., Bautista, K. T., Campos, H., Daza, S., & Vargas, G. (2024). Biofuel production in Latin America: A review for Argentina, Brazil, Mexico, Chile, Costa Rica and Colombia. *Energy Reports, 11*, 28–38. https://doi.org/10.1016/j.egyr.2023.10.060

Velvizhi, G., Sarkar, O., Rovira-Alsina, L., Puig, S., & Mohan, S. V. (2022). Conversion of carbon dioxide to value added products through anaerobic fermentation and electro fermentation: A comparative approach. *International Journal of Hydrogen Energy, 47*(34), 15442–15455. https://doi.org/10.1016/j.ijhydene.2021.12.205

Villano, M., Aulenta, F., Ciucci, C., Ferri, T., Giuliano, A., & Majone, M. (2010). Bioelectrochemical reduction of CO2 to CH4 via direct and indirect extracellular electron transfer by a hydrogenophilic methanogenic culture. *Bioresource Technology, 101*(9), 3085–3090. https://doi.org/10.1016/j.biortech.2009.12.077

Villano, M., Paiano, P., Palma, E., Miccheli, A., & Majone, M. (2017). Electrochemically driven fermentation of organic substrates with undefined mixed microbial cultures. *Chemsuschem, 10*(15), 3091–3097. https://doi.org/10.1002/cssc.201700360

Wang, A., Liu, W., Cheng, S., Xing, D., Zhou, J., & Logan, B. E. (2009). Source of methane and methods to control its formation in single chamber microbial electrolysis cells. *International Journal of Hydrogen Energy, 34*(9), 3653–3658. https://doi.org/10.1016/j.ijhydene.2009.03.005

Xu, L., Fan, J., & Wang, Q. (2019). Omics application of bio-hydrogen production through green alga Chlamydomonas reinhardtii. *Frontiers in Bioengineering and Biotechnology, 7*. https://doi.org/10.3389/fbioe.2019.00201

Yarkent, Ç., Oral, İ, Öncel, D. Ş, & Oncel, S. S. (2024). Focusing on a new biohydrogen production strategy (HS-HS) using Chlamydomonas reinhardtii mutants: The effects of mutation points on biohydrogen production capacity. *International Journal of Hydrogen Energy, 52*, 88–102. https://doi.org/10.1016/j.ijhydene.2023.02.034

Yu, S., Lai, B., Plan, M. R., Hodson, M. P., Lestari, E. A., Song, H., & Krömer, J. O. (2018). Improved performance of *Pseudomonas putida* in a bioelectrochemical system through overexpression of periplasmic glucose dehydrogenase. *Biotechnology and Bioengineering, 115*(1), 145–155. https://doi.org/10.1002/bit.26433

Zhou, M., Chen, J., Freguia, S., Rabaey, K., & Keller, J. (2013). Carbon and electron fluxes during the electricity driven 1, 3-propanediol biosynthesis from glycerol. *Environmental Science & Technology, 47*(19), 11199–11205.

Zhou, M., Freguia, S., Dennis, P. G., Keller, J., & Rabaey, K. (2015). Development of bioelectrocatalytic activity stimulates mixed-culture reduction of glycerol in a bioelectrochemical system. *Microbial Biotechnology, 8*(3), 483–489. https://doi.org/10.1111/1751-7915.12240

Chapter 6
Application of Microbial Electrosynthesis and Electro-Fermentation Using Carbon Dioxide for the Production of Useful Products

Sadaf Shakeel, Mohammad Zain Khan, Sadiya Amanullah, and Khushneda Jabin

1 Introduction

The urgent need to address climate change and global warming has driven research into innovative technologies capable of reducing atmospheric CO_2 levels while simultaneously producing valuable chemicals. Among these emerging solutions, microbial electrosynthesis (MES) and electro-fermentation (EF) stand out as promising biotechnological processes that combine microbial metabolism with electrochemical principles to convert CO_2 into useful products (Jiang et al., 2019a, 2019b).

MES and EF are part of the broader microbial electrosynthesis technologies' (METs) platform, which harnesses the power of electroactive microorganisms to produce chemicals and fuels. MES primarily focuses on the direct conversion of CO_2 into organic compounds using electrical energy and specific microbial communities in bioelectrochemical systems. Electroactive microbes, particularly electroautotrophs, can directly utilize electrons from cathodes or through mediation by hydrogen or other energy carriers (Jiang et al., 2019a, 2019b). Complementing MES, electro-fermentation enhances traditional microbial fermentation processes through electrochemistry. EF aims to overcome metabolic constraints by adjusting the fermentation process, improving the conversion of organic matter or wastes into valuable products. This approach employs electrodes as electron sources or sinks to facilitate product synthesis via anodic or cathodic EF, thereby modulating oxidation–reduction potentials and metabolic fluxes using electricity (Moscoviz et al., 2016).

S. Shakeel (✉) · M. Z. Khan · S. Amanullah · K. Jabin
Industrial Chemistry Research Laboratory, Department of Chemistry, Faculty of Sciences, Aligarh Muslim University, Aligarh, India
e-mail: Sadaf.icb20@gmail.com

© The Author(s), under exclusive license to Springer Nature Singapore Pte Ltd. 2025
S. Sevda (ed.), *Electro-Fermentation for Synthesis of Chemicals, Biochemicals and Biofuels*, https://doi.org/10.1007/978-981-96-6358-3_6

As research in this field progresses, MES and EF show great promise for the sustainable production of chemicals and fuels, harnessing renewable energy sources for a greener future. Their potential to mitigate CO_2 emissions while advancing chemical production makes them attractive solutions in the face of global environmental challenges. This chapter explores the applications of MES and EF in useful product production, challenges, and future prospects. By understanding the potential of these technologies, we can appreciate their transformative role in tackling some of the most urgent issues facing our world today, such as global warming and the exhaustion of natural resources, while paving the way for more sustainable industrial practices.

2 Application of Microbial Electrosynthesis and Electro-Fermentation

2.1 Application in Biofuel Production

MES and EF are innovative techniques showing promise for sustainable biofuel production, offering new ways to tackle climate change and reduce reliance on fossil fuels. A range of biofuels—including biomethanol, bioethanol, and biogas—are crucial in society's shift toward decarbonization. These biofuels can be utilized either as additives or in their pure forms, while hydrogen can be converted to electricity through fuel cell technology. These fossil fuel alternatives have the potential to decrease greenhouse gas emissions and mitigate global warming, contributing to the achievement of sustainable development objectives (Mahapatra et al., 2021). This section explores the applications of these technologies in producing key biofuels such as ethanol, butanol, methanol, and methane.

Alcohols are economically important value-added chemicals with high demand in the global market (Bajracharya et al., 2017a). While traditionally produced using inorganic catalysts under costly, high-temperature, and high-pressure conditions with toxic reagents, bioalcohol production through MES has become an attractive alternative due to their lower hygroscopicity and higher energy density (Vassilev et al., 2019). In MES, bioalcohols are primarily obtained during the solvatogenesis phase, where medium and small chain carboxylates are converted to their corresponding alcohols under mildly acidic pH (Anwer et al., 2021a, 2021b; Bajracharya et al., 2017b).

Bioethanol demand is rising, and it is produced globally at an annual rate of 130 billion liters. It is a very suitable alternative to petroleum-based fuels as it can be derived from renewable resources, less toxic, and can be directly used in vehicles, mixed with various fuel blends (Joyia et al., 2024). Ethanol production in MES has been widely reported and is considered a promising future biofuel (Bajracharya et al., 2017b). At high acetate accumulation, decreasing pH triggers the reduction of acetate to ethanol. Sufficient hydrogen availability acts as a reducing equivalent, promoting ethanol production from acetate (Bajracharya et al., 2017a). Homoacetogens like *C. ljungdahlii* and *C. autoethanogenum* can reduce CO_2 to ethanol along with volatile fatty acids (VFAs) at lower pH when sufficient reducing equivalents (H + or H2) are available (Steinbusch et al., 2010). The selectivity of products can be influenced by limiting micronutrients and trace elements, such as pantothenic acid and cobalt, which can enhance ethanol production over acetate (Bajracharya et al., 2017a).

Li et al., (2024) used machine learning (ML) algorithms like random forest, gradient-boosted decision trees, and XGBoost which were used to predict acetate and ethanol concentrations based on factors like cathode material, pH, applied potential, temperature, and inorganic carbon (IC) levels. After hyperparameter optimization, XGBoost showed the best accuracy, with R2 values of 0.877 for acetate and 0.647 for ethanol. A two-stage XGBoost model further improved ethanol predictions, achieving R2 values of 0.998 for training and 0.727 for testing. Key features influencing ethanol production were current, pH, IC, and total organic carbon (TOC), while temperature and applied potential had less impact. This research highlights the potential of ML to optimize MES and offers insights for improving sustainable chemical production (Li et al., 2024).

Butanol, a promising renewable transportation fuel, is mainly obtained in MES through the further reduction of butyrate under slightly acidic conditions (Bajracharya et al., 2017a). Li et al. (2012) reported the production of isobutanol and 3-methyl-1-butanol (3 MB) from CO2 using *Ralstonia eutropha H16*. Propanol has also been reported in several MES studies (Kondaveeti & Min, 2015; Sharma et al., 2013; Gavilanes et al., 2019). Mogollón et al., 2023 have demonstrated how the electroactive properties of the bacterial strain and bioelectrochemical systems can optimize fermentative processes. The research investigated enhancing acetone, butanol, and ethanol (ABE) production by modifying extracellular redox potential in *Clostridium saccharoperbutylacetonicum* N1-4 using a dual-chamber Microbial Fuel Cell. By adding NADH and adjusting cathode potential, researchers achieved significant improvements in solvent production. NADH supplementation at 200 mM doubled acetone yields to 2.4 g L-1, while cathode poising at −600 mV increased butanol production to 5.8 g L-1 (Mogollón et al., 2023).

Methanol, an important biofuel with potential to reduce greenhouse gas emissions, has been produced in MES through various pathways. It can be synthesized by reducing acetic acid and butyric acid, or through enzymatic routes involving methanol dehydrogenase, formate dehydrogenase, and carbonic anhydrase for CO_2 conversion (Addo et al., 2011). Formic acid can also be converted to methanol via alcohol dehydrogenase and formaldehyde dehydrogenase (Srikanth et al., 2018). Planktonic cells may also participate in methanol production in MES (Gavilanes et al., 2019).

Methane production in MES, known as electro-methanogenesis, occurs primarily at the cathode through methanogenic archaea that use hydrogen as an electron donor and CO_2 as a terminal electron acceptor (Bajracharya et al., 2017a). This process can occur through both direct and indirect electron transfer mechanisms, with indirect transfer via hydrogen supply typically requiring less energy (Eerten-jansen et al., 2015). Recent studies have explored various approaches to enhance methane production in MES, including the integration of oxy-combustion (Diego-García et al., 2022), sulfur oxidation at the anode (Dinh et al., 2022), and the use of novel cathode materials like hollow porous stainless steel/graphene foam (Fu et al., 2022). The performance of methane-producing MES systems has been reported with cathode potentials ranging from −0.55 to −1.15 V vs. SHE, with production rates varying from 0.12 L m^{-2} d^{-1} to 24 L m^{-2} d^{-1} (Bajracharya et al., 2017a). Coulombic efficiencies are generally higher (95–99%) at less negative cathode potentials (−0.7 to −0.8 V vs SHE), suggesting direct electron transfer to microbes in CO_2 reduction (Cheng et al., 2009).

EF has shown promising results in enhancing biofuel production, particularly for ethanol, butanol, and methane. Various studies have demonstrated the effectiveness of using electrodes as electron sinks or sources to improve production yields and efficiency. An engineered *Shewanella oneidensis* strain successfully converted glycerol to ethanol with the electrode serving as an electron acceptor (Flynn et al. 2010). Similarly, an engineered Escherichia coli strain with modified respiratory chains, incorporating three cytochromes (STC, CymA, and MtrA), showed increased ethanol production from glycerol (Sturm-Richter et al., 2015). A co-culture system combining *Clostridium cellobioparum* and *Geobacter sulfurreducens* achieved ethanol production of up to 10 g/L from glycerol (Speers et al., 2014). The addition of mediators like neutral red, methylene blue, and potassium ferricyanide has been shown to improve ethanol production in some EF studies with *E. coli* and *Klebsiella pneumoniae* (Harrington et al., 2015). In anodic EF, Awate et al. demonstrated the effectiveness of using *Cellulomonas uda* to produce ethanol from cellobiose, with a genetically modified consortium of *Geobacter sulfurreducens* strains removing non-ethanol fermentation byproducts. This process achieved 2.7 and 3.7-times higher yields and productivity, respectively, compared to standard fermentation (Awate et al., 2017).

Cathodic EF has been particularly effective in stimulating butanol production. Studies with *Clostridium tyrobutyricum BAS 7* and *Clostridium pasteurianum* showed that electrons supplied by the electrode altered NADH formation, leading to enhanced butanol production from sucrose and glucose, respectively (Choi et al., 2012). The mechanism involves increasing the availability of reducing equivalents, favoring NADH-dependent pathways crucial for butanol biosynthesis (Choi et al., 2014). Khosravanipour et al. (2016) achieved a maximum butanol concentration of 13.31 g/L using an initial glucose concentration of 120 g/L, electro-fermented at 33.51 °C with a fixed potential of +1.32 V vs. Ag/AgCl. This study also investigated the effects of electrode materials, glucose concentration, temperature, and applied potential on butanol production optimization. Zhang et al., 2023 introduced a novel cathodic electro-fermentation CEF system to enhance butanol production by solventogenic *Clostridia.* By optimizing conditions including cathode potential, methyl viologen, and glucose concentration, researchers achieved significant improvements in butanol yield. Initial production reached 7.17 g/L, which was further boosted to 13.14 g/L by adding butyric acid. The polarized electrode and methyl viologen inhibition enabled more efficient butanol formation, with a yield of 0.43 g/g and 90.44% selectivity, presenting a promising approach for cost-effective bio-butanol production.

In electro-methanogenesis, direct electron transfer from cathodes to archaeal biofilms has been observed, primarily in species like *Methanobacterium palustre* (Cheng et al., 2009). Research has shown that methane synthesis can occur even without supplied hydrogen when using thermophilic bacteria or species such as *Methanococcus maripaludis*, with electrons from the cathode being utilized directly (Beese-Vasbender et al., 2015). The applied voltage plays a crucial role, with studies showing that a voltage of 0.3 V can promote the growth of hydrogenotrophic methanogens, leading to a 22.4% increase in methane production compared to conventional anaerobic digestion. One study demonstrated that the addition of conductive graphite in an EF system increased methane yield by 54.3% and maximum methane production by 72.2%. Recent research has explored the introduction of electrodes into anaerobic digesters, successfully enriching *Geobacter* species and increasing methane production, with up to 50% of the synthesized methane attributed to unknown metabolic pathways. EF technology has also shown potential for producing other valuable chemicals and biofuels. For instance, Zhou et al. studied the cathodic EF of glycerol to 1,3-propanediol (1,3-PDO) using a mixed population of fermenting glycerol microbes. They demonstrated that current enhances 1,3-PDO production from 24.8% to 50.1% at −0.9 V vs. SHE, resulting from a metabolic shift from propionate to 1,3-PDO (Zhou et al., 2013).

Both MES and EF have shown significant potential in enhancing biofuel production. However, while these technologies offer great potential, but these are still at an early development stage and require significant research effort for its practical implementation and widespread use in industry (Bhagchandanii et al., 2020).

2.2 Application in High-Value Chemical Production

MES has emerged as a promising technology for the production of various high-value chemicals, particularly using CO_2 as a substrate. The major product obtained from MES is acetic acid (mainly referred to as acetate), which can be produced with high selectivity (>90%) (Nevin et al., 2010). However, production is currently limited by low current densities, necessitating further research to develop cathodes with improved performance. Acetate serves as a key intermediate in MES systems and can be further converted into other valuable chemicals (Bajracharya et al., 2017a). The global demand for acetic acid is projected to reach 16.8 million tons by 2022 (Gadkari et al., 2022). Conventional production methods, such as acetaldehyde oxidation, ethylene oxidation, and the Cativa process, rely on non-renewable feedstocks and corrosive catalysts, resulting in high carbon footprints (Merli et al., 2021). These processes also generate harmful wastes and face issues with non-recyclable catalysts, catalyst loss, and limited catalyst solubility (Murali et al., 2018).

MES offers advantages over conventional methods, including the use of biological catalysts that show poisoning resistance and produce fewer pollutants (Merli et al., 2021). However, MES still lags behind in terms of cost and energy consumption (Christodoulou et al., 2017). Recent studies have addressed these challenges, with Bian et al., (2022) using nickel-coated ceramic hollow fiber membrane cathodes to enhance microbial catalyst enrichment and acetate separation. Song et al., (2022) introduced CuO/g-C_3N_4 as a photoelectric material in the cathode chamber, achieving 2.6 times higher acetic acid production compared to control systems. The concentration of acetic acid produced by MES is close to the concentration of acetic acid used in Chinese vinegar, indicating potential application of MES in the food industry (Bian et al., 2020). Fluidized bed reactors have known to improve acetate production from MES in many studies (Caizán-Juanarena et al., 2024) where Dong et al., (2018) reported that addition of 16 g L^{-1} GAC to the MES system significantly boosted the volumetric acetate production rate, achieving a 2.8-fold increase compared to the control setup without GAC. This enhancement resulted in a final acetate concentration of 3.92 g L^{-1} after a 24-day period. Shakeel et al., 2020, reported that a system utilizing nitric acid-treated graphite granules demonstrated superior performance in acetate production. It achieved a volumetric production rate which was 1.4 times greater than the system with untreated graphite granules and 4.2 times higher than the system lacking any graphite granules in the cathode chamber. Shakeel and Kheel, 2022, used a packed-fluidized cathode and reported higher acetate production as compare to non-packed cathode. Vassilev et al., 2024, recently compared fixed and fluidized bed cathodes for acetate production and found that fixed bed reactors outperformed fluidized bed reactors in terms of acetate production and current density, showing 42% and 47% higher rates, respectively. The fixed GAC bed contained less biomass, suggesting higher specific acetate production rates. Fixed GAC beds also exhibited superior charge efficiency, although they showed lower charge recovery in carboxylates. This indicates enhanced hydrogen evolution in fixed beds, but also suggests a need for improved CO_2 feeding to boost acetate production rates. Furthermore, acetate can be chain-elongated to produce bioplastics, contributing to a sustainable environment (Mistry et al., 2019). The ability of MES to produce diverse products positions it as an important approach for creating a sustainable and carbon-circular economy.

Recently, the range of products from CO_2 in MES has expanded to include more valuable compounds such as butyrate and caproate, alongside the primary product, acetic acid (Dessi et al., 2021; Tahir et al., 2021). Acetate can undergo chain elongation to form higher carbon compounds such as caproate (C6), butyrate (C4), and propionate (C3) (Agler et al., 2011). Butyric acid has been observed as a byproduct alongside acetic acid in various MES studies. Recently, Hui et al., 2024, investigated the use of methanol as a co-substrate with CO_2 to improve butyrate production in microbial electrosynthesis. The combination of CO_2 and methanol feeding yielded superior results, with butyrate production rates reaching 0.36 ± 0.01 g L −1 d −1 and final concentrations of 8.6 ± 0.2 g L −1. These figures significantly surpassed those achieved in reactors fed with CO_2 alone, which produced 0.20 ± 0.03 g L − 1 d −1 and reached concentrations of 5.2 ± 0.1 g L −1. Methanol served a dual purpose, functioning as both an electron donor and a carbon source, with each role contributing approximately half of the carbon found in the final products.

Research employing pure cultures and metabolic engineering has led to the production of chemical commodities like formate (Tashiro et al., 2018) and 3-hydroxybutyrate (Chen et al., 2018) from CO_2 as the exclusive carbon source, though scalability remains a significant challenge. Batlle-Vilanova et al., (2017) demonstrated enhanced butyrate production under conditions of hydrogen partial pressure above 1 atm and pH near 5. CO_2 can be reduced to butyrate at a cathode potential of −0.37 V vs. SHE under biological conditions (Batlle-Vilanova et al., 2017). Indirect CO_2 conversion to butyric acid has also been reported, with ethanol serving as an electron donor in the chain elongation process (Jourdin et al., 2019). In contrast, Jourdin et al., (2018) observed butyrate production without ethanol, suggesting that the cathode may act as an electron donor. Near-neutral pH conditions have been shown to promote butyric and caproic acid formation through chain elongation (Vassilev et al., 2018).

Recent advancements include the use of gas diffusion membrane composite biocathodes for high butyrate production (Wu et al., 2022b) and in situ chain elongation for caproate production (Li et al., 2022). Wang et al. (2022) found that mixotrophic pre-enriched acetogens enhanced chain elongation to C4 and C6 products. Zhao et al., (2023) utilized Chinese liquor brewing wastewater to obtain higher amounts of medium chain fatty acids. Other products reported include propionate and iso-butyrate from bicarbonate (Annie Modestra et al., 2015), small quantities of 2-oxo-butyrate (Nevin et al., 2011), and caproic acid (Vassilev et al., 2018). Wu et al., (2022a) recently reported lycopene production using engineered *Cupriavidus necator*, further expanding the range of valuable compounds obtainable through MES.

Glycerol production via CO_2 reduction has been achieved using *Geobacter sulfurreducens* with succinate, reduction current, and gas diffusion electrodes over a 90-day period (Soussan et al., 2013). Mixed microbial cultures have also shown the capability to produce carboxylic acids and alcohols with up to six carbon atoms (Vassilev et al., 2018). Key microorganisms involved in MES processes, including *Clostridium aceticum, Clostridium ljungdahlii*, and *Moorella thermoacetica*, are known to generate acetate, 2-oxo-butyrate, and formate. These organisms are frequently used in MES research for their efficiency in reducing CO_2 to these organic compounds with the aid of an external electron source, such as a BES cathode (Das et al., 2018). Additionally, acetogenic bacteria like *Sporomusa ovata, Rhodobacter sphaeroides, Clostridium ljungdahlii*, and *Moorella thermoacetica* have demonstrated the ability to convert CO_2 into acetate using electrons from MES cathode (Nevin et al., 2010).

EF has emerged as a promising technology for producing a variety of value-added chemicals from organic and inorganic materials, including carbon dioxide and waste effluents. This process harnesses the power of electricity and microbial catalysis to drive the synthesis of a wide range of compounds, utilizing acidogenesis to convert substrates into intermediate forms of fatty acids. One of the key advantages of EF is its ability to produce volatile fatty acids (VFAs) or short-chain fatty acids (SCFAs) with carbon chain lengths from C2 to C5 (Logan & Rabaey, 2012; Bhagchandanii et al., 2020). The electrodes in EF systems play a crucial role in controlling reaction kinetics and influencing the acidification of monosaccharides by anaerobic bacteria, resulting in increased VFA productivity (Wong et al., 2016). SCFAs have diverse applications, serving as electron donors in various bioprocesses such as bioplastic production, diesel synthesis, and bioelectricity generation (Bhagchandanii et al., 2020).

A notable example of EF's potential is the production of acetate from CO_2. Nevin et al. (2011) demonstrated the reduction of CO_2 to acetate using electro-fermentation systems with different acetogenic bacteria at potentials of −0.4 V and −0.6 V vs. SHE. Gyldemyn et al., 2017, further improved this process by using anionic exchange membranes, achieving a 32% higher production rate and recovery compared to systems without product recovery. The production of n-butyrate through EF has been explored via two pathways. The first involves reducing carbon dioxide to acetate using hydrogen as an electron donor via the acetyl-CoA pathway (Nikhil et al., 2015). Alternatively, n-butyrate can be produced through the elongation of acetate using the reverse beta-oxidation process, with ethanol or lactic acid serving as electron donors (Pasupuleti et al., 2014). Choi et al., (2012), examined butyrate production from sucrose using *Clostridium tyrobutyricum BAS 7* at −0.4 V vs. Ag/AgCl with neutral red as an electron mediator, achieving increases of 1.7 and 1.3 times in productivity and yield, respectively. Ganigué et al., (2015), demonstrated the bioelectrochemical production of butyrate directly from CO_2 at −0.8 V vs. SHE, showing that the process was hydrogen-driven and could also produce ethanol and butanol at low pH values.

Medium-chain fatty acids (MCFAs) represent another valuable product class obtainable through EF. These compounds are naturally hydrophobic with a low oxygen/carbon ratio, making them energy-dense and easily separable from fermentation broths (Pasupuleti et al., 2014). In bioelectrochemical systems, MCFAs can be produced in the cathodic chamber using acetate as a regenerative source (Zhao et al., 2016). Research has shown that the bacterium *Clostridium kluyveri* can produce caproate (C6) and caprylate (C8) from acetate when the cathode is set at −0.9 V against a normal hydrogen electrode (NHE), using hydrogen generated in the system as an electron source (Schröder et al., 2015). The production of MCFAs through EF offers impressive energy efficiency, as the energy content of one mole of caproic acid (3452 kJ) exceeds that of the two moles of ethanol (2638 kJ) required for its production (Cheng et al., 2009).

EF has also been applied to the production of other organic acids. Schuppert et al., (1992) conducted EF of lactate using *Propionibacterium acidipropionici* with cobalt sepulchrate as an electron donor at −0.47 V vs. SHE, resulting in propionate as the sole fermentation product. Xu et al., (2018), demonstrated the enhancement of lactic acid productivity and optical activity from activated sludge and food waste using a mixed microbial community at +0.1 V vs. SHE, showing a 4.73-fold increase in productivity and improved optical activity toward L-lactic production. Researchers have also focused on modifying microorganisms to enhance electroactivity and direct metabolic pathways toward target products. Wu et al., (2019) engineered an *E. coli strain* for succinate production, achieving a yield of 1.10 mol of succinate per mol of glucose at −0.65 V vs. Ag/AgCl using neutral red as an electron mediator. Kim et al., (2017), genetically modified *Klebsiella pneumoniae L17* to produce 3-hydroxypropionic acid from glycerol at +0.5 vs. Ag/AgCl, reporting a 1.7-fold increase in production compared to controls without applied potential. Acetoin has been produced from lactate using engineered *S. oneidensis strains*, and 2,3-butanediol has been synthesized from acetoin through cathodic EF.

These examples highlight the versatility and potential of EF in producing a wide range of valuable chemicals, from simple organic acids to more complex compounds like MCFAs. The ability to control reaction kinetics and influence metabolic pathways through electrode potentials offers unique advantages over traditional fermentation processes, making EF a promising technology for sustainable chemical production with diverse applications in renewable fuels, antimicrobials, and flavor and fragrance intermediates.

Both EF and MES show great promise for producing a wide range of value-added chemicals from waste materials and CO_2. However, further research is needed to improve production efficiencies, develop mediator-less systems, and reduce energy consumption and costs for large-scale applications. The continued development of these technologies could lead to more sustainable and environmentally friendly chemical production methods, contributing to the establishment of a carbon-circular economy.

2.3 Application in Bioplastics Production

MES has demonstrated significant potential in the sustainable production of bioplastics, particularly polyhydroxybutyrate (PHB). Acetate, a common product of these processes, can serve as a feedstock for biomass growth, including the production of bioplastics such as polyhydroxybutyrate (PHB) (Wood et al., 2021). Dai et al., (2007) demonstrated that acetate could be used as the sole substrate for biomass growth, resulting in polyhydroxybutyrate (PHB) accumulation up to 41% of the dry cell weight. This "Power-to-Plastics" approach offers a promising route for converting electrical energy and CO_2 into valuable biopolymers through MES. A recent study showcased an effective two-step process for bioplastic production using MES. In the first step, acetate is generated through MES at a specific concentration. This acetate then serves as a substrate for PHB production by *Ralstonia eutropha* in the second step. The study compared acetate produced in double-chamber versus triple-chamber MES reactors, finding that the triple-chamber setup yielded the highest PHB production, achieving 0.460 g/L of PHB from 5 g/L of acetate. These advancements demonstrate the effective conversion of CO_2 into valuable bioplastics through MES, underlining its potential as a sustainable technology for bioplastic production and contributing to circular economy principles (Pepè Sciarria et al., 2018). However, the economic viability of bioplastic production via MES faces challenges. PHB production from acetate requires costly cell disruption processes for recovery, resulting in a levelized cost of production (LCC) of approximately US$7100/t, significantly higher than the market price of conventional plastics at around US$1000/t (Kopke & Simpson, 2020). To address these economic hurdles, researchers are exploring alternative approaches, such as engineering acetogens to produce 3-hydroxybutyrate (3HB), the monomer unit of PHB, extracellularly (Kopke & Simpson, 2020). This development could potentially improve process economics by eliminating the need for cell disruption and simplifying downstream processing in MES systems.

While the current costs of bioplastic production through MES high, ongoing research and technological advancements may enhance the economic feasibility of these processes. The ability to produce bioplastics from renewable resources using electrical energy and CO_2 as inputs aligns with the goals of creating a sustainable and carbon-circular economy, driving continued interest in improving these technologies for large-scale application.

2.4 Application in Storing Electrical Energy

MES offers an innovative approach to energy storage, particularly in the context of renewable energy integration and grid stability. As countries worldwide increase their reliance on renewable sources like wind and solar, the intermittent nature of these energy sources poses challenges for grid management. MES presents a solution by converting excess electricity into more manageable forms of chemical energy, effectively storing surplus energy during low-demand periods (Vidales et al., 2019). One of the most promising applications of MES in energy storage is the production of methane (CH_4) from carbon dioxide (CO_2), aligning with the concept of Bioelectrochemical Power-to-Gas (BEP2G) technology (Lehner et al., 2014). This process offers several advantages over traditional Power-to-Gas systems that produce hydrogen (Jeremiasse et al., 2010). Methane has a higher volumetric energy density and can utilize existing natural gas infrastructure for transport and storage, making it a more practical option for large-scale energy storage (Ding et al., 2016).

MES systems also demonstrate potential for short-term energy storage through the development of MES supercapacitors. The substantial internal capacitance of electroactive biofilms in bioelectrochemical systems allows for rapid, on-demand energy return to the electrical grid. This dual capability—long-term storage through methane production and short-term storage via biofilm capacitance—makes MES a versatile technology for addressing various energy storage needs (Vidales et al., 2019). Furthermore, MES offers impressive energy conversion efficiencies. When coupled with photovoltaic systems, MES can potentially achieve a combined efficiency of about 4.5%, which is nine times higher than conventional bio-based processes. This increased efficiency suggests that MES could dramatically improve bioproduction rates when powered by solar energy.

The economic viability of MES for energy storage is closely tied to electricity market dynamics. As renewable energy sources gain market share, periods of electricity surplus may lead to depressed wholesale prices, potentially improving the economic feasibility of MES technologies. This evolving energy landscape creates new opportunities for MES implementation and profitability in energy storage applications (Zhang & Tremblay, 2019). By harnessing renewable electricity to produce useful organic compounds or methane, MES represents a promising pathway toward more efficient and sustainable energy storage solutions. It not only offers a means to enhance renewable energy utilization but also addresses challenges in grid stability and carbon dioxide mitigation, positioning itself as a key technology in the transition to a more sustainable energy future.

EF shows promising applications in energy storage and production of valuable compounds from various carbon-rich substrates. One significant approach involves combining EF with Microbial Fuel Cells (MFCs) to create self-sufficient systems that can generate electricity while simultaneously producing valuable products (Salar-García et al., 2024). Chandrasekhar and colleagues (2015) illustrated this concept by creating an innovative self-powered, single-chamber electro-fermenter without a membrane and featuring an air-cathode. Their system effectively transformed solid food waste into three valuable products: bioelectricity, biohydrogen, and bioethanol, achieving notable results with a peak power output of 162.4 $mW.m^{-2}$, hydrogen generation at 21.9 $mL.h^{-1}$, and a bioethanol yield of 4.85% w/v (Chandrasekhar et al., 2015). Similarly, Nikhil et al. (2015), developed a hybrid MFC-EF system that reached a maximum power output of 72 $mW.m^{-2}$ and produced 343 mL of biohydrogen using synthetic wastewater as a substrate (Nikhil et al., 2015). Another approach to energy storage through EF involves submersible Microbial Electrolysis Cells (MECs). Van Eerten-Jansen et al., (2013) coupled two cathodic chambers to an anaerobic digester, with one chamber primarily producing electricity and the other generating hydrogen. This configuration achieved hydrogen production rates between 5 and 487 mL. d^{-1} (Van Eerten-Jansen et al., 2013). EF also shows potential in enhancing the sustainability of other energy-related processes. For instance, it can serve as a hydrogen supplier for biomass pyrolysis, offering a more environmentally friendly alternative to fossil fuel-derived hydrogen (Bhagchandanii et al., 2020). Conversely, EF can be employed to valorize the bio-oil aqueous phase (BOAP) obtained during pyrolysis, converting it into hydrogen or other value-added products (Borole et al., 2018).

These applications demonstrate the versatility of EF in energy storage and production, highlighting its potential to contribute to more sustainable and integrated energy systems. By converting waste streams into valuable energy carriers and chemicals, EF offers a promising approach to closing the loop in carbon-based energy cycles.

2.5 Application in Syngas Fermentation

Microbial electrosynthesis (MES) shows promise for enhancing syngas fermentation processes, particularly in industries with significant CO_2 emissions. Syngas, containing varying ratios of CO, CO_2, and H_2, can be derived from multiple sources including industrial flue gases, natural gas reformation, and gasification of various waste materials (Griffin & Schultz, 2012). Several companies have developed technologies for converting syngas to ethanol at demonstration scales (Heijstra et al., 2017). The composition of syngas varies depending on its source and production method. For instance, biomass-derived syngas can have significantly different ratios of N_2, CO, CO_2, and H_2 based on the gasification process used (Munasinghe & Khanal, 2010). Steel mill flue gases typically contain 20–25% CO, 20–25% CO_2, and about 5% H_2 (Romano et al., 2013).

However, the CO and H_2 fractions in these gas feedstocks are often insufficient for optimal conversion of CO_2 into desired products. For example, producing ethanol from steel mill emissions would require a significant increase in the H_2 fraction relative to CO and CO_2 (Zhang & Tremblay, 2019). This is where MES could play a crucial role. By providing additional reducing equivalents to the syngas fermentation process, MES could substantially increase production yields and reduce greenhouse gas emissions (Liew et al., 2016). This application could be particularly impactful in the steel industry, which accounted for 5.5% of global CO_2 emissions in 2015 (Zhang & Tremblay, 2019).

The electricity required for MES could come from renewable sources or be generated on-site during syngas production. For instance, steel mills produce other gaseous byproducts with high calorific values that could be used to generate electricity for powering MES. Additionally, in the case of municipal solid waste gasification, organic carbon in low-strength wastewater could be oxidized at the MES anode, contributing to wastewater treatment and reducing the overall energy requirements for CO_2 reduction (Zhang & Tremblay, 2019).

By integrating MES with syngas fermentation, industries could potentially achieve higher conversion efficiencies, reduce their carbon footprint, and create value from waste streams, making the overall process more economically viable and environmentally sustainable.

2.6 Application in Biogas Upgradation

MES presents a promising solution for upgrading biogas, a valuable product of anaerobic digestion of various waste streams. Biogas typically contains 50–70% methane, 30–50% CO_2, and trace amounts of other compounds such as H2S and siloxanes (Götz et al., 2016). Its high CO_2 and water content make it corrosive and less compatible with existing natural gas infrastructure, necessitating upgrading to increase its methane content for improved energy density, economic value, and compatibility (Chen et al., 2015; Mamun et al., 2016). While traditional upgrading methods focus on removing CO_2, MES offers a unique approach by directly converting CO_2 in biogas to methane, enhancing productivity and value from biogas plants (Marshall et al., 2012).

In a MES-based biogas upgrading system, biogas is fed into the cathodic chamber of a MES reactor, where CO_2 is reduced to methane using electricity. The process can be optimized by feeding liquid effluent from anaerobic fermentation to the anodic chamber, where microbial catalysts oxidize organic carbon molecules to CO_2, generating electrons that are transferred to the anode. This integrated approach offers multiple benefits, including direct conversion of CO_2 to methane, further reduction of chemical oxygen demand in wastewater, and potential recovery of other valuable components from wastewater (Zhang & Tremblay, 2019). Additionally, using appropriate ion-exchange membranes, the system can concentrate and recycle other wastewater pollutants such as ammonia, sulfate, and phosphate (Tice & Kim, 2014).

The integration of anaerobic digestion (AD) with MES further enhances biogas upgrading and wastewater treatment. In this combined AD-MES system, complex organic substrates are first processed through AD, producing volatile fatty acids (VFAs), H_2, and CO_2. The MES process then utilizes the VFA-rich effluent from AD as its influent, further fermenting VFAs to CH_4 and reducing CO_2 to CH_4 either directly or indirectly via H_2. This integrated approach offers several advantages over traditional AD, including enhanced hydrolysis and acidification of waste materials, accelerated conversion of VFAs, increased CH_4 content in the final biogas product, improved overall rate of CH_4 production, and more efficient removal of VFAs from the wastewater (Kong et al., 2020).

While this multifaceted approach addresses biogas upgrading, wastewater treatment, and resource recovery simultaneously, thorough life cycle and techno-economic assessments are necessary to evaluate its environmental impact and economic feasibility under various conditions. These analyses will be crucial in determining the practical applicability and scalability of MES-based biogas upgrading systems in different contexts (Zhang & Tremblay, 2019).

EF presents a promising approach for upgrading biogas and enhancing the production of valuable chemicals. EF can stabilize and optimize fermentation processes by managing redox and pH imbalances, thus improving substrate purity and controlling byproduct formation (Schievano et al., 2016). One key application of EF is in biogas upgrading, particularly in increasing the methane (CH_4) content of biogas. Chu et al., (2020) provided a critical assessment of microbial electrochemical biogas upgrading strategies, highlighting the potential of this approach. The development of improved electrode materials is crucial for enhancing CH_4 production in biogas systems (Arunasri et al., 2020).

The integration of anaerobic fermentation with microbial electrolysis cells (MECs) represents a significant advancement in biogas technology through electromethanogenesis (Zhao et al., 2015). In this process, electroactive microorganisms efficiently utilize electrons under low-voltage conditions (Marshall et al., 2013). CH_4 formation can occur via two pathways near the cathode: direct electron utilization or hydrogenotrophic methanogenesis using cathode-generated hydrogen.

$$CO_2 + 8H^+ + 8e^- \rightarrow CH_4 + 2H_2O \ (E_\circ = -240 \text{ mV vs SHE})$$

$$CO_2 + 4H_2 \rightarrow CH_4 + 2H_2O$$

The electrochemical reduction of CO_2 at the cathode, facilitated by hydrogenotrophic methanogens, can significantly increase the CH_4 content in biogas. Simultaneously, oxidation of substrates by anodic exoelectrogens provides electrons for cathodic reduction, enhancing the overall electrolysis process and improving end-product yields.

In EF systems, exoelectrogenic bacteria can utilize various feedstocks such as H_2, volatile fatty acids (VFAs), and alcohols for CH_4 production. The process typically involves initial oxidation of glucose into intermediate acid metabolites by anaerobic fermentation microorganisms, followed by further oxidation by exoelectrogens in the anode compartment. This two-step process helps mitigate feedback inhibition often encountered in glucose fermentation. To further enhance CH_4 production through EF, it is crucial to deepen our understanding of microbial–electrode interactions and the microbiome's role in modifying CH_4 production (Barbosa et al., 2021). This knowledge will be instrumental in optimizing EF strategies for biogas upgrading and maximizing the production of high-value chemicals from waste streams (Chandrasekhar et al., 2021).

2.7 Application in Hybrid Fermentation

MES and EF show promising applications in hybrid systems for sustainable energy production and waste management. These hybrid configurations combine MES or EF with other bioelectrochemical or industrial processes to enhance efficiency and expand the range of products.

In MES, the integration of electrocatalysts has been proposed to address limitations such as low production rates. A study evaluating metal-based hybrid cathodes demonstrated improved performance, with zinc-based cathodes achieving notable increases in acetic acid production at higher current densities. This zinc-based electrical-biological hybrid system shows potential for improving CO_2 valorization efficiency through MES (Jiang et al., 2019a, 2019b). Hybrid systems in MES also include combinations of autotrophic and heterotrophic microorganisms. Autotrophs at the biocathode reduce CO_2, while heterotrophs at the bioanode can treat organic waste streams, addressing multiple environmental challenges simultaneously. Integration of MES with established biotechnologies like anaerobic digestion and fermentation is another promising approach (Jiang et al., 2019a, 2019b). The OxyMES system, combining an oxy-combustion plant with MES, demonstrates the potential for energy storage without CO_2 emissions. This hybrid approach achieved a Power-to-Gas performance of 60%, surpassing standalone MES systems. Another concept, the Microbial Rechargeable Battery (MRB), combines MES and Microbial Fuel Cell (MFC) technologies for cyclical energy storage and generation (Diego-García et al., 2022).

2.8 Application in Biorefinery

Both MES and EF contribute to waste management and resource recovery, particularly in the context of organic waste and wastewater treatment. Employing MES and

EF in waste biorefineries allows organic waste to be transformed into biofuels and valuable chemicals.

MES holds significant promise for biorefineries by transforming CO_2 emissions into valuable chemicals, thus contributing to a sustainable and circular economy. In the context of biorefineries, MES can be integrated into existing industrial processes to enhance the production of bio-based chemicals and fuels. This integration can help reduce reliance on fossil fuels and lower greenhouse gas emissions. MES utilizes microorganisms as biocatalysts to convert CO_2 into organic compounds such as acetic acid, alcohols, and bioplastics. This process can be powered by renewable energy sources, making it an environmentally friendly alternative. The primary advantage of MES in biorefineries is its ability to utilize waste CO_2, a common byproduct in many industrial processes, thereby turning a waste product into a valuable resource (Kamm et al., 2005).

In wastewater treatment, MES can clean effluents while generating upgraded biogas and recycling pollutants. When integrated with bioethanol production from lignocellulosic biomass, MES can process liquid stillages containing glycerol, carbohydrates, cell debris, and residual bioethanol. These substances serve as substrates for biological oxidation reactions at the MES anode, producing reducing equivalents and CO_2 that can be used at the cathode to synthesize valuable compounds. This integration increases the overall energetic efficiency of bioethanol production by recycling energy from liquid waste (Blasco-Gómez et al., 2017; Jadhav et al., 2017; Sadhukhan et al., 2016). In biodiesel plants, which are energy-intensive and generate significant glycerol byproduct and oily waste, MES can recover energy from these waste streams. Glycerol can be converted to bioethanol in the anodic chamber, with the resulting CO_2 and reducing equivalents used at the cathode to produce valuable products. By incorporating MES, biorefineries can improve energy efficiency, reduce waste, generate additional valuable products, and enhance the sustainability of biofuel production processes (Sadhukhan et al., 2016).

These applications demonstrate MES's potential as a crucial component in developing advanced, more sustainable biorefineries, contributing to the circular economy concept in biofuel production by recovering energy and resources from waste streams (Zhang & Tremblay, 2019). However, several challenges need to be addressed to fully realize the potential of MES in biorefineries. These include optimizing the efficiency of CO_2 conversion, improving the purity of the produced chemicals, and scaling up the technology for industrial applications. Flexible reactor designs that can be easily integrated into existing biorefinery infrastructure are crucial for reducing costs and enhancing efficiency. Future research and development efforts should focus on increasing the production rates of MES, developing efficient methods for product extraction, and exploring the production of higher-value chemicals. By overcoming these challenges, MES can play a pivotal role in the transition to a more sustainable and circular bioeconomy, providing a viable pathway for the decarbonization of various industries (Dessì et al., 2021).

EF is a transformative technology for biorefineries, combining the principles of electrochemistry with microbial fermentation to enhance the production of bio-based chemicals and fuels. This approach harnesses electrodes to control the redox environment and pH levels within fermentation processes, optimizing microbial activity and improving the efficiency of bioconversion. In biorefineries, EF can significantly enhance the utilization of low-grade carbon substrates and waste materials, transforming them into valuable products (Roy et al., 2016). By stabilizing the fermentation environment, EF reduces the need for chemical additives, which can lower production costs and minimize environmental impact. This method also allows for greater control over microbial metabolism, leading to higher yields and selectivity of desired products. One of the key advantages of EF is its ability to address common challenges in traditional fermentation processes, such as redox imbalances and pH fluctuations. By providing a more stable and controlled environment, EF can improve the overall efficiency and sustainability of biorefineries. Additionally, the integration of renewable energy sources to power the electrochemical processes further enhances the environmental benefits of EF.

Future research in EF should focus on scaling up the technology for industrial applications, optimizing electrode materials and configurations, and exploring the production of a wider range of bio-based chemicals. By overcoming these challenges, EF has the potential to revolutionize biorefineries, making them more efficient, sustainable, and economically viable. Moreover, EF can be integrated with other biorefinery processes, such as anaerobic digestion and microbial electrosynthesis, to create a more comprehensive and efficient system for converting biomass into a variety of valuable products. This integration can help maximize the use of available resources and reduce waste, further contributing to the sustainability of biorefineries (Rago et al., 2019).

3 Challenges and Limitations

While MES and EF hold significant promise for producing useful chemicals from CO_2, several challenges and limitations must be addressed for successful implementation. MES faces several significant challenges that need to be addressed for its successful application. One of the primary challenges is ensuring high product yield and stability over long-term operations. This involves optimizing the efficiency of CO_2 conversion and maintaining consistent production rates. Additionally, the economic feasibility of MES is a concern, as balancing the capital and operating costs with the revenue generated from waste treatment can be difficult. Scaling up MES systems from laboratory to industrial scale presents another major hurdle, requiring the development of flexible reactor designs that can be integrated into existing industrial processes. Furthermore, efficient methods for product separation and storage are essential to enhance the overall viability of MES (Quraishi et al., 2021).

EF also encounters several challenges that must be overcome to optimize its application. One of the key issues is the optimization of critical factors such as applied voltage, hydraulic retention time, and organic loading rate to achieve better performance. EF systems need to be scaled up effectively to handle industrial-scale operations, which involve addressing technical and economic barriers. Another significant challenge is the integration of EF with existing biorefinery processes, ensuring that it can operate efficiently alongside other technologies. Long-term stability and consistent performance are also critical, requiring continuous monitoring and adjustment of the fermentation environment. Additionally, developing efficient methods for product extraction and purification is essential to maximize the economic benefits of EF (Quraishi et al., 2021).

4 Regulatory and Policy Barriers

The development and implementation of MES and EF technologies face potential obstacles due to the absence of clear regulatory frameworks. To facilitate adoption, it's crucial to establish policies that support research, development, and commercialization. This includes developing standardized safety protocols for large-scale bioelectrochemical systems, creating guidelines for using genetically modified organisms, and establishing product quality standards for chemicals and fuels produced via these processes. Regulatory incentives, such as carbon credits, subsidies, tax breaks, and grants, could encourage investment and promote market growth. It's also important to ensure that these technologies comply with existing environmental regulations while advocating for updates that may unintentionally hinder their development.

A balanced approach to intellectual property rights is necessary to encourage innovation while promoting knowledge sharing and collaborative research. Research collaboration and implementing policies to support education and training in bioelectrochemical systems, such as funding specialized university programs and apprenticeships, are crucial for developing a skilled workforce. Finally, developing policies that promote public understanding and acceptance of these new technologies through mandated consultations and funded education initiatives is essential for their successful integration into society. These regulatory and policy considerations are vital for overcoming barriers and fostering the growth of MES and EF technologies in the sustainable energy and chemical production sectors.

By addressing these regulatory and policy barriers, governments and international bodies can create a more supportive environment for the development and implementation of MES and EF technologies, ultimately accelerating their integration into sustainable biorefinery processes and the broader bioeconomy.

5 Future Perspectives

The future of MES and EF holds immense promise for revolutionizing sustainable processes. These innovative technologies are poised to play a crucial role in the circular economy by converting CO_2 and waste materials into valuable chemicals and fuels. Future research in MES is expected to focus on expanding the range of high-value products beyond simple carbon compounds, enhancing electroactive microorganisms through genetic engineering, improving electron transfer mechanisms, increasing production yield and rate, and developing efficient product separation methods. For EF, key areas of development include optimizing critical operational factors, advancing electrode materials, integrating with other biorefinery processes, ensuring long-term system stability, and improving product extraction and purification techniques. Both technologies share common future goals, such as integration with renewable energy sources, scaling up for industrial applications, enhancing economic viability, and garnering policy and regulatory support. Advances in synthetic biology offer powerful tools for optimizing microbial strains and maximizing product yields. The integration of MES and EF with renewable energy systems presents an opportunity to address energy storage challenges while contributing to carbon capture efforts. Increased investment in research and development initiatives, along with collaborative efforts among academia, industry, and government, can accelerate innovation and promote commercialization. Furthermore, raising public awareness and education about these technologies can drive acceptance and support for their adoption. By addressing these challenges and focusing on these future directions, MES and EF have the potential to significantly enhance their application in biorefineries, leading to more efficient and sustainable processes for converting biomass and waste into valuable products, ultimately contributing to a more sustainable and circular economy (Jiang et al., 2019a, 2019b).

6 Conclusion

Microbial electrosynthesis and electro-fermentation represent transformative approaches for converting CO_2 into valuable chemicals across various industries. Their applications in chemical production, energy storage, waste management, and other innovative areas highlight their potential to contribute to sustainable development and address pressing environmental challenges. While challenges remain, advancements in synthetic biology, renewable energy integration, and supportive policies can pave the way for the widespread adoption of these technologies. Continued research and collaboration will be essential in unlocking the full potential of MES and EF for a more sustainable future.

References

Addo, P. K., Arechederra, R. L., Waheed, A., Shoemaker, J. D., Sly, W. S., & Minteer, S. D. (2011). Methanol production via bioelectrocatalytic reduction of carbon dioxide: Role of carbonic anhydrase in improving electrode performance. *Electrochemical and Solid-State Letters, 14*(9), E9–E13.

Agler, M. T., Wrenn, B. A., Zinder, S. H., & Angenent, L. T. (2011). Waste to bioproduct conversion with undefined mixed cultures: The carboxylate platform. *Trends in Biotechnology, 29*(2), 70–78.

Annie Modestra, J., Navaneeth, B., & Venkata Mohan, S. (2015). Bio-electrocatalytic reduction of CO2: Enrichment of homoacetogens and pH optimization towards enhancement of carboxylic acids biosynthesis. *Journal of CO2 Utilization, 10*, 78–87.

Anwer, A. H., Khan, N., Khan, M. D., Shakeel, S., & Khan, M. Z. (2021a). Redox mediators as cathode catalyst to boost the microbial electro-synthesis of biofuel product from carbon dioxide. *Fuel, 302*, Article 121124.

Anwer, A. H., Khan, N., Umar, M. F., Rafatullah, M., & Khan, M. Z. (2021b). Electrodeposited hybrid biocathode-based CO2 reduction via microbial electro-catalysis to biofuels. *Membranes, 11*(3), 223.

Arunasri, K., Yeruva, D. K., Vamshi Krishna, K., & Venkata Mohan, S. (2020). Monitoring metabolic pathway alterations in *Escherichia coli* due to applied potentials in microbial electrochemical system. *Bioelectrochemistry, 134*, Article 107530. https://doi.org/10.1016/j.bioelechem.2020.107530

Awate, B., Steidl, R. J., Hamlischer, T., & Reguera, G. (2017). Stimulation of electro-fermentation in single-chamber microbial electrolysis cells driven by genetically engineered anode biofilms. *Journal of Power Sources, 356*, 510–518.

Bajracharya, S., Yuliasni, R., Vanbroekhoven, K., Buisman, C. J., Strik, D. P., & Pant, D. (2017a). Long-term operation of microbial electrosynthesis cell reducing CO2 to multi-carbon chemicals with a mixed culture avoiding methanogenesis. *Bioelectrochemistry, 113*, 26–34.

Bajracharya, S., Srikanth, S., Mohanakrishna, G., & Zacharia, R. (2017b). Biotransformation of carbon dioxide in bioelectrochemical systems: State of the art and future prospects. *Journal of Power Sources, 356*, 256–273.

Barbosa, S. G., Peixoto, L., Alves, J. I., & Alves, M. M. (2021). Bioelectrochemical systems (BESs) towards conversion of carbon monoxide/syngas: A mini-review. *Renewable and Sustainable Energy Reviews, 135*, Article 110358. https://doi.org/10.1016/j.rser.2020.110358

Batlle-Vilanova, P., Ganigué, R., Ramió-Pujol, S., Bañeras, L., Jiménez, G., Hidalgo, M., Balaguer, M. D., Colprim, J., & Puig, S. (2017). Microbial electrosynthesis of butyrate from carbon dioxide: Production and extraction. *Bioelectrochemistry, 117*, 57–64.

Beese-Vasbender, P. F., Grote, J. P., Garrelfs, J., Stratmann, M., & Mayrhofer, K. J. (2015). Selective microbial electrosynthesis of methane by a pure culture of a marine lithoautotrophic archaeon. *Bioelectrochemistry, 102*, 50–55.

Bhagchandanii, D. D., Babu, R. P., Sonawane, J. M., Khanna, N., Pandit, S., Jadhav, D. A., Khilari, S., & Prasad, R. (2020). A comprehensive understanding of electro-fermentation. *Fermentation, 6*(3), 92.

Bian, B., Bajracharya, S., Xu, J., Pant, D., & Saikaly, P. E. (2020). Microbial electrosynthesis from CO2: Challenges, opportunities and perspectives in the context of circular bioeconomy. *Bioresource Technology, 302*, Article 122863.

Bian, B., Singh, Y., Rabaey, K., & Saikaly, P. E. (2022). Nickel-coated ceramic hollow fiber cathode for fast enrichment of chemolithoautotrophs and efficient reduction of CO2 in microbial electrosynthesis. *Chemical Engineering Journal, 450*, Article 138230.

Blasco-Gómez, R., Batlle-Vilanova, P., Villano, M., Balaguer, M. D., Colprim, J., & Puig, S. (2017). On the edge of research and technological application: A critical review of electromethanogenesis. *International Journal of Molecular Sciences, 18*(4), 874.

Borole, A. P., Tsouris, C., Pavlostathis, S. G., Yiacoumi, S., Lewis, A. J., Zeng, X., & Park, L. (2018). Efficient conversion of aqueous-waste-carbon compounds into electrons, hydrogen, and chemicals via separations and microbial electrocatalysis. *Frontiers in Energy Research, 6*, 94.

Cai, W., Cui, K., Liu, Z., Jin, X., Chen, Q., Guo, K., & Wang, Y. (2022). An electrolytic-hydrogen-fed moving bed biofilm reactor for efficient microbial electrosynthesis of methane from CO2. *Chemical Engineering Journal, 428*, Article 132093.

Caizán-Juanarena, L., Sleutels, T., Kokko, M., Berenguer, R., & ter Heijne, A. (2024). Fluidized and fixed granular beds of activated carbon as electrodes in microbial electrochemical technologies. In *Emerging trends and advances in microbial electrochemical technologies* (pp. 425–478). Elsevier.

Chaitanya, N. K., Nair, P. S., Rajpurohit, A., & Chatterjee, P. (2024). Impact of cell voltage on synthesis of caproic acid from carbon dioxide and ethanol in direct current powered microbial electrosynthesis cell. *Bioresource Technology, 412*, Article 131383.

Chandrasekhar, K., Amulya, K., & Mohan, S. V. (2015). Solid phase bio-electrofermentation of food waste to harvest value-added products associated with waste remediation. *Waste Management, 45*, 57–65.

Chandrasekhar, K., Kumar, A. N., Kumar, G., Kim, D. H., Song, Y. C., & Kim, S. H. (2021). Electro-fermentation for biofuels and biochemicals production: Current status and future directions. *Bioresource Technology, 323*, Article 124598.

Chen, X. Y., Vinh-Thang, H., Ramirez, A. A., Rodrigue, D., & Kaliaguine, S. (2015). Membrane gas separation technologies for biogas upgrading. *RSC Advances, 5*(31), 24399–24448.

Chen, X., Cao, Y., Li, F., Tian, Y., & Song, H. (2018). Enzyme-assisted microbial electrosynthesis of poly (3-hydroxybutyrate) via CO2 bioreduction by engineered Ralstonia eutropha. *ACS Catalysis, 8*(5), 4429–4437.

Cheng, S., Xing, D., Call, D. F., & Logan, B. E. (2009). Direct biological conversion of electrical current into methane by electromethanogenesis. *Environmental Science & Technology, 43*(10), 3953–3958.

Choi, O., Um, Y., & Sang, B. I. (2012). Butyrate production enhancement by Clostridium tyrobutyricum using electron mediators and a cathodic electron donor. *Biotechnology and Bioengineering, 109*(10), 2494–2502.

Choi, O., Kim, T., Woo, H. M., & Um, Y. (2014). Electricity-driven metabolic shift through direct electron uptake by electroactive heterotroph Clostridium pasteurianum. *Scientific Reports, 4*, 6961.

Christodoulou, X., Okoroafor, T., Parry, S., & Velasquez-Orta, S. B. (2017). The use of carbon dioxide in microbial electrosynthesis: Advancements, sustainability and economic feasibility. *Journal of CO2 Utilization, 18*, 390–399.

Chu, N., Liang, Q., Jiang, Y., & Zeng, R. J. (2020). Microbial electrochemical platform for the production of renewable fuels and chemicals. *Biosensors and Bioelectronics, 150*, Article 111922. https://doi.org/10.1016/j.bios.2019.111922

Dai, Y., Yuan, Z., Jack, K., & Keller, J. (2007). Production of targeted poly(3-hydroxyalkanoates) copolymers by glycogen accumulating organisms using acetate as sole carbon source. *Journal of Biotechnology, 129*(3), 489–497.

Das, S., Chatterjee, P., & Ghangrekar, M. M. (2018). Increasing methane content in biogas and simultaneous value-added product recovery using microbial electrosynthesis. *Water Science and Technology, 77*, 1293–1302. https://doi.org/10.2166/wst.2018.002

Dessì, P., Rovira-Alsina, L., Sánchez, C., Dinesh, G. K., Tong, W., Chatterjee, P., Tedesco, M., Farras, P., Hamelers, H. M., & Puig, S. (2021). Microbial electrosynthesis: towards sustainable biorefineries for production of green chemicals from CO2 emissions. *Biotechnology Advances, 46*, 107675.

Diego-García, R., Morán, A., & Romeo, L. M. (2022). Integration of oxycombustion and microbial electrosynthesis for sustainable energy storage. *Energy Conversion and Management, 269*, Article 116002.

Ding, A., Yang, Y., Sun, G., & Wu, D. (2016). Impact of applied voltage on methane generation and microbial activities in an anaerobic microbial electrolysis cell (MEC). *Chemical Engineering Journal, 283*, 260–265.

Dinh, H. T., Kambara, H., Matsushita, S., Aoi, Y., Kindaichi, T., Ozaki, N., & Ohashi, A. (2022). Biological methane production coupled with sulfur oxidation in a microbial electrosynthesis system without organic substrates. *Journal of Environmental Sciences, 116*, 68–78.

Dong, Z., Wang, H., Tian, S., Yang, Y., Yuan, H., Huang, Q., Song, T. S., & Xie, J. (2018). Fluidized granular activated carbon electrode for efficient microbial electrosynthesis of acetate from carbon dioxide. *Bioresource Technology, 269*, 203–209.

Eerten-Jansen, M. C., Jansen, N. C., Plugge, C. M., de Wilde, V., Buisman, C. J., & ter Heijne, A. (2015). Analysis of the mechanisms of bioelectrochemical methane production by mixed cultures. *Journal of Chemical Technology & Biotechnology, 90*(5), 963–970.

Flynn, J. M., Ross, D. E., Hunt, K. A., Bond, D. R., & Gralnick, J. A. (2010). Enabling unbalanced fermentations by using engineered electrode-interfaced bacteria. *mBio, 1*(5), e00190–10.

Fu, Q., He, Y., Li, Z., Li, J., Zhang, L., Zhu, X., & Liao, Q. (2022). Direct CO2 delivery with hollow stainless steel/graphene foam electrode for enhanced methane production in microbial electrosynthesis. *Energy Conversion and Management, 268*, Article 116018.

Gadkari, S., Mirza Beigi, B. H., Aryal, N., & Sadhukhan, J. (2021). Microbial electrosynthesis: Is it sustainable for bioproduction of acetic acid. *RSC advances, 11*(17), 9921–9932.

Ganigué, R., Puig, S., Batlle-Vilanova, P., Balaguer, M. D., & Colprim, J. (2015). Microbial electrosynthesis of butyrate from carbon dioxide. *Chemical Communications, 51*(15), 3235–3238.

Gavilanes, J., Reddy, C. N., & Min, B. (2019). Microbial electrosynthesis of bioalcohols through reduction of high concentrations of volatile fatty acids. *Energy and Fuels, 33*(5), 4264–4271.

Gildemyn, S., Verbeeck, K., Jansen, R., & Rabaey, K. (2017). The type of ion selective membrane determines stability and production levels of microbial electrosynthesis. *Bioresource Technology, 224*, 358–364.

Götz, M., Lefebvre, J., Mörs, F., McDaniel Koch, A., Graf, F., Bajohr, S., Reimert, R., & Kolb, T. (2016). Renewable Power-to-Gas: A technological and economic review. *Renewable Energy, 85*, 1371–1390.

Griffin, D. W., & Schultz, M. A. (2012). Fuel and chemical products from biomass syngas: A comparison of gas fermentation to thermochemical conversion routes. *Environmental Progress & Sustainable Energy, 31*(2), 219–224.

Heijstra, B. D., Leang, C., & Juminaga, A. (2017). Gas fermentation: Cellular engineering possibilities and scale up. *Microbial Cell Factories, 16*(1), 60.

Harrington, T. D., Mohamed, A., Tran, V. N., Biria, S., Gargouri, M., Park, J. J., Gang, D. R., & Beyenal, H. (2015). Neutral red-mediated microbial electrosynthesis by *Escherichia coli*, Klebsiella pneumoniae, and Zymomonas mobilis. *Bioresource Technology, 195*, 57–65.

Jadhav, D. A., Ray, S. G., & Ghangrekar, M. M. (2017). Third generation in bio-electrochemical system research—a systematic review on mechanisms for recovery of valuable by-products from wastewater. *Renewable and Sustainable Energy Reviews, 76*, 1022–1031.

Jeremiasse, A. W., Hamelers, H. V., & Buisman, C. J. (2010). Microbial electrolysis cell with a microbial biocathode. *Bioelectrochemistry, 78*(1), 39–43.

Jiang, Y., May, H. D., Lu, L., Liang, P., Huang, X., & Ren, Z. J. (2019a). Carbon dioxide and organic waste valorization by microbial electrosynthesis and electro-fermentation. *Water Research, 149*, 42–55.

Jiang, Y., Chu, N., Zhang, W., Ma, J., Zhang, F., Liang, P., & Zeng, R. J. (2019b). Zinc: A promising material for electrocatalyst-assisted microbial electrosynthesis of carboxylic acids from carbon dioxide. *Water Research, 159*, 87–94.

Jourdin, L., Winkelhorst, M., Rawls, B., Buisman, C. J. N., & Strik, D. P. B. T. B. (2019). Enhanced selectivity to butyrate and caproate above acetate in continuous bioelectrochemical chain elongation from CO2: Steering with CO2 loading rate and hydraulic retention time. *Bioresource Technology Reports, 7*, Article 100284.

Jourdin, L., Raes, S. M. T., Buisman, C. J. N., & Strik, D. P. B. T. B. (2018). Critical biofilm growth throughout unmodified carbon felts allows continuous bioelectrochemical chain elongation from CO2 up to caproate at high current density. *Frontiers in Energy Research, 6*, 1–15.

Joyia, M. A. K., Ahmad, M., Chen, Y. F., Mustaqeem, M., Ali, A., Abbas, A., & Gondal, M. A. (2024). Trends and advances in sustainable bioethanol production technologies from first to fourth generation: A critical review. *Energy Conversion and Management, 321*, Article 119037.

Kamm, B., Kamm, M., Gruber, P. R., & Kromus, S. (2005). Biorefinery systems—an overview. In B. Kamm, P. R. Gruber, & M. Kamm (Eds.), *Biorefineries-Industrial Processes and Products: Status Quo and Future Directions* (pp. 1–40). Wiley-VCH Verlag GmbH.

Khosravanipour Mostafazadeh, A., Drogui, P., Brar, S. K., Tyagi, R. D., Le Bihan, Y., Buelna, G., & Rasolomanana, S. D. (2016). Enhancement of biobutanol production by electromicrobial glucose conversion in a dual chamber fermentation cell using C. pasteurianum. *Energy Conversion and Management, 130*, 165–175.

Kim, C., Kim, M. Y., Michie, I., Jeon, B. H., Premier, G. C., Park, S., & Kim, J. R. (2017). Anodic electro-fermentation of 3-hydroxypropionic acid from glycerol by recombinant Klebsiella pneumoniae L17 in a bioelectrochemical system. *Biotechnology for Biofuels, 10*, 1–11.

Kondaveeti, S., & Min, B. (2015). Bioelectrochemical reduction of volatile fatty acids in anaerobic digestion effluent for the production of biofuels. *Water Research, 87*, 137–144.

Kong, F., Ren, H. Y., Pavlostathis, S. G., Nan, J., Ren, N. Q., & Wang, A. (2020). Overview of value-added products bioelectrosynthesized from waste materials in microbial electrosynthesis systems. *Renewable and Sustainable Energy Reviews, 125*, Article 109816.

Kopke, M., & Simpson, S. D. (2020). Pollution to products: Recycling of 'above ground' carbon by gas fermentation. *Current Opinion in Biotechnology, 65*, 180–189. https://doi.org/10.1016/j.copbio.2020.02.017

Lehner, M., Tichler, R., Steinmüller, H., & Koppe, M. (2014). *Power-to-gas: Technology and business models* (pp. 7–17). Springer International Publishing.

Li, H., Opgenorth, P. H., Wernick, D. G., Rogers, S., Wu, T., Higashide, W., Malati, P., Huo, Y., Cho, K. M., & Liao, J. C. (2012). Integrated electromicrobial conversion of CO2 to higher alcohols. *Science, 335*(6076), 1596–1596.

Li, Z., Cai, J., Gao, Y., Zhang, L., Liang, Q., Hao, W., Jiang, Y., & Zeng, R. J. (2022). Efficient production of medium chain fatty acids in microbial electrosynthesis with simultaneous bio-utilization of carbon dioxide and ethanol. *Bioresource Technology, 352*, Article 127101.

Li, C., Li, H., Li, P., Dang, Y., Sun, D., & Guo, D. (2024). Enhancing predictions of acetate and ethanol production from microbial electrosynthesis using optimized machine learning models. *ACS Sustainable Chemistry & Engineering, 12*(10), 4264–4275.

Liew, F., Martin, M. E., Tappel, R. C., Heijstra, B. D., Mihalcea, C., & Köpke, M. (2016). Gas fermentation-a flexible platform for commercial scale production of low-carbon-fuels and chemicals from waste and renewable feedstocks. *Frontiers in Microbiology, 7*, 694.

Logan, B. E., & Rabaey, K. (2012). Conversion of wastes into bioelectricity and chemicals by using microbial electrochemical technologies. *Science, 337*(6095), 686–690.

Mahapatra, S., Kumar, D., Singh, B., & Sachan, P. K. (2021). Biofuels and their sources of production: A review on cleaner sustainable alternative against conventional fuel, in the framework of the food and energy nexus. *Energy Nexus, 4*, Article 100036.

Mamun, M. R. A., Karim, M. R., Rahman, M. M., Asiri, A. M., & Torii, S. (2016). Methane enrichment of biogas by carbon dioxide fixation with calcium hydroxide and activated carbon. *Journal of the Taiwan Institute of Chemical Engineers, 58*, 476–481.

Marshall, C. W., LaBelle, E. V., & May, H. D. (2013). Production of fuels and chemicals from waste by microbiomes. *Current Opinion in Biotechnology, 24*(3), 391–397. https://doi.org/10.1016/j.copbio.2013.03.016

Marshall, C. W., Ross, D. E., Fichot, E. B., Norman, R. S., & May, H. D. (2012). Electrosynthesis of commodity chemicals by an autotrophic microbial community. *Applied and Environmental Microbiology, 78*(23), 8412–8420.

Merli, G., Becci, A., Amato, A., & Beolchini, F. (2021). Acetic acid bioproduction: The technological innovation change. *Science of the Total Environment, 798*, Article 149292.

Mistry, A. N., Ganta, U., Chakrabarty, J., & Dutta, S. (2019). A review on biological systems for CO2 sequestration: Organisms and their pathways. *Environmental Progress & Sustainable Energy, 38*(1), 127–136.

Mogollón, C. A. G., Díaz, J. C. Q., & Posada, J. O. G. (2023). Production of acetone, butanol, and ethanol by electro-fermentation with Clostridium saccharoperbutylacetonicum N1–4. *Bioelectrochemistry, 152*, Article 108414.

Moscoviz, R., Toledo-Alarcón, J., Trably, E., & Bernet, N. (2016). Electro-fermentation: How to drive fermentation using electrochemical systems. *Trends in Biotechnology, 34*(11), 856–865.

Mostafazadeh, A. K., Drogui, P., Brar, S. K., Tyagi, R. D., Le Bihan, Y., Buelna, G., & Rasolomanana, S. D. (2016). Enhancement of biobutanol production by electromicrobial glucose conversion in a dual chamber fermentation cell using C. pasteurianum. *Energy Conversion and Management, 130*, 165–175.

Munasinghe, P. C., & Khanal, S. K. (2010). Biomass-derived syngas fermentation into biofuels: Opportunities and challenges. *Bioresource Technology, 101*(13), 5013–5022.

Murali, N., Srinivas, K., & Ahring, B. K. (2017). Biochemical production and separation of carboxylic acids for biorefinery applications. *Fermentation, 3*(2), 22.

Nevin, K. P., Hensley, S. A., Franks, A. E., Summers, Z. M., Ou, J., Woodard, T. L., Snoeyenbod-West, O. L., & Lovley, D. R. (2011). Electrosynthesis of organic compounds from carbon dioxide is catalyzed by a diversity of acetogenic microorganisms. *Applied and Environmental Microbiology, 77*(9), 2882–2886.

Nevin, K. P., Woodard, T. L., Franks, A. E., Summers, Z. M., & Lovley, D. R. (2010). Microbial electrosynthesis: Feeding microbes electricity to convert carbon dioxide and water to multicarbon extracellular organic compounds. *mBio, 1*(2), e00103–10.

Nikhil, G. N., Venkata Subhash, G., Yeruva, D. K., & Venkata Mohan, S. (2015). Synergistic yield of dual energy forms through biocatalyzed electrofermentation of waste: Stoichiometric analysis of electron and carbon distribution. *Energy, 88*, 281–291.

Pasupuleti, S. B., Sarkar, O., & Mohan, S. V. (2014). Upscaling of biohydrogen production process in semi-pilot scale biofilm reactor: Evaluation with food waste at variable organic loads. *International Journal of Hydrogen Energy, 39*(14), 7587–7596.

Pepè Sciarria, T., Batlle-Vilanova, P., Colombo, B., Scaglia, B., Balaguer, M. D., Colprim, J., Puig, S., & Adani, F. (2018). Bio-electrorecycling of carbon dioxide into bioplastics. *Green Chemistry, 20*(17), 4058–4066.

Quraishi, M., Wani, K., Pandit, S., Gupta, P. K., Rai, A. K., Lahiri, D., Jadhav, D. A., Ray, R. R., Jung, S. P., Thakur, V. K., & Prasad, R. (2021). Valorisation of CO2 into value-added products via microbial electrosynthesis (MES) and electro-fermentation technology. *Fermentation, 7*(4), 291.

Rago, L., Pant, D., & Schievano, A. (2019). Electro-fermentation—microbial electrochemistry as new frontier in biomass refineries and industrial fermentations. In *Advanced bioprocessing for alternative fuels, biobased chemicals, and bioproducts* (pp. 265–287). Woodhead Publishing.

Romano, M. C., Anantharaman, R., Arasto, A., Ozcan, D. C., Ahn, H., Dijkstra, J. W., Carbo, M., & Boavida, D. (2013). Application of advanced technologies for CO2 capture from industrial sources. *Energy Procedia, 37*, 7176–7185.

Roy, S., Schievano, A., & Pant, D. (2016). Electro-stimulated microbial factory for value added product synthesis. *Bioresource Technology, 213*, 129–139. https://doi.org/10.1016/j.biortech.2016.03.052

Sadhukhan, J., Lloyd, J. R., Scott, K., Premier, G. C., Yu, E. H., Curtis, T., & Head, I. M. (2016). A critical review of integration analysis of microbial electrosynthesis (MES) systems with waste biorefineries for the production of biofuel and chemical from reuse of CO2. *Renewable and Sustainable Energy Reviews, 56*, 116–132.

Salar-García, M. J., Ortiz-Martínez, V. M., Sánchez-Segado, S., Valero Sánchez, R., Sáez López, A., Lozano Blanco, L. J., & Godínez-Seoane, C. (2024). Sustainable production of biofuels and biochemicals via electro-fermentation technology. *Molecules, 29*(4), 834.

Schievano, A., Pepè Sciarria, T., Vanbroekhoven, K., De Wever, H., Puig, S., Andersen, S. J., Rabaey, K., & Pant, D. (2016). Electro-fermentation—merging electrochemistry with fermentation in industrial applications. *Trends in Biotechnology, 34*(11), 866–878. https://doi.org/10.1016/j.tibtech.2016.04.007

Schröder, U., Harnisch, F., & Angenent, L. T. (2015). Microbial electrochemistry and technology: Terminology and classification. *Energy & Environmental Science, 8*(2), 513–519.

Schuppert, B., Schink, B., & Trösch, W. (1992). Batch and continuous production of propionic acid from whey permeate by Propionibacterium acidi-propionici in a three-electrode amperometric culture system. *Applied Microbiology and Biotechnology, 37*, 549–553.

Shakeel, S., Anwer, A. H., & Khan, M. Z. (2020). Nitric acid treated graphite granular cathode for microbial electro reduction of carbon dioxide to acetate. *Journal of Cleaner Production, 269*, Article 122391.

Shakeel, S., & Khan, M. Z. (2022). Enhanced production and utilization of biosynthesized acetate using a packed-fluidized bed cathode based MES system. *Journal of Environmental Chemical Engineering, 10*(4), Article 108067.

Shang, G., Cui, K., Cai, W., Hu, X., Jin, P., & Guo, K. (2023). A 20 L electrochemical continuous stirred-tank reactor for high rate microbial electrosynthesis of methane from CO2. *Chemical Engineering Journal, 451*, Article 138898.

Sharma, M., Aryal, N., Sarma, P. M., Vanbroekhoven, K., Lal, B., Benetton, X. D., & Pant, D. (2013). Bioelectrocatalyzed reduction of acetic and butyric acids via direct electron transfer using a mixed culture of sulfate-reducers drives electrosynthesis of alcohols and acetone. *Chemical Communications, 49*(58), 6495–6497.

Song, T. S., Li, T., Tao, R., Huang, H. F., & Xie, J. (2022). CuO/g-C3N4 heterojunction photocathode enhances the microbial electrosynthesis of acetate through CO2 reduction. *Science of the Total Environment, 818*, Article 151820.

Soussan, L., Riess, J., Erable, B., Delia, M.-L., & Bergel, A. (2013). Electrochemical reduction of CO2 catalysed by Geobacter sulfurreducens grown on polarized stainless-steel cathodes. *Electrochemistry Communications, 28*, 27–30. https://doi.org/10.1016/j.elecom.2012.11.033

Speers, A. M., Young, J. M., & Reguera, G. (2014). Fermentation of glycerol into ethanol in a microbial electrolysis cell driven by a customized consortium. *Environmental Science & Technology, 48*(11), 6350–6358.

Srikanth, S., Singh, D., Vanbroekhoven, K., Pant, D., Kumar, M., Puri, S. K., & Ramakumar, S. S. V. (2018). Electro-biocatalytic conversion of carbon dioxide to alcohols using gas diffusion electrode. *Bioresource Technology, 265*, 45–51.

Steinbusch, K. J., Hamelers, H. V., Schaap, J. D., Kampman, C., & Buisman, C. J. (2010). Bioelectrochemical ethanol production through mediated acetate reduction by mixed cultures. *Environmental Science & Technology, 44*(1), 513–517.

Sturm-Richter, K., Golitsch, F., Sturm, G., Kipf, E., Dittrich, A., Beblawy, S., Kerzenmacher, S., & Gescher, J. (2015). Unbalanced fermentation of glycerol in *Escherichia coli* via heterologous production of an electron transport chain and electrode interaction in microbial electrochemical cells. *Bioresource Technology, 186*, 89–96.

Tahir, K., Miran, W., Jang, J., Woo, S. H., & Lee, D. S. (2021). Enhanced product selectivity in the microbial electrosynthesis of butyrate using a nickel ferrite-coated biocathode. *Environmental Research, 196*, Article 110907.

Tashiro, Y., Hirano, S., Matson, M. M., Atsumi, S., & Kondo, A. (2018). Electrical-biological hybrid system for CO2 reduction. *Metabolic Engineering, 47*, 211–218.

Tice, R. C., & Kim, Y. (2014). Energy efficient reconcentration of diluted human urine using ion exchange membranes in bioelectrochemical systems. *Water Research, 64*, 61–72.

Van Eerten-Jansen, M. C. A. A., Heijne, A. T., Grootscholten, T. I. M., Steinbusch, K. J. J., Sleutels, T. H. J. A., Hamelers, H. V. M., & Buisman, C. J. N. (2013). Bioelectrochemical production of caproate and caprylate from acetate by mixed cultures. *ACS Sustainable Chemistry & Engineering, 1*(5), 513–518.

Vassilev, I., Rinta-Kanto, J. M., & Kokko, M. (2024). Comparing the performance of fluidized and fixed granular activated carbon beds as cathodes for microbial electrosynthesis of carboxylates from CO2. *Bioresource Technology*, 130896.

Vassilev, I., Hernandez, P. A., Batlle-Vilanova, P., Freguia, S., Krömer, J. O., Keller, J., Ledezma, P., & Virdis, B. (2018). Microbial electrosynthesis of isobutyric, butyric, caproic acids, and corresponding alcohols from carbon dioxide. *ACS Sustainable Chemistry & Engineering, 6*(7), 8485–8493.

Vassilev, I., Kracke, F., Freguia, S., Keller, J., Krömer, J. O., Ledezma, P., & Virdis, B. (2019). Microbial electrosynthesis system with dual biocathode arrangement for simultaneous acetogenesis, solventogenesis and carbon chain elongation. *Chemical Communications, 55*(30), 4351–4354.

Vidales, A. G., Omanovic, S., & Tartakovsky, B. (2019). Combined energy storage and methane bioelectrosynthesis from carbon dioxide in a microbial electrosynthesis system. *Bioresource Technology Reports, 8*, Article 100302.

Wang, D., Liang, Q., Chu, N., Zeng, R. J., & Jiang, Y. (2023). Deciphering mixotrophic microbial electrosynthesis with shifting product spectrum by genome-centric metagenomics. *Chemical Engineering Journal, 451*, 139010.

Wong, J. W.-C., Tyagi, R. D., & Pandey, A. (2016). *Current developments in biotechnology and bioengineering: Solid waste management*. Elsevier Science.

Wood, J. C., Grové, J., Marcellin, E., Heffernan, J. K., Hu, S., Yuan, Z., & Virdis, B. (2021). Strategies to improve viability of a circular carbon bioeconomy-A techno-economic review of microbial electrosynthesis and gas fermentation. *Water Research, 201*, Article 117306.

Wu, Y., Li, W., Wang, L., Wu, Y., Wang, Y., Wang, Y., & Meng, H. (2022a). Enhancing the selective synthesis of butyrate in microbial electrosynthesis system by gas diffusion membrane composite biocathode. *Chemosphere, 308*, Article 136088.

Wu, H., Pan, H., Li, Z., Liu, T., Liu, F., Xiu, S., Wang, J., Wang, H., Hou, Y., Yang, B., & Lei, L. (2022b). Efficient production of lycopene from CO2 via microbial electrosynthesis. *Chemical Engineering Journal, 430*, Article 132943.

Wu, Z., Wang, J., Liu, J., Wang, Y., Bi, C., & Zhang, X. (2019). Engineering an electroactive Escherichia coli for the microbial electrosynthesis of succinate from glucose and CO 2. *Microbial Cell Factories, 18*, 1–14.

Xue, G., Lai, S., Li, X., Zhang, W., You, J., Chen, H., Qian, Y., Gao, P., Liu, Z., & Liu, Y. (2018). Efficient bioconversion of organic wastes to high optical activity of L-lactic acid stimulated by cathode in mixed microbial consortium. *Water Research, 131*, 1–10.

Yao, H., Rinta-Kanto, J. M., Vassilev, I., & Kokko, M. (2024). Methanol as a co-substrate with CO2 enhances butyrate production in microbial electrosynthesis. *Applied Microbiology and Biotechnology, 108*(1), 372.

Zhang, T., & Tremblay, P. L. (2019). Possible industrial applications for microbial electrosynthesis from carbon dioxide. In *Microbial electrochemical technology* (pp. 825–842). Elsevier.

Zhang, Y., Li, J., Yong, Y. C., Fang, Z., Yan, H., Li, J., & Meng, J. (2023). Highly selective butanol production by manipulating electron flow via cathodic electro-fermentation. *Bioresource Technology, 374*, Article 128770.

Zhao, Z., Zhang, Y., Wang, L., & Quan, X. (2015). Potential for direct interspecies electron transfer in an electric-anaerobic system to increase methane production from sludge digestion. *Scientific Reports, 5*, 1–12. https://doi.org/10.1038/srep11094

Zhao, Z., Sun, J., Yu, Q., Ma, W., Quan, X., & Zhang, Y. (2016). Enhanced decomposition of waste activated sludge via anodic oxidation for methane production and bioenergy recovery. *International Biodeterioration & Biodegradation, 106*, 161–169.

Zhao, J., Ma, H., Wu, W., Bacar, M. A., Wang, Q., Gao, M., Wu, C., Xia, C., & Qian, D. (2023). Conversion of liquor brewing wastewater into medium chain fatty acids by microbial electrosynthesis: Effect of cathode potential and CO2 supply. *Fuel, 332*, Article 126046.

Zhou, M., Chen, J., Freguia, S., Rabaey, K., & Keller, J. (2013). Carbon and electron fluxes during the electricity driven 1, 3-propanediol biosynthesis from glycerol. *Environmental Science & Technology, 47*(19), 11199–11205.

Chapter 7
Applications of Electro-Fermentation of Food Waste for Biofuel Production

Gunaswetha Kuraganti and Sujatha Edla

1 Introduction

Electro-fermentation is a well-known biofuel that combines electromagnetic stimulation via microbial metabolism to improve the generation of ethanol, a widely recognized biofuel generated through fermentation processes (Salar-García et al., 2024; Saini and Mishra, (2024); . This process makes use of certain bacteria that are skilled in breaking down carbohydrates like glucose and fructose in anaerobic environments to create ethanol, and these include yeast (*Saccharomyces cerevisiae*) and *Zymomonas mobilis*. This combination contributes to the ability of these bacteria to capture and store carbon efficiently, which is important for reducing greenhouse gas emissions (Thakur et al. 2024). The electronic design helps capture and store carbon dioxide (CO_2) more efficiently. Electron fermentation is carried out under favorable pH conditions such as acidic or basic conditions and a redox atmosphere, i.e., transfer of electrons in a chemical reaction, these materials can be modified to aim for higher yields (Roy & Das 2023). Electron fermentation controls the formation or breakdown of large carbon chain products to enable the production of different products depending on the needs of the products, electron fermentation is also used in large-scale biomass production by optimizing fermentation conditions. Electron fermentation also helps to maintain complex microbial assemblages, i.e., coexistence of different microbial populations (Qingshan et al., 2019).

Fermenters, another name for electrochemical cells, aid in the creation of an environment that allows microbes—tiny living organisms—to increase and generate useful goods. Devices like these electrochemical cells can change chemical energy into electrical energy or vice versa. Microbial metabolism and fermentation conditions can be influenced by electrochemical cells (Nuohan et al., 2023a, 2023b). It is possible to modify the circumstances under which fermentation occurs by employing

G. Kuraganti · S. Edla (✉)
Department of Microbiology, Kakatiya University, Hanamkonda, Telangana, India
e-mail: sujathaedla_1973@kakatiya.ac.in

© The Author(s), under exclusive license to Springer Nature Singapore Pte Ltd. 2025

S. Sevda (ed.), *Electro-Fermentation for Synthesis of Chemicals, Biochemicals and Biofuels*, https://doi.org/10.1007/978-981-96-6358-3_7

electrochemical cells. They can assist in regulating the quantities of oxygen, acidity, or other crucial elements. This implies that we can improve the conditions in which the bacteria operate, potentially resulting in more effective fermentation. This strategy can potentially enhance present fermentation technologies by employing electrochemical cells to enhance current fermentation processes. This means that we can make them more efficient and faster, with less substrate consumption and product inhibition effects. Sometimes creating objects during production can slow down or stop the process. By controlling the environment with electrolyte cells, we can minimize these negative effects, allowing it to heat up more efficiently. Substrates are substances (such as sugars) that microorganisms feed on to grow. By doing the process right we can use less of these materials, which is good for conservation (Noordam et al., 2024).

An electro-fermentor is an integrated bioelectrochemical device that uses both biological and electrochemical processes to simultaneously produce compounds with higher value and electricity. The electrodes will function as sinks or sources of electrons (Chandrasekhar et al., 2015; Gambino & Cammarata, 2017; Modestra et al., 2020). The bulk to investigate on electro-fermentation was first concentrated on increasing the metabolic rate of microbes (of the conventional fermentation process). Subsequently, the research attempts were directed toward the generation of value-added chemicals and fuels (such as ethanol, hydrogen (H_2), methane (CH_4), biopolymers, and others) through bioelectrosynthesis. Electromagnetic technology is used in the medical field to improve electron transfer, which is a novel way of combining electricity and synthesis for efficient recycling of biological waste, especially when low-energy or high-energy electro-fermentation (EF) fermentation systems are used, cell proliferation, microbial metabolism to faster and thus the fermentation efficiency is improved (Sravan et al., 2021). Compared to the conventional electromagnetic method, this method uses less electricity. No toxic or harmful substances are produced, the reactor is simple and accessible and has a large control capacity.

2 Types of Organic Food Waste in Electro-Fermentation

The application of microbes to the recycling of readily degrading organic waste has garnered significant interest in the Green Age, which is centered on renewable energy and sustainable economic growth. Recycling organic waste is a successful application of anaerobic fermentation. In an anaerobic setting, organic waste could be converted by a range of microbes into organic acid, biogas, and various other products with added value (Lutfi et al., 2024). Currently, many researches focus on improving the electron transfer speed, increasing the electron transfer between different materials, improving the electron transport, and adding conductive materials, but the impact is often limited Current management strategies suffer from limited resources and low prices (Dar et al., 2024). Value-added product advancement is now a necessary trend for continued expansion. Through the application of low voltage or current to the fermentation system, electrical fermentation (EF) is a method that stimulates cell growth, promotes microbial metabolism, and enhances the generation of value-added products. This approach stands for a fresh line of inquiry at the nexus of biology and electrochemistry (Vira, 2024).

Food waste contains a lot of organic resources that could be turned into other things through recovery systems, as it is a substantial source. A large volume of food waste generated worldwide presents great opportunities for resource recovery. Different types of fermentation can yield different products, and the product can be affected by changing specific fermentation conditions. Key parameters include the type of inoculum, pH levels, ORP, OLR as well as nutrient availability. The main products from fermentation of food wastes are lactic acid and volatile fatty acids, with an emphasis on high yield and purity (Nuohan et al., 2023a, 2023b; Wang et al., 2020).

Different feed wastes are used to enhance bioelectrochemical processes in electro-fermentation. Studies show that feed wastewater is a particularly promising product after pre-fermentation, which increases the concentration of short-chain organic acids and ethanol, thereby improving microbial electrolytic cell (MEC) performance for hydrogen production in addition, waste-activated The cotreatment of sludge and food waste has been shown to effectively produce volatile fatty acids (VFAs), especially with electrochemical pretreatment with metal oxide electrotreatment specificity serves to enhance it (Virdis et al., 2022). The acidogenic fermentation of food waste leads to electron acceptors and donors' production in addition to the generation of medium-chain carboxylic acids (Qingshan et al., 2024; Wang et al., 2024). Furthermore, MECs can use fermentations that are rich in simple organic acids and ethanol as good source of electrons, whereby the choice of substrate varies the amount of hydrogen generated. All these studies underline the numerous applications of food wastage in electro-fermentation, optimizing energy recovery, and valorizing resources (Li et al., 2024).

Electro-fermentation (EF) is an innovative bioprocess that uses food waste to produce valuable chemicals, fuels, and energy by combining electrical stimulation and fermentation techniques. Using various food wastes in electro-fermentation dealing with, promoting renewable energy is an emerging area of research and development with f objectives in. This comprehensive discussion examines the various types of food waste used in EF programs, their potential, and the processes for converting them into valuable by-products (Jun-Sil et al., 2024; Upadhyay, (2024). Different types of food waste used in biofuel production are given in Table 1.

2.1 *Vegetable and Fruit Waste*

Fruit and vegetable waste is an important part of global food waste, affecting both developing and developed countries. This waste includes fruit and vegetable peels, skins, seeds, and other inedible or excess fruit and vegetable parts, which are often discarded rather than used Biochemical composition of fruit and vegetable waste Carbohydrates, including sugars, cellulose, pectin, vitamins, and minerals, together,

Table 1 The table shows types of food waste, major components, microorganisms involved, biofuel production, and applications of these biofuels. It highlights the versatility and potential of electro-fermentation in converting various food wastes into valuable energy sources

Type of food waste	Major components	Microorganisms used	Biofuels produced	Application of biofuel
Fruit and vegetable waste	Sugars (glucose, fructose), cellulose, pectin	*Saccharomyces cerevisiae*, *Rhodopseudomonas palustris*	Ethanol, Hydrogen	Transportation fuel, electricity generation
Starchy food waste	Starch, polysaccharides	*Clostridium acetobutylicum*, *Zymomonas mobilis*	Ethanol, Butanol	Blending with gasoline, industrial solvents
Dairy waste	Lactose, proteins	*Saccharomyces cerevisiae*, *Lactobacillus*	Ethanol, Lactic Acid	Fuel, feedstock for chemical industries
Fats, oils, and grease (FOG) waste	Lipids, fatty acids, glycerol	*Methanosarcina*, *Methanosaeta*	Methane, Biodiesel	Biogas for energy, biodiesel for vehicles
Bakery and confectionery waste	Sugars, starch	*Clostridium butyricum*, *Saccharomyces cerevisiae*	Ethanol, Hydrogen	Fuel for transportation, hydrogen for fuel cells
Meat and fish waste	Proteins, fats	*Methanosarcina*, *Clostridium*	Methane, Hydrogen	Biogas for heating and power, hydrogen for clean energy

so that wastes can be converted into valuable products. Microorganisms metabolize sugars in fruit and vegetable waste during electro-fermentation under controlled electrochemical conditions. The use of electrical inputs can increase fermentation efficiency and increase organic acids (such as lactic and acetic acids) and wine production (Yıldız & Öztekin 2024). This method helps to destroy waste and contributes to renewable energy. The by-products of electrolysis of fruit and vegetable waste can be used for a variety of applications. For example, they can produce biofuels such as ethanol and hydrogen, which are renewable energy sources. In addition, organic acids produced can be used in the food industry as preservatives or as pharmaceuticals.

Despite the potential benefits, managing fruit and vegetable waste is a challenge. The large amount of waste generated, especially in urban areas, requires efficient collection and processing systems. Furthermore, due to the heterogeneous nature of the waste, the involvement of mixed microbial consortia is important for proper breakdown, which can complicate the fermentation process (Petrescu-Mag et al., 2024). The future of fruit and vegetable waste management through electro-fermentation depends on further research and development. There is a need for scalable and economically viable programs that can effectively turn food waste into a sustainable resource. This includes optimizing microbial habitat and providing electrical conditions optimized to maximize desired results (Abate et al., 2024; Bhattacharya, 2024).2024

2.2 *Starchy Food Waste: Sources and Implications*

There are many starchy foods that are wasted from the agricultural sector and food processing industry. They are created when farmers grow, process, and eat starchy crops like potatoes, rice, maize, wheat, and others. This type of waste includes potato skins; rice bran; remnants of pasta; leftover bread, or other products from these raw ingredients that can be found everywhere in the world's markets and homes contributing to its current state of crisis on food wastage globally (Wang et al., 2023).

During the harvesting of crops, a lot of starchy food waste is generated. This includes crops that had been damaged or remained unharvested but were still left in the fields. A case in point is the small or crooked potatoes that are often viewed as trash. A lot of waste is produced especially during the processing of starchy crops into other food products. For example, rice bran is a by-product of milling rice while removing its outside bran layer. Likewise, trimmings and scraps arise as pasta-making progresses (Sharma et al., 2020). At the consumer level, starchy food waste occurs when foods are not consumed on time. Stale bread and leftover pasta are examples of foods that are thrown away rather than eaten often. Starchy foods consist primarily of polysaccharides, especially starch, which can be hydrolyzed into simple sugars such as glucose. These natural properties make starchy food wastes a valuable feedstock, as these sugars can be converted into biofuels and various chemicals (Ezejiofor et al., 2014).

Many people can make a lot of mistakes about this; it's best to properly explore the application of starchy food waste in electro-fermentation. As microbial electrolysis cells (MECs) convert starch into glucose which then enters into microbial metabolism under controlled electrochemical conditions, it allows its hydrolysis by electricity, thereby controlling their metabolism by controlling their metabolic pathways (Sharma et al., 2020). In doing so they can produce biohydrogen, methane, or ethanol from glucose released by starchy waste for instance (Sharma et al., 2020). These processes not only help with waste management but also produce renewable energy. The by-products from the electro-fermentation of starchy food waste can have many uses. For instance, biofuels such as ethanol serve as renewable energy sources. Moreover, during fermentation, organic acids are formed and can find applications in preservation or added to foods (Economou et al., 2024; Sharma et al., 2020).

2.3 Dairy Waste: Sources and Implications

Dairy waste is a significant by-product of the dairy industry, arising from the production and processing of milk and various milk products. This waste includes whey, a by-product of cheese making, as well as leftover materials from butter and yogurt production, and spoiled or unused dairy products (Jayasree Joshi et al., 2024). The management and utilization of dairy waste present both challenges and opportunities for sustainability. The dairy industry generates waste at multiple stages, starting from milk production on farms. This includes milk that is not collected due to spoilage or contamination, as well as milk that does not meet quality standards for processing. During the processing of milk into products like cheese, butter, and yogurt, significant amounts of waste are produced. For instance, whey is generated as a by-product when cheese is made, accounting for a large volume of waste in cheese production facilities (Hameed et al., 2023). Additionally, during butter production, buttermilk is produced, which can also be considered waste if not utilized effectively. Dairy products have a limited shelf life, leading to spoilage and waste. Products that are not sold or consumed in time, such as yogurt or milk, often end up being discarded (Alzahrani et al., 2024).

Lactose, proteins (casein and whey proteins), and fats are abundant in dairy waste. The whey particularly contains high amounts of lactose which serves as a good substrate for fermentation processes. The biochemical composition of dairy waste makes it a valuable resource for converting waste into useful products. In electro-fermentation lactate can be fermented by microbes under electrically controlled conditions (Vira, 2024). The application of electrical inputs helps to modulate the microbial environment hence resulting in higher yields of valuable compounds like lactic acid and ethanol. This process not only aids in waste management but also contributes to renewable energy generation and the creation of valuable co-products. The fermentation of dairy waste generates by-products that have various applications such as lactic acid produced from lactose fermentation which contains different uses in food preservation; bio-based plastics manufacturing; and food additives. Also, recovering residual whey protein would mean less wastage and improved resource efficiency because they can be used as animal feed instead (Hameed et al., 2023).

2.4 Bakery and Confectionery Waste: A Valuable Resource

Bakery and confectionery waste is a major by-product of the food industry, with products such as bread, cakes, biscuits, candies, and other sweets, contaminated, unused, or left behind during production and consumption These wastes represent not only a loss of resources but through microbial fermentation, They also provide opportunities for sustainable practices (Kaur et al., 2023; Kumar et al., 2023). Different processing factors can lead to waste degradation. For example, incomplete baked goods, sliced cakes and breads, and items that do not meet proper standards are often discarded. In addition, overproduction to meet demand can lead to excess products that may be contaminated before consumption. For consumers, bakery and confectionery products generally have shorter shelf lives. Lack of timely sales or availability of products can contribute significantly to food waste. Stale bread and cake leftovers are examples of common waste. Bakery and confectionery waste contain a lot of sugar, especially the sugar used in these products, syrups, sweeteners, and other products. In addition to sugar, this waste can also contain starches, fats, and sometimes preservatives or artificial flavors. These natural sweeteners make bread waste a good source of microbial fermentation (Siddiqu et al., 2023).

The high sugar content contained in bakery and confectionery waste can be effectively used in electro-fermentation. In this process, microbes metabolize sugars and produce valuable by-products such as ethanol, hydrogen, or volatile fatty acids under electrochemical control. The application of electrical stimulation can enhance selectivity and efficiency of the fermentation process, reducing the formation of unwanted by-products. This innovative approach helps in waste management and contributes to the renewable production of energy. The primary output from electro-fermentation of bakery and confectionery waste is ethanol which may be used as a biofuel or within the food and beverage industry. Ethanol obtained from wastes has the potential to replace fossil fuels on a sustainable basis thereby reducing greenhouse gas emissions. Furthermore, hydrogen gas production from these wastes provides a renewable energy source that can be tapped for various purposes (Kaur et al., 2023; Kumar et al., 2023).

2.5 Brewery and Distillery Waste: Composition and Utilization

Waste from refineries and breweries is a significant side-effect of the methodology used to make lager and distill spirits. This waste comprises trub (sediment), spent grains, and spent yeast, all of which, if inappropriately made due, can create ecological issues. These side effects do, notwithstanding, likewise offer opportunities for reasonable practices and asset recuperation.

2.6 Sources of Brewery and Distillery Waste:

Spent Grains: The process of beer production is usually characterized by the use of grains including barley, wheat, and corn which are used during the extraction stage to acquire sugars that are essential in fermentation. At the end of the brewing process, any remaining grains known as spent grains are often thrown away. Such grains can therefore represent a large part of total waste from breweries (Dymchenko et al., 2023).

Spent Yeast: Yeast is an important fermentation agent, converting sugar into alcohol. Yeast that has completed its activity after fermentation is called spent yeast. This complement contains valuable nutrients and can be a source of protein and other compounds (Dymchenko et al., 2023).

Trub: The sediment that collects at the bottom of fermentation tanks is known as trub. It contains solid components that are not required for the finished product, such as hop remnants and dead yeast cells. Trub is frequently taken out of beer or spirit to make it clearer, although it's still a waste (Dymchenko et al., 2023).

The industrial waste from breweries and distilleries is rich in protein, carbohydrates, and remaining alcohol. Spent grains contain fiber and protein that make them valuable substances for different uses. Spent yeast includes important amino acids, vitamins, and minerals whereas trub contains organic matter that can be utilized as well. The conversion of carbohydrates in spent grains and leftover sugars in spent yeast can be carried out by microbes under electrochemical conditions to yield products of great economic importance such as ethanol, methane, or hydrogen (Gallón-Bedoya et al., 2024). When fermentation processes apply electrical stimulation, microbial activity increases resulting in higher recoveries of biofuels and other by-products. Also, spent yeast doubles up as a bio-catalyst or nutrient source which enhances fermentation processes even further. The expanse of these side products is monumental. Animal fodder may be made from spent grains, which is a healthy option for animals kept by farmers; they can also be turned into biofuels like ethanol or biogas that are part of renewable energy sources. Spent yeast however have their uses in formulating protein supplements or as a commencement for the production of other foods that undergo fermentation. Trub is always considered rubbish but at times this is not the case because it can serve the purpose of composting material to produce biogas and associated products (Caro et al., 2023). Despite these advantages, however, it is difficult to deal with waste from breweries and distilleries. The high-water content in spent grains means they would spoil if no further action on them is taken rapidly enough. Moreover, proper fermentation leads to specific microbial strains which implies that complexity comes during conversion processes due to the heterogeneous nature of waste materials.

2.7 Fats, Oils, and Grease Waste: Environmental Impacts and Management

Food preparation processes such as frying, cooking, and processing of oils and fatty foods generate fat, oil, and grease (FOG) waste. With its ability to plug wastewater systems, FOGs waste represents an immense threat to the environment because it leads to expensive infrastructure challenges and environmental degradation (Ahmad et al., 2023). Commercial kitchens, restaurants, and food manufacturing units primarily produce FOG waste. Some common sources include frying oils from deep fryers in restaurants, cooking oils used in food preparation at homes or restaurants as well as residual fats from the meat industry. Even at the household level, we contribute to FOG through our cooking practices that involve either liquid fats or oils. FOG waste consists mainly of triglycerides together with fatty acids and glycerol. The nature of these components varies according to the kind of food being cooked or even depending on which oil is employed. FOG wastes possess a high content of lipids making them a source of major ecological problems because they can be solidified and adhere to pipes of both drainage systems and sewers. The accumulation of FOG within wastewater systems can lead to blockages that result in sewer overflows and backflows. Such situations disrupt wastewater treatment processes by contaminating water bodies with untreated sewage besides posing health threats and environmental hazards. Biofilms can develop in an abundance of FOG in the environment, which can make clogging problems worse (Thomas et al., 2020; Wallace et al., 2017).

Despite environmental challenges, FOG waste can be efficiently utilized by electro-fermentation processes. Under controlled electrical conditions, FOG can be produced into biogas (methane) or biodiesel. The use of electrodes helps to optimize microbial activity, thereby increasing the breakdown of lubricants into volatile lubricants and other organic by-products These changes do not necessarily reduce the impact of FOG waste not only under the effects on the environment but also contribute to the production of renewable energy. The conversion of FOG waste into biodiesel presents a sustainable alternative to fossil fuels. FOG-treated biodiesel can be used in diesel engines, reducing reliance on renewable energy sources. In addition, methane produced from FOG fermentation can be used for energy production, providing renewable energy sources for various applications (Elvis & Karlis, 2018).

2.8 *Municipal Food Waste: Composition and Challenges*

Urban food waste is a complex mixture of food waste from various sources including households, restaurants, and institutions This waste flow poses significant challenges to waste management and the environmental impact is enormous (Wen-Shuenn & Yi-Chen, 2023). Understanding the nature and associated issues is important for developing effective waste management strategies. Urban food waste often includes a variety of organic products. The most common are:

Fruit Peels: These can include the peels from bananas, apples, oranges, and other fruits, and are frequently thrown away after eating. Despite being high in nutrients, they are usually thrown away (Mehdi et al., 2024).

Vegetable Trimmings: Leftovers from vegetables, including stems, leaves, and skins, fall into this group. Though they are frequently deemed inedible, these remnants can be a good source of nutrients and vitamins (Mehdi et al., 2024).

Leftover Cooked Food: After meals, a lot of households and dining establishments produce a lot of leftover food. This can include leftover food, which is frequently thrown out even though it might still be healthy to eat (Mehdi et al., 2024).

Spoiled Food: Municipal food waste is a result of spoilage from faulty storage or depreciation of food. This consists of rotting fruits, moldy bread, and past-due dairy goods (Mehdi et al., 2024).

Discarded Packaging: Food scraps are not inherently waste, but when they are, packaging materials like cardboard, plastic, and metallic containers frequently end up in waste streams with them, making recycling more difficult (Mehdi et al., 2024).

Urban food waste disposal comes with significant environmental impacts. When food waste is sent to landfills, it decomposes anaerobically, releasing methane, a potent greenhouse gas that contributes to climate change. In addition, the presence of food waste in landfills creates sedimentation, which can contaminate groundwater and soil. The heterogeneous nature of urban food waste poses challenges in effective management. Different materials require different treatment methods, making them more difficult to treat properly. For example, some materials can be composted, while others require anaerobic digestion or incineration. Furthermore, the recycling process is complicated by the presence of packaging materials and other inorganic materials. Adding plastic or metal to organic waste can interfere with composting efforts and reduce the quality of the final compost product (Saiphet & Kunta 2023).

Bolster your perseverance and have your vision on! When broken down, municipal food waste can yield very useful resources. Electro-fermentation processes can take this waste and convert it into renewable energy. On the other hand, municipal solid waste streams are heterogeneous and contain different types of substrates which require the use of mixed microbial consortia for efficient breakdown. Different microbial species' metabolic pathways can be optimized through electric stimulation to produce biogas (methane), hydrogen, ethanol, or organic acids from these complex waste streams. The potential applications of municipal food waste are enormous. Local governments will increase their supply of renewable energy by turning scraps of food into biogas thereby reducing dependence on fossil fuels. Organic acids and biofuels such as ethanol could also be removed from it using electro-fermentation methods enhancing circular economy (Saiphet & Kunta 2023; Ghanbarzadeh et al., 2024).

3 Electro-Fermentation in Biofuel Production

EF for biofuel production is a forward-looking process. It combines microbial fermentation with electricity to convert organic matter into such biofuels as ethanol, butanol, hydrogen, methane, and biodiesel. The main advantage of using EF in biofuel production is that it employs electric fields in optimizing the fermentation process. Therefore, this biotechnological approach could be key to achieving sustainable energy production from renewable resources like food waste, agricultural residues, and industrial by-products. Electro-fermentation integrates metabolic pathways involving microorganisms that can be controlled electrochemically. In conventional fermentation, microorganisms break down organic materials (sugars, fats, and proteins) to produce biofuels among others. This process goes through redox reactions through which electron transfer is essential for metabolism. Additionally, adding an external electric field or electrodes enhances the process including control over microbial metabolism, product selectivity, and energy efficiency among others (Bahry et al., 2017; Bender et al., 2022). The biofuel production process using electro-fermentation is shown in Fig. 1.

Providing an external electron donor or acceptor increases microbial metabolism, leading to increased biofuel production. The electric field can divert the production pathways of desired products (e.g., biofuels) and reduce the formation of undesirable by-products, and using electricity to direct fermentation reduces overall energy consumption on, stabilizing biofuel production (Choudhari et al., 2024; Parambath et al., 2023).

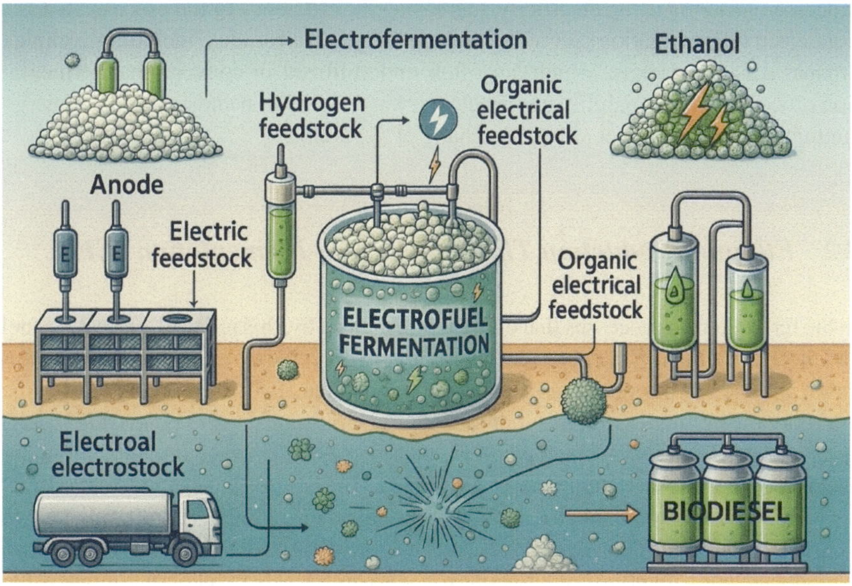

Fig. 1 It shows basic components like a fermentation reactor with organic waste, anode and cathode electrodes of the reactor, yeast and bacteria, etc. The figure also highlights the use of external electrodes to stimulate microbial activity and the production of different biofuels such as ethanol, hydrogen, methane, and biodiesel from the system

3.1 Types of Biofuels Produced by Electro-Fermentation

Electro-fermentation (EF) is a creative biomolecule high-tech process that uses a microbe's metabolic abilities to allow different kinds of substrates such as food wastage to be converted into biofuels effectively. Electrical inputs are incorporated into fermentation processes so that they improve not only the output of biofuels but also enable the production of various types of bioenergy products (Salar-García et al., 2024). The basic principle behind electro-fermentation involves exposing microorganisms to controlled electrochemical conditions that enable their metabolic processes. The application of electric currents can increase electron transport processes within cells hence improving fermentative efficiency. Specifically, this technique allows for the optimization of microbial metabolism leading toward the production of certain types of biodiesels and their by-products. Also, it is possible to optimize the growing environment for bacteria with the help of electricity which translates to transforming substrates of desired compounds (Dhungana et al., 2023).

The use of electro-fermentation for biofuel production offers many advantages over conventional fermentation methods. Higher yields can be achieved and processed faster by using electricity, increasing the efficiency of the process. In addition, electro-fermentation can utilize a wide range of substrates, including complex organic waste materials, which are often underutilized in conventional fermentation processes This capability contributes to waste utilization and encourages energy continuous renewal (Li et al., 2023; Zhang et al., 2023).

3.2 Ethanol Production Through Electro-Fermentation (EF)

In the fermentation processes that ethanol is produced by, this is a well-known biofuel and electro-fermentation enhances its production processes through electrical stimulation and the use of microbial metabolism (Dhungana et al., 2023). Several specific microorganisms are used in this method including *Saccharomyces cerevisiae* (yeast) and *Zymomonas mobilis* which can convert glucose and fructose into ethanol under anaerobic conditions (Dhungana et al., 2023). The microbes selected for electro-fermentation are responsible for transforming sugars into ethanol. Commonly known as baker's yeast, *Saccharomyces cerevisiae* is widely utilized in fermentation because it produces large quantities of ethanol even with high levels of alcohol tolerance. On the other hand, *Zymomonas mobilis* is a bacterium whose fermentation ability is particularly good in terms of converting sugars to ethanol. Glycolysis –the metabolic pathway through which these microorganisms liberate energy from glucose- yields pyruvate that can either be oxidized to acetic acid or reduced to form ethanol when there are no oxygen molecules present (Zhang et al., 2023).

- Anaerobic conditions are important for ethanol production, as they inhibit the formation of sugars that would produce carbon dioxide and water instead of ethanol. Under anaerobic conditions, microbial metabolism continues to ferment, resulting in the accumulation of ethanol as a major by-product (Dhungana et al., 2023).

- The introduction of electricity in electro-fermentation serves several purposes. It helps to maintain the redox balance in bacterial cells, which is essential for maintaining optimal metabolic function. The electric field enhances electron transfer processes, leading to the efficient conversion of substrates to ethanol. Through efficient electron flow, microorganisms can achieve higher ethanol yields compared to conventional fermentation methods (Dhungana et al., 2023).
- Fermentation kinetics can be improved by using electrical stimulation. The electric field can influence microbial community structure, resulting in more efficient microbial specificity in ethanol production. In addition, the controlled electrochemical environment allows for better control of the exogenous products, minimizing the formation of unwanted compounds that can inhibit fermentation (Dhungana et al., 2023).

3.3 Butanol Production Through Electro-Fermentation (EF)

However, butanol is a potential liquid biofuel with a higher energy content than ethanol, thus making it more favorable in several areas such as transport and industry. Butanol can be produced via electro-fermentation (EF) through special microorganisms such as *Clostridium acetobutylicum* known for its ability to carry out the acetone-butanol-ethanol (ABE) fermentation process (Nailwal et al., 2024). *Clostridium acetobutylicum* is an anaerobic, gram-positive bacteria. This organism ferments carbohydrates like glucose into mixtures of acetone, butanol, and ethanol during the ABE fermentation process. This fermentation pathway starts with the breakdown of glucose into pyruvate followed by the conversion of pyruvate into different fermentation products among which is butanol production. The usual ratios for these products differ but one aim remains which involves enhancing butanol generation (Mogollón et al., 2023).

The role of an electric current in electro-fermentation is essential for butanol production enhancement, as it determines the metabolic pathways of *Clostridium acetobutylicum*. The oxidation-reduction potential (ORP) in the fermentation system can be altered by the use of an electric field. Moreover, further importance lies in that it alters the metabolic pathways within *Clostridium acetobutylicum*, since by applying an electric current metabolic balance may shift toward the production of butanol instead of acetone and ethanol. The ORP measures how many electrons are available in a system and therefore influences microbial metabolism. For instance, during ABE fermentation when ORP is higher it results in more conversion of intermediate products to butanol as compared to acetone or ethanol. To increase the amount of butanol produced one needs to control the electric current to have favorable fermentation conditions that would result in a higher yield. This fine-tuning of ORP creates a better environment for the pathway used for the production of Butanol which leads to enhanced efficiency of biofuel production (Nailwal et al., 2024; Zhang et al., 2023).

3.4 Hydrogen Production Through Microbial Electrolysis Cells

The paramount notion of using hydrogen gas (H2) as an energy carrier free of pollution is constantly gaining prominence among individuals who desire cleaner methods of generating power. Biological means through which hydrogen can be produced are presently being looked into; one such method is by employing microbial electrolysis cells (MECs) within electro-fermentation processes. It relies on certain types of bacteria that produce hydrogen to transform organic wastes into gaseous form hence offering alternative renewable energy solutions so much required today. In actual sense, MECS can serve as bioreactors that assist different species of microorganisms in producing hydrogen hence requiring organisms to enter into their metabolic phases. For this reason, organic refuse is used as a raw material that bacteria decompose (Aboelela and Soliman 2024). When external voltage inputs are given, the rate at which organic matters are turned into gaseous hydrogen increases. Some bacteria responsible for producing hydrogen include *Rhodopseudomonas palustris* and *Clostridium butyricum*. These are anaerobic fermenters that convert organic compounds into H2 gases. A key characteristic used to classify *Rhodopseudomonas palustris* is its photosynthesis which utilizes light energy; *Clostridium butyricum* on the other hand, can ferment other compounds such as carbohydrates or carboxylic acids (Gautam et al., 2023).

The application of external energy in MECs is important for increasing hydrogen production. This electrical induction helps to optimize electron transfer in microbial cells. By providing additional energy, the external voltage increases the reducing agents needed to produce hydrogen gas, thereby increasing the total hydrogen yield the electrical supply of MECs provides electrons a movement between bacteria and electrode is increased. This development is important because efficient electron transfer is a key factor in hydrogen enrichment. The presence of external energy allows bacteria to efficiently metabolize organic compounds to hydrogen, yielding higher yields than conventional treatment methods (Gautam et al., 2023; Aboelela & Soliman, 2024).

3.5 Methane Production Through Electro-Fermentation

Methane, or CH4, is an important part of biogas which is a renewable source of energy. Methane production happens mainly through the work of methanogenic archaea and is vital to anaerobic digestion processes. The integration of electrical fields into electro-fermentation has increased the rate of methane production, especially in microbial electrolysis cells (MECs) and microbial fuel cells (MFCs). Some specialized microorganisms known as methanogenic archaea such as *Methanosarcina* and *Methanosaeta* convert organic matter to produce methane. They thrive in anaerobic conditions where substrates like acetate, carbon dioxide, and hydrogen are used for methane production. These organisms' metabolic pathways are very important for an anaerobic digestion process which breaks complex organic materials into simple compounds. While undergoing anaerobic digestion, organic matter will be decomposed by a consortium of microorganisms including hydrolytic bacteria, acidogenic bacteria, and methanogens. The conversion starts from breaking down complex organic materials into simpler types such as organic acids or alcohols that are subsequently converted to methane by methanogenic archaea This multi-step procedure is key in effective biogas production (Liu et al., 2019).

The electric field applied in electro-fermentation supports the electron transfer processes necessary for the production of methane. The induction of electricity helps to optimize the metabolism of methanogenic archaea, facilitating the conversion of substrates to methane. By increasing the electron density, the power unit accelerates the overall production of methane, resulting in higher yields in microbial electrolyte cells (MECs) and microbial fuel cells (MFCs) This new system utilizes microbial activity to gain strength. In MECs, an external energy source is used to drive the production of methane from organic substrates, while in MFCs, microbial metabolism generates electricity directly from organic matter Both systems benefit from enhanced electron transfer simply by electrolysis, thereby efficient methane production (Jia et al., 2020).

3.6 *Biodiesel Production Through Electro-Fermentation*

In place of standard diesel, biodiesel is a green fuel that can be employed. It is derived from lipids, which constitute fats and oils. These lipids may originate from various sources like plants, animals, and even algae. Lipids resemble the elements in the construction of fats; just as bricks are used to construct an edifice. Algae or other fatty substances frequently serve as the sources of lipids for biodiesel production purposes. Tiny aquatic flora known as algae have been discovered to grow heartily with the capacity for producing huge quantities of oil that could be harvested for biodiesel production purposes. If one were to visualize it, then algae might come off as a kind of organic factory that has been specifically engineered to generate oil, which may be converted into fuel (Bhagchandanii et al., 2020). The process of electro-fermentation uses specialized microorganisms. These microorganisms help break down complex fatty acids into simpler components called free fatty acids (FFAs) and glycerol. You can think of microorganisms as little workers that help break down fats into smaller pieces, making them easier to use. Electrical stimulation is used to make fat breakdown more efficient. This means that less electricity is used for the microorganisms to function properly. It appears that some efficiency has been increased to help employees do their jobs faster and more efficiently (Zhang et al., 2023).

The electrical stimulation makes the process more efficient, producing a very stable lubricant. This means that more beneficial parts are derived from fat. Higher yields are important because they mean that more biodiesel can be produced with the same amount of starting material. Once the free fats are produced, they must be converted into biodiesel through a process called transesterification. This is a chemical reaction in which a fatty acid is mixed with an alcohol (such as methanol) and a catalyst (an agent that accelerates the reaction) Think of transesterification as a culinary process in which you mix ingredients new food in this case biodiesel (Choudhari et al., 2024; Parambath et al., 2023).

3.7 Advantages of Electro-Fermentation in Biofuel Production

- When you utilize an outer Electric Field (EF), more productivity is experienced in microbial Metabolism which yields higher biofuels than those derived from normal fermentation processes. The use of an Electrolytic process (EF) ensures superior control over the fermentation conditions such as redox potential, microbial activity, and PH. The approach minimizes unwanted by-products leading to more specificity for the required product; in this case, biofuels (Chandrasekhar et al., 2021).
- Reduced unwanted by-product formation including alcohols or organic acids has been recorded due to the use of electrical stimulation hence increasing process efficiency and reducing costs associated with overall fermentation processes. Electro-fermentation can be scaled easily enough to integrate it with alternative energy sources like wind power and solar energy making it environmentally friendly over a long-term period. Also, it offers an effective method of recycling organic waste thus reducing pollution levels reaching nearby towns. Furthermore, the application of EF through external electricity enhances energy productivity since there would be minimal heat requirements as well as chemical additives that are usually needed in traditional fermentations (Saiphet & Kunta, 2023).
- The addition of electro-fermentation to biofuel production offers many advantages. This not only increases ethanol production but also shortens the fermentation time, increasing the efficiency of the process. Furthermore, the ability to use a wide range of feedstocks, including organic solid waste, makes biofuel production more sustainable. This process helps to destroy waste by converting food scraps and other organic materials into valuable energy sources (Choudhari et al., 2024).
- Compared to ethanol, butanol has many advantages as a biofuel. It has high efficiency, which means it can deliver more energy per unit volume. In addition, butanol is less volatile and has a lower vapor pressure than ethanol, making it easier to handle and store. These properties make butanol a suitable gasoline blend

for use in existing internal engines without substantial modification (Dhungana et al., 2023).

- Hydrogen gas is considered a clean energy carrier because its combustion produces only water as a by-product, making it environmentally friendly. In addition, hydrogen can be used to generate electricity in fuel cells, providing versatile energy solutions. Biological hydrogen production through MEC not only contributes to renewable energy production but also helps in biomass waste management (Nailwal et al., 2024).
- Methane is a valuable renewable energy source that can be used for heating, electricity, and fuel for vehicles. When it burns, it releases carbon dioxide and water, making it cleaner than fossil fuels. The production of methane from natural sources through electro-fermentation not only contributes to the production of renewable energy but also to biomass waste management (Nailwal et al., 2024).

4 Applications of Electro-Fermentation for Biofuel Production from Food Waste

4.1 Renewable Energy Generation from Electro-Fermentation

Electro-fermentation is a promising technology for producing biofuels that can be used in generating renewable energy. Some of the biofuels produced from this process, including ethanol, butanol, and hydrogen are cleaner than conventional fossil fuels. Through electro-fermentation, ethanol, and butanol can be obtained from various organic wastes (Bhattacharya et al., 2024). They can be mixed with gasoline for use in internal combustion engines. This mixture burns cleaner than traditional gas thus reducing emissions into the atmosphere as opposed to the conventional fossil fuels. Thus, it helps to enhance sustainable development by utilizing waste materials and decreasing reliance on fossil resources. Furthermore, hydrogen can also be generated through electro-fermentation which later generates electricity in fuel cells. It is indeed a highly efficient process since water is typically the only product of hydrogen combustion thus making it an eco-friendly energy source. The inclusion of hydrogen alongside electrical systems contributes to building up green hydrogen economies meant to enable an overall lowering in carbon emissions as well as seeking/upholding sustainable energy approaches.

4.2 *Waste Management and Circular Economy Through Electro-Fermentation*

Electro-fermentation plays an important role in waste management and supports the principles of circular economy by converting food waste into valuable biofuels Fruit and vegetable residues, starchy waste, urban food waste used to produce biofuels like ethanol, butanol, and hydrogen in electro-fermentation, etc. Different types of food waste are used This method effectively reduces the amount of waste that would otherwise end up in landfills. Power generation not only solves the critical issue of food waste but also contributes to a sustainable circular economy by converting waste into valuable products, reducing environmental pollution our environment and encouraging more efficient use of resources This new approach is essential for future waste management and renewable energy (Qingshan et al., 2024; Wang et al., 2024).

4.3 *Decentralized Biofuel Production Through Electro-Fermentation*

Decentralized biofuel generation schemes can benefit greatly from electro-fermentation. Food waste can be treated directly at the place of businesses, towns, and farms through electro-fermentation. Thus, this transportation is eliminated thereby reducing costs incurred especially during transportation which are unsafe for the environment. By processing food waste on their premises, these institutions would thus reduce expenses related to transferring the wastes to one plant situated far away which moreover can be economically unsustainable (Wang et al., 2020; Nuohan et al., 2023a, 2023b). Therefore, it becomes less expensive in general to produce biofuels. The carbon footprint attributed to waste transportation would also be reduced with decentralized systems and thereby less landfill usage along with reductions in greenhouse gases resulting from degrading garbage.

Recycling our trash into useful products is one of the functions of decentralized systems. This leads to additional jobs within the area, improvement of quality of life, and diversification of income sources. Decentralized biofuel production can support local economies by creating jobs in waste management and biofuel production; also, farmers can use it as a technique to cope with waste. It stimulates new opportunities for economic growth by developing employment in localities related to solid waste disposal and biofuel creation. Furthermore, it fosters sustainable farming strategies since it enables agrarians to effectively manage their refuse (Choudhari et al., 2024; Parambath et al., 2023).

4.4 Industrial Applications of Biofuels from Electro-Fermentation

Biofuels produced through electro-fermentation have various applications in different industries. These biofuels are valuable in many areas as solvents and fertilizers. It has been used in medicine and as an ingredient in pharmaceuticals. Acts as an ingredient in personal care products, increasing product stability and effectiveness. Acts as a feedstock for various chemicals, and contributes to the production of plastics and other materials (Virdis et al., 2022). Methane can be burned to produce electricity, providing a sustainable energy source compared to coal. Hydrogen is essential in cleaning processes such as water splitting and can also be used for heating, increasing energy efficiency. The use of these biofuels helps reduce greenhouse gas emissions and dependence on fossil fuels promotes a cleaner environment and supports efforts to do things that have remained constant across industries (Rouf et al., 2024).

4.5 Environmental Sustainability Through Electro-Fermentation

Electro-fermentation that turns food waste products into biofuels has massive benefits for the environment. The strategy prevents methane output through the food waste because it is never sent to a dump. For this reason, the overall greenhouse gas emissions are decreased which is vital in reducing global warming effects. The biofuels that are formed from these processes include such items as ethanol and methane which act as renewable sources of energy (Yıldız & Öztekin, 2024). They replace fossil fuels resulting in an energy landscape that is more sustainable decreasing reliance on non-renewable resources. With its utilization of waste materials, electro-fermentation reduces the environmental pollution of energy generation. Such a system not only aids in waste management work but also reduces its ecological impact as compared to usual means of generating electricity. In doing so, it provides a low-carbon energy source that fits into an attempt made at the world level toward climate change combating strategy. It embraces programs that aim at cutting down carbon dioxide emission rates while promoting less polluted forms of electric power generation (Economou et al., 2024; Petrescu-Mag et al., 2024). The applications of electro-fermentation of food waste for biofuel generation in shown in Fig. 2.

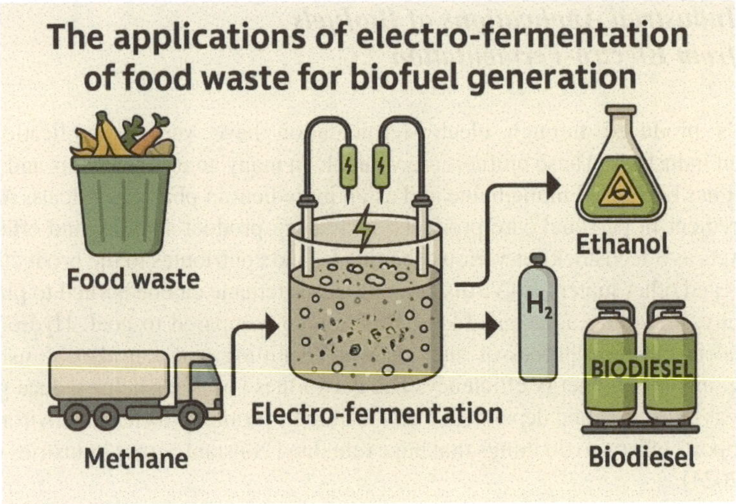

Fig. 2 Illustration of electro-fermentation of food waste for biofuel production. The visualization visually connects the electro fermentation process to a variety of applications, using bright and vivid colours to establish the flow from food waste to useful biofuels and other by-products

5 Conclusion

The electrolysis of food waste into biofuels represents an unprecedented advance in waste management and renewable energy production. This innovative process harnesses the synergy between microbial metabolism and electrical stimulation, providing an efficient and sustainable method for converting organic waste into biofuels valuable as the world faces increasing pressure from the accumulation of waste and depletion of fossil fuels, electrolysis appears as a promising solution with principles reconciling the circular economy with environmental sustainability. The constraints imposed on microorganisms by their metabolic limitations characteristically restrict conventional fermentation techniques. In contrast, electro-fermentation bypasses these restrictions through the application of an external electric field to alter redox balance inside microorganisms' environments. This control allows for the optimization of several metabolism pathways hence leading to larger amounts or yields of necessary biofuels such as ethanol, butanol, hydrogen, and methane. Through the directional manipulation of microbial metabolism toward defined end products, electro-fermentation has turned out to be one versatile and effective means of producing biofuel. One major advantage associated with this technique is its ability to employ diverse forms of food waste including vegetable scraps, fruits, fatty matter such as cooking oil, and FOG waste. This flexibility in raw material use not only enhances biosynthesis but also helps in lessening waste disposal challenges. The most important aspect is how electro-fermentation transforms organic refuse into power-laden gas or liquid which diminishes strains on landfills while at the same time curbing climate change through greenhouse gas release from organic materials degradation. Biofuels such as ethanol, hydrogen, methane, and butanol can replace fossil fuels to reduce pollution and provide clean energy for transportation and electricity generation electro-fermentation of food waste helps the environment through garbage and methane reduction, waste is bio-useful. Oil also provides economic opportunities. Electro-fermentation systems can be developed for diverse scaling from small decentralized units to larger industrial installations. This scalability allows different contexts such as rural areas; urban centers or industry to adopt the technology. In this sense, decentralized biofuel production systems are very much needed in remote and underserved areas with locally available renewable energy sources that can also address waste management problems. While electro-fermentation has tremendous promise it requires the resolution of some challenges before mass acceptance including optimization of energy efficiency for the process; price reduction; as well as an understanding of intricate relationships between microbial communities and electrochemical inputs. Therefore, more research in this field helps to solve these problems to exploit electro-fermentation fully. To sum up, the conversion of food waste through electrical fermentation into biocides relates to waste management; eco-friendliness, and renewable energy production. By transforming unusable food remnants into valuable liquid fuels, such a revolutionary approach provides a map for a sustainable future where less waste will exist at last more renewable sources should be advocated. As the technology matures, electrolysis will play a key role in the global transition to a circular economy and a zero-carbon energy system.

References

Ayele, A., Masi, C., Abda, E. M., & Korsa, G. (2024). Wastes from fruits and vegetables processing industry for value-added products. https://doi.org/10.1007/978-3-031-48143-7_7

Aboelela, D., & Soliman, M. A. (2022). Hydrogen production from microbial electrolysis cells powered with microbial fuel cells. *Journal of King Saud University-Engineering Sciences*. https://doi.org/10.1016/j.jksues.2022.05.008

Ahmad, I., Abdullah, N., Koji, I., Yuzir, A., Ahmad, M. D., Rachmadona, N., Al-Dailami, A., Show, P. L., & Khoo, K. S. (2023). Micro and macro analysis of restaurant wastewater containing fat, oil, grease (FOG): An approach based on prevention, control, and sustainable management. *Chemosphere, 325*, 138236. https://doi.org/10.1016/j.chemosphere.2023.138236

Al-Gheethi, A. (2019). Establish in-house: A pre-treatment method of fat, oil and grease (FOG) in kitchen wastewater for safe disposal. *International Journal of Integrated Engineering, 11*(2). https://doi.org/10.30880/ijie.2019.11.01.018

Alzahrani, F., Akanbi, T. O., Scarlett, C. J., & Aryee, A. N. (2024). The use of immobilised enzymes for lipid and dairy processing and their waste products: A review of current progress. *Processes, 12*(4), 634. https://doi.org/10.3390/pr12040634

Bahry, H., Pons, A., Abdallah, R., Pierre, G., Delattre, C., Fayad, N., Taha, S., & Vial, C. (2017). Valorization of carob waste: Definition of a second-generation bioethanol production process. *Bioresource Technology, 235*, 25–34. https://doi.org/10.1016/j.biortech.2017.03.056

Bender, L. E., Lopes, S. T., Gomes, K. S., Devos, R. J. B., & Colla, L. M. (2022). Challenges in bioethanol production from food residues. *Bioresource Technology Reports, 19*, Article 101171. https://doi.org/10.1016/j.biteb.2022.101171

Bhagchandanii, D. D., Babu, R. P., Sonawane, J. M., Khanna, N., Pandit, S., Jadhav, D. A., Khilari, S., & Prasad, R. (2020). A comprehensive understanding of electro-fermentation. *Fermentation, 6*(3), 92. https://doi.org/10.3390/fermentation6030092

Bhattacharya, I. N., & Sahiba, Suhag, M. (2024). Sustainable agricultural waste management practices and approaches: An overview. https://doi.org/10.58532/v3bcag2p2ch3

Caro, I., Portales, S., & Gómez, M. (2023). Microbial characterization of discarded breads. *LWT, 173*, 114291. https://doi.org/10.1016/j.lwt.2022.114291

Cerciello, M., Agovino, M., & Garofalo, A. (2019). Estimating food waste under the FUSIONS definition: What are the driving factors of food waste in the Italian provinces? *Environment, Development and Sustainability, 21*, 1139–1152. https://doi.org/10.1007/s10668-017-0080-0

Chandrasekhar, K., Amulya, K., & Mohan, S. V. (2015). Solid phase bio-electrofermentation of food waste to harvest value-added products associated with waste remediation. *Waste Management, 45*, 57–65. https://doi.org/10.1016/j.wasman.2015.06.001

Chandrasekhar, K., Kumar, A. N., Kumar, G., Kim, D. H., Song, Y. C., & Kim, S. H. (2021). Electro-fermentation for biofuels and biochemicals production: Current status and future directions. *Bioresource Technology, 323*, Article 124598. https://doi.org/10.1016/j.biortech.2020.124598

Choudhari, J. K., Anandkumar, J., Choubey, J., & Sahariah, B. P. (2024). Molecular Mechanisms behind the plant/microbe induced biofuel production: Systems biology approach. In *Emerging Sustainable Technologies for Biofuel Production* (pp. 371–399). Cham: Springer Nature Switzerland. https://doi.org/10.1007/978-3-031-52167-6_16

Dar, R. A., Tsui, T.-H., Zhang, L., Tong, Y. W., Sharon, S., Shoseyov, O., & Liu, R. (2024). Fermentation of organic wastes through oleaginous microorganisms for lipid production—Challenges and opportunities. *Renewable & Sustainable Energy Reviews*. https://doi.org/10.1016/j.rser.2024.114328

Dhungana, P., Prajapati, B., Bhatt, P., Regmi, D., Yadav, M., Maharjan, S., Lamsal, U., Kathariya, S., Chaudhary, P., & Joshi, J. (2023). Production of bioethanol from Saccharum spontaneum by simultaneous saccharification and electro-fermentation using mixed culture of microbes. *Biofuels, 14*(2), 191–199. https://doi.org/10.1080/17597269.2022.2126083

Dymchenko, A., Geršl, M., & Gregor, T. (2023). Trends in bread waste utilisation. Trends in Food Science & Technology, 132, 93–102.132:93–102. https://doi.org/10.1016/j.tifs.2023.01.004

Economou, F., Voukkali, I., Papamichael, I., Phinikettou, V., Loizia, P., Naddeo, V., Sospiro, P., Liscio, M. C., Zoumides, C., Tirca, D. M., & Zorpas, A. A. (2024). Turning food loss and food waste into watts: A review of food waste as an energy source. *Energies, 17*(13), 3191. https://doi.org/10.3390/en17133191

Elvis, K., & Karlis, S. (2018). Problems with fat, oil, and grease (FOG) in food industry wastewaters and recovered FOG recycling methods using anaerobic co-digestion: A short review. *Key Engineering Materials, 762*, 61–68. https://doi.org/10.4028/WWW.SCIENTIFIC.NET/KEM.762.61

Ezejiofor, T. I. N., Enebaku, U. E., & Ogueke, C. (2014). Waste to wealth-value recovery from agro-food processing wastes using biotechnology: A review. *British Biotechnology Journal, 4*(4), 418–481. https://doi.org/10.9734/BBJ/2014/7017

Gallón-Bedoya, M., Cortés-Rodríguez, M., Gil-González, J. H., Guil-Guerrero, J. L., & Ortega-Toro, R. (2024). Valorization of Andean berries through whole-fruit powder development. *Applied Food Research, 4*(2), Article 100447. https://doi.org/10.1016/j.afres.2024.100447

Gambino, S., & Cammarata, L. (2017). Tilt measurements on volcanoes: More than a hundred years of recordings. *Italian Journal of Geosciences, 136*(2), 275–295. https://doi.org/10.3301/IJG.2017.07

Gautam, R., Nayak, J. K., Ress, N. V., Steinberger-Wilckens, R., & Ghosh, U. K. (2023). Biohydrogen production through microbial electrolysis cell: Structural components and influencing factors. *Chemical Engineering Journal, 455*, Article 140535. https://doi.org/10.1016/j.cej.2022.140535

Ghanbarzadeh Lak, M., Ghaffariraad, M., & Jahangirzadeh Soureh, H. R. (2024). Characteristics and impacts of municipal solid waste (MSW). Springer water. https://doi.org/10.1007/978-3-031-52633-6_2

Hameed, A., Anwar, M. J., Perveen, S., Amir, M., Naeem, I., Imran, M., Hussain, M., Ahmad, I., Afzal, M. I., Inayat, S., & Awuchi, C. G. (2023). Functional, industrial and therapeutic applications of dairy waste materials. *International Journal of Food Properties, 26*(1), 1470–1496. https://doi.org/10.1080/10942912.2023.2213854

Jayasree Joshi, T., Harsha, V., Nandagopal, P., Ashok, A., & Pokhrel, S. (2024). Value addition and sustainable management of dairy industry byproducts. In *Value added products from food waste* (pp. 19–38). Cham: Springer Nature Switzerland. https://doi.org/10.1007/978-3-031-48143-7_2

Jia, X., Li, M., Wang, Y., Wu, Y., Zhu, L., Wang, X., & Zhao, Y. (2020). Enhancement of hydrogen production and energy recovery through electro-fermentation from the dark fermentation effluent of food waste. *Environmental Science and Ecotechnology, 1*, Article 100006. https://doi.org/10.1016/j.ese.2019.100006

Jun-Sil, C., Yunjeong, C., Huiju, P., Danbee, K., & Gahyun, B. (2024). Combining pre-fermentation and microbial electrolysis for efficient hydrogen production from food wastewater. *Process Safety and Environmental Protection*. https://doi.org/10.1016/j.psep.2024.05.062

Kaur, M., Singh, A. K., & Singh, A. (2023). Bioconversion of food industry waste to value added products: Current technological trends and prospects. *Food Bioscience,* 102935. https://doi.org/10.1016/j.fbio.2023.102935

Kumar, V., Brancoli, P., Narisetty, V., Wallace, S., Charalampopoulos, D., Dubey, B. K., Kumar, G., Bhatnagar, A., Bhatia, S. K., & Taherzadeh, M. J. (2023). Bread waste–a potential feedstock for sustainable circular biorefineries. *Bioresource Technology, 369*, 128449. https://doi.org/10.1016/j.biortech.2022.128449

Li, K., Zhang, J. W., Li, F., Zhao, X. Q., Bai, F. W., Song, H., & Liu, C. G. (2023). Electrochemical control of cell metabolism improves ethanol production of Zymomonas mobilis in an electro-fermentation system. *ACS Sustainable Chemistry & Engineering, 11*(6), 2364–2374. https://doi.org/10.1021/acssuschemeng.2c06188

Li, W., Han, Y., Zhang, Z., Cai, T., Wang, J., Gao, T., Lu, Z., & Zhen, G. (2024). Arousing the bioelectroactivity and syntrophic metabolic functionality of microorganisms in a recirculated two-phase anaerobic digestion bioreactor for enhanced biohythane recovery from high-solids biowaste. *Chemical Engineering Journal,* 154321. https://doi.org/10.1016/j.cej.2024.154321

Liu, S., Deng, Z., Li, H., & Feng, K. (2019). Contribution of electrodes and electric current to process stability and methane production during the electro-fermentation of food waste. *Bioresource Technology, 288*, Article 121536. https://doi.org/10.1016/j.biortech.2019.121536

Lutfi, K., Maryudi, M., & Erna, A. (2024). Utilization of tofu liquid waste as liquid organic fertilizer using the fermentation method with activator effective microorganisms 4 (EM-4): A review. *Ekuilibrium*. https://doi.org/10.20961/equilibrium.v8i1.84056

Modestra, J. A., Reddy, C. N., Krishna, K. V., Min, B., & Mohan, S. V. (2020). Regulated surface potential impacts bioelectrogenic activity, interfacial electron transfer and microbial dynamics in microbial fuel cell. *Renewable Energy, 149*, 424–434. https://doi.org/10.1016/j.renene.2019.12.018

Mogollón, C. A. G., Díaz, J. C. Q., & Posada, J. O. G. (2023). Production of acetone, butanol, and ethanol by electro-fermentation with Clostridium saccharoperbutylacetonicum N1–4. *Bioelectrochemistry, 152*, Article 108414. https://doi.org/10.1016/j.bioelechem.2023.108414

Nailwal, V., Mukherjee, T., & Mohan, S. V. (2024). Decoupling acidogenic cascade via bio-electrogenic induced ABE pathway for enhanced n-butanol synthesis in Clostridium acetobutylicum ATCC-824: Bioenergetics and gene expression analysis. *Chemical Engineering Journal, 486*, Article 150015. https://doi.org/10.1016/j.cej.2024.150015

Noordam, S. J., Madjarov, J., Louro, R. O., Paquete, C. M. (2024). Electroactive (Micro) organisms (pp. 51–77). https://doi.org/10.1007/978-3-031-54306-7_4

Nuohan, W., Ming, G., Shuo, L., Xiaona, W., & Qunhui, W. (2023a). Electrochemical promotion of organic waste fermentation: Research advances and prospects. *Environmental Research*. https://doi.org/10.1016/j.envres.2023.1174224

Nuohan, W., Ming, G., Shuo, L., Xiaona, W., & Qunhui, W. (2023b). Electrochemical promotion of organic waste fermentation: Research advances and prospects. *Environmental Research*. https://doi.org/10.1016/j.envres.2023.117422

Özcan, B., Tetik, N., & Aloğlu, H. Ş. (2024). Polysaccharides from fruit and vegetable wastes and their food applications: A review. *International Journal of Biological Macromolecules*. https://doi.org/10.1016/j.ijbiomac.2024.134007

Parambath, J. B., Abla, F., Ahmad, A. A., Kanan, S. M., & Mohamed, A. A. (2023). Biocatalysts for biofuels production. In *Bioenergy engineering* (pp. 409–425). Woodhead Publishing. https://doi.org/10.1016/B978-0-323-98363-1.00005-3

Petrescu-Mag, R. M., Petrescu, D. C., Ajtai, I, Roba, C., Gică, O. A., Cuibus, L., Tenter, A. R., Toader, V., Negruşa, A., & Bican-Brişan, N. (2024). Causes and solutions for fruit and vegetable waste: A participatory approach with Romanian farmers for sustainable agriculture. *International Journal of Agricultural Sustainability*. https://doi.org/10.1080/14735903.2024.2329391

Lin, Q., Xi, S., Cheng, B., Jiang, J.-C., Zan, F., Tang, Y., Li, Y., Khanal, S. K., Wang, Z., Chen, G., & Guo, G. (2024). Electrogenerated singlet oxygen and reactive chlorine species enhancing volatile fatty acids production from co-fermentation of waste activated sludge and food waste: The key role of metal oxide coated electrodes. *Water Research*. https://doi.org/10.1016/j.watres.2024.121953

Rago, L., Pant, D., & Schievano, A. (2019). Electro-fermentation—microbial electrochemistry as new frontier in biomass refineries and industrial fermentations. In *Advanced bioprocessing for alternative fuels, biobased chemicals, and bioproducts* (pp. 265–287). Woodhead Publishing. https://doi.org/10.1016/B978-0-12-817941-3.00014-0Get rights and content

Roy, M., & Das, S. (2023). Microbial electrosynthesis: A biobased pathway for the production of value-added chemicals through carbon sequestration. https://doi.org/10.1002/9783527839001.ch15

Saini, P., & Mishra, P. (2024). Microbial nanowires: Future of bioenergy applications. In *Bioprospecting of microbial resources for agriculture, environment and bio-chemical industry* (pp. 193–208). Cham: Springer Nature Switzerland. https://doi.org/10.1007/978-3-031-63844-2_12

Saiphet, A., & Kunta, K. (2023). Municipal food waste management and its greenhouse gas emission in Thailand: A Case study of 8 municipalities. *Environmental Research, Engineering and Management*. https://doi.org/10.5755/j01.erem.79.3.33014

Salar-García, M. J., Ortiz-Martínez, V. M., Sánchez-Segado, S., Valero Sánchez, R., Sáez López, A., Lozano Blanco, L. J., & Godínez-Seoane, C. (2024). Sustainable production of biofuels and biochemicals via electro-fermentation technology. *Molecules, 29*(4), 834. https://doi.org/10.3390/molecules29040834

Sharma, P., Gaur, V. K., Kim, S. H., & Pandey, A. (2020). Microbial strategies for bio-transforming food waste into resources. *Bioresource Technology, 299*, Article 122580. https://doi.org/10.1016/j.biortech.2019.122580

Siddiqui, S. A., Erol, Z., Rugji, J., Taşçı, F., Kahraman, H. A., Toppi, V., Musa, L., Di Giacinto, G., Bahmid, N. A., Mehdizadeh, M., & Castro-Muñoz, R. (2023). An overview of fermentation in the food industry-looking back from a new perspective. *Bioresources and Bioprocessing, 10*(1), 85. https://doi.org/10.1186/s40643-023-00702-y

Sravan, J. S., Tharak, A., Modestra, J. A., Chang, I. S., & Mohan, S. V. (2021). Emerging trends in microbial fuel cell diversification-Critical analysis. *Bioresource Technology, 326*, Article 124676. https://doi.org/10.1016/j.biortech.2021.124676

Thakur, S., Das, B., & Das, S. (2024). Microbial electrosynthesis of fatty acids and industrial chemicals through carbon sequestration (pp. 279–292). https://doi.org/10.1016/b978-0-443-18820-6.00014-x

Thomas, C., Rachel, C., Bruce, J., Raffaella, V., & Raffaella, V. (2020). Characterisation and energy assessment of fats, oils and greases (FOG) waste at catchment level. *Waste Management, 103*, 399–406. https://doi.org/10.1016/J.WASMAN.2019.12.040

Todd, E. C. D., & Faour-Klingbeil, D. (2024). Impact of food waste on society, specifically at retail and foodservice levels in developed and developing countries. *Foods, 13*(13), 2098. https://doi.org/10.3390/foods13132098

Upadhyay, D. (2024). Food fermentation as a biotechnological process. https://doi.org/10.58532/v3bkbt21p1ch4

Vira, H. (2024). The effect of fermentation modes on the efficiency of organic waste treatment in batch bioreactors. *Archives of Environmental Protection*. https://doi.org/10.24425/aep.2024.149434

Virdis, B., Hoelzle, R. D., Marchetti, A., Boto, S. T., Rosenbaum, M. A., Blasco-Gómez, R., Puig, S., Freguia, S., & Villano, M. (2022). Electro-fermentation: sustainable bioproductions steered by electricity. *Biotechnology Advances, 59*, 107950. https://doi.org/10.1016/j.biotechadv.2022.107950

Wallace, T., Gibbons, D., O'Dwyer, M., & Curran, T. P. (2017). International evolution of fat, oil and grease (FOG) waste management—A review. *Journal of Environmental Management, 187*, 424–435. https://doi.org/10.1016/j.jenvman.2016.11.003

Wang, Q., Li, H., Feng, K., & Liu, J. (2020). Oriented fermentation of food waste towards high-value products: A review. *Energies, 13*(21):5638. https://doi.org/10.3390/EN13215638

Wang, N., Gao, M., Liu, S., Zhu, W., Zhang, Y., Wang, X., Sun, H., Guo, Y., & Wang, Q. (2023). Electrochemical promotion of organic waste fermentation: Research advances and prospects. *Environmental Research*, 117422. https://doi.org/10.1016/j.envres.2023.117422

Wang, C., Feng, D., Xia, A., Nizami, A. S., Huang, Y., Zhu, X., Liao, Q., & Murphy, J. D. (2024). A comparative life cycle assessment of electro-anaerobic digestion to evaluate biomethane generation from organic solid waste. *Renewable and Sustainable Energy Reviews, 196*, 114347. https://doi.org/10.1016/j.rser.2024.114347

Wen-Shuenn, D., & Yi-Chen, L. (2023). (1) Determinants of municipal-level household food waste. *Agribusiness*. https://doi.org/10.1002/agr.21878

Yıldız, A. Y., & Öztekin, S. (2024). Utilization of fruit and vegetable wastes in food packaging. https://doi.org/10.1201/9781003453277-9

Zhang, Y., Li, J., Yong, Y. C., Fang, Z., Yan, H., Li, J., & Meng, J. (2023). Highly selective butanol production by manipulating electron flow via cathodic electro-fermentation. *Bioresource Technology, 374*, Article 128770. https://doi.org/10.1016/j.biortech.2023.128770

Chapter 8
Electro-Fermentation for Treatment of Solid Waste for the Synthesis of High Value-Added Product Generation

Priyanka, Devendra Pratap Singh, Tanya Raheja, Deepka Umrao, Sharanjeet kaur, Mushkaan Duggal, Khusboo Gupta, Surajbhan Sevda, and Meena Krishania

1 Overview of Solid Waste Management Challenges

Solid waste management is a critical aspect of urban planning and environmental sustainability. The increasing population in developing countries has resulted in substantial increase in municipal solid waste (MSW) generation. The improper collection, transportation, disposal, and treatment are the major issues faced by India for municipal waste management (Prajapati et al., 2021). MSW is categorized into degradable and non -degradable waste. Paper, textiles, food scraps, straw, and yard waste are examples of degradable waste; non-biodegradable waste includes things like leather, plastic waste, rubbers, metallic substances, ash of coal, wood, and electronic waste. Traditional methods of waste disposal like open dumping and burning have led to various environmental problems. In India only 21% of MSW is treated while the remaining waste is disposed of in unsanitary landfill sites without recycling and treatment technologies. The improper waste disposal can lead to pollution of air, water and soil. By 2050, there will be 3.4 billion tons of municipal waste produced worldwide, a rise of 70% over 2016 levels (Kaza et al., 2018). According to estimations by 2050, trash created on daily per capita is likely to climb by 19% in developed nations and by 40% or more in developing and underdeveloped countries. According to a study India ranks among the top ten countries that contribute to major MSW worldwide. The recent data on solid waste from the center pollution control

Priyanka · D. P. Singh · T. Raheja · D. Umrao · S. kaur · M. Duggal · K. Gupta · M. Krishania (✉)
Center of Innovative and Applied Bioprocessing (DBT-CIAB), Mohali, India
e-mail: meena@ciab.res.in

S. Sevda
Environmental Bioprocess Laboratory, Department of Biotechnology, National Institute of Technology, Warangal, Telangana, India
e-mail: sevdasuraj@nitw.ac.in

© The Author(s), under exclusive license to Springer Nature Singapore Pte Ltd. 2025

S. Sevda (ed.), *Electro-Fermentation for Synthesis of Chemicals, Biochemicals and Biofuels*, https://doi.org/10.1007/978-981-96-6358-3_8

board, New Delhi, shows that 160.038 TPD (tons per day) of waste is produced by nation with overall estimated efficiency of 95.4% of collected waste with 152.74 kTPD (kilotons per day). The 79.95 kTPD of collected waste is treated and 29.42 kTPD is dumped into landfills. In India, the highest rates of waste are generated by residents of Delhi following Lakshadweep and Mizoram (Sk et al., 2023).

Solid waste is categorized into different types like municipal solid waste, industrial waste, biomedical waste, construction waste, electronic waste, hazardous waste, and agricultural waste based on the source of generation. Municipal waste consists of organic waste, plastics, papers, glass and metals generated from households, hotels, markets, offices, institutions, and other non-industrial sources. Industrial waste comprises hazardous and non-hazardous waste like solvents, chemicals, metals, and byproducts generated from industries. Construction waste consists of materials such as bricks, concrete, wood, metals, plastics, and packaging waste from construction sites. Electronic waste known as e-waste includes discarded or damaged electronic devices like computers, mobile phones, televisions, and electrical gadgets. Hazardous waste includes toxic chemicals, pesticides, and radioactive materials that interferes with human health and poses serious risks to the environment contributing to its physical and chemical properties. Agricultural wastes include waste generated from agricultural activities like crop residues, animal waste, agrochemicals, and packaging materials used in farming (Devi et al., 2014).

India needs comprehensive waste management techniques for garbage reduction, collection, separation, treatment, and disposal. In addition to resource recovery, energy generation, and employment, the use of sustainable waste management techniques has helped to lower pollution and health concerns. Present-day solid waste management strategies focus on reducing trash production, optimizing resource recovery, and appropriately disposing of leftover garbage (Sharma et al., 2021). The major challenges faced by India for solid waste management are the lack of infrastructure and adequate resource constraints. The inadequate waste collection vehicles, treatment, and disposal facilities pose a major setback to solid waste management. Solid waste management procedures are hampered by inadequate infrastructure for effective collection, transportation, and treatment. Another significant obstacle to solid waste management is the absence of adequate infrastructure for recycling. Financial limitations also prevent waste management infrastructure and techniques from being developed and maintained. Due to a lack of funding, waste management programs cannot be carried out effectively. Another factor impeding the solid waste management procedures is technical limitations. The lack of specialized technological knowledge and skills hinders solid waste management practices. Moreover, the shortage of skilled labor and experts' personnel in the waste management fields pose a significant challenge in adoption and implementation of innovative waste management practices. Insufficient human resources are also a major challenge in solid waste management. The lack of efficient trained personnel like waste collectors, supervisors, and administrative staff affects the efficient waste management practices. Thus, insufficient infrastructure and resources directly impact the efficacy of waste management practices (Mir et al., 2021).

The institutional and governances also play a role in solid waste management. The lack of proper coordination among roles and responsibilities among various stakeholders, government agencies, municipalities and other private entities leads to poor waste management practices (Robins et al., 2011). Improper waste management has adverse environment and social effects. The improper waste treatment methods like dumping and landfilling adversely affects public health.

2 Electro-Fermentation as a Novel Approach

Electro-fermentation (EF) is a revolutionary method, uses electrodes to control the electrical metabolism of microbial fermentation. The process is a hybrid technology that utilizes electrodes during fermentation process to influence microbial metabolism resulting in enhanced production rates and modified product profiles. The worldwide release of greenhouse gases has been caused by the growing need for energy and the extraction of fossil fuels for energy generation. Alternately, low-value organic substrates like acetate, lactate caproate, and valerate can be utilized to generate precursor fuels like C_2H_5OH, CH_3OH, CH_4, H_2, and other compounds using biocatalyst-like enzymes and microorganisms (bacteria, fungi, and algae) (Mohan et al., 2016). The conventional method used for fermentation that is acidic fermentation has the limitation of feedback inhibition for the production of carbon neutral precursors like hydrogen from various organic substrates. In feedback inhibition, the physiological buffering capacity of microbial cells gets altered due to the accumulation of products from substrates that affects the growth and metabolism of microorganism (Kengen et al., 2009). The researchers have proposed a new integrated approach in which various microorganism work synergistically within a complex biological system to improve the metabolic and thermodynamic stability of the process. For example, H_2 was produced by integration of acidic fermentation with photo-fermentation, purple non-sulfur bacteria and electroactive bacteria in a bioelectrochemical system to facilitate the production of organic acid (Sriram et al., 2022).

EF utilizes the electrode as electron scavenger that facilitates the production of oxidized and reduced products simultaneously differentiating it from conventional fermentation process (Ikeda et al., 2021a, 2021b).The integration of EF into microbial electrolysis cells facilitated the production of hydrogen and other compounds from various waste streams leading to improved bioethanol production, waste recycling, and environment protection through the conversion of lignocellulosic materials into biofuels (Wang et al., 2023a, 2023b). Anode electro-fermentation and cathode electro-fermentation are two additional categories for EF. By oxidizing organic materials at the anode, electrons in anode EF create a proton gradient that improves biomass production and ATP synthesis while also controlling the redox state of cells. The oxidized products like acetate, 2-ketoglutonate, and hydrogen are mainly produced by anode electro-fermentation where anode acts as electron acceptor that drives electrodes toward product synthesis at cathode (Selvasmbian et al., 2022).

In cathode electro-fermentation, cathode donates the received electrons from anode to produce a pool of electrons for synthesis of redox cofactors like NADH and additional ATP (Kracke et al., 2014, Choi et al., 2014). Although the conventional method of anaerobic digestion for process of hydrolysis and methanogenesis has been successful in industrial-based applications, but due to restriction in intracellular redox imbalance and low product yield in large-scale applications, it remains a critical process. To overcome the restrictions in acidic fermentation integrating it with electric potential could be a promising approach to improve intracellular metabolism and redox balance for desired product yield in large-scale applications. The research gaps in electro-fermentation need continuous efforts to make it commercially and economically viable technology for real-life applications.

In order to improve metabolic pathways and creation of products, microbial EF technology integrates microbial activities with electrochemical principles. Using exoelectrogenic bacteria that produce energy and other substances, this innovative technique transfers electrons outside of cells. According to Bhagchandanii et al., (2020), the application of electric field via EF can enhance metabolic processes, regulate redox and pH imbalance, and promote the creation of certain products, ultimately resulting in increased output and biomass yield. Additionally, by combining electrochemistry with biotechnology, microbial electrosynthesis effectively produces chemicals and fuels using renewable resources and regenerative energies (Fruehauf et al., 2020).

2.1 Electrochemical and Microbial Processes in Electro-Fermentation

The EF enables direct control of redox states and metabolic pathways by fusing the ideas of electrochemistry with the fermentative metabolism of microbes (Bhagchandanii et al., 2020; Tribhuvan et al., 2022). By employing electrodes as sources or sinks of electrons, EF optimizes microbial activity and enhances biomass yield and product selectivity (Haidar et al., 2022). This technique offers a sustainable alternative to existing chemical processes by reducing their environmental impact and carbon footprint (Tribhuvan et al., 2022). Understanding the fundamentals of electron and mass transfer in electrochemistry, which is necessary to carry out electrochemical transformations in organic synthesis, provides the theoretical foundation for EF applications (Scheremetjew et al., 2021). Furthermore, bioelectrochemical techniques like microbial fuel cells and microbial electrolysis cells highlight the potential for efficient pollutant removal and chemical-free treatment methods, demonstrating the practical applications of redox reactions in industrial fermentation and wastewater treatment (Beschkov & Razkazova-Velkova, 2021). Understanding electrochemical processes—which involve many electrodes and reactions—is crucial for understanding fundamental ideas. Platinum and alloys containing platinum are strongly catalyzed reactions, including oxygen reduction and hydrogen evolution.

Single-crystal electrode preparation methods facilitate the research of surface reactivity and interfacial behavior of gold is easier to comprehend (Climent & Herrero, 2020). Charge transfer reactions are explained theoretically, and important variables like half-wave potential and limiting current are calculated by this (González et al., 2023). Voltammetry and impedance measurements are used to study topotactic solid-to-solid interconversions and phase changes that occur in solid-state processes related to cultural heritage (Doménech-Carbó & Doménech-Carbó, 2023).

Some value-added compounds have been produced commercially in the food and chemical sectors by microbial fermentation techniques. Fermentation processes can still only produce a few number of chemical species, despite the fact that metabolic engineering and genome editing allow for the free creation of microbial metabolic pathways at the genomic level (Yamada et al., 2022). By combining the principles of electrochemical cells with microbial biological reactions, microbial fermentation in EF allows for improved control over pH and redox states and improves product yields and quality by regulating fermentative metabolism under an electric potential (Tribhuvan et al., 2022). Without appreciable increase in the yield of ethanol and 2, 3-butanediol, this procedure can be used for a variety of microorganisms, such as *Clostridium autoethanogenum,* where external potentials and redox mediators affect biomass and metabolic pathways, including lactate and acetate (Martínez-Ruano et al., 2023). In order to optimize desired product outputs, EF models suggest dynamically adjusting electrode potentials (Haidar et al., 2022). However, in order to implement successful control techniques, a greater comprehension of the rates at which fermentation and EF behavior transition is requires. Further evidence of the potential of EF to optimize complex microbial processes like rumen fermentation comes from the use of an electric potential during rumen fermentation, which boosts microbial biomass, dry matter absorption, production of volatile fatty acids, and drops the NADH/NAD ratio (López-Hernández et al., 2023).

Clostridium autoethanogenum (Martínez-Ruano et al., 2023), *Zymomonas mobilis* (Li et al., 2023), *Saccharomyces cerevisiae, Aspergillus niger, Ganoderma sessile, Bacillus licheniformis, Bacillus pulmilus, Bacillus sp.*, and *Aeribacillus sp.* are significant microorganisms that are employed in EF. Research has looked into these species' ability to exploit EF processes to produce a range of chemicals, such as ethanol, alcohols, organic acids, and thermostable lipases (Dhungana & Rajbhandari, 2022; Garuba et al., 2020). Bhagchandanii et al., (2020) state that EF improves yields of products and metabolic pathways by regulating microbial metabolism by the application of external potentials. By examining the relationship between intracellular redox parameters and upregulating genes linked to particular metabolites, these microbial species are capable of producing valuable substances like ethanol and lipases in EF systems, indicating the possible use of this technology for industrial applications.

In an effort to consume less energy and produce less sludge, microbial fuel cells (MFCs) are BESs that employ organic waste to produce bioelectricity (Naha et al., 2023). MFCs are a helpful instrument for applications including bioremediation and the production of renewable energy. MFCs have shown encouraging results in the removal of organic matter and pollutants from wastewater while concurrently producing bioelectricity. Typically composed of an anode, cathode, and a proton exchange membrane dividing the two chambers (Iqbal et al., 2023; Mohyudin et al., 2022). Increasing power generation efficiency is the aim of MFC design modifications and developments. This will make it possible to integrate it into the wastewater treatment facilities that are already in place and contribute to the creation of a cleaner, greener ecology, particularly in developing and impoverished countries (Naha et al., 2023).

2.2 Synergistic Effects of Combining Fermentation with Electrochemistry

EF, which blends industrial fermentation with electrochemistry, maximizes fermentation technologies by boosting carbon efficiency, growth of microbial species, and product recovery (Rago et al., 2019). This approach can lead to reduced substrate utilization, increased carbon efficiency, and improved product recovery when using renewable electricity sources (Rago et al., 2019). Furthermore, employing electrochemical acetate as a substrate for fermentation offers monetary benefits, reduced manufacturing expenses, and improved market pricing stability, underscoring the possibility of combining electrochemistry and precision fermentation for environmentally friendly chemical and food production (Crandall et al., 2023).

Enhancing microbial metabolism in EF technology requires electrochemical stimulation, which modifies fermentation conditions and facilitates electron transfer. Electrical energy can be used to stimulate microbial cells to create more valuable chemicals or fuels. Using electron mediators, like phenazine-1-carboxylic acid, can alter metabolism to create more reduced metabolites and boost the efficiency of extracellular electron transfer (Feng et al., 2020). Moreover, the combination of microbial catalysis and bioelectrochemistry in a two-dimensional biofilm electrode reactor can promote the growth and multiplication of electroactive decolorization bacteria, resulting in high decolorization rates for the treatment of dye wastewater (Liu et al., 2022). Additionally, the design of n-type conjugated polyelectrolytes can boost current intake from electrodes, allowing for continuous chemical production, by enhancing the electrical communication between bacteria and electrodes (Quek et al., 2022).

Combining electrochemistry and fermentation has been shown in numerous studies to boost product yields. For instance, research on *Clostridium autoethanogenum* revealed that employing an external potential during EF raised biomass, acetate, and lactate yields (Martínez-Ruano et al., 2023). Likewise, studies on *Clostridium pasteurianum* have demonstrated that altering the oxidation–reduction potential through electrochemical control significantly increased the molar yield for 1, 3-propanediol, indicating that electrochemistry may be used to increase product yield (Arbter et al., 2022). Furthermore, when redox mediators like Neutral Red and Brilliant Blue were utilized in bioelectrochemical system (BES) with glycerol metabolism, *C. pasteurianum* demonstrated a concentration-dependent change in product synthesis, which increased the yields of 1,3-propanediol and n-butanol.

The electro-fermentation technology has shown a great deal of promise in adjusting the electrochemical parameters that impact the fermentation process. EF can enhance the efficiency of bioprocesses by directly exchanging electrons with electrode surfaces or indirectly activating targeted pathways with the application of electrical energy (Li et al., 2023). According to Rago et al., (2019) & Singh et al., (2021), this approach can maximize product yields, carbon efficiency, and optimized microbial development while minimizing substrate utilization and product-inhibition impacts. The use of electrochemical cells in EF systems promotes efficient alcohol generation and metabolic alterations in microbial systems, which can impact microbial metabolism, redox balance, and even gene expression (Li et al., 2023). Moreover, mathematical models have been proposed for the study and control of fermentations in EF settings, highlighting the need to understand optimal techniques to maximize desired product yields via dynamic electrode control (Haidar et al., 2022).

One significant problem in conventional fermentation methods is balancing the redox equivalents of substrates and products, which restricts the variety of compounds that may be effectively generated (Yamada et al., 2022). Furthermore, optimization attempts are hampered by the lack of theoretical framework for understanding and managing the EF processes which require the dynamic electrode potential regulation to maximize targeted product yield (Haidar et al., 2022). As in BES, where the anode reaction is primarily the oxygen evolution reaction, the problem of oxygen penetration to the cathode results in creative reactor designs to keep oxygen out and guarantee high efficiency in microbial electrosynthesis processes (Abdollahi et al., 2022).

3 Description of Reactor Configurations and Operational Parameters

Running costs, manufacturing costs, electrode equipment type, redox mediator use, separator type, and any new addition to an existing fermenter supporting its electrochemical needs are all determined by the architectural layout of an electric fermenter. The industry uses reactors with an external power source to drive fermentation on the one hand, and anaerobic fermenters in different configurations on the other.

The "H-type" arrangement is the most widely used for bench-scale electrochemical systems. It consists of two sections that are separated by a charged exchange membrane and form an electrochemical half-cell that facilitates the anode and cathode independently. However, because of their overall ease of use and fabrication, singled chamber reactors have also been used (Paiano et al., 2021).The reactor's design is heavily influenced by the characteristics of the EF process.

The BESs are the name given to the reactors utilized for MES. They are essentially divided into two types: single and double-chambered. Anodic and cathodic responses persist in the electrolyte configuration of the microbial medium in one-chambered BES, whereas in two-chambered reactors, the cathode and anode are genuinely segregated yet ionically connected. Although this isn't true in two-chambered reactors, obstruction caused by the membrane separator may limit overall performance. Typically, it is assumed that compounds framed at one anode could impede the process at one, along with the other way around (Abdollahi et al., 2022).Enhancing reaction rates and yields to bring EF to technological readiness levels (TRLs) is still a primary goal of reactor and process design (Jadhav et al., 2022).

3.1 Electrodes, Microbial Inoculum, and Substrate Selection

3.1.1 Electrodes

Anodic electron discharge can be effectively coupled with microbial production, as demonstrated by some research studies. One of the earliest proof-of-principle investigations paired heterologous glycerol to ethanol fermentation with *S. oneidensis's* native electron transfer pathway. Later, using heterologous *S. oneidensis* cytochromes and an engineered redox mediator, an analogous pathway for an imbalanced conversion of glycerol into acetic acid via electron discharge was discovered in E. coli. More recently, anodic electron discharge was used to regenerate NAD^+, which increased the breakdown of glycerol to 3-hydroxypropionate in *Klebsiella pneumoniae*. A more thorough examination of the metabolic shift under higher oxidative anodic electron discharge circumstances in comparison with traditional fermentation was conducted after this work (Virdis et al., 2022). An alternative method couples the anode with the discharge of excess reducing equivalents by means of artificial redox mediators. The impact of several external redox mediators on *Clostridium cellulolyticum's* fermentation profile demonstrated that the variety of mediators significantly affects the overall quantity and range of fermentation products.

3.1.2 Microbial Inoculum

Due to electrochemical activities, certain microorganisms are used in EF. These can be categorized as electrochemically flexible microbes, exoelectrogen, electricigen, or anode-respiring microbes. This implies that there may be another method by which microbes can transport electrons outside of their cell membranes. Due to their ability to reduce metals, microorganisms classified as exoelectrogens can generate electrons at the anode and provide power. With endoelectrogen, the situation is the opposite; it consumes electric flow at the cathode by oxidizing metals (Choi et al., 2016, Nikhil et al., 2016). They can use membrane proteins, pili, fibers, cytochromes, or expelled mediators to transfer electrons. For example, *Geobacter sulfurreducens* uses pili as conductive wires to transfer electrons between cells. However, *Shewanella* species include mediators like riboflavin in addition to conductive fibers. Remarkably, sulfate-lessening microscopic organisms (*Desulfovibrio sp.)* have been reported to include three different potential arbiters: formate dehydrogenase, hydrogenase, and rubredoxin (Choi et al., 2016). It is theoretically possible to convert any biodegradable substance into organic molecules and energy (Logan et al., 2012). Modern wastewater effluents, waste biomass, inorganic substrates like alkali, corrosive mining seepage, and so on are among the waste substrates that are most frequently utilized (Chandrasekhar et al., 2017). In any event, it is anticipated that microbial co-activity will achieve the goal. For this reason, mixed cultures are typically used (Kiran Kumar et al., 2012). Microbes are primarily arranged with the ability to hydrolyze complicated waste materials into simpler substrates. The second group of electrochemically active bacteria can use the latter to oxidize them in the anode chamber. Cathode-based EFs differ from the one shown above in that they require specialized systems, which may need an external electric source, in order to reduce the amount of oxidized chemicals.

3.1.3 Substrate

The variety of intricate metabolic pathways that organisms have evolved to regulate electron loads dictates the range of biochemicals that can be produced by fermentation processes. The substrates required for creating these pathways are dictated by the unique microbial population in the process, since some microorganisms, due to their genetic composition, only contain a portion of these pathways. Polarized electrodes are used as extra electron sources in EF technology, which ferments energy-rich substrates including alcohol and carbohydrates (Chandrasekhar et al., 2017). Depending on the type of substrate, multiple metabolic methods are used by fermenting organisms to produce energy. Since they are metabolized by central carbon metabolism pathways, carbohydrates are generally readily available as substrates for a variety of microbial species. Some organisms use glycerol fermentation, a branch of carbohydrate fermentation, to produce greatly reduced compounds. The later stages of either the ED pathway or EMP glycolysis allow certain species of *Lactobacillus, Enterobacter, Clostridium,* and *Klebsiella* to produce energy by consuming glycerol; at the same time, they balance the electron load by directly reducing a portion of the glycerol to 1,3-propanediol. This procedure only takes place when there are no carbs present since they prevent the synthesis of 1, 3-propanediol (Virdis et al., 2022). Using "clean" substrates, specific nutrient additions, and strict process controls in fermentation increases operational expenses. In contrast, basic systems that do not required sterilization are often suitable with mixed-culture systems because microbial diversity generally encouraged in such systems. According to Shanthi Saravan et al., (2018), "dirty" substrates like agricultural byproducts, municipal waste streams, and other organic wastes can also be utilized as feedstock with little to no pretreatment. Owing to these equipment and metabolic choices, defined-culture systems are usually used to produce high-value products that justify higher construction and operating costs, while mixed-culture processes are used to produce lower-value products or when it is desirable to use a variety of substrates or products (Virdis et al., 2022).

4 Types of Solid Waste Suitable for Electro-Fermentation

Traditionally waste has been viewed as a burden on society, requiring significant amounts of land, energy, and money for treatment and disposal (Jiang et al., 2019). A huge amount of organic waste is generated, consisting of a large complex. Every year, anthropogenic CO_2emissions into the atmosphere amount to almost 35.5 gigatonnes (Gt), and are thought to be the primary cause of climate change (Mac Dowell et al., 2017). Around the world, 2.01 billion tons of MSW and 1000 km^3 of wastewater are released into the environment, posing serious dangers to human health and the environment, especially in developing countries. A subgroup of easily degradable organic waste in this category has great potential for resource exploitation and future development (Savvidou et al., 2022). The conversion of organic waste (solid, liquid, and gaseous) into biomaterials, biochemicals, and biofuels is causing a paradigm shift toward the creation of a closed-loop approach for utilizing waste because these materials have inherent energy that can be reused to create products with added value. In the green era, which revolves around sustainable development and renewable energy, using microbes to recycle organic waste that breaks down quickly has gained a lot of interest (Wang et al., 2020).One efficient way to recycle organic waste is by anaerobic fermentation. In an anaerobic environment, organic waste can be converted into organic acid, biogas, and other value-added products using various microbes (Uddin et al., 2021). Conventional fermentation requires a variety of waste feedstock to form bio-based products. Typically, fermentation reactions exhibit redox imbalance and many thermodynamic constraints, limiting their productivities and overall yields. However, at the industrial scale, these limitations need to be overcome to improve product and platform chemical recovery through cutting-edge hybrid bioprocesses and biotechnologies (Roy et al., 2016). MES have attracted a lot of focus lately due to its potential applications in numerous sectors. MES is a hybrid method that turns waste carbon resources into bioenergy and bioproducts by using electroactive microbes to catalyze bioelectrochemical processes (Zou & He, 2018).When fermentation and MES are integrated, a hybrid EF bioprocess is formed that combines the advantages of both processes and can direct metabolism in the direction of specific production. EF is a novel technique for combining the electrochemical and fermentation processes to enhance the recycling of organic waste, in which electrochemical technology is used to augment the electron transfer process in fermentation (Shanthi Sravan et al., 2021). Electrode insertion accelerates processes and increases electron kinetics in EF (Moscoviz et al., 2016a, 2016b).This regulates the rate of kinetics by improving bioelectrocatalysis and the synthesis of products from a particular substrate (Stamenkovic et al., 2016). By reducing the energy of activation and heat of activation, the variations in the electrode properties in conventional approaches can get around the rate restrictions (Villano et al., 2017).

Various types of solid wastes are suitable for EF. Some of them are municipal organic waste, industrial waste, and agriculture waste. These wastes are suitable as they are high in organic content, which bacteria can decompose during fermentation. Through the use of an electrical current during EF process, microbial activity can be increased and waste materials can be more effectively converted into useful products including biofuels, biochemicals, and energy. Food trash is one kind of organic waste that is formed during the preparation and consumption of food. It accounts for 30–60% of the mass of municipal solid garbage (Li et al., 2016a, 2016b). A third of all food produced for human consumption globally is lost or squandered each year. Food waste is perishable during storage and transport because it is typically high in lipids, sugar, cellulose, water, and protein. Food waste can produce foul-smelling fumes, contaminate water sources, and attract flies and harmful microbes if it is not properly treated (Cuéllar & Webber, 2010). Numerous methods for utilizing food waste have been developed in order to address this problem. These techniques include growing insects for protein feed, applying heat to reduce waste volume, anaerobic digestion for biogas production, and composting for use as fertilizer. Among these alternatives, anaerobic digestion is a frequently employed technique for producing biogas since food waste is a feedstock that is rich in nutrients and carbohydrates (Ahamed et al., 2016). Since typical anaerobic digestion is a slower process, some technologies have been added to the anaerobic digestion of food waste to increase process stability and speed up methanogenesis. Electrochemical techniques have been verified to be efficient for enhancing anaerobic digestion, and EF combines anaerobic fermentation with a MEC (Park et al., 2018). Electrodes in the EF system can control the metabolism and interactions of the microbial population. This is due to the fact that the energized electrodes' two powerful electron sources and sinks have the ability to significantly change the extracellular electron transfer process of the relevant microorganisms. This enhances the ability of fermentation of bacteria to breakdown organic substrates and methanogens turn intermediate products into biogas (Du et al., 2007a, 2007b).

4.1 Pretreatment Steps Required for Effective Conversion of Solid Waste into Fermentable Substrates

The pretreatment process reduces biomass's complex organic structures to simpler molecules, making them more vulnerable to microbial breakdown (Barik, 2019). The selection of pretreatment technique has a major impact on the sustainability, economy, and efficiency of fermentation (Vieira et al., 2020). So different solid waste also requires pretreatment before converting them into fermentable substrates. Before being used in the fermentation process, the agri-food waste needs to be preprocessed. The order in which these procedures are completed varies depending on the kind of waste material and the fermentation method. These preprocessing steps include pretreatment, hydrolysis, and size reduction (such as chipping and grinding). Apart from the preprocessing step, the downstream processing of fermentation products is also necessary. This includes the separation and concentration of the product from the fermentation medium. Pretreatment is a crucial phase in the biochemical conversion of agricultural and food waste during the preprocessing stage, particularly for lignocellulosic waste because of their fermentation (Doddapaneni & Kikas, 2021).

Pretreatment of lignocellulosic materials by physical, chemical, or biological processes is a thoroughly studied process that generates ethanol. Furthermore, considerable research has been conducted on pretreating waste materials in order to produce biogas. By making the materials more accessible to the enzymes and making the wastes more bio-digestible for the production of ethanol and biogas, pretreatment can increase the output of C_2H_5OH or biogas from the wastes (Taherzadeh & Karimi, 2008).

Pretreatment of biomass is needed for enzymes to change cellulosic biomass into fermentable sugars. This involves the release of cellulose from cell walls, the breakdown of lignin, and the pre-extraction of hemicellulose. In the biorefinery process, this is one of the most expensive steps. The most often utilized biomass pretreatment processes are hot water pretreatment, alkaline solution pretreatment, and diluted acid pretreatment. These methods have benefits as well as drawbacks. The kind of biomass to be processed largely determines the pretreatment technique selected. The technique of alkaline pretreatment is employed to elevate cellulose fermentation. Cellulose is hydrated using the hot water pretreatment technique before processing. Biomass is treated with acid to extract hemicelluloses (Wasewar, 2023).

Anaerobic digestion is regarded as the best method for treating organic waste and producing valuable goods (biogas and biofertilizer). This process is advantageous since it produces less sludge, uses less energy, and recovers green energy. The decomposition of proteins, starches, fatty, and cellulosic components into basic soluble molecules is the rate-limiting step in the anaerobic digestion process. Thus, different pretreatment methods for organic solid waste were used in order to enhance organic material solubilization, thereby raising methane generation. The processes which stick among them are biological, thermochemical or acid hydrolysis, microwave, ultrasonication, and thermochemical processes (Mlaik et al., 2019).

4.2 Factors Influencing the Efficiency of Solid Waste Treatment in Electro-Fermentation Systems

Since the functional microorganisms in EF are a succession of native microbes, the initial environmental factors—temperature, pH, oxygen level, moisture content, growth factors, conductivity, nutrient ions, types, and concentrations of contaminants—affect them (Lu et al., 2014).

4.2.1 PH

Protons and electrons are consumed by the cathodic reaction simultaneously, and when they combine with oxygen, they form water, for instance in an air cathode (Du et al., 2007b). A specific range of pH is needed for the survival of microbes, so the difference of acid–base between the cathode and anode severely affects the activity and growth of microbes (Yu et al., 2024).For example, *Clostridium autoethanogenum* grows heterotrophically utilizing fructose as a substrate, forming alcohols (ethanol and 2, 3-butanediol) and organic acids (formic, lactic, and acetic acid). According to the literature, EF improves yield with *Clostridium* strain. Through the use of solid-state electrodes and an external potential—with or without redox mediators—EF modifies the metabolism of microorganisms. However, the reduction in pH value can restrict substrate conversion and cellular viability during fermentation (Martínez-Ruano et al., 2023). Microbes in EF system adapt to varying pH conditions, and it is of great value to know the pH values that are simultaneously appropriate for both electrogenic microorganisms and degrading microorganisms (Zhang et al., 2019).

4.2.2 Temperature

Temperature has a major impact on the composition of microbial populations and metabolites, making it a crucial element in the fermentation process. Temperature changes affect the microbial growth rate, activity, and distribution in EF system (Michie et al., 2013). The ideal temperature range for most microorganisms to thrive, activate their enzymes, and form stable biofilm is 35–40 °C. According to the literature, many times elevating the temperature will increase the yield of EF process. It has shown the relationship between the anode current output and temperature in the 30 to 45 °C region as per the Arrhenius law (Liu et al., 2011). However, when the temperature reaches a certain level, the microbial enzyme's activity and structure are compromised, lowering the EF's performance. As the operating temperature increases, the total microbial activity reduces, which slows down the microbial production of protons and electrons (Liu et al., 2005). Choosing the optimum temperature conditions can lead to satisfactory EF performance, depending on the varying temperature sensitivity of different microbes (Zhang et al., 2019).

4.2.3 Electrode Materials

Carbon-based materials are frequently used as EF electrode materials because of their high specific surface areas (Yaqoob et al., 2020). However, the materials that are frequently used for this have some drawbacks. The mechanical strength of carbon cloth is low, the thick carbon felt substance would reduce the diffusion and mass transfer efficiency, and the smooth surface of graphene makes it difficult for electric-producing bacteria to stick to it. To overcome these drawbacks, these materials need to be changed to add materials such as conductive polymers, metal oxides, etc. (Hindatu et al., 2017).The conductive carbon materials, such as activated carbon, biochar, and carbon fiber, facilitate electron transfer in the media and also give microbes a favorable attached space in the EF system (Lu et al., 2014). Activated carbon is also less expensive and has better catalytic performance than noble metals (such as platinum). Its catalytic efficiency is dependent on its specific surface area, pore structure, functional groups, and graphitization content (Zhang et al., 2019). Carbon fiber efficiently shape functional microbial groups and are easily extracted from the remediated soil and reused (Li et al., 2016a, 2016b). Biochar is advantageous for the growth of electrochemically active microorganisms and substrate transfer as an anode material. Additionally, the rough surface structure and large specific surface area following high temperature carbonization can aid in electron transport. All these characteristics make its cost-effective as compared to other materials (Yin et al., 2019).

4.2.4 Substrate Composition

Substrates in the EF include co-metabolizing substrates and contaminants, which are classified into fermentable substrates and non-fermentable substrates. According to earlier research, functional microorganisms in EF are strongly linked with the substrates, and the kinds and concentrations of the substrates have a significant impact on how the microbial communities are structured (Chae et al., 2009). According to a study, higher substrate concentrations have a significant effect on the synergistic interaction between the microbes and the adapted electrode, which caused the fermentation pattern to be redirected (Jiang et al., 2020). Some studies showed that EF originating from diverse carbon sources possess varying capacities to adjust to alterations in their substrate. While EF enriched with acetate cannot use glucose immediately after the substrate is changed, EF enriched with glucose performed better than EF supplemented with butyrate or acetate (Zhang et al., 2011). Additionally, substrates with various bioavailabilities and morphologies could affect the performance of EF system. Additional studies have demonstrated an inverse relationship between the system performance and the complexity of the substrate, which in turn is connected to the substrate's bioavailability (Qin et al., 2014).

5 Identify the Different Types of High-Value Products that Can Be Synthesized via Electro-Fermentation of Solid Waste

An electro-fermenter is an integrated bioelectrochemical device that simultaneously produces molecules with added value and power through electrochemical and biological processes. Electrodes will act as electron sources or sinks (Chandrashekhar et al.,

(2015), Gambino et al., (2017), and Modestra et al., 2020). Improving the microbial metabolism of the traditional fermentation process was the primary focus of the majority of the initial research on EF. Later on, though, research focused on using bioelectrosynthesis to produce fuels and chemicals with added value, like ethanol, biopolymers, (H_2), and (CH_4). (Jia et al., 2020; Sravan et al., 2019).

5.1 Production of Biofuels from Electro-Fermentation of Solid Waste

A major challenge remains in the direct upcycling of waste materials in an aqueous electrolyte via electro-reforming because of the tremendous complex chemistry and high level of polymerization of solid waste. Therefore, before electro-reforming, it is important to sort and depolymerize biomass or plastic streams, particularly polyolefin and lignin with C–C linkages, in order to gain monomeric units that are regarded as the platform agents for the electro-reforming reaction (Ma et al., 2023).

To run on self-driven bioelectrogenic activity, a novel solid-state bio-EF system (SBES) was constructed. Food waste served as the SBES's operational substrate, and it was evaluated if bioelectricity, (H_2), and bioethanol could all be produced simultaneously. The system demonstrated a highest open-channel voltage and power density of 443 mV and 162.4 mW/m^2, respectively, on the ninth day of operation. The rate of H_2 generation increased to 21.9 ml/h on the 19th day of operation. The SBES system additionally measured 4.85% w/v bioethanol production on the twentieth day of operation. The analysis of the end products confirmed that the typical fermentation process that produced H_2 was a mixed acetate/butyrate process (Chandrasekhar et al., 2015).

Process instability makes it difficult to constantly anaerobic digest (AD) submerged aquatic plants, like Egeria densa, in order to recover renewable energy. A single-chamber membrane-free MEC equipped with a pair of Ti/RuO_2 mesh electrodes (i.e., the combined MEC-AD system) was used to test the potential effects of bioelectrochemical activation on methane production and process stability of *E. densa* fermentation at various applied voltages (0–1.0 V). Application of MEC improved overall process performance, particularly solid matter removal, and stabilized *E. densa* fermentation successfully. Throughout the bioelectrochemical process, the *E. densa* AD process was run consistently and without any indications of imbalance. The average methane yield at 1.0 V was around 248.2 ± 21.0 mL L^{-1} d^{-1}, as applied voltage increases, there is a steady increase in the solubilization removal of solid matter and the efficiency of methane conversion. However, as soon as the external power was cut off, the process's stability rapidly deteriorated, resulting in the removal of weaker solid matter and methane production, demonstrating the beneficial and essential function that methane plays in preserving process stability (Zhen et al., 2016).

During the EF process, it is also possible to reduce the amount of volatile fatty acids intermediates produced by releasing electrons in the anodic compartment. The biocathode, which in turn regulates the entire electro-fermentation process, depends on the kinetics of these reactions. Using *Klebsiella pneumoniae* L17 as a potential pure culture, Kim et al., (2020) investigated the impact of the cathode (applied potential of – 900 mV versus Ag/AgCl) mediated breakdown of glycerol (Plant sources such as soybeans or palm are common. Tallow obtained from animals is another source) to 1, 3-Propanediol (1, 3-PDO) using several electron mediators. The cathode's external potential led to a rise in 1, 3-PDO (35.5 mm) production, which was equivalent to 23.7 mm of fermentation in a non-BES. 1, 3-PDO is one industrial solvent with significant added value that may be produced via glycerol fermentation.

Exoelectrogenic bacteria in MEC will break down the carbon-rich substance into protons and electrons due to the voltage of the in situ electrode. The bacteria then move the protons and electrons they have produced to the cathodic compartment. Protons are also transferred to the cathode, which drops to accommodate biological H_2. According to Marshall et al., (2013), electroactive microorganisms exhibit enhanced efficiency in using released electrons during electro methanogenesis when operating at low voltage settings. The H_2 from the cathodic terminal can also help electric carriers produce high-CH_4. Methane can be produced in two ways: either directly using electrons as a source of power loss or by hydrogenotrophic methanogenesis via EF via two pathways near the cathodic terminal.

The study combined anaerobic digestion with wastewater-mixed bioelectrolysis of agricultural waste to produce biogas. The bioelectrochemical digesters were run in anoxic conditions at 20, 40, 80, 120 mV, and Control. At 40 mV, an increased methane output of 175.17 ± 81.39 ml/g COD was recorded, whereas under control conditions (no voltage supply), it was only 105.36 ± 40.73 ml/g COD. At different applied voltages, the rate of total COD removal as well as the organic conversion to methane were both improved (Prajapati & Singh, 2020). In another study conducted by Sravan et al., 2018, in order to compare the impact of electrical stimulation on increasing the rate of fermentation, three single-chambered EF systems (Applied potential (AP: – 0.6 V), Closed circuit (CC: 100 Ω), and Control (C)) were separately operated utilizing canteen waste (COD-10 g/l). Compared to CC (3593 mg/l) and C (2666 mg/l), AP (4595 mg/l) produced more total volatile fatty acids. Extended chain length was demonstrated by electron flux mechanism-induced fermentation. Along with waste cleanup, electrochemical analysis revealed a clear difference in electron transfer rates (Kapp). The crucial conclusion is that electrical stimulation plays a crucial function in fermentation to improve the conversion of traditional substrates to specific carboxylic acids. A further sustainable biofuel known as biohythane was also created when the right amounts of bioH_2 and bioCH_4 were combined.

In this work, a two-chambered microbial fuel cell (MFC) seeded with a combination of cellulose-degrading bacteria (called CDBs) was used to create power from rice straw without any prior preparation. The power density was 145 mW/m^2, and the coulombic efficiencies (CEs) ranged from 54.3 to 45.3% with a baseline concentration of 1 g/L of rice straw. These values matched to initial concentrations of 0.5–1 g/

Table 1 Biofuel production from EF of solid waste

Value-added products by electro-fermentation of solid waste	Substrate	Production value	Ref
Biohydrogen	Food waste	H_2 production rate (21.9 ml/h) on 19th day	Chandrasekhar et al., 2015
Bioethanol	Food waste	4.85% w/v bioethanol production on 20th day	Chandrasekhar et al., 2015
Methane	Aquatic plant (*E. densa*) powder	11.1 ± 1.0 mmol/L/d	Zhen et al., 2016
1, 3-Propanediol	Glycerol	39.3 ± 0.8 mM,	Kim et al., 2020
Biogas	Agricultural waste (wheat straw)	175.17 ± 81.39 ml/g COD at 49 mV	Prajapati & Singh, 2020
Biohydrogen	Food waste	VFA: 4595 $mg.L^{-1}$ H_2 (26%) > CH_4 (4%)	Sravan et al., 2018

L. Stackable MFCs in series and parallel produced open circuit voltages of 2.17 and 0.723 V, respectively, using hexacyanoferrate as the catholyte. The highest power of three stacked MFCs linked serially might be 490 mW/m^2 (0.5 mA). About three times (1.5 mA) as much current was created in MFCs stacked in parallel as compared to those generated via the serial connection (Hassan et al., 2014) (Table 1).

5.2 *Biopolymer Production from Electro-Fermentation of Solid Waste*

Biopolymers, including cellulose, starch, pectin, and derivatives, are primarily derived from agri waste. Synthetic biodegradable polymers include polyesters, polysaccharides, and protein-based ones. Natural biodegradable polymers include polyglycolide, PLA, and polyurethanes. Polymers like P(3HB-co-3HV) and PHA are produced through microbial fermentation. The demand for polylactic acid (PLA) and polyhydroxyalkanoates (PHA) in the biodegradable plastics market is expected to double by 2023 (Ranganathan et al., 2020). Because of their capacity to oxidize organic substances through metabolic processes, bacteria are essential to BES (bioelectrochemical system). PHA is a kind of biopolyester mostly made up of monomers hydroxyalkanoates. Examples of its copolymers are polyhydroxybutyrate (PHB), polyhydroxypropionate (PHP), and polyhydroxyvalerate (PHV). According to Morgan-Sagastume et al., (2014), these biopolymers have mechanical, physicochemical, and thermoplastic properties that make them comparable to petroleum-derived plastics like polypropylene.

The benefit of bioelectrochemical system is that microorganisms may be used for both oxidation and reduction. In MFC, an oxidation process typically occurs on the anode, while a reduction reaction on the cathode produces useful goods or biofuel. Some of the mechanisms involved in BES are as follows: (i) anaerobic redox reaction occur in the anode; (ii) anaerobic redox reaction occur in the cathode; (iii) adsorption or absorption on electrodes; and (iv) depending on the type of BES, a combination of the aforementioned mechanisms (Banu et al., 2019).

The process of producing PHA from organic wastes involves two steps: the first is the hydrolysis of carbohydrates, followed by the fermentation of the hydrolysis products to produce carboxylic acids. These carboxylic acids are transformed into PHAs in the second step (Dionisi et al., 2005). In the second stage, PHA accumulates and PHA-producing microbes multiply, for example, when feast/famine circumstances are applied.

5.3 Value-Added Chemicals Production from Bioelectrochemical System

Short-chain fatty acids (SCFA) are intermediate metabolites produced during acidogenic fermentation (Rai & Singh, 2016). The carbon-rich organic waste and CO_2 may be utilized as an energy source in the EF process to make SCFA, according to Mohan & Pandey, (2013). Anaerobic/dark fermentation combined with MEC technology is a very lucky choice for increased feedstock reduction and the transformation of VFA-rich sewage into H_2 (Marone et al., 2017). For bacterial metabolism to occur, the biological redox co-factors NAD^+ and NADH are essential. In the past ten years, microorganisms have gained the capacity to actively or implicitly share electrons with cathodes, which could result in the development of novel reductive processes using electrodes (Mostafazadeh et al., 2016). Since SCFAs are very promising, low in weight, pollution-free, and easily biodegradable, they are a major commodity for the apparel, fruit, beverage, polymer, and plastics sectors. By use of acetogenic processes, feedstock or CO_2 is converted via a series of metabolic events mediated by microorganisms to produce SCFA.

Foods rich in saturated or unsaturated fatty acids (FA), like coconut oil, include a lot of *medium-chain fatty acids* (MCFA), of which lauric acid (C12:0) makes up around half of the fat. The oxygen-to-carbon ratio of these naturally hydrophobic MCFA is quite low. To improve energy density, it is feasible to separate from the fermentation broth due to its hydrophobic nature (Rigouin et al., 2018).In the cathode chamber during the bioelectrochemical MCFA synthesis process, acetate is utilized as a carbon source. Using acetic acid for the generation of MCFA, having a high market value and several industrial applications, would be a wise choice as opposed to using it for the electrochemical production of CH_4 (Rigouin et al., 2018).The most readily accessible compounds in fermentation effluents are acetate, propionate, and butyrate, which may be used as precursor molecules in BES under applied power source conditions to synthesize MCFA (Tharak & Mohan, 2021).Several experiments have tried to produce n-caprylate by using ethanol as an electron source. Van Eerten-Jansen et al., (2013) investigated the bioelectrochemical conversion of acetate to ethanol and MCFA (butyrate, caproate, and caprylate) while preserving − 900 mV Vs. SHE (standard hydrogen electrode) using mixed microbial culture as a biocatalyst and no electron mediators. MCFA caproate is a highly important chemical molecule due to its numerous medical applications. MCFA is used in airplane fuels for a variety of purposes, including waste management, the production of green diesel precursors, antimicrobials, flavors, and fragrance-related intermediates.

6 Applications and Case Studies

The novel technique known as EF which combines electrochemical systems with conventional fermentation processes, has drawn a lot of interest recently due to its potential uses in a variety of waste streams. It is a promising bioprocess that controls the metabolism of microorganisms that are electrochemically active by supplying electrodes that serve as electron acceptors during fermentation, assisting in the creation and movement of protons and electrons, and ultimately generating bioelectricity and chemicals with added value. Redox imbalance during metabolic activities is a typical occurrence, and the standard fermentation technique has limited economic feasibility and applicability (Bhagchandanii et al., 2020). According to Chandrashekhar et al., (2021a, 2021b), EF has become a hybrid technology that can control redox instability and improve the metabolic process in general to produce high biomass yields and products. This method produces higher production rates and altered product profiles by manipulating microbial metabolism during fermentation operations with electrodes (Moscoviz et al., 2016a, 2016b). The ability to drive fermentation processes through electrochemical systems, like polarized electrodes, which can affect fermentative bacteria and their metabolism, is one of the main benefits of EF. Due to this influence, important molecules like hydrogen and 1, 3-propanediol can be produced at a rapid rate from substrates like glucose (Tian et al., 2023). Furthermore, EF differs from traditional fermentation methods in that it has been demonstrated to enable the generation of both oxidized and reduced products at high yields by employing electrodes as electron sources or scavengers (Ikeda et al., 2021a, 2021b).

Research has demonstrated the feasibility of EF as a means of creating biofuels and biopolymers via electro-oxidative processes, underscoring the adaptability of this strategy in the production of compounds with added value from biomass feedstocks (Chernysh et al., 2022). EF circumvents the redox constraints usually associated with conventional fermentation by modifying the internal redox state of electrochemically active bacteria, providing a viable route for sustained bioproduction (Yamada et al., 2022). This technology's applicability to waste valorization and circular economy initiatives has been examined in the context of biorefineries (Mohan & Mohan, 2022).

Moreover, the effective synthesis of hydrogen and other chemicals from a variety of waste streams has been made possible by the incorporation of EF into microbial electrolysis cells. Researchers have developed bioelectrochemical systems for energy production and waste management by utilizing the synergies between electrochemical and microbiological processes (Muguruma et al., 2014). By turning lignocellulosic materials into biofuels, this integration has enhanced the production of bioethanol, recycled trash, and protected the environment (Wang et al., 2023a, 2023b).

EF provides a sustainable way to use carbon-rich off-gases from industrial operations in syngas fermentation, a potential method for producing biofuel. By using electrochemical systems, syngas may be fermented by microorganisms to produce biofuels; this illustrates how electro-fermentation can be used to alleviate energy issues and lower greenhouse gas emissions (Dang et al., 2021). This use case demonstrates how versatile electro-fermentation is for a range of industrial processes and waste streams. Furthermore, research has been done on EF as a sustainable and effective method of producing bio-based products from biomass feedstocks in order to produce valuable compounds. Researchers have shown that EF can be used to change microbial metabolism and increase the synthesis of target molecules by utilizing BES (Yamada et al., 2022). This strategy has potential to support the growth of biorefineries and the shift to a more sustainable bioeconomy. EF is an innovative method that efficiently drives bioproduction from various waste streams by combining classic fermentation processes with electrochemical devices. Through leveraging the complementary qualities of both electricity-dependent microbial fermentation and electrochemical processes, scientists are creating new avenues for waste valorization, energy production, and sustainable biomanufacturing. The wide range of uses for EF highlight its potential to revolutionize bioprocessing and open the door to a more sustainable future.

7 Summary of the Potential of Electro-Fermentation for Solid Waste Treatment

The two most important concerns facing humanity now are the power crisis and climate change. Because of this, the scientific community is concentrating on creating sustainable processes to create a circular economy as well as searching for biofuels that can replace traditional fossil fuels. It has been shown that bioelectrochemical methods are effective in converting various waste kinds into bioenergy and value-added goods. In recent years, EF has attracted a lot of attention because of its potential to produce biofuel and biochemicals. The benefits of electrocatalysis and biocatalysis are dynamically coupled in bioelectrocatalysis. High selectivity, high activity, a broad range of substrates, and gentle reaction conditions are some of the benefits of biocatalysis. High energy conversion efficiency and the potential to use renewable electricity as an electron source are two benefits of electrocatalysis. These characteristics work together to produce a variety of products, effective energy conversion, and selective biosensing (Chen et al., 2020). In the past, linear economic frameworks have been used to generate chemicals and energy, which require finite resources to produce valuable items and then dispose of the leftover wastes. Because of the "take-make-disposal" linear model, an enormous quantity of waste was produced. Because of this, the waste was deemed inevitable and detrimental to society and the environment, which led to its incineration and disposal in landfills or wastewater (Jung et al., 2020). Anaerobic microbial reactions in MFCs break down organic wastes to produce electricity. As a result, the overall amount of solid waste is reduced, resulting in the production of energy. Because of these benefits, MFCs are included in the potential circular bioeconomy model. On the other hand, the wastewater treatment system's linear model aerobic digestion system uses energy continually to create air flow, which breaks down organic waste. Given the usage of fossil fuels and greenhouse gas emissions from the generation of power, BESs have the potential to develop into a more viable circular bioeconomy model that is good for the environment. The EF has the potential to make circular bioeconomic processes from the solid waste, without compromising climate and environmental conditions (Jung et al., 2020).

7.1 Implication for Sustainable Waste Management and Resource Recovery

An effective and planned solid waste management have ability to reduce negative effects on the environment and human health, saves resources, and makes cities more livable. But unsustainable solid waste management methods have a detrimental effect on environmental sustainability and public health, which is made worse by fast urbanization as well as institutional and financial constraints. An essential component of sustainable source management is resource recovery. It entails a number of tasks, such as identifying, gathering, and processing wastewater streams and waste products in order to recover and extract materials, chemicals, and resources with added value. Since additional processing procedures are frequently required to improve the purity of water or other natural assets, resource recovery may include a quality-cost trade-off. Higher capital expenses associated with additional processing have an impact on the affordability and recovery of resources. The existence of pollutants linked to the recovered products throughout their life cycle raises safety issues. When it comes to urban wastewater, safety is particularly important because of the presence of pathogens, substances like heavy metals, and more recently, "emerging contaminants" such as waste of pharma industries, and genes from drug-resistant bacteria in the effluents that can be used to recover resources (Parada et al., 2022). Environmental, social, economical as well as human safety concerns are associated with resource recovery from waste, but these challenges can be addressed by the utilization of advance techniques which are safe in all aspects of human and environmental consideration.

References

Abdollahi, M., Al Sbei, S., Rosenbaum, M. A., & Harnisch, F. (2022). The oxygen dilemma: The challenge of the anode reaction for microbial electrosynthesis from CO2. *Frontiers in Microbiology, 13*. https://doi.org/10.3389/fmicb.2022.947550

Ahamed, A., Yin, K., Ng, B. J. H., Ren, F., Chang, V. W. C., & Wang, J. Y. (2016). Life cycle assessment of the present and proposed food waste management technologies from environmental and economic impact perspectives. *Journal of Cleaner Production, 131*, 607–614. https://doi.org/10.1016/J.JCLEPRO.2016.04.127

Arbter, P., Widderich, N., Utesch, T., Hong, Y., & Zeng, A. P. (2022). Control of redox potential in a novel continuous bioelectrochemical system led to remarkable metabolic and energetic responses of Clostridium pasteurianum grown on glycerol. *Microbial Cell Factories, 21*(1). https://doi.org/10.1186/s12934-022-01902-5

Banu, J. R., Kumar, M. D., Gunasekaran, M., & Kumar, G. (2019). Biopolymer production in bio electrochemical system: Literature survey. *Bioresource Technology Reports, 7*, Article 100283.

Barik, D. (2019). Chapter 3–Energy extraction from toxic waste originating from food processing industries. In D. Barik (Ed.), *Energy from Toxic Organic Waste for Heat and Power Generation* (pp. 17–42). Woodhead Publishing. https://doi.org/10.1016/B978-0-08-102528-4.00003-1

Beschkov, V., & Razkazova-Velkova, E. (2021). Bioelectrochemical processes in industrial biotechnology. *Energy storage battery systems–fundamentals and applications*. https://doi.org/10.5772/intechopen.98582

Bhagchandanii, D. D., Babu, R. P., Sonawane, J. M., Khanna, N., Pandit, S., Jadhav, D. A., Khilari, S., & Prasad, R. (2020). A comprehensive understanding of electro-fermentation. *Fermentation, 6*(3). https://doi.org/10.3390/FERMENTATION6030092

Chae, K.-J., Choi, M.-J., Lee, J.-W., Kim, K.-Y., & Kim, I. (2009). Effect of different substrates on the performance, bacterial diversity, and bacterial viability in microbial fuel cells. *Bioresource Technology, 100*, 3518–3525. https://doi.org/10.1016/j.biortech.2009.02.065

Chandrasekhar, K., & Ahn, Y.-H. (2017). Effectiveness of piggery waste treatment using microbial fuel cells coupled with elutriated-phased acid fermentation. *Bioresource Technology, 244*(Pt 1), 650–657.

Chandrasekhar, K., Amulya, K., & Mohan, S. V. (2015). Solid phase bio-EFof food waste to harvest value-added products associated with waste remediation. *Waste Management, 45*, 57–65.

Chandrasekhar, K., Kumar, A. N., Kumar, G., Kim, D. H., Song, Y. C., & Kim, S. H. (2021a). Electro-fermentation for biofuels and biochemicals production: Current status and future directions. *Bioresource Technology, 323*, Article 124598.

Chandrasekhar, K., Naresh Kumar, A., Kumar, G., Kim, D. H., Song, Y. C., & Kim, S. H. (2021b). Electro-fermentation for biofuels and biochemicals production: Current status and future directions. *Bioresource Technology, 323*. https://doi.org/10.1016/j.biortech.2020.124598

Chen, H., Simoska, O., Lim, K., Grattieri, M., Yuan, M., Dong, F., Lee, Y. S., Beaver, K., Weliwatte, S., Gaffney, E. M., & Minteer, S. D. (2020). Fundamentals, applications, and future directions of bioelectrocatalysis. *Chemical Reviews, 120*(23), 12903–12993.

Chernysh, Y., Bataltsev, Y., Shen, X., Bohdanovych, O., & Yakhnenko, O. (2022). Electro-fermentation for biopolymers production: trends determination with bioinformatics data analysis. *Journal of Engineering Sciences, 9*(2), H1-H8. https://doi.org/10.21272/jes.2022.9(2).h1

Choi, O., & Sang, B.-I. (2016). Extracellular electron transfer from cathode to microbes: Application for biofuel production. *Biotechnology for Biofuels, 9*, 11.

Choi, O., Kim, T., Woo, H. M., & Um, Y. (2014). Electricity-driven metabolic shift through direct electron uptake by electroactive heterotroph Clostridium pasteurianum. *Scientific Reports, 4*(1), 6961.

Climent, V., & Herrero, E. (2020). Electrochemical behavior of single crystal electrodes on model processes. *In Springer Handbooks*. https://doi.org/10.1007/978-3-030-46906-1_34

Crandall, B. S., Overa, S., Shin, H., & Jiao, F. (2023). Turning Carbon Dioxide into sustainable food and chemicals: How Electrosynthesized acetate is paving the way for fermentation innovation. *Accounts of Chemical Research, 56*(12). https://doi.org/10.1021/acs.accounts.3c00098

Cuéllar, A. D., & Webber, M. E. (2010). Wasted food, wasted energy: The embedded energy in food waste in the United States. *Environmental Science and Technology, 44*(16), 6464–6469. https://doi.org/10.1021/ES100310D/SUPPL_FILE/ES100310D_SI_001.PDF

Dang, J., Wang, N., & Atiyeh, H. (2021). Review of dissolved co and h2 measurement methods for syngas fermentation. *Sensors, 21*(6), 2165. https://doi.org/10.3390/s21062165

Devi, K. S. (2014). Literature review on solid waste management practices in India.

Dhungana, P., & Rajbhandari, J. (2022). Integrated method of pretreatment, Saccharification and electro-fermentation of Saccharum Spontaneum to Produce Bioethanol Using microbial consortium. *ECS Meeting Abstracts, MA2022–02*(64). https://doi.org/10.1149/ma2022-02642383mtgabs

Dionisi, D., Beccari, M., Di Gregorio, S., Majone, M., Papini, M. P., & Vallini, G. (2005). Storage of biodegradable polymers by an enriched microbial community in a sequencing batch reactor operated at high organic load rate. *Journal of Chemical Technology & Biotechnology: International Research in Process, Environmental & Clean Technology, 80*(11), 1306–1318.

Krishna C. Doddapaneni, T. R., & Kikas, T. (2021). Chapter 39–Thermochemical and biochemical treatment strategies for resource recovery from agri-food industry wastes. In R. Bhat (Ed.), *Valorization of Agri-Food Wastes and By-Products* (pp. 787–807). Academic Press. https://doi.org/10.1016/B978-0-12-824044-1.00007-6

Doménech-Carbó, A., & Doménech-Carbó, M. T. (2023). *Electrochemical Processes and Techniques*. https://doi.org/10.1007/978-3-031-31945-7_2

Du, Z., Li, H., & Gu, T. (2007a). A state of the art review on microbial fuel cells: A promising technology for wastewater treatment and bioenergy. *Biotechnology Advances, 25*(5), 464–482. https://doi.org/10.1016/J.BIOTECHADV.2007.05.004

Du, Z., Li, H., & Gu, T. (2007b). A state of the art review on microbial fuel cells: A promising technology for wastewater treatment and bioenergy. *Biotechnology Advances, 25*(5), 464–482. https://doi.org/10.1016/j.biotechadv.2007.05.004

Feng, J., Lu, Q., Li, K., Xu, S., Wang, X., Chen, K., & Ouyang, P. (2020). Construction of an electron transfer mediator pathway for Bioelectrosynthesis by Escherichia coli. *Frontiers in Bioengineering and Biotechnology, 8*. https://doi.org/10.3389/fbioe.2020.590667

Fruehauf, H. M., Enzmann, F., Harnisch, F., Ulber, R., & Holtmann, D. (2020). Microbial electrosynthesis—an inventory on technology readiness level and performance of different process variants. *Biotechnology Journal, 15*(10). https://doi.org/10.1002/biot.202000066

Gambino, E., Toscanesi, M., Del Prete, F., Flagiello, F., Falcucci, G., Minutillo, M., & Jannelli, E. (2017). Polycyclic aromatic hydrocarbons (PAHs) degradation and detoxification of water environment in single-chamber air-cathode microbial fuel cells (MFCs). *Fuel Cells, 17*(5), 618–626.

Garuba, E. O., Ajunwa, O. M., Oguntomi, O. S., & Onilude, A. A. (2020). Electro-fermentative production of thermotolerant lipases by electrogenic bacteria isolated from palm oil mill effluents. *Novel Research in Microbiology Journal, 4*(1). https://doi.org/10.21608/nrmj.2020.73436

González, J., Laborda, E., & Molina, Á. (2023). Voltammetric kinetic studies of electrode reactions: Guidelines for detailed understanding of their fundamentals. *Journal of Chemical Education, 100*(2). https://doi.org/10.1021/acs.jchemed.2c00944

Haidar, I., Desmond-Le Quéméner, E., Barbot, J. P., Harmand, J., & Rapaport, A. (2022). Modeling and optimal control of an electro-fermentation process within a batch culture. *Processes, 10*(3). https://doi.org/10.3390/pr10030535

Hassan, S. H., El-Rab, S. M. G., Rahimnejad, M., Ghasemi, M., Joo, J. H., Sik-Ok, Y., & Oh, S. E. (2014). Electricity generation from rice straw using a microbial fuel cell. *International Journal of Hydrogen Energy, 39*(17), 9490–9496.

Hindatu, Y., Annuar, M. S. M., & Gumel, A. M. (2017). Mini-review: Anode modification for improved performance of microbial fuel cell. *Renewable and Sustainable Energy Reviews, 73*, 236–248. https://doi.org/10.1016/j.rser.2017.01.138

Ikeda, S., Takamatsu, Y., Tsuchiya, M., Suga, K., Tanaka, Y., Kouzuma, A., & Watanabe, K. (2021a). Shewanella oneidensismr-1 as a bacterial platform for electro-biotechnology. *Essays in Biochemistry, 65*(2), 355–364.

Ikeda, S., Takamatsu, Y., Tsuchiya, M., Suga, K., Tanaka, Y., Kouzuma, A., & Watanabe, K. (2021b). Shewanella oneidensis mr-1 as a bacterial platform for electro-biotechnology. *Essays in Biochemistry, 65*(2), 355–364. https://doi.org/10.1042/ebc20200178

Iqbal, A., Khan, B. B., Iqbal, S., Khan, N., Rehman, M. I., & M. (2023). A quick review on the mechanism and applications of microbial mediated fuel cells. *Journal of Bacteriology & Mycology: Open Access, 11*(1). https://doi.org/10.15406/jbmoa.2023.11.00335

Jadhav, D. A., Park, S.-G., Pandit, S., Yang, E., Ali Abdelkareem, M., Jang, J. K., Chae, K. J. (2022). Scalability of microbial electrochemical technologies: applications and challenges. *Bioresour Technol*ogy, *345*, 126498. https://doi.org/10.1016/j.biortech.2021.126498

Jia, X., Li, M., Wang, Y., Wu, Y., Zhu, L., Wang, X., & Zhao, Y. (2020). Enhancement of hydrogen production and energy recovery through electro-fermentation from the dark fermentation effluent of food waste. *Environmental Science and Ecotechnology, 1*, Article 100006.

Jiang, Y., May, H. D., Lu, L., Liang, P., Huang, X., & Ren, Z. J. (2019). Carbon dioxide and organic waste valorization by microbial electrosynthesis and electro-fermentation. *Water Research, 149*, 42–55. https://doi.org/10.1016/j.watres.2018.10.092

Jiang, Y., Chu, N., Zhang, W., Zhang, L., & Jianxiong Zeng, R. (2020). Electro-fermentation regulates mixed culture chain elongation with fresh and acclimated cathode. *Energy Conversion and Management, 204*. https://doi.org/10.1016/j.enconman.2019.112285

Jung, S., Lee, J., Park, Y. K., & Kwon, E. E. (2020). Bioelectrochemical systems for a circular bioeconomy. *Bioresource Technology, 300*, Article 122748.

Junho, B., Ahn, J., Lee, J., Hwang, C., Kim, G., Lee, J., & Lee, S. (2021). Synthetic formatotrophs for one-carbon biorefinery. *Advanced Science, 8*(12). https://doi.org/10.1002/advs.202100199

Kaza, S., Yao, L., Bhada-Tata, P., & Van Woerden, F. (2018). *What a waste 2.0: a global snapshot of solid waste management to 2050*. World Bank Publications.

Kengen, S. W., Goorissen, H. P., Verhaart, M., van Niel, E., Claassen, P., & Stams, A. (2009). Biological hydrogen production by anaerobic microorganisms. *Biofuels; John Wiley Sons: Hoboken, NJ, USA*.

Kim, C., Lee, J. H., Baek, J., Kong, D. S., Na, J. G., Lee, J., & Kim, J. R. (2020). Small current but highly productive synthesis of 1, 3-propanediol from glycerol by an electrode-driven metabolic shift in klebsiella pneumoniae L17. *Chemsuschem, 13*(3), 564–573.

Kiran Kumar, A., Venkateswar Reddy, M., Chandrasekhar, K., Srikanth, S., & Venkata Mohan, S. (2012). Endocrine disruptive estrogens role in electron transfer: Bio-electrochemical remediation with microbial mediated electrogenesis. *Bioresource Technology, 104*, 547–556.

Kracke, F., & Krömer, J. O. (2014). Identifying target processes for microbial electrosynthesis by elementary mode analysis. *BMC Bioinformatics, 15*, 1–14.

Li, Y., Jin, Y., Li, J., Chen, Y., Gong, Y., Li, Y., & Zhang, J. (2016b). Current Situation and development of kitchen waste treatment in China. *Procedia Environmental Sciences, 31*, 40–49. https://doi.org/10.1016/J.PROENV.2016.02.006

Li, X., Wang, X., Zhao, Q., Wan, L., Li, Y., & Zhou, Q. (2016a). Carbon fiber enhanced bioelectricity generation in soil microbial fuel cells. *Biosensors and Bioelectronics, 85*, 135–141. https://doi.org/10.1016/j.bios.2016.05.001

Li, K., Zhang, J. W., Li, F., Zhao, X. Q., Bai, F. W., Song, H., & Liu, C. G. (2023). Electrochemical control of cell metabolism improves ethanol production of Zymomonas mobilis in an electro-fermentation system. *ACS Sustainable Chemistry and Engineering, 11*(6). https://doi.org/10.1021/acssuschemeng.2c06188

Liu, H., Cheng, S., & Logan, B. E. (2005). Power generation in fed-batch microbial fuel cells as a function of ionic strength, temperature, and reactor configuration. *Environmental Science & Technology, 39*(14), 5488–5493. https://doi.org/10.1021/es050316c

Liu, Y., Climent, V., Berná, A., & Feliu, J. M. (2011). Effect of temperature on the catalytic ability of electrochemically active biofilm as anode catalyst in microbial fuel cells. *Electroanalysis, 23*(2), 387–394. https://doi.org/10.1002/ELAN.201000499

Liu, S., Xue, H., Feng, X., & Pyo, S. H. (2022). Electrostimulation for promoted microbial community and enhanced biodegradation of refractory azo dyes. *Journal of Environmental Chemical Engineering, 10*(6). https://doi.org/10.1016/j.jece.2022.108626

Logan, B. E., & Rabaey, K. (2012). Conversion of wastes into bioelectricity and chemicals by using microbial electrochemical technologies. *Science, 337*, 686–690.

López-Hernández, R., Cercado-Quezada, B., Gómez-Velázquez, H. D. J., Robles-Rodríguez, C., González-Dávalos, L., Varela-Echavarría, A., Shimada, A., & Mora, O. (2023). Single-chamber EFof Rumen fluid increases microbial biomass and volatile fatty acid production without major changes in diversity. *Fermentation, 9*(6). https://doi.org/10.3390/fermentation9060502

Lu, L., Yazdi, H., Jin, S., Zuo, Y., Fallgren, P. H., & Ren, Z. J. (2014). Enhanced bioremediation of hydrocarbon-contaminated soil using pilot-scale bioelectrochemical systems. *Journal of Hazardous Materials, 274*, 8–15. https://doi.org/10.1016/j.jhazmat.2014.03.060

Ma, Y., Zhang, Y., Yuan, W., Du, M., Kang, S., & Qiu, B. (2023). Electroreforming injects a new life into solid waste. *EES Catalysis, 1*(6), 892–920.

Mac Dowell, N., Fennell, P. S., Shah, N., & Maitland, G. C. (2017). The role of CO2 capture and utilization in mitigating climate change. *Nature Climate Change, 7*(4), 243–249. https://doi.org/10.1038/nclimate3231

Marone, A., Ayala-Campos, O. R., Trably, E., Carmona-Martínez, A. A., Moscoviz, R., Latrille, E., & Bernet, N. (2017). Coupling dark fermentation and microbial electrolysis to enhance bio-hydrogen production from agro-industrial wastewaters and by-products in a bio-refinery framework. *International Journal of Hydrogen Energy, 42*(3), 1609–1621.

Marshall, C. W., LaBelle, E. V., & May, H. D. (2013). Production of fuels and chemicals from waste by microbiomes. *Current Opinion in Biotechnology, 24*(3), 391–397.

Martínez-Ruano, J., Suazo, A., Véliz, F., Otálora, F., Conejeros, R., González, E., & Aroca, G. (2023). Electro-fermentation with clostridium autoethanogenum: Effect of pH and neutral red addition. *Environmental Technology and Innovation, 31*. https://doi.org/10.1016/j.eti.2023.103183

Michie, I. S., Kim, J. R., Dinsdale, R. M., Guwy, A. J., & Premier, G. C. (2013). Factors affecting microbial fuel cell acclimation and operation in temperate climates. *Water Science and Technology : A Journal of the International Association on Water Pollution Research, 67*(11), 2568–2575. https://doi.org/10.2166/WST.2013.159

Mir, I. S., Cheema, P. P. S., & Singh, S. P. (2021). Implementation analysis of solid waste management in Ludhiana city of Punjab. *Environmental Challenges*, 2, 100023.

Mlaik, N., Khoufi, S., Hamza, M., Masmoudi, M. A., & Sayadi, S. (2019). Enzymatic pre-hydrolysis of organic fraction of municipal solid waste to enhance anaerobic digestion. *Biomass and Bioenergy, 127*, 105286. https://doi.org/10.1016/j.biombioe.2019.105286

Modestra, J. A., Katakojwala, R., & Mohan, S. V. (2020). CO2 fermentation to short chain fatty acids using selectively enriched chemolithoautotrophic acetogenic bacteria. *Chemical Engineering Journal, 394*, Article 124759.

Mohan, S., & Mohan, S. (2022). Hybrid electrosynthesis as non-genetic approach for regulating microbial metabolism towards waste valorization in circular framework. *Microbial Biotechnology, 16*(2), 184–189. https://doi.org/10.1111/1751-7915.14143

Mohan, S. V., Butti, S. K., Amulya, K., Dahiya, S., & Modestra, J. A. (2016). Waste biorefinery: A new paradigm for a sustainable bioelectro economy. *Trends in Biotechnology, 34*(11), 852–855.

Mohan, S. V., & Pandey, A. (2013). Biohydrogen production: an introduction. *Biohydrogen,* 1–24.

Mohyudin, S., Farooq, R., Jubeen, F., Rasheed, T., Fatima, M., & Sher, F. (2022). Microbial fuel cells a state-of-the-art technology for wastewater treatment and bioelectricity generation. *Environmental Research, 204*. https://doi.org/10.1016/j.envres.2021.112387

Morgan-Sagastume, F., Valentino, F., Hjort, M., Cirne, D., Karabegovic, L., Gerardin, F., & Werker, A. (2014). Polyhydroxyalkanoate (PHA) production from sludge and municipal wastewater treatment. *Water Science and Technology, 69*(1), 177–184.

Moscoviz, R., Toledo-Alarcón, J., Trably, E., & Bernet, N. (2016b). Electro-fermentation: How to drive fermentation using electrochemical systems. *Trends in Biotechnology, 34*(11), 856–865. https://doi.org/10.1016/j.tibtech.2016.04.009

Moscoviz, R., Toledo-Alarcón, J., Trably, E., & Bernet, N. (2016a). Electro-Fermentation: How to drive fermentation using electrochemical systems. *Trends in Biotechnology, 34*(11), 856–865. https://doi.org/10.1016/j.tibtech.2016.04.009

Mostafazadeh, A. K., Drogui, P., Brar, S. K., Tyagi, R. D., Le Bihan, Y., Buelna, G., & Rasolomanana, S. D. (2016). Enhancement of biobutanol production by electromicrobial glucose conversion in a dual chamber fermentation cell using C. pasteurianum. *Energy Conversion and Management, 130*, 165–175.

Muguruma, H., Hoshino, T., & Nowaki, K. (2014). Electronically type-sorted carbon nanotube-based electrochemical biosensors with glucose oxidase and dehydrogenase. *Acs Applied Materials & Interfaces, 7*(1), 584–592. https://doi.org/10.1021/am506758u

Naha, A., Debroy, R., Sharma, D., Shah, M. P., & Nath, S. (2023). Microbial fuel cell: A state-of-the-art and revolutionizing technology for efficient energy recovery. *Cleaner and Circular Bioeconomy, 5*. https://doi.org/10.1016/j.clcb.2023.100050

Nikhil, G. N., Yeruva, D. K., Venkata Mohan, S., & Swamy, Y. V. (2016). Assessing potential cathodes for resource recovery through wastewater treatment and salinity removal using non-buffered microbial electrochemical systems. *Bioresource Technology, 215*, 247–253.

Paiano, P., Premier, G., Guwy, A., Kaur, A., Michie, I., Majone, M., & Villano, M. (2021). Simplified reactor design for mixed culture-based EF toward butyric acid production. *Processes, 9*, 417. https://doi.org/10.3390/pr9030417

Parada, M. P., Kehrein, P., Xevgenos, D., Asveld, L., & Osseweijer, P. (2022). Societal values, tensions and uncertainties in resource recovery from wastewaters. *Journal of Environmental Management, 319*, Article 115759.

Park, J., Lee, B., Tian, D., & Jun, H. (2018). Bioelectrochemical enhancement of methane production from highly concentrated food waste in a combined anaerobic digester and microbial electrolysis cell. *Bioresource Technology, 247*, 226–233. https://doi.org/10.1016/J.BIORTECH.2017.09.021

Peekhaus, N., & Conway, T. (1998). What's for Dinner?:Entner-Doudoroff metabolism in Escherichia coli. *Journal of Bacteriology, 180*, 3495–3502.

Prajapati, K. B., & Singh, R. (2020). Enhancement of biogas production in bio-electrochemical digester from agricultural waste mixed with wastewater. *Renewable Energy, 146*, 460–468.

Prajapati, K. K., Yadav, M., Singh, R. M., Parikh, P., Pareek, N., & Vivekanand, V. (2021). An overview of municipal solid waste management in Jaipur city, India-current status, challenges and recommendations. *Renewable and Sustainable Energy Reviews, 152*, Article 111703.

Qin, B., Dong, X., Wang, J., Sun, C., Guo, D., & Li, Y. (2014). The Influence on MFC voltage of different types of oilfield produced water as anode substrate. *Applied Mechanics and Materials, 675–677*, 592–595. https://doi.org/10.4028/www.scientific.net/AMM.675-677.592

Quek, G., Vázquez, R. J., McCuskey, S. R., Kundukad, B., & Bazan, G. C. (2022). Enabling electron injection for microbial Electrosynthesis with n-type conjugated polyelectrolytes (Adv. Mater. 37/2022). *Advanced Materials, 34*(37). https://doi.org/10.1002/adma.202270262

Rago, L., Pant, D., & Schievano, A. (2019). Electro-fermentation-microbial electrochemistry as new frontier in biomass refineries and industrial fermentations. *In advanced bioprocessing for alternative fuels, biobased chemicals, and bioproducts: technologies and approaches for scale-up and commercialization*. https://doi.org/10.1016/B978-0-12-817941-3.00014-0

Rai, P. K., & Singh, S. P. (2016). Integrated dark-and photo-fermentation: Recent advances and provisions for improvement. *International Journal of Hydrogen Energy, 41*(44), 19957–19971.

Ranganathan, S., Dutta, S., Moses, J. A., & Anandharamakrishnan, C. (2020). Utilization of food waste streams for the production of biopolymers. *Heliyon, 6*(9).

Rigouin, C., Croux, C., Borsenberger, V., Ben Khaled, M., Chardot, T., Marty, A., & Bordes, F. (2018). Increasing medium chain fatty acids production in Yarrowialipolytica by metabolic engineering. *Microbial Cell Factories, 17*, 1–12.

Robins, G., Bates, L., & Pattison, P. (2011). Network governance and environmental management: Conflict and cooperation. *Public Administration, 89*(4), 1293–1313.

Roy, S., Schievano, A., & Pant, D. (2016). Electro-stimulated microbial factory for value added product synthesis. *Bioresource Technology, 213*, 129–139. https://doi.org/10.1016/J.BIORTECH.2016.03.052

Savvidou, M. G., Pandis, P. K., Mamma, D., Sourkouni, G., & Argirusis, C. (2022). Organic waste substrates for bioenergy production via microbial fuel cells: a key point review. *Energies, 15*(15), 5616. https://doi.org/10.3390/EN15155616

Scheremetjew, A., Meyer, T. H., Lin, Z., Massignan, L., & Ackermann, L. (2021). Fundamental principles of organic electrochemistry. In *Science of Synthesis* (Vol. 2021, no. 5). https://doi.org/10.1055/sos-SD-236-00002

Selvasembian, R., Mal, J., Rani, R., Sinha, R., Agrahari, R., Joshua, I., & Pradhan, N. (2022). Recent progress in microbial fuel cells for industrial effluent treatment and energy generation: Fundamentals to scale-up application and challenges. *Bioresource Technology, 346*, Article 126462.

Shanthi Sravan, J., Tharak, A., Annie Modestra, J., Seop Chang, I., & Venkata Mohan, S. (2021). Emerging trends in microbial fuel cell diversification-Critical analysis. *Bioresource Technology, 326*, Article 124676. https://doi.org/10.1016/J.BIORTECH.2021.124676

ShanthiSravan, J., Butti, S. K., Sarkar, O., Vamshi Krishna, K., Venkata Mohan, S. (2018). EFof food waste—regulating acidogenesis towards enhanced volatile fatty acids production. *Chemical Engineering Journal, 334*, 1709–1718. https://doi.org/10.1016/j.cej.2017.11.005

Sharma, H. B., et al. (2021). Circular economy approach in solid waste management system to achieve UN-SDGs: solutions for post-COVID recovery. *Science of the Total Environment 800*, 149605.

Singh, M., Joshi, G., & Ali, M. A. (2021). Electro-fermentation technology: emerging platform. *Bioelectrochemical Systems: Principles and Processes*. https://doi.org/10.1007/978-981-15-6872-5_4

Sk, M. M., Qamar, S., & Sethy, T. (2023). Solid waste management in Indian perspectives: a comprehensive review.

Sravan, J. S., Butti, S. K., Sarkar, O., Krishna, K. V., & Mohan, S. V. (2018). EFof food waste–regulating acidogenesis towards enhanced volatile fatty acids production. *Chemical Engineering Journal, 334*, 1709–1718.

Sravan, J. S., Butti, S. K., Sarkar, O., & Mohan, S. V. (2019). Electrofermentation: chemicals and fuels. In *Microbial Electrochemical Technology* (pp. 723–737). Elsevier.

Sriram, S., Wong, J. W. C., & Pradhan, N. (2022). Recent advances in electro-fermentation technology: A novel approach towards balanced fermentation. *Bioresource Technology, 360*, 127637.

Stamenkovic, V. R., Strmcnik, D., Lopes, P. P., & Markovic, N. M. (2016). Energy and fuels from electrochemical interfaces. *Nature Materials, 16*(1), 57–69. https://doi.org/10.1038/nmat4738

Taherzadeh, M. J., & Karimi, K. (2008). Pretreatment of lignocellulosic wastes to improve ethanol and biogas production: a review. *International Journal of Molecular Sciences, 9*(9), 1621–1651. https://doi.org/10.3390/IJMS9091621

Tharak, A., & Mohan, S. V. (2021). Electrotrophy of biocathodes regulates microbial-electro-catalyzation of CO2 to fatty acids in single chambered system. *Bioresource Technology, 320*, Article 124272.

Tian, J., Lacroix, R., Yaqoob, A., Bureau, C., Midoux, C., Quéméner, E., & Bouchez, T. (2023). Study of a pilot scale microbial electrosynthesis reactor for organic waste biorefinery. *Energies, 16*(2), 591. https://doi.org/10.3390/en16020591

Tribhuvan, D., Vinay, V., Gite, S., & Ahmed, S. (2022). Electro-fermentation technology: synthesis of chemicals and biofuels. *In Bio-Electrochemical Systems*. https://doi.org/10.1201/9781003225430-6

Uddin, M. N., Siddiki, S. Y. A., Mofijur, M., Djavanroodi, F., Hazrat, M. A., Show, P. L., Ahmed, S. F., & Chu, Y. M. (2021). Prospects of bioenergy Production from organic waste using anaerobic digestion technology: A mini review. *Frontiers in Energy Research, 9*. https://doi.org/10.3389/FENRG.2021.627093

Utesch, T., Sabra, W., Prescher, C., Baur, J., Arbter, P., & Zeng, A. P. (2019). Enhanced electron transfer of different mediators for strictly opposite shifting of metabolism in Clostridium pasteurianum grown on glycerol in a new electrochemical bioreactor. *Biotechnology and Bioengineering, 116*(7). https://doi.org/10.1002/bit.26963

Van Eerten-Jansen, M. C., Ter Heijne, A., Grootscholten, T. I., Steinbusch, K. J., Sleutels, T. H., Hamelers, H. V., & Buisman, C. J. (2013). Bioelectrochemical production of caproate and caprylate from acetate by mixed cultures. *ACS Sustainable Chemistry & Engineering, 1*(5), 513–518.

Vieira, S., Schneider, J., Martinez Burgos, W. J., Magalhães, A., Medeiros, A. B. P., de Carvalho, J. C., de Souza Vandenberghe, L. P., Soccol, C. R., & Sydney, E. B. (2020). Pretreatments of Solid Wastes for Anaerobic Digestion and Its Importance for the Circular Economy. In C. Baskar, S. Ramakrishna, S. Baskar, R. Sharma, A. Chinnappan, & R. Sehrawat (Eds.), *Handbook of Solid Waste Management: Sustainability through Circular Economy* (pp. 1–27). Springer Singapore. https://doi.org/10.1007/978-981-15-7525-9_5-1

Villano, M., Paiano, P., Palma, E., Miccheli, A., & Majone, M. (2017). Electrochemically driven fermentation of organic substrates with undefined mixed microbial cultures. *Chemsuschem, 10*(15), 3091–3097. https://doi.org/10.1002/CSSC.201700360

Virdis, B., Hoelzle, R. D., Marchetti, A., Boto, S. T., Rosenbaum, M. A., Blasco-Gómez, R., Puig, S., Freguia, S., & Villano, M. (2022). Electro-fermentation: Sustainable bioproductions steered by electricity. *Biotechnology Advances, 59*, Article 107950.

Wang, Q., Li, H., Feng, K., & Liu, J. (2020). Oriented fermentation of food waste towards high-value products: A review. *Energies, 13*(21), 5638. https://doi.org/10.3390/EN13215638

Wang, C., Chang, D., Zhang, Q., & Yu, Z. (2023a). Enhanced bioethanol production in a microbial electrolysis cell by evolved escherichia coli lge2-h.

Wang, C., Chang, D., Zhang, Q., & Yu, Z. (2023b). Enhanced bioethanol production in a microbial electrolysis cell by evolved escherichia coli lge2-h. https://doi.org/10.21203/rs.3.rs-3208004/v1

Wasewar, K. L. (2023). Chapter 31–Sustainability of biorefineries for waste management. In P. Singh, P. Verma, R. Singh, A. Ahamad, & A. C. S. Batalhão (Eds.), *Waste Management and Resource Recycling in the Developing World* (pp. 721–754). Elsevier. https://doi.org/10.1016/B978-0-323-90463-6.00006-3

Yamada, S., Takamatsu, Y., Ikeda, S., Kouzuma, A., & Watanabe, K. (2022). Towards application of electro-fermentation for the production of value-added chemicals from biomass feedstocks. *Frontiers in Chemistry, 9*. https://doi.org/10.3389/fchem.2021.805597

Yaqoob, A. A., Ibrahim, M. N. M., & Rodríguez-Couto, S. (2020). Development and modification of materials to build cost-effective anodes for microbial fuel cells (MFCs): An overview. *Biochemical Engineering Journal, 164*, 107779. https://doi.org/10.1016/j.bej.2020.107779

Yin, C., Shen, Y., Yuan, R., Zhu, N., Yuan, H., & Lou, Z. (2019). Sludge-based biochar-assisted thermophilic anaerobic digestion of waste-activated sludge in microbial electrolysis cell for methane production. *Bioresource Technology, 284*, 315–324. https://doi.org/10.1016/j.biortech.2019.03.146

Yu, L., An, Z., Xie, D., Yin, D., Xie, G., Gao, X., Xiao, Y., Liu, J., & Fang, Z. (2024). From waste to protein: A new strategy of converting composted distilled grain wastes into animal feed. *Frontiers in Microbiology, 15*, 1405564. https://doi.org/10.3389/FMICB.2024.1405564/BIBTEX

Zhang, Y., Min, B., Huang, L., & Angelidaki, I. (2011). Electricity generation and microbial community response to substrate changes in microbial fuel cell. *Bioresource Technology*, *102*(2), 1166–1173. https://doi.org/10.1016/j.biortech.2010.09.044

Zhang, X., Li, X., Zhao, X., & Li, Y. (2019). Factors affecting the efficiency of a bioelectrochemical system: A review. *RSC Advances, 9*(34), 19748–19761. https://doi.org/10.1039/c9ra03605a

Zhen, G., Kobayashi, T., Lu, X., Kumar, G., & Xu, K. (2016). Biomethane recovery from Egeria densa in a microbial electrolysis cell-assisted anaerobic system: Performance and stability assessment. *Chemosphere, 149*, 121–129.

Zou, S., & He, Z. (2018). Efficiently "pumping out" value-added resources from wastewater by bioelectrochemical systems: A review from energy perspectives. *Water Research, 131*, 62–73. https://doi.org/10.1016/J.WATRES.2017.12.026

Chapter 9
Advanced Electro-Fermentation Strategies for Enhanced Bioproduct Synthesis in Anaerobic Environments

Vishal Ahuja, Himani Deepak, Neha Saxena, and Manik Sharma

1 Introduction

Microbial fermentation is one of the most prominent and oldest forms of bioprocess, known to have a direct relation with industries. Initially, it was only restricted to food fermentation, i.e., production of curd, wine and alcohol but now it has been expanded the dimension and covers both edibles, i.e., production of bread, soy, yogurt, probiotics, and vitamins as well as biofuels, and biopolymers, etc., (Bhagchandanii et al., 2020). The majority of the fermentation processes have suffered from several bottlenecks including capital investment (CAPEX), operational cost (OPEX), long time duration, economic feasibility, high substrate cost, and low product yield. In addition, downstream processing alone can cost more than 50% of the total cost (Flynn et al., 2010; Soccol et al., 2013). Several strategies have been defined to reduce the process cost like the use of lignocellulosic and agricultural residues as main feedstock. However, the collection and processing of such feedstock is itself a challenge due to cost, efficiency, and unwanted by-products.

The fermentation process has also shown a wide range of adaptability due to operational conditions like aerobic (presence of oxygen), anaerobic (absence of

V. Ahuja (✉)
Department of Biotechnology, University Centre for Research and Development, Chandigarh University, Mohali, Punjab, India
e-mail: vishal.e14483@cumail.in

V. Ahuja · M. Sharma
University Institute of Biotechnology, a University, Mohali, Punjab, India

H. Deepak
Jigyasa University, Sherpur, Vikasnagar, Uttarakhand, India

N. Saxena
School of Basic Sciences and Technology, IIMT University, Meerut, India

© The Author(s), under exclusive license to Springer Nature Singapore Pte Ltd. 2025
S. Sevda (ed.), *Electro-Fermentation for Synthesis of Chemicals, Biochemicals and Biofuels*, https://doi.org/10.1007/978-981-96-6358-3_9

oxygen), solid-state, and submerged fermentation. Among different variants, anaerobic fermentation gets special attention due to the production of alcohol, biogas, and other energy-related products. United States Environmental Protection Agency (US EPA) defined anaerobic digestion as the process that breaks down the complex organic materials from agricultural residues, plant litter, animal fecal matter and slaughterhouse waste, municipal and industrial waste in the absence of oxygen. The major products of the process are alcohol, gases, and fatty acids (Bolzonella et al., 2005). The available literature has shown the valorization of organic waste for ethanol (Li et al., 2021), ethanol, and lactic acid waste from starch and sugar waste (Li et al., 2021), methane (Li et al., 2011) and butanol from food waste (Huang et al., 2015; Li et al., 2011; Wang et al., 2023).

Anaerobic environment, one the most common and conventional approaches, has shown widespread applicability in fuel generation from waste biomass digestion. Besides, substrate selection, feed C/N ratio, moisture content in substrate, electron transfer, and redox balance are the major issues. Poor electron transfers also make the system sensitive to redox disturbance and affect the yield. The chapter elaborates on the advancements in fermentation and the evolution of electro-fermentation to overcome the bottlenecks of conventional anaerobic fermentation or digestion.

2 Anaerobic Digestion and Fermentation: Scope and Challenges

Anaerobic digestion and fermentation are a bio-catalytic process that represents the degradation and conversion of complex organic compounds into simpler compounds. Anaerobic digestion is one of the most explored approaches for the production of biogas. The process is mainly comprised of four stages, i.e., hydrolysis, acidogenesis, acetogenesis, and methanogenesis which convert substrate to product. Biogas is mainly comprised of methane (CH_4; > 70%) along with a minor fraction of SOx, NOx, COx, and water vapor. The exact composition of biogas, the concentration of CH_4, and other constituents depend upon the type of feedstock, C/N, moisture, microbial community, and process conditions (Kumar et al., 2018; Wang et al., 2022). Similarly, for other fermentation processes as well, the feedstock is converted to ethanol and butanol in multiple stages, i.e., biomass processing and pretreatment, inoculum preparation and optimization, fermentation, and product recovery.

Anaerobic fermentation and electron transfer mechanism: All organisms utilize the available substrate as food and generate cellular energy (ATP) via cellular respiration. In the electron transport system, terminal electron acceptor species depends upon the type of respiration, i.e., in aerobic respiration, the electron acceptor is oxygen, while during anaerobic respiration, exogenous substrates act as terminal electron acceptors. However, ATP synthesis follows oxidative respiration for ATP generation regardless of the electron acceptor. The process refers to the transfer of electrons via electron carriers of the electron transport chain. In contrast to aerobic

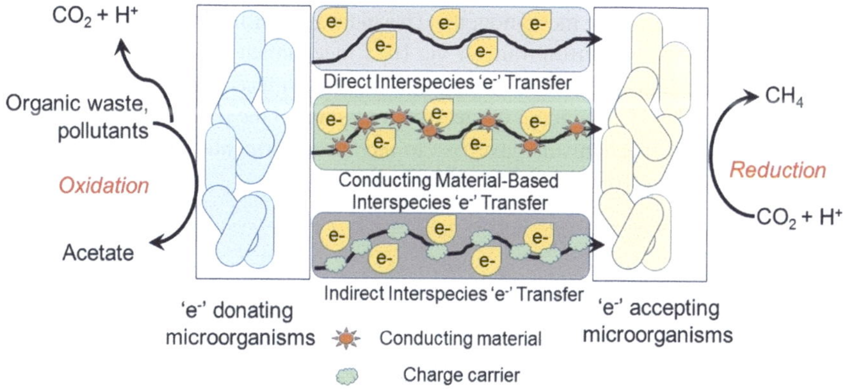

Fig. 1 Types of interspecies electron 'e' transfer (IET) in fermentation

fermentation, microbial cells rely upon internal electron acceptors during anaerobic conditions hence ATP generation is substrate-level phosphorylation which refers to the transfer of phosphate moieties to ADP from metabolic intermediates (Tejedor-Sanz et al., 2022). During anaerobic fermentation, the availability of electron acceptors and intermediates is a matter of concern which slows down digestion as well as energy generation. Therefore, the synchronization between oxidizing and reducing reactions must be in balance with the help of diverse microbial species. Hence, the process referred to as Interspecies electron transfer and reactions include oxidation of organic feedstock and reduction of CO_2 to CH_4 (Batstone et al., 2006; Baek et al., 2016). Two types of Interspecies electron transfer (IET) have been reported (Fig. 1):

Direct Interspecies Electron Transfer (DIET): DIET is the electron transfer mechanism, mainly used by syntrophic microorganisms during the anaerobic degradation of volatile fatty acids (VFAs). In this, exoelectrogenic microorganisms release electrons which will be carried via electrical pilli, and membrane-associated cytochromes and finally captured by electron-capturing microorganisms (Sharma et al., 2024a, 2024b; Valentin et al., 2023; Yuan et al., 2021).

Indirect interspecies electron transfer or mediated interspecies electron transfer (IIET/MIET): IIET is the mechanism by which microorganisms use molecules for electron transfer. Based on the molecule involved, IIET can be classified on interspecies formate transfer and interspecies hydrogen transfer (Wang & Lee, 2021). Besides these two, some other molecules have also shown the potential to act as electron transmitters. These molecules are small molecules including phenazines, riboflavins, and quinones that are secreted by the cell itself and compounds of natural or synthetic origin such as neutral reds and humic substances (Huang et al., 2020; Nozhevnikova et al., 2020).

In comparison with IIET, DIET is a faster and more efficient method of electron transmission. In addition to microorganisms, it can also be mediated by the addition of abiotic conductive materials (Cheng and Call, 2016). In the case of methanogenesis, high organic loading led to the accumulation of VFAs due to rapid acetogenesis which is toxic and inhibitory for methanogens. The main reason for VFAs accumulation and inhibition is slow electron transfer via IIET mechanism. The limitation can be countered by the addition of conducting materials that induce DIET (Su et al., 2023; Wang & Lee, 2021). The concept of electron transfer and the possible integration of additional transfer mechanisms gave rise to advanced fermentation systems called electro-fermentation (EF).

3 Electro-Fermentation

Electro-fermentation (EF) is an emerging technology that can regulate electrogenic bacteria. In order to produce electricity and valuable products, the electrofermentors are made up of electrodes that function as a source and sink for the electrons. Conventional fermentation methods involve a number of drawbacks that restrict their applicability and economic value. Recently discovered 'Electro-fermentation' (EF) can electrochemically regulate the metabolic pathways of various microbial strains (Fig. 2).

In order to maximize the biomass yield and production of desired products, EF can be used to regulate pH and redox imbalance, promote carbon chain elongation, and control other metabolic processes occurring in the fermenter (Bhagchandanii

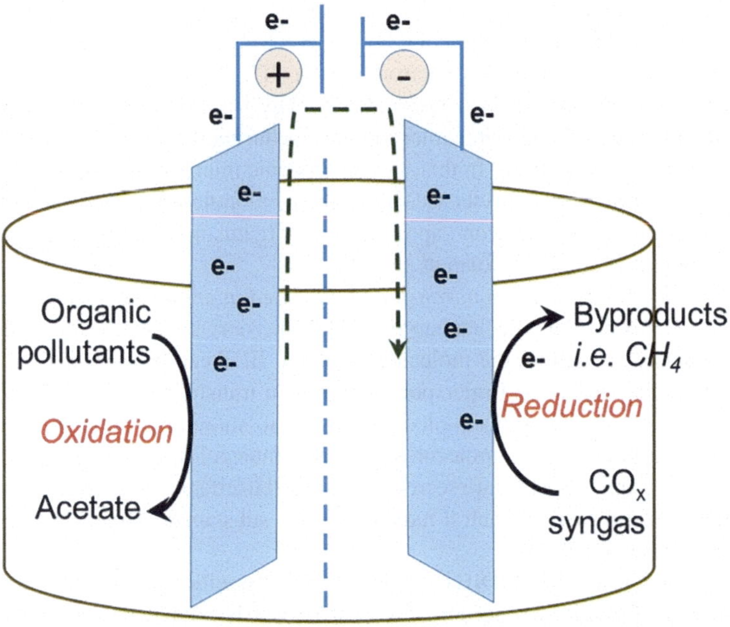

Fig. 2 Electro-fermentation mechanisms for waste valorization and product generation

et al., 2020). The mechanism of EF relies on a polarized electrode interface that helps in the modulation of fermentation rate, biomass yield, titer, and distribution of by-products. The electrode in electrofermentors can either act as electron donor or electron acceptor, which is also known as cathodic electro-fermentation (CEF) or anodic electro-fermentation (AEF), respectively. Both AEF and CEF can be utilized to regulate the oxidation–reduction potential of the fermentation system (Virdis et al., 2022).

EF combines industrial fermentation with electrochemistry which suggests that the electrical field has a direct impact on fermentation and metabolic pathways in microbes. It offers opportunities to improve carbon efficiency, reduce additives, enhance microbial growth, and boost product recovery in various fermentation processes (Moscoviz et al., 2016). Electrodes in EF alter the redox balance by affecting the $NADH/NAD^+$ ratio, influencing fermentation outcomes (Tejedor-Sanz et al., 2022). This can enhance product yield through adjustments in the pattern of produced compounds (Ramanaiah et al., 2023). Research also reveals that external potential can boost lactic acid production by *Clostridium autoethanogenum* DSM 10,061. Overall, manipulating electron flow can optimize fermentation processes and product formation in various bacterial species (Martínez-Ruano et al., 2023).

Bioelectrochemical reactions with bacteria produce various metabolites like ethanol and acetate. These reactions involve microbial cells interacting with electrodes to regulate redox imbalance and enhance biosynthetic processes. Electroactive microorganisms' potential to transfer electrons extracellularly and improve the process performance (Shang et al., 2024). Some of the microorganisms like *Shewanella oneidensis* and *Geobacter sulfurreducens* have high electrochemical activities (Shi et al., 2019), while another group of microorganisms including *Escherichia coli (E. coli), Saccharomyces cerevisiae*, and *Bacillus subtilis* have poor electroactivity, and hence proved insufficient for practical EF applications therefore addition of redox mediators can be considered. However, the addition of exogenous redox mediators increases environmental burden and cost. This highlights the importance of understanding the potential limitations and challenges in implementing EF processes efficiently (Salar-García et al., 2024). 3-Hydroxypropionic acid (3-HP) is a key substrate for several industrial processes. *Klebsiella pneumoniae,* an efficient biocatalyst, has been engineered (recombinant *K. pneumoniae* L17) by overexpressing aldehyde dehydrogenase (AldH) in a bioelectrochemical system (BES). This led to a 1.7-fold increase in 3-HP production compared to the control group. The study also explored the use of HNQ for electron shuttling to improve electrode interaction and redox state control for higher yields of 3-HP and other bioconversion processes (Kim et al., 2017). A study on *Shewanella oneidensis* MR-1, a facultative anaerobe, revealed that it represses the expression of carbon metabolic genes in the absence of electron acceptors, which hinders fermentative growth on glucose. The findings suggest that MR-1 requires essential nutrients like amino acids for fermentative growth. Understanding the genetic regulation in MR-1 could lead to the development of new fermentation technologies, like EF, and provide insights into bacteria's survival strategies in diverse environments (Ikeda et al., 2023).

EF system consists of two electrodes (anode and cathode) fitted in a vessel in which biocatalysts (enzymes and cells) are incubated in optimal reaction conditions. In the system, both electrodes are isolated via an ion-selective membrane that allows the alteration of intracellular redox balance. The electrodes modulate the overall process and allow optimal operation to attain maximum product yield and offer better process economics as observed in acetogenic bacteria like *Clostridium beijerinckii* and *Clostridium pasteurianum*, which produce more reduced compounds like butanol and 1,2-propanediol when provided with an external electron source. Similarly, *C. autoethanogenum* has been found to increase lactic acid production under an external potential (Liberato et al., 2019). *C. autoethanogenum*, a spore-forming anaerobe, can metabolize industrial gases like $CO/H_2/CO_2$ and sugars like xylose via the Wood–Ljungdahl pathway. This makes it a promising candidate for producing microbial biomass along with alcohols and organic acids from hemicellulose-rich biomass including hardwood, softwood, and agricultural residues like sugarcane bagasse and corn husk (Martínez-Ruano et al., 2023). However, factors like culture medium composition, pH, temperature, age of the active inoculum, and inhibitory agents play critical roles in the fermentative production of organic acids and alcohols (Martínez-Ruano et al., 2023). The pH regulation becomes crucial in the case of pH-sensitive species like *Clostridium*, where a shift from acidogenesis to solventogenesis is influenced by the pH of the medium. Controlling pH can enhance xylose uptake and butyric acid production. Additionally, pH changes in the anodic and cathodic chambers in EF processes can impact the growth and activity of microorganisms (Herzog, 2024).

4 Types of Electro-Fermenters and Bioelectrochemical Systems

EF combines industrial fermentation with electrochemistry, offering a novel biotechnological transformation route. The use of electrochemical cells can optimize fermentation processes by reducing substrate usage and product inhibition. EF enables improved carbon efficiency, eliminates the need for additives, and enhances microbial growth. Renewable energy sources can power pure-culture fermentations for higher-value products and also benefit agro-food residue utilization for added-value biorefineries (Rago et al., 2019). Various issues including redox mediators, construction of electrodes, and design parameters for cells led to the development of novel electrofermentors (Maureira et al., 2023).

4.1 Bioelectrochemical Systems

BESs are unique in converting organic waste into electricity or hydrogen/chemical products (Bajracharya et al., 2016; del Pilar Anzola Rojas et al., 2018). BES operates under mild conditions, using diverse organic substrates without expensive metals as catalysts. Recent developments in BESs have expanded their applications in electricity generation, including microbial electrosynthesis (Pant et al., 2012). Innovations in electrode materials, separators, and catalysts have transformed BES is among the promising technologies with improved performance and potential for various applications.

One such example of BES includes the anaerobic technologies (AT) for generating energy from waste water treatment, despite facing challenges such as ammonia and VFA accumulation. Introduction of syntrophic fatty acid oxidizing bacteria can improve the performance of anaerobic units of BES by working with hydrogenotrophic methanogenic organisms (Gupta et al., 2023). BES can enrich syntrophic fatty acid oxidizing bacteria populations and remove ammonium, enhancing AT performance for sustainable recovery. BESs can also be classified as microbial fuel cell (MFC) (Slate et al., 2019), microbial electrolysis cell (MEC) (Zhang & Angelidaki, 2014), enzymatic fuel cell (EFC) (Leech et al., 2012), microbial desalination cell (MDC) (Ali et al., 2024), and microbial solar cells (MSC) (Imoro & Buamah, 2021; Zuo et al., 2016) based on their application. Among all the BES, the concept of MDCs has gained attention as a way to simultaneously desalinate water and recover energy or hydrogen. Stacked MDCs with exchange membranes have shown promise in increasing desalination rates. Additionally, the idea of microbial electrochemical snorkels (MES) has been introduced for efficient organic matter oxidation in urban wastewater treatment (Pant et al., 2012). Different types of BES are summarized below:

4.1.1 Microbial Fuel Cells

MFCs use bacteria to convert organic substances into electricity (Yaqoob et al., 2021). MFC systems are designed for environmental monitoring, pollutant removal, and power generation. MFCs systems utilize microbial metabolism to produce bioelectricity in anaerobic environments by oxidizing organic matter. The electrodes in EF capture the current created by electron movement, with the anode in contact with biofilm and the cathode accepting electrons for oxygen reduction (Mohd Noor et al., 2024). In MFCs, both the electrodes are separated by proton exchange membranes. A biocompatible molecule called chondroitin has been used in creating sulfate-based proton exchange membranes for MFCs. Chondroitin membranes exhibited low internal resistance, high power density, and good coulomb efficiency, with high proton conductivity, the membranes show potential as effective proton exchange materials for MFCs, indicating their suitability as an alternative energy source (Erenler et al., 2024). In addition to energy/ electricity generation, the decomposition of organic materials by microbes reduces the organic load in wastewater, aiding in the removal of biochemical oxygen demand and chemical oxygen demand, indicators of organic pollution levels. By combining microbial metabolism with electricity generation, MFCs offer a sustainable and energy-efficient approach to wastewater treatment (Mkilima et al., 2024).

4.1.2 Microbial Electrolysis Cells

MECs are innovative technologies for sustainable power generation, providing biohydrogen production and wastewater treatment (Arun et al., 2024). The potential of hydrogen gas as an eco-friendly vehicle fuel is significant. The process for electrochemically assisted hydrogen production using microbial electrochemical cells, BES, bio-catalyzed electrolysis, and MES, is defined as electro-hydrogenesis (Cardeña et al., 2019). MECs function as reactors by combining MFC with electrolysis to produce biohydrogen efficiently and cleanly. Several critical factors such as microbial culture, electrodes, membrane, and substrate impact the hydrogen production in MECs (Gautam et al., 2023). However, challenges like high production costs, internal resistance, methanogenesis, and biofouling hamper effective scale-up.

A conventional MEC consists of two chambers (one anaerobic and one cathodic chamber). Bacteria on the anode surface oxidize organic waste to generate oxidized products along with electrons and protons that acidify the anolyte solution. Protons are then absorbed in the cathode, resulting in hydrogen gas production and alkalization of the catholyte solution (Arun et al., 2024). During anaerobic digestion of nitrogen-rich substrates, methane production is hindered by ammonia's toxic effects on microorganisms. Previous studies aimed to improve ammonia tolerance in microorganisms without addressing high ammonia levels via a semi-continuous anaerobic digestor, creating a microbial electrolysis cell system (Sharma et al., 2024a, 2024b). Results showed methane yields increased by 40–110% with voltage application at 2000 and 3000 mg/L total ammonia nitrogen. Voltage promoted nitrogen-metabolism microorganisms and functional gene expression for ammonia oxidation and denitrification, reducing ammonia levels. It also maintained microbial structure and enhanced methanogenesis gene expression. This approach offers a solution for efficient anaerobic digestion under high ammonia conditions. Fabrication of cathodic catalysts for hydrogen production (HER) via MECs can become a boon for the energy sector. The use of carbon materials in fabrication can offer high conductivity, a stable system, a large specific surface area, and cost-effectiveness (Xu et al., 2023).

4.1.3 Microbial Desalination Cells

MDCs are considered one of the most emerging technologies in bioelectrochemical generation. The technique has been proven for various sustainable technologies including waste water treatment, seawater desalination, and eco-energy production (Guang et al., 2020). The working mechanism of MDC largely depends upon the generation of electricity via the bioelectrochemical oxidation of organic compounds present in waste water and the release of electrons which are first captured by the anode and transferred to the cathode (Mohammadi et al., 2024). The generated bio electricity is further utilized for the migration of ions throughout the microbial desalination system which will further result in the formation of valuable by-products. However, the factors including pH in cathodic and anodic fluids, higher internal resistance, presence of the catalyst, type of membrane, and type of desalination cell have a high impact on the MDC configurations (Al-Mamun et al., 2018). The type of resistors in MDCs emphasizes the impact of acclimation resistance. Higher acclimation resistance exhibits higher power performance (Wang et al., 2024). MDCs play a critical role in soil salination and crop production. A halotolerant *Citrobacter* sp. strain KUT (CKUT) has shown exceptional salt tolerance along with the production of extracellular polymeric substances (0.04 mg/ml). Inoculation of CKUT in *Vigna radiata* seedlings can enhance the chlorophyll content, plant growth, and overall plant characteristics (Chakraborty and Mondal, 2024). However, integration of CKUT into MDCs might have more possibilities for fresh water generation from seawater along with enhanced crop yield.

4.1.4 Microbial Solar Cells

MSCs are recently discovered bioelectrochemical systems that utilize solar cells in order to generate electricity and biochemical production. The working principle of MSCs involves the use of photoautotrophic microorganisms and higher plants to generate electric current from solar energy using electrochemically active microorganisms (Strik et al., 2011). Traditional soil microbial fuel cells (SFMC) offer various limitations including lower efficiency and high resistance. To overcome these limitations, MSCs have been included with SFMCs, making it suitable for in situ soil remediation. The maximum output power and best performance occurred at voltage 2.0 V which may be due to faster phenol degradation rates and high concentration limits, with excellent stability over multiple cycles (Xie et al., 2023).

Phenol is a persistent primary pollutant that is difficult to degrade, posing a challenge for environmental cleanup. MFC including MSC offer a more efficient method of phenol removal compared to other techniques like photocatalysis and chemical oxidation (Zou et al., 2024). MSC includes ‘Dye-Sensitized Solar Cells’ (DSSCs), which is a third-generation solar cells and aim to improve solar efficiency and lower costs. DSSCs were first developed in 1991 by Gratzel, which are made up of non-toxic materials, and are bifacial, collecting light from both sides (Alhamed et al., 2012). Utilizing microbial pigments from *Talaromyces atroroseus* GH2, *Arthrobacter bussei* CP30, and *Paracoccus bogoriensis* BOG6 for innovative DSSCs is essential for sustainable technology (Spadaro et al., 2024). DSSCs are a promising technology due to their low-cost, and eco-friendly production. The key components of DSSCs include conducting substrate, photoanode, dye, electrolyte, and catalyst with the photoanode being crucial for energy conversion efficiency, where TiO_2 and Zn are commonly used photoanode materials (Shakeel Ahmad et al., 2017). Another microbial solar cell includes a biological photovoltaic cell (BPV) which can harness solar energy via biological organisms to generate clean electricity. Cyanobacteria sp. including *Leptolyngbia* sp. along with BPV cells conjugated with silver nanoparticles to efficiently generate photocurrent and hydrogen gas under visible light unveiling the potential of BPV solar cell technology for future electricity generation (Carbas et al., 2023).

5 Electro-Fermentation Under Anaerobic Environment

EF is known to improve the efficiency of conventional anaerobic digestion by enhancing the movement of electrons under the influence of an electric field. Based on the electrode participating critically in anaerobic digestion, i.e., anode and cathode, the fermentation process has been classified into the following two types:

5.1 Anodic Electrode Fermentation

AEF is an emerging biotechnology that endows anaerobic respiration using anode as an unended terminal electron acceptor (Gu et al., 2023; Moscoviz et al., 2016). The technology can offer positive outcomes for aerobic as well as anaerobic processes. AEF can modulate the carbon flux, reroute microbial metabolisms, and stabilize microbial redox without any continuous oxygen supply (Vassilev et al., 2018). One of its biggest challenges is to establish an efficient extracellular electron transfer (EET) system between microbe and anode and connect it to central carbon metabolism. It can operate in aerobic processes under anoxic conditions by using an anode as an alternative to oxygen as an electron acceptor (Vassilev et al., 2018; Zhen et al., 2024). *Lactococcus lactis* is a versatile bacterium used in food production for various purposes. By utilizing aerobic conditions, lactate production can be suppressed, allowing for more efficient growth and production of beneficial compounds. Some strains of *L. lactis* can even respire with the presence of specific compounds, enhancing their growth and yield. However, aerated culturing has its challenges, including the use of animal blood as a heme source (Zhao et al., 2021). On the other hand, anaerobic conditions limit product diversity due to the high lactate production. To overcome these limitations, alternative electron acceptors can be used to modify fermentation products and improve redox balance. A *Lactococcus lactis* strain, unable to regenerate NAD^+ , successfully utilized ferricyanide as an electron acceptor for rapid growth. However, high ferricyanide concentrations limited practical use. An adaptively evolved mutant with enhanced ferricyanide respiration capacity outperformed the parent strain when utilizing an anode as an electron acceptor in a BES. The mutant efficiently interacted with the anode, achieving a high current density and demonstrating the potential of AEF with Lactic Acid Bacteria (Gu et al., 2023).

5.2 Cathodic Electrode Fermentation

Biochemical reactions in the anodic chamber generate electricity, while reactions in the cathodic chamber consume current to produce valuable biochemical products. Cathodic modifications, like adding chitosan to increase surface area, enhance current consumption. Microbe-EET in a cathodic chamber catalyzes diverse bioelectrochemical reactions. The anodic methanogenesis pathway includes electron-proton transfer, carbon dioxide reduction, and hydrogen formation with a cathode as a hydrogen sink (Bhagchandanii et al., 2020; Kadier et al., 2014). EF applies BES in anaerobic fermentation, using inexpensive cathodes instead of traditional electron donors like hydrogen or ethanol. These systems also produce hydrogen on-site for electron donation (Jiang et al., 2020). Chain elongation is a new process using secondary fermentation to produce medium-chain fatty acids from organic waste. EF can impact the production process by using a fresh carbon cathode increased caproate specificity by 28% compared to other methods. The bacteria involved in the process were mainly Firmicutes, Bacteroidetes, and Proteobacteria (Jiang et al., 2020).

Biobutanol, a green alternative to fossil fuels produced by *Clostridium beijerinckii*, faces challenges due to low productivity in ABE fermentation. Cathodic-electro-fermentation can improve butanol synthesis by modifying NAD^+/NADH ratios. A study on glucose-based ABE-EF at a lab-scale showed that poising E_{cat} at -0.8 V vs Ag/AgCl resulted in optimal conditions. Compared to traditional ABE fermentation, cathodic electrode fermentation with a redox mediator led to higher butanol productivity (152.9 mg/L/h) and lower energy consumption (0.02 kWh/kg-butanol). This technique offers advancements in butanol production, selectivity, and energy efficiency for sustainable fuel applications (Molognoni et al., 2023). Butyric acid is a key precursor used in various industries, including food additives, chemicals, and pharmaceuticals. Current production methods rely on non-renewable fossil resources, but there is growing interest in using microorganism fermentation with different substrates for more sustainable production. Lignocellulosic butyric acid production with mixed cultures, especially using rice straw, has shown promise due to high productivity and low cost (Guo et al., 2020). Cellular metabolism during fermentation is influenced by the redox couple of NAD^+/NADH, and exogenous electron mediators can enhance NADH availability. BES are being explored to manipulate electron flow and enhance butyric acid production. Both MFC and CEF systems have the potential to generate biofuels while treating organic waste. MFC enhances NAD^+ regeneration, while CEF enhances NADH formation, both leading to improved butyric acid production (Vassilev et al., 2018; Zhang et al., 2021).

6 Anaerobic Electro-Fermentation for Energy Generation

Anaerobic fermentation is known to produce energy products like alcohol and biogas. The presence of conducting materials in the external field to improve electron mobility might enhance the product yield. Some of the researches for anaerobic EF are summarized below:

6.1 Biogas/Methane Production

The energy crises that plague the overwhelming majority of countries, particularly the developed nations, are characterized by a high reliance on fossil fuels (Gürsan & de Gooyert, 2021; Mohammad et al., 2021). The gap between demand and availability fueled the shift to alternative greener energy sources that will also aid in countering environmental pollution. Additionally, numerous legislations, including Agenda 21 and the Kyoto Protocol, encouraged the shift toward renewable and low-carbon emitting energy sources to lower greenhouse gas emissions by fossil fuels (Moreau & Vuille, 2018; Sahota et al., 2018). The deployment of renewable residual feedstock and waste for electricity, fuel, chemicals, and hydrogen has been the subject of increasing interest due to the increasing environmental and policy consciousness and activities (Stevens, 2001). Fossil sources, such as coal, oil, and gas, account for approximately 60% of global electricity generation. In the first quarter of 2020, renewable sources contributed only 28%, while variable renewable sources contributed only 9% (IEA50, 2024). It has been noted that the proportion of renewable energy in global electricity generation has been increasing at a rate of 2% per year, in contrast to the 1.8% annual growth in electricity demand that began at the beginning of 1990. Biogas exhibited the third-fastest annual development in capacity of this renewable source generation, at 11.5%, trailing only solar PV at 36.5% and wind at 24.0%. Biofuels have been expanding at an average annual growth rate of 9.7% since 1990 (IEA50, 2020), which is another indicator of the centrality of biomass in global electricity generation. This suggests that biogas is a critical component of the energy balance and should be promoted in terms of production and utilization.

Biogas is probably one of the most primitive and conventional models of anaerobic digestion model for energy generation. It can be produced through controlled anaerobic decomposition of chicken dung, agricultural crop residues, and livestock slurry (Abanades et al., 2022). The raw biogas generated through the microbial digestion process can be utilized directly for the generation of thermal energy, power, gaseous fuels like biomethane, and other added-value compounds (Abanades et al., 2022). The utilization of biogas in renewable energy has the potential to mitigate the effects of increasing greenhouse gas emissions (Chen et al., 2007; Jeremiah et al., 2020a, b). For example, biogas was utilized in Swaziland in 2014, contributing 0. 29% of the total energy consumption and nearly 8% of the total renewable energy, excluding hydropower (Holliger et al., 2017).

Biogas has the potential to supplant solid biomass, such as firewood, which is frequently employed as a cooking fuel in our camps. By 2040, biogas has the potential to provide pure cooking fuel to approximately 200 million individuals, with a particular emphasis on Africa and Asia. This implies that biogas contributes to the attainment of social development objectives, which are referred to as the Sustainable Development Goals. Biogas is converted to biomethane, which is a superior fuel compared to unprocessed biogas (IEA, 2020). This establishes biogas as a sustainable and reliable energy source within the decarbonization process of the energy and electricity matrix (Tabatabaei et al., 2015; Jeremiah et al., 2020b, c). As per an estimate, annual biogas-based electricity generation has crossed the mark of 7000 MW (Topolewski & Radulescu, 2021).

Biogas is generated through the anaerobic digestion (AD) process. In addition to its capacity to generate a renewable energy source, the process can also treat the feedstock and generate a digestate that can serve as a substitute for chemical fertilizers in organic cultivation (Lebuhn et al., 2014; Jeremiah et al., 2020b). As a component of the natural carbon cycle, microorganisms are crucial in the production of biogas. Microbial anaerobic biodegradation is estimated to release between 590 and 800 million tonnes of methane into the atmosphere (Joale et al., 2020). Electrochemical approaches have recently been confirmed as efficient means of enhancing AD. The integration of AD with MEC is the base concept of EF (Park et al., 2018). Electrodes in EF systems can regulate microbial metabolism and the interplay between different microbial communities (Moscoviz et al., 2016). The energized electrodes possess strong electron sources/sinks that can greatly influence the EET of microorganisms. This can enhance the microbial oxidation of organic matter followed by its conversion to methane by methanogens (Zhou et al., 2015). Unlike MEC, EF can maintain oxidation–reduction potential in digesters, stabilize the digestion process, and regulate, or trigger the environment for better microbial metabolism (Kracke et al., 2015). Essentially, an electric current catalyzes initiating fermentation in an unstable setting and promoting thermodynamic unfavorable processes. Therefore, EF requires minimal energy expenditure, which is determined by the substrates and target products (Kracke & Krömer, 2014).

Several investigations have verified that applying a low voltage (less than 1.0 V) can enhance the production of desired products and decrease the generation of unwanted by-products in conventional fermentation procedures (Hassanein et al., 2017). Nevertheless, there is a scarcity of studies that have explored the application of EF to enhance methanogenic fermentation. Within EF systems utilized for methane production, the process of electron transfer (EET) plays a crucial part. This involves the exchange of electrons between microorganisms and electrodes, as well as the transfer of electrons between different species of microbes (IET) that work together in a mutually beneficial manner. Furthermore, apart from interspecies hydrogen transfer, DIET also plays a role in the breakdown of organic matter and the generation of methane (Schievano et al., 2016). The main groups involved in syntrophic interactions are fermentative bacteria, electroactive microorganisms, and methanogens.

Among these, electroactive microorganisms are particularly important in the process known as DIET (Schievano et al., 2016). Scientists discovered that the population of hydrogenotrophic methanogens in EF was 17 times higher than in CAD, but the presence of acetotrophic methanogens remained relatively constant (Gajaraj et al., 2017). Sangeetha et al., 2017 discovered the presence of electroactive *Geobacter* and *Desulfovibrio* in the cathode biofilm (Sangeetha et al., 2017). They also observed that hydrogenotrophic methanogens made up 80% of the biodiversity in this biofilm, whereas acetotrophic methanogens, specifically *Methanosarcina*, accounted for 40% of the anode biofilm (Sangeetha et al., 2017). The microbial community's variability leads to the significance of the CO_2 to CH_4 reduction as a mechanism for methanogenesis. *Geobacter* plays a role in this process by decomposing organic acids through DIET or by gathering electrons from the cathodes (Zhao et al., 2015). In addition, the efficiency of internal hydrogen transfer is improved by the production of hydrogen at the cathodes and its subsequent use by hydrogenotrophic methanogens (Tian et al., 2018; Villano et al., 2010). Overall, the activity of functional microorganisms in EF systems led to an enhancement in non-electrode methanogenesis (Hirano & Matsumoto, 2018). Additionally, electrode reactions accounted for only 40.6% of the overall methane production (Kim et al., 2017).

6.2 Bio-Alcohol Production

Fossil fuels discharge lethal gases and suspended particulates, causing environmental damage. Bioethanol is a cleaner alternative that limits SOx and particulate matter emissions. It is commonly used as a mixture in gasoline and has the potential for sustainable aviation fuels (Perera, 2018; Shindell & Smith, 2019). Bioethanol can replace gasoline completely or be blended as E15 or E85 (Jain & Kumar, 2024). Unlike petroleum-based ethanol, bioethanol is derived from biomass, making it eco-friendly with a lower carbon footprint. Although the farming and production processes may introduce some carbon footprint, bioethanol still has a significantly lower carbon footprint compared to fossil-based ethanol (Jain & Kumar, 2024; Sanni et al., 2022). Bioalcohols are alcohols produced from biological resources or biomass like crops, lignocellulosic residues, and food waste. Based on the feedstock used, alcohols are categorized into generations, i.e., first generation is produced from crops, second generation is derived from forestry or agricultural residues (Melikoglu et al., 2016).

Sugarcane molasses and bagasse, a by-product of the sugar industry, are used as a carbon source for microorganisms to produce bioethanol. In high temperatures, the industrial yield of bioethanol from *Saccharomyces cerevisiae* decreases, especially in tropical regions. Boosting bioethanol production on a large scale is necessary for commercial profitability. A study analyzed the impact of three supplements (thiamine, pyridoxine, and biotin) on alcohol production and showed a positive impact on alcohol production and yeast growth with lower glucose concentration, despite a longer lag phase with higher vitamin doses. Besides, biomass complexity is also a challenge to improve the ethanol yield. To overcome the challenge, biomass must be hydrolyzed with physical, chemical, and enzymatic treatments that reduce the crystallinity of biomass, and release fermentable sugars from biomass. Efficient pretreatment is crucial for converting corncob biomass to bioethanol, a vital stage for economic success. Various physical, chemical, or combined pretreatment methods are used, with physicochemical processes using binary acids considered the most effective for commercial application (Selvakumar et al., 2022).

Bio-oil extraction from lignocellulose via pyrolysis can be a good option for fuel preparation. However, the extracted oil also contains a significant amount of glucosans which can be used for ethanol production via microbial fermentation. The presence of aldehyde, phenols, and acids in oil also inhibits microbial growth. In a study, bio-oil was electrically treated and supplemented in growth media for further use in ethanol production by *E. coli*-H strain. The strain has shown comparatively higher levoglucosan consumption as well as ethanol production potential over the control group. In undetoxified bio-oil media with 1.0% (w/v) levoglucosan offered an ethanol yield of 0.54 $g_{ethanol}/g_{levoglucosan}$ (around 94% of theoretical ethanol yield) (Wang et al., 2024). The microbial strain has shown ethanol production potential at the industrial scale however needs scale-up studies and cost-analysis. Gu et al. (2022) have explored the effect of electrode potentials on metabolic products. For analysis, microorganisms (*Ethanoligenens harbinense*) inoculated on anode. *E. harbinense*-inoculated anodic EF or cathodic EF with 0 V poised potential offered maximum H_2 production of 1888–1986 mL/L_{medium} which was 23–26% higher than open-circuit EF. Along with it, ethanol production was also increased by 30.7 ± 13.3% in comparison with the open-circuit system. In comparison with open-circuit systems, glucose consumption and cell growth were also increased. Gene expression analysis showed the overexpression of, rubredoxin, ferredoxin, and [FeFe]-hydrogenase genes, upregulation of reduced NADH regeneration and electron transfer-related genes along with downregulation of lower expression of ATP-binding cassette transporter system-related genes which ultimately increased the ethanol yield. Li et al., (2023) also reported the increased in ethanol yield by more than 12% by *Zymomonas mobilis* in comparison with control fermentation trials (without electricity input) (Li et al., 2023).

6.3 Biochemicals

EF has been known to improve the yield by improving the movement of electrons. Besides ethanol and biogas, some of the commodity chemicals are also known to be produced from anaerobic fermentation and hence EF might be a boon for such processes. Medium-chain fatty acids are fatty acids with $C_{6\text{-}12}$ containing a carboxylic acid functional group. These compounds have a high energy density of 3492.4–4798.7 kJ/mol and a low O/C ratio ranging from 1:3 to 1:6. Besides energy, MFAs also used in food, biomedical, and chemical industries. As per an estimate, the global market size for MFAs has crossed the market demand of $8 billion by the end of

2023 from \$5.32 billion in 2014. Palm or coconut oil are conventional feedstock for MFAs production (Sun et al., 2024). Waste materials from sewage sludge can be used for MFAs. In comparison to the control, EF increased the short-chain fatty acid by 0.5 folds and attained a concentration of 82.4 mmol C/L. The combination of EF at acidification, anaerobic fermentation, and chain elongation has helped in attaining MFAs yield of 27.9 mmol C/L. The concentration was around 20%–86.6% higher than other groups (Sun et al., 2024). It was suggested that the increased performance was due to the improved electrochemical performance of system which enhanced the product yield. Anaerobic EF has supported the growth and diversity of electrochemically active bacteria, and improve the production of functional enzymes related to the production of Acetyl-CoA and fatty acid biosynthesis. An imbalanced electrochemical condition during the process lowers the end-products yield and increases the process period. In contrast, EF approach can modulate the metabolic process, stabilize the redox balance and increase the conversion process. Adilkhanova et al., 2024 reported that EF enhanced poly-γ-glutamic acid (γ-PGA) production by *Bacillus subtilis* ATCC 6051 by almost 3 times and attained 9.13 ± 1.4 g/ L in the presence of electrode potential of 0.4 V versus Ag/AgCl over control (3.3 ± 0.99 g/ L) (Adilkhanova et al., 2024). The overall benefit of EF can be interpreted in terms of cost–benefit. A comparison of acetic acid production via microbial electrosynthesis, anaerobic fermentation, and industrial process (ethane direct oxidation and methanol carbonylation) via reduction of CO_2, CO, and H_2O revealed that both elector-synthesis and anaerobic fermentation have high production and investment costs, i.e., 1.44 £/kg and 1770 £/t (microbial electrosynthesis) and 4.14 £/kg and 1598 £/t (anaerobic fermentation) respectively. In contrast, methanol carbonylation has production and investment costs of around 0.26 £/kg and 261 £/t while the same will be 0.11 £/kg and 258 £/t for ethane direct oxidation. Combining both anaerobic fermentation and microbial electrosynthesis reduced the investment costs by 9% and resulted in a production cost of 0.24 £/kg which was much lower than the current market price of 0.48 £/kg (Christodoulou & Velasquez-Orta, 2016).

The process not only reduced the cost but also utilized CO_2 in production which improved the environmental feasibility of the process. It also suggested possible ways to lower down the environmental pollution. Although, EF has proved efficient and advanced commercialization and scale-up need to overcome the challenges and bottlenecks like the requirement of high voltage which might be lethal for cells. In addition, electrode materials, toxicity, and cost of fabrication might also act as constrain.

7 Conclusion

Electro-fermentation is an advanced approach that can be integrated into conventional anaerobic fermentation models. Normally anaerobic fermentation is used for the production of biogas, alcohols, and commodity chemicals but it offers low product yield due to slow electron transfer. Some of the conductive materials can be added to the system to improve the transfer. Further advancements lead to the use of electrodes and processes run under the influence of electric fields. This approach combined two different approaches, i.e., fermentation and electrochemical fuel cell. Under the influence of an electric field, the flux of biological pathways changes and stabilizes the redox balance which might reduce the production time and product cost. However, the fabrication cost and stability of microbial cells at high voltage are still a challenge. At higher voltage, the viability of cells is reduced. Therefore, it still needs upgradation to increase the stability of the model with consistent productivity.

References

Abanades, S., et al. (2022). A critical review of biogas production and usage with legislations framework across the globe. *International Journal of Environmental Science and Technology, 19*(4), 3377–3400. https://doi.org/10.1007/S13762-021-03301-6/FIGURES/6

Adilkhanova, A., et al. (2024). Electrofermentation increases concentration of poly γ-glutamic acid in Bacillus subtilis biofilms. *Microbial Biotechnology, 17*(3), e14426. https://doi.org/10.1111/1751-7915.14426

Alhamed, M., Issa, A., & Doubal, A. (2012). Studying of natural dyes properties as photo-sensitizer for Dye Sensitized Solar Cells (DSSC). *Chemistry, Environmental Science, Materials Science, 16*, 1370–1383.

Ali, Q. A., et al. (2024). Enhancing microbial desalination cell performance for water desalination and wastewater treatment: experimental study and modelling of electrical energy production in open and closed-circuit modes. *Journal of Chemical Technology & Biotechnology, 99*(1), 294–306. https://doi.org/10.1002/JCTB.7538

Al-Mamun, A., et al. (2018). A review of microbial desalination cell technology: Configurations, optimization and applications. *Journal of Cleaner Production, 183*, 458–480. https://doi.org/10.1016/J.JCLEPRO.2018.02.054

Arun, J., et al. (2024). New insights into microbial electrolysis cells (MEC) and microbial fuel cells (MFC) for simultaneous wastewater treatment and green fuel (hydrogen) generation. *Fuel, 355*, 129530. https://doi.org/10.1016/J.FUEL.2023.129530

Baek, G., Kim, J., & Lee, C. (2016). A long-term study on the effect of magnetite supplementation in continuous anaerobic digestion of dairy effluent—Enhancement in process performance and stability. *Bioresource Technology, 222*, 344–354. https://doi.org/10.1016/J.BIORTECH.2016.10.019

Bajracharya, S., et al. (2016). An overview on emerging bioelectrochemical systems (BESs): Technology for sustainable electricity, waste remediation, resource recovery, chemical production and beyond. *Renewable Energy, 98*, 153–170. https://doi.org/10.1016/J.RENENE.2016.03.002

Batstone, D. J., Picioreanu, C., & van Loosdrecht, M. C. M. (2006). Multidimensional modelling to investigate interspecies hydrogen transfer in anaerobic biofilms. *Water Research, 40*(16), 3099–3108. https://doi.org/10.1016/J.WATRES.2006.06.014

Bhagchandanii, D. D., et al. (2020). A Comprehensive Understanding of Electro-Fermentation. *Fermentation, 6*(3), 92. https://doi.org/10.3390/FERMENTATION6030092

Bolzonella, D., et al. (2005). Anaerobic Fermentation of Organic Municipal Solid Wastes for the Production of Soluble Organic Compounds. *Industrial and Engineering Chemistry Research, 44*(10), 3412–3418. https://doi.org/10.1021/IE048937M

Carbas, B. B., et al. (2023). Construction of novel cyanobacteria-based biological photovoltaic solar cells: Hydrogen and photocurrent generated via both photosynthesis and respiratory system. *Journal of Photochemistry and Photobiology A: Chemistry, 442*, 114764. https://doi.org/10.1016/J.JPHOTOCHEM.2023.114764

Cardeña, R., Cercado, B., & Buitrón, G. (2019). Microbial electrolysis cell for biohydrogen production. In *Biomass, Biofuels, Biochemicals: Biohydrogen* (2nd ed., pp. 159–185). https://doi.org/10.1016/B978-0-444-64203-5.00007-1

Chakraborty, S., & Mondal, S. (2024). Halotolerant Citrobacter sp. remediates salinity stress and promotes the growth of Vigna radiata (L) by secreting extracellular polymeric substances (EPS) and biofilm formation: a novel active cell for microbial desalination cell (MDC). *International Microbiology, 27*(1), 291–301. https://doi.org/10.1007/S10123-023-00386-6/METRICS

Chen, S. T., Kuo, H. I., & Chen, C. C. (2007). The relationship between GDP and electricity consumption in 10 Asian countries. *Energy Policy, 35*(4), 2611–2621. https://doi.org/10.1016/J.ENPOL.2006.10.001

Cheng, Q., & Call, D. F. (2016). Hardwiring microbes via direct interspecies electron transfer: mechanisms and applications. *Environmental Science: Processes & Impacts, 18*(8), 968–980. https://doi.org/10.1039/C6EM00219F

Christodoulou, X., & Velasquez-Orta, S. B. (2016). Microbial electrosynthesis and anaerobic fermentation: An economic evaluation for acetic acid production from CO_2 and CO. *Environmental Science and Technology, 50*(20), 11234–11242. https://doi.org/10.1021/ACS.EST.6B02101/SUPPL_FILE/ES6B02101_SI_001.PDF

Erenler, S. A., et al. (2024). Development of microbial chondroitin sulfate-based proton exchange membranes for microbial fuel cells. *Fuel, 363*, 130976. https://doi.org/10.1016/J.FUEL.2024.130976

Flynn, J. M., et al. (2010). Enabling unbalanced fermentations by using engineered electrode-interfaced bacteria. *MBio, 1*(5), 190–200. https://doi.org/10.1128/MBIO.00190-10/ASSET/9F914ED2-4554-4621-A4AB-8AB477BB1897/ASSETS/GRAPHIC/MBO0041010550004.JPEG

Gajaraj, S., et al. (2017). Methane production improvement and associated methanogenic assemblages in bioelectrochemically assisted anaerobic digestion. *Biochemical Engineering Journal, 117*, 105–112. https://doi.org/10.1016/J.BEJ.2016.11.003

Gautam, R., et al. (2023). Bio-hydrogen production through microbial electrolysis cell: Structural components and influencing factors. *Chemical Engineering Journal, 455*, 140535. https://doi.org/10.1016/J.CEJ.2022.140535

Gu, J., et al. (2022). Electro-fermentation enhances H2 and ethanol co-production by regulating electron transfer and substrate transmembrane transport. *Chemical Engineering Journal, 429*, 132223. https://doi.org/10.1016/J.CEJ.2021.132223

Gu, L., et al. (2023). Superior anodic electro-fermentation by enhancing capacity for extracellular electron transfer. *Bioresource Technology, 389*, 129813. https://doi.org/10.1016/J.BIORTECH.2023.129813

Guang, L., et al. (2020). Performance of exoelectrogenic bacteria used in microbial desalination cell technology. *International Journal of Environmental Research and Public Health, 17*(3), 1121. https://doi.org/10.3390/IJERPH17031121

Guo, X., et al. (2020). Direct conversion of untreated cane molasses into butyric acid by engineered Clostridium tyrobutyricum. *Bioresource Technology, 301*, 122764. https://doi.org/10.1016/J.BIORTECH.2020.122764

Gupta, R., et al. (2023). A perspective on Bio-Electrochemical System (BES) as a tool for boosting the performance of the nonperforming anaerobic units. *Cleaner and Circular Bioeconomy, 6*, 100066. https://doi.org/10.1016/J.CLCB.2023.100066

Gürsan, C., & de Gooyert, V. (2021). The systemic impact of a transition fuel: Does natural gas help or hinder the energy transition? *Renewable and Sustainable Energy Reviews, 138*, 110552. https://doi.org/10.1016/J.RSER.2020.110552

Hassanein, A., et al. (2017). Next generation digestion: Complementing anaerobic digestion (AD) with a novel microbial electrolysis cell (MEC) design. *International Journal of Hydrogen Energy, 42*(48), 28681–28689. https://doi.org/10.1016/J.IJHYDENE.2017.10.003

Herzog, J. (2024). Co-culture of Acetobacterium woodii and Clostridium drakei using CO_2 and in situ generated H_2 to produce caproic acid via lactic acid. https://doi.org/10.15480/882.9002

Hirano, S., & Matsumoto, N. (2018). Analysis of a bio-electrochemical reactor containing carbon fiber textiles for the anaerobic digestion of tomato plant residues. *Bioresource Technology, 249*, 809–817. https://doi.org/10.1016/J.BIORTECH.2017.09.206

Holliger, C., de Laclos, H. F., & Hack, G. (2017). Methane production of full-scale anaerobic digestion plants calculated from substrate's biomethane potentials compares well with the one measured on-site. *Frontiers in Energy Research, 5*(JUN), 263802. https://doi.org/10.3389/FENRG.2017.00012/BIBTEX

Huang, H., Singh, V., & Qureshi, N. (2015). Butanol production from food waste: a novel process for producing sustainable energy and reducing environmental pollution. *Biotechnology for Biofuels, 8*(1), 1–12. https://doi.org/10.1186/S13068-015-0332-X/FIGURES/8

Huang, L., et al. (2020). Evidence for the coexistence of direct and riboflavin-mediated interspecies electron transfer in Geobacter co-culture. *Environmental Microbiology, 22*(1), 243–254. https://doi.org/10.1111/1462-2920.14842

IEA. (2020). *An introduction to biogas and biomethane—Outlook for biogas and biomethane: Prospects for organic growth—Analysis—IEA, International Energy Agency*. Retrieved September 7, 2024 from https://www.iea.org/reports/outlook-for-biogas-and-biomethane-prospects-for-organic-growth/an-introduction-to-biogas-and-biomethane

IEA50. (2020). *Global Energy Review 2020—Analysis—IEA, International Energy Agency*. Retrieved September 7, 2024 from https://www.iea.org/reports/global-energy-review-2020

IEA50. (2024). *Renewables Information—Data product—IEA, International Energy Agency*. Retrieved September 7, 2024 from https://www.iea.org/data-and-statistics/data-product/renewables-information

Ikeda, S., et al. (2023). Supplementation with amino acid sources facilitates fermentative growth of Shewanella oneidensis MR-1 in Defined Media. *Applied and Environmental Microbiology, 89*(7). https://doi.org/10.1128/AEM.00868-23/SUPPL_FILE/AEM.00868-23-S0002.XLSX

Imoro, A. Z., Mensah, M., & Buamah, R. (2021). Developments in the microbial desalination cell technology: A review. *Water-Energy Nexus, 4*, 76–87. https://doi.org/10.1016/J.WEN.2021.04.002

Jain, S., & Kumar, S. (2024). A comprehensive review of bioethanol production from diverse feedstocks: Current advancements and economic perspectives. *Energy, 296*, 131130. https://doi.org/10.1016/J.ENERGY.2024.131130

Jeremiah, M., Kabeyi, B., & Olanrewaju, O. A. (2020a). *Development of a Biogas Plant with Electricity Generation, Heating and Fertilizer Recovery Systems*, pp. 146–159.

Jeremiah, M., Kabeyi, B., & Olanrewaju, O. A. (2020b). Optimization of biogas production for optimal abattoir waste treatment with bio-methanation as solution to Nairobi slaughterhouses waste disposal. In *Proceedings of the 2nd African International Conference on Industrial Engineering and Operations Management*, pp. 160–170.

Jeremiah, M., Kabeyi, B., & Olanrewaju, O. A. (2020c). The potential of grid power generation from municipal solid waste for Nairobi City. In *Proceedings of the 2nd African International Conference on Industrial Engineering and Operations Management*, pp. 132–148.

Jiang, Y., et al. (2020). Electro-fermentation regulates mixed culture chain elongation with fresh and acclimated cathode. *Energy Conversion and Management, 204*, 112285. https://doi.org/10.1016/J.ENCONMAN.2019.112285

Joale, P. C., Ricardo, P. F., & Medina, T. G. I. (2020). Sizing of electric power generation systems from biogas through the generator staggering methodology. *SN Applied Sciences, 2*(2), 1–11. https://doi.org/10.1007/S42452-020-1968-0/TABLES/15

Kadier, A., et al. (2014). A review of the substrates used in microbial electrolysis cells (MECs) for producing sustainable and clean hydrogen gas. *Renewable Energy, 71*, 466–472. https://doi.org/10.1016/J.RENENE.2014.05.052

Kim, C., et al. (2017). Anodic electro-fermentation of 3-hydroxypropionic acid from glycerol by recombinant Klebsiella pneumoniae L17 in a bioelectrochemical system. *Biotechnology for Biofuels, 10*(1), 1–11. https://doi.org/10.1186/S13068-017-0886-X/FIGURES/6

Kim, K. R., Kang, J., & Chae, K. J. (2017). Improvement in methanogenesis by incorporating transition metal nanoparticles and granular activated carbon composites in microbial electrolysis cells. *International Journal of Hydrogen Energy, 42*(45), 27623–27629. https://doi.org/10.1016/J.IJHYDENE.2017.06.142

Kracke, F., & Krömer, J. O. (2014). Identifying target processes for microbial electrosynthesis by elementary mode analysis. *BMC Bioinformatics, 15*(1), 1–14. https://doi.org/10.1186/S12859-014-0410-2/FIGURES/6

Kracke, F., Vassilev, I.,& Krömer, J. O. (2015). Microbial electron transport and energy conservation—The foundation for optimizing bioelectrochemical systems. *Frontiers in Microbiology, 6*(JUN). https://doi.org/10.3389/FMICB.2015.00575/FULL

Kumar, P., et al. (2018). Electro-fermentation in aid of bioenergy and biopolymers. *Energies, 11*(2), 343. https://doi.org/10.3390/EN11020343

Lebuhn, M., Munk, B., & Effenberger, M. (2014). Agricultural biogas production in Germany—from practice to microbiology basics. *Energy, Sustainability and Society, 4*(1), 1–21. https://doi.org/10.1186/2192-0567-4-10/TABLES/3

Leech, D., Kavanagh, P., & Schuhmann, W. (2012). Enzymatic fuel cells: Recent progress. *Electrochimica Acta, 84*, 223–234. https://doi.org/10.1016/J.ELECTACTA.2012.02.087

Li, K., et al. (2023). Electrochemical control of cell metabolism improves ethanol production of zymomonas mobilis in an electro-fermentation system. *ACS Sustainable Chemistry and Engineering, 11*(6), 2364–2374. https://doi.org/10.1021/ACSSUSCHEMENG.2C06188/SUPPL_FILE/SC2C06188_SI_002.XLSX

Li, Y., et al. (2021). High-efficiency ethanol yield from anaerobic fermentation of organic wastes via stimulating growth of ethanol-producing Fe(III)-reducing bacteria with magnetite. *ACS Sustainable Chemistry and Engineering, 9*(3), 1246–1253. https://doi.org/10.1021/ACSSUSCHEMENG.0C07348/SUPPL_FILE/SC0C07348_SI_001.PDF

Li, Y., Park, S. Y., & Zhu, J. (2011). Solid-state anaerobic digestion for methane production from organic waste. *Renewable and Sustainable Energy Reviews, 15*(1), 821–826. https://doi.org/10.1016/J.RSER.2010.07.042

Liberato, V., et al. (2019). Clostridium sp. as bio-catalyst for fuels and chemicals production in a biorefinery context. *Catalysts, 9*(11), 962. https://doi.org/10.3390/CATAL9110962

Martínez-Ruano, J., et al. (2023). Electro-fermentation with Clostridium autoethanogenum: Effect of pH and neutral red addition. *Environmental Technology & Innovation, 31*, 103183. https://doi.org/10.1016/J.ETI.2023.103183

Maureira, D., et al. (2023). Industrial bioelectrochemistry for waste valorization: State of the art and challenges. *Biotechnology Advances, 64*, 108123. https://doi.org/10.1016/J.BIOTECHADV.2023.108123

Melikoglu, M., et al. (2016). Biochemical production of bioalcohols. In *Handbook of Biofuels Production: Processes and Technologies* (2 ed., pp. 237–258). https://doi.org/10.1016/B978-0-08-100455-5.00009-6

Mkilima, T., et al. (2024). Carwash wastewater treatment through the synergistic efficiency of microbial fuel cells and metal-organic frameworks with graphene oxide integration. *Case Studies in Chemical and Environmental Engineering, 9*, 100582. https://doi.org/10.1016/J.CSCEE.2023.100582

Mohammad, N., et al. (2021). Natural gas as a key alternative energy source in sustainable renewable energy transition: A mini review. *Frontiers in Energy Research, 9*, 625023. https://doi.org/10.3389/FENRG.2021.625023/BIBTEX

Mohammadi, M., Fazeli, F., & Baghsheikhi, M. (2024). Enhancing water desalination and power generation in microbial desalination cells: A comprehensive review of effective parameters and structural types for optimal performance. *Journal of Power Sources, 602*, 234267. https://doi.org/10.1016/J.JPOWSOUR.2024.234267

Mohd Noor, N. N., Oktavitri, N. I., & Kim, K. (2024). Effect of submerged and floating cathodes on sustainable bioelectricity generation and benthic nutrient removal in sediment microbial fuel cells. *Fuel, 367*, 131438. https://doi.org/10.1016/J.FUEL.2024.131438

Molognoni, D., et al. (2023). Enhancing butanol production by Clostridium beijerinckii through cathodic electrofermentation approach. *Journal of Chemical Technology & Biotechnology, 98*(4), 856–864. https://doi.org/10.1002/JCTB.7324

Moreau, V., & Vuille, F. (2018). Decoupling energy use and economic growth: Counter evidence from structural effects and embodied energy in trade. *Applied Energy, 215*, 54–62. https://doi.org/10.1016/J.APENERGY.2018.01.044

Moscoviz, R., et al. (2016). Electro-fermentation: How to drive fermentation using electrochemical systems. *Trends in Biotechnology, 34*(11), 856–865. https://doi.org/10.1016/J.TIBTECH.2016.04.009

Nozhevnikova, A. N., et al. (2020). Syntrophy and interspecies electron transfer in methanogenic microbial communities. *Microbiology (Russian Federation), 89*(2), 129–147. https://doi.org/10.1134/S0026261720020101/METRICS

Pant, D., et al. (2012). Bioelectrochemical systems (BES) for sustainable energy production and product recovery from organic wastes and industrial wastewaters. *RSC Advances, 2*(4), 1248–1263. https://doi.org/10.1039/C1RA00839K

Park, J., et al. (2018). Bioelectrochemical enhancement of methane production from highly concentrated food waste in a combined anaerobic digester and microbial electrolysis cell. *Bioresource Technology, 247*, 226–233. https://doi.org/10.1016/J.BIORTECH.2017.09.021

Perera, F. (2018). Pollution from fossil-fuel combustion is the leading environmental threat to global pediatric health and equity: Solutions exist. *International Journal of Environmental Research and Public Health, 15*(1). https://doi.org/10.3390/IJERPH15010016

del Pilar Anzola Rojas, M., et al. (2018). Microbial electrosynthesis (MES) from CO2 is resilient to fluctuations in renewable energy supply. *Energy Conversion and Management, 177*, 272–279. https://doi.org/10.1016/J.ENCONMAN.2018.09.064

Rago, L., Pant, D., & Schievano, A. (2019). Electro-fermentation—microbial electrochemistry as new frontier in biomass refineries and industrial fermentations. In *Advanced Bioprocessing for Alternative Fuels, Biobased Chemicals, and Bioproducts: Technologies and Approaches for Scale-Up and Commercialization*, pp. 265–287. https://doi.org/10.1016/B978-0-12-817941-3.00014-0

Ramanaiah, S. V., et al. (2023). Bioelectrochemical systems (BESs) for agro-food waste and wastewater treatment, and sustainable bioenergy—A review. *Environmental Pollution, 325*, 121432. https://doi.org/10.1016/J.ENVPOL.2023.121432

Sahota, S., et al. (2018). Review of trends in biogas upgradation technologies and future perspectives. *Bioresource Technology Reports, 1*, 79–88. https://doi.org/10.1016/J.BITEB.2018.01.002

Salar-García, M. J., et al. (2024). Sustainable production of biofuels and biochemicals via electro-fermentation technology. *Molecules, 29*(4), 834. https://doi.org/10.3390/MOLECULES29040834

Sangeetha, T., et al. (2017). Energy recovery evaluation in an up flow microbial electrolysis coupled anaerobic digestion (ME-AD) reactor: Role of electrode positions and hydraulic retention times. *Applied Energy, 206*, 1214–1224. https://doi.org/10.1016/J.APENERGY.2017.10.026

Sanni, A., et al. (2022). Sustainability analysis of bioethanol production from grain and tuber starchy feedstocks. *Scientific Reports, 12*(1), 1–15. https://doi.org/10.1038/s41598-022-24854-7

Schievano, A., et al. (2016). Electro-fermentation—Merging electrochemistry with fermentation in industrial applications. *Trends in Biotechnology, 34*(11), 866–878. https://doi.org/10.1016/J.TIBTECH.2016.04.007

Selvakumar, P., et al. (2022). Optimization of binary acids pretreatment of corncob biomass for enhanced recovery of cellulose to produce bioethanol. *Fuel, 321*, 124060. https://doi.org/10.1016/J.FUEL.2022.124060

Shakeel Ahmad, M., Pandey, A. K., & Abd Rahim, N. (2017). Advancements in the development of TiO2 photoanodes and its fabrication methods for dye sensitized solar cell (DSSC) applications. A review. *Renewable and Sustainable Energy Reviews, 77*, 89–108. https://doi.org/10.1016/J.RSER.2017.03.129

Shang, G., et al. (2024). A cathodic electro-fermentation system for enhanced methane production from high-concentration potato starch industrial wastewater. *Journal of Water Process Engineering, 59*, 105006. https://doi.org/10.1016/J.JWPE.2024.105006
Sharma, A., et al. (2024a). Factors affecting hydrogen production in microbial electrolysis cell (MEC): A review. *International Journal of Hydrogen Energy, 61*, 1473–1484. https://doi.org/10.1016/J.IJHYDENE.2024.02.193
Sharma, D., et al. (2024b). Simultaneous production of biogas and electricity from anaerobic digestion of pine needles: Sustainable energy and waste management. *BioTech, 13*(3), 35. https://doi.org/10.3390/BIOTECH13030035
Shi, M. M., Jiang, Y. G., & Shi, L. (2019). Electromicrobiology and biotechnological applications of the exoelectrogens Geobacter and Shewanella spp. *Science China Technological Sciences, 62*(10), 1670–1678. https://doi.org/10.1007/S11431-019-9509-8/METRICS
Shindell, D., & Smith, C. J. (2019). Climate and air-quality benefits of a realistic phase-out of fossil fuels. *Nature, 573*(7774), 408–411. https://doi.org/10.1038/s41586-019-1554-z
Slate, A. J., et al. (2019). Microbial fuel cells: An overview of current technology. *Renewable and Sustainable Energy Reviews, 101*, 60–81. https://doi.org/10.1016/J.RSER.2018.09.044
Soccol, C. R., Pandey, A., & Larroche, C. (2013). *Fermentation processes engineering in the food industry*. CRC Press. Retrieved September 7, 2024 from https://www.routledge.com/Fermentation-Processes-Engineering-in-the-Food-Industry/Soccol-Pandey-Larroche/p/book/9781138198678
Spadaro, D., et al. (2024). Development of innovative dye sensitized solar cells (DSSCs) based on co-sensitization of natural microbial pigments. *Dyes and Pigments, 229*, 112311. https://doi.org/10.1016/J.DYEPIG.2024.112311
Stevens, D. J. (2001). Hot gas conditioning: recent progress with larger-scale biomass gasification systems; update and summary of recent progress. *Natinal Renewable energy laboratory*. (NREL/SR-510–29952), p. 103. https://doi.org/10.2172/786288.
Strik, D. P. B. T. B., et al. (2011). Microbial solar cells: Applying photosynthetic and electrochemically active organisms. *Trends in Biotechnology, 29*(1), 41–49. https://doi.org/10.1016/J.TIBTECH.2010.10.001
Su, K., et al. (2023). A review on the interspecies electron transfer of methane production in anaerobic digestion system. *Fermentation, 9*(5), 467. https://doi.org/10.3390/FERMENTATION9050467
Sun, X., et al. (2024). Enhanced medium-chain fatty acid production from sewage sludge by combined electro-fermentation and anaerobic fermentation. *Bioresource Technology, 404*, 130917. https://doi.org/10.1016/J.BIORTECH.2024.130917
Tabatabaei, M., et al. (2015). Renewable energy and alternative fuel technologies. *BioMed Research International, 1*, 245935. https://doi.org/10.1155/2015/245935
Tejedor-Sanz, S., et al. (2022). Extracellular electron transfer increases fermentation in lactic acid bacteria via a hybrid metabolism. *eLife, 11*. https://doi.org/10.7554/ELIFE.70684
Tian, T., et al. (2018). Bio-electrochemically assisting low-temperature anaerobic digestion of low-organic strength wastewater. *Chemical Engineering Journal, 335*, 657–664. https://doi.org/10.1016/J.CEJ.2017.11.016
Topolewski, Ł, & Radulescu, M. (2021). Relationship between energy consumption and economic growth in European countries: evidence from dynamic panel data analysis. *Energies, 14*(12), 3565. https://doi.org/10.3390/EN14123565
Valentin, M. T., et al. (2023). Direct interspecies electron transfer mechanisms of a biochar-amended anaerobic digestion: a review. *Biotechnology for Biofuels and Bioproducts, 16*(1), 1–21. https://doi.org/10.1186/S13068-023-02391-3
Vassilev, I., et al. (2018). Anodic electro-fermentation: Anaerobic production of L-Lysine by recombinant Corynebacterium glutamicum. *Biotechnology and Bioengineering, 115*(6), 1499–1508. https://doi.org/10.1002/BIT.26562

Villano, M., et al. (2010). Bioelectrochemical reduction of CO_2 to CH_4 via direct and indirect extracellular electron transfer by a hydrogenophilic methanogenic culture. *Bioresource Technology, 101*(9), 3085–3090. https://doi.org/10.1016/J.BIORTECH.2009.12.077

Virdis, B., et al. (2022). Electro-fermentation: Sustainable bioproductions steered by electricity'. *Biotechnology Advances, 59*, 107950. https://doi.org/10.1016/J.BIOTECHADV.2022.107950

Wang, C., et al. (2024). Enhanced bioethanol production by evolved Escherichia coli LGE2-H in a microbial electrolysis cell system. *Bioresources and Bioprocessing, 11*(1), 1–10. https://doi.org/10.1186/S40643-023-00717-5/FIGURES/5

Wang, C. T., Dwivedi, K. A., & Lui, W. M. (2024). Exploring the role of domesticated resistors in batch-mode microbial desalination cell. *Chemosphere, 358*, 142173. https://doi.org/10.1016/J.CHEMOSPHERE.2024.142173

Wang, W., Chang, J. S., & Lee, D. J. (2022). Integrating anaerobic digestion with bioelectrochemical system for performance enhancement: A mini review. *Bioresource Technology, 345*, 126519. https://doi.org/10.1016/J.BIORTECH.2021.126519

Wang, W., & Lee, D. J. (2021). Direct interspecies electron transfer mechanism in enhanced methanogenesis: A mini-review. *Bioresource Technology, 330*, 124980. https://doi.org/10.1016/J.BIORTECH.2021.124980

Wang, X., et al. (2023). Feasibility of efficient, direct, butanol production from food waste without nutrient supplement by clostridium saccharoperbutylacetonicum N1–4. *Sustainability (Switzerland), 15*(7), 6061. https://doi.org/10.3390/SU15076061/S1

Xie, W., et al. (2023). In situ degradation of organic pollutants by novel solar cell equipped soil microbial fuel cell. *Environmental Science and Pollution Research, 30*(11), 30210–30220. https://doi.org/10.1007/S11356-022-24356-Z/METRICS

Xu, L., et al. (2023). Carbon-based materials as highly efficient catalysts for the hydrogen evolution reaction in microbial electrolysis cells: Mechanisms, methods, and perspectives. *Chemical Engineering Journal, 471*, 144670. https://doi.org/10.1016/J.CEJ.2023.144670

Yaqoob, A. A., Ibrahim, M. N. M., & Guerrero-Barajas, C. (2021). Modern trend of anodes in microbial fuel cells (MFCs): An overview. *Environmental Technology & Innovation, 23*, 101579. https://doi.org/10.1016/J.ETI.2021.101579

Yuan, H., et al. (2021). Disentangling the syntrophic electron transfer mechanisms of Candidatus geobacter eutrophica through electrochemical stimulation and machine learning. *Scientific Reports, 11*(1), 1–14. https://doi.org/10.1038/s41598-021-94628-0

Zhang, Y., et al. (2021). A cathodic electro-fermentation system for enhancing butyric acid production from rice straw with a mixed culture. *Science of The Total Environment, 767*, 145011. https://doi.org/10.1016/J.SCITOTENV.2021.145011

Zhang, Y., & Angelidaki, I. (2014). Microbial electrolysis cells turning to be versatile technology: Recent advances and future challenges. *Water Research, 56*, 11–25. https://doi.org/10.1016/J.WATRES.2014.02.031

Zhao, G., et al. (2021). Efficient production of Nisin a from low-value dairy side streams using a nonengineered dairy lactococcus lactis strain with low lactate dehydrogenase activity. *Journal of Agricultural and Food Chemistry, 69*(9), 2826–2835. https://doi.org/10.1021/ACS.JAFC.0C07816

Zhao, Z., et al. (2015). Potential for direct interspecies electron transfer in an electric-anaerobic system to increase methane production from sludge digestion. *Scientific Reports, 5*(1), 1–12. https://doi.org/10.1038/srep11094

Zhen, J., et al. (2024). Extracellular electron transfer (EET) enhanced anammox process for progressive nitrogen removal: A review. *Chemical Engineering Journal, 482*, 148886. https://doi.org/10.1016/J.CEJ.2024.148886

Zhou, M., et al. (2015). Development of bioelectrocatalytic activity stimulates mixed-culture reduction of glycerol in a bioelectrochemical system. *Microbial Biotechnology, 8*(3), 483–489. https://doi.org/10.1111/1751-7915.12240/SUPPINFO

Zou, J., et al. (2024). Dodecahedral Ag3PO4 photocatalysis and biodegradation synergistically remove phenol and generate electricity. *Renewable Energy, 231*, 120994. https://doi.org/10.1016/J.RENENE.2024.120994

Zuo, K., et al. (2016). A novel multi-stage microbial desalination cell for simultaneous desalination and enhanced organics and nitrogen removal from domestic wastewater. *Environmental Science: Water Research & Technology, 2*(5), 832–837. https://doi.org/10.1039/C6EW00196C

Chapter 10
Electro-Fermentation for Circular Carbon Economy

Saswata Goswami, Harshdeep Rana, Shreosi Chatterjee, Anjana, and Amita Sharma

1 Introduction

Fermentation technology is a crucial leading-edge biotechnological platform for producing a wide array of products. These include microbial whole cells in form of probiotic microbial culture and starters used in fermentation of foods, fermented dairy products like yogurt, beverages like beer and wine, bread, soy sauce, aromas for the food and perfume industries, antibiotic and pharmaceuticals, organic solvents (e.g., ethanol, acetone, butanol, etc.), amino acids, vitamins like ascorbic acid, riboflavin and cobalamin etc., enzymes, and biodegradable polymers (e.g., PHA and PHB) (Albuquerque et al., 2007). However, industrial fermentation technologies' applicability and economic viability are sometimes restricted due to several technical limitations. For instance, the limited availability of pure and defined substrates (for example, white sugars) are often associated with several environmental and economic sustainability challenges, since substrate specificity significantly enhances the degree of selectivity and purity of the target bioproducts during fermentation. The overall downstream and upstream processes including pure substrate procuring from defined and faithful agro-industrial sources, sustaining sterile environment for inoculum and fermenter, and harvesting products aseptically encompass up to 50% of total operational cost (Rago et al., 2019). Additionally, the majority of the time, the culture medium is augmented with optimal amount of macro and micronutrient precisely designed for distinct microbial strain and metabolic pathway since precise types and amounts of vitamins, minerals, buffers, chelating factors, nitrogen content, and

S. Goswami (✉) · H. Rana · S. Chatterjee · Anjana
Chemical Engineering Division, Department of Biotechnology, Center of Innovative and Applied Bioprocessing (CIAB), Government of India, Sector-81 (Knowledge City), Mohali, Punjab, India
e-mail: saswatagoswami2015@gmail.com; saswata@ciab.res.in

A. Sharma
Department of Chemical Engineering, Erode Senguthar Engineering College, Thudupatti, Erode, Tamil Nadu, India

© The Author(s), under exclusive license to Springer Nature Singapore Pte Ltd. 2025
S. Sevda (ed.), *Electro-Fermentation for Synthesis of Chemicals, Biochemicals and Biofuels*, https://doi.org/10.1007/978-981-96-6358-3_10

antifoaming agents supplemented in the production media significantly influence microbial metabolism greatly contributing overall production expenses (Rago et al., 2019).

The metabolic pathways in typical fermentations are frequently redox unbalanced, which drastically restricts the amount of product selectivity from a given substrate, which is primary drawback associated with conventional fermentation. To encounter such redox imbalance, several conventional strategies have been employed, for instance controlling pH, sparging of air, O_2, N_2, H_2, etc., and supplement of co-substrates like heme-containing proteins (Amaro-Reyes et al., 2024). As an electron acceptor, the culture media is supplemented with sterile air/O_2 to maintain the balance the redox equivalents facilitating the microbial metabolism. In order to prevent the gaseous metabolites (such as hydrogen and ammonia) buildup, sparging of di-nitrogen in production media is required in anaerobic fermentations. Alternatively, H_2 or other low-molecular-soluble substances like formate or acetate can be sparged as electron donors to improve reductive metabolism (Kim et al., 2006; Lyu et al., 2021). However, micro-bubbling has major disadvantages allied with mass-transfer efficacy as a consequence of reduced solubility of the gaseous phase. Therefore, to increase the gaseous solubilization, fermenters are further kept under pressure, which in turn causes inflation of reactor engineering costs (Raghavendran et al., 2020).

Additionally, chemical parameters such as supplement of acid/base, chelating agents, buffers, antifoam agents in production media is difficult to maintain in conventional fermentation. Typically pH of the media can maintained by either adding buffer or acid (HCl)/alkaline (NaOH) solution to the culture media. However, common constraint often occurs during the maintenance of the pH. It is quite difficult to achieve optimum pH without hampering the osmotic balance in the media because of ionic inflation such as Na^+ and Cl^- (Schievano et al., 2016; Singh et al., 2017). Separation and collection of all soluble metabolites from culture media, secondary chemicals, and other by-products is a highly challenging task. Furthermore, microbial growth and metabolism may be inhibited due to the presence of products or co-products in large quantities in the medium. Nowadays, research in biorefineries exploring microorganism have exclusively revolved around fermentation of agro-waste and low-grade carbon substrate (Roy et al., 2016).

Microbes have a natural ability to develop pathways and enzymes to convert carbon sources derived from biomass into a variety of novel compounds. Even if the substrate is impure, this ability can be efficiently cloned into engineered systems and carry out various techniques to manage culture purity (waste). In this regard, extremophiles such as hyperhemophile *Thermotoga neapolitana* can be a good choice (d'Ippolito et al., 2010). Moreover, culture purity can be achieved with modification of condition such as extreme temperature, pH, salinity, and other physico-chemical parameters, despite of substrate impurity. This could be crucial steps for determining specific metabolic pathways and improving the probability of regulating metabolic processes. Nevertheless, microbial transformation of impure substrates (waste, byproducts) have major challenges in managing the microbial population and redox conditions, leading to suboptimal process stability and product selectivity (Manzini et al., 2015).

To optimize conventional pure-culture fermentations and improvement of microbial biomass, electrochemically driven microbial metabolism may be a disruptive possibility. In this way, the application of electrodes can regulate and optimize fermentation settings to produce products of greater purity, to encourage the proliferation and density of microbial cells, to attain chain elongation, etc., which is known as electro-fermentation (EF) (Cardeña et al., 2024; Lu et al., 2024; Schievano et al., 2016). This idea integrates electrochemistry and industrial fermentation to create an innovative biotechnology perspective. In EF, an applied electrical field can have an oxidative or reductive effect on the intracellular metabolic pathways of microorganisms as well as the fermentation environment. Opportunities to improve carbon efficiency, reduce the amount of chemicals used for pH or redox balance control, strengthen microbial resilience and growth, or improve product recovery could lead to new developments in many fermentation platforms (Moscoviz et al., 2016). It is possible to directly drive microorganisms to employ electrons as reductant sources, as well as for integration or substitution of sugar (Zhang et al., 2024). Furthermore, it may be possible to stimulate biosynthesis from cheap carbon sources, maintain redox equilibrium in fermentation processes, and encourage cell development by electrically controlling metabolic pathways. Electrodes allow for the control and optimization of fermentation conditions, leading to the production of highly pure products, the growth and density of microbial cells, and the elongation of chains. As a result, electrical current can be utilized as a more environmentally friendly substitute for traditional substrates made from crops, like glucose, when reducing or oxidizing power is needed (Liu et al., 2023).

EF offers advantages over conventional fermentation which includes: (1) adjusting pH and redox imbalances caused by factors such as substrate purity and byproduct accumulation for optimization and stabilization of microbial metabolism during fermentation, (2) enhancing synthesis of ATP thus improving total microbial biomass yield, (3) facilitating either chain elongation or breakdown of carbon substances under diverse oxido-reductive environment, (4) extraction of desired value added bio products via selective membrane usage, (5) Direct fermentation pathways primarily toward the synthesis of a specific and distinct product (Schievano et al., 2016).

2 Principle of Electro-fermentation

The electro-fermentation (EF) process harnesses solid-state electrodes as electron mediators to strategically influence microbial metabolism. In this system, the anodic terminal acts as a powerful electron acceptor, drawing electrons away, while the cathodic terminal serves as an electron donor, actively supplying electrons to drive biochemical reactions. This dynamic electron exchange enables precise metabolic control, optimizing microbial activity for enhanced bioprocess efficiency (Roy et al., 2016). The distinctive function of EF approach integrates a microbial system with electrochemistry that allows all electrochemical redox reactions to be catalyzed (Fig. 1). The synergistic integration of electrochemical and biological reactions helps mitigate the thermochemical limitations typically associated with conventional fermentation, enabling more efficient and controlled bioprocessing (Bhagchandanii et al., 2020). Incorporating electrodes into a bioreactor facilitates precise control over electron transfer kinetics, thus enabling balanced redox states and synchronized electro-metabolic pathway regulation (Boucher et al., 2023). This approach drives the production of specific biobased products. The EF method sets itself apart from traditional fermentation techniques such as stirred-tank bioreactors, photobioreactors, and airlift bioreactors by leveraging electrochemical interactions to selectively develop microbial strains and broaden metabolic pathways. This strategic enhancement of microbial processes leads to a significant boost in the production of fatty acids, biogas, and platform chemicals, driving greater efficiency and yield in bioprocessing (Salar-García et al., 2024; Sravan et al., 2019). Significant increase in product yield is primarily caused by controlled use of reducing ions at the cathode, facilitated by integrated electron metabolism. EF can aid in the stabilization and optimization of metabolic pathway via balancing redox and pH constraints, such as the generation of byproducts and substrate purity (Schievano et al., 2016). It alters different oxidative and reductive conditions or upregulates the disintegration of the carbon chain. Additionally, EF promotes the production of more adenosine triphosphate, which raises the biomass output. EF significantly governs the fermentation process to maximize the yield of a certain product, such as ethanol produced from glycerol (Virdis et al., 2022).

2.1 Anodic Electro-fermentation

During fermentation processes, redox reactions typically facilitate oxidation and reduction of electron carriers, e.g., reduction of nicotinamide adenine dinucleotide (NAD^+) to NADH aiding simultaneous oxidation of substrate oxidation. This phenomenon leads to excretion of end products such as ethanol and CO_2 from the microbial cell into the environment. In biochemical pathways, the production of various end products depends on the flow of electrons through the system. By altering the conditions that affect electron flow, such as hydrogen partial pressure or pH levels, one can influence the balance and proportions of the products formed, thus enhancing yield of final product (Cheah et al., 2019; Zhou et al., 2017). In addition to conventional approaches, intracellular NAD^+/NADH ratio can be altered to achieve cellular redox balancing with a polarized electrode, which is considered as fundamental aspect of the EF process. Introduction of electrodes in fermentation system facilitates increased the production of either oxidized or reduced products. EF technology uses polarized electrodes to ferment energy-rich substrates, such as alcohols and carbohydrates, by serving as additional sources or sinks for electrons. In anodic electro-fermentation (AEF), the working electrode acts as an anode, helping to disintegrate surplus electrons. This process enhances the fermentation efficiency

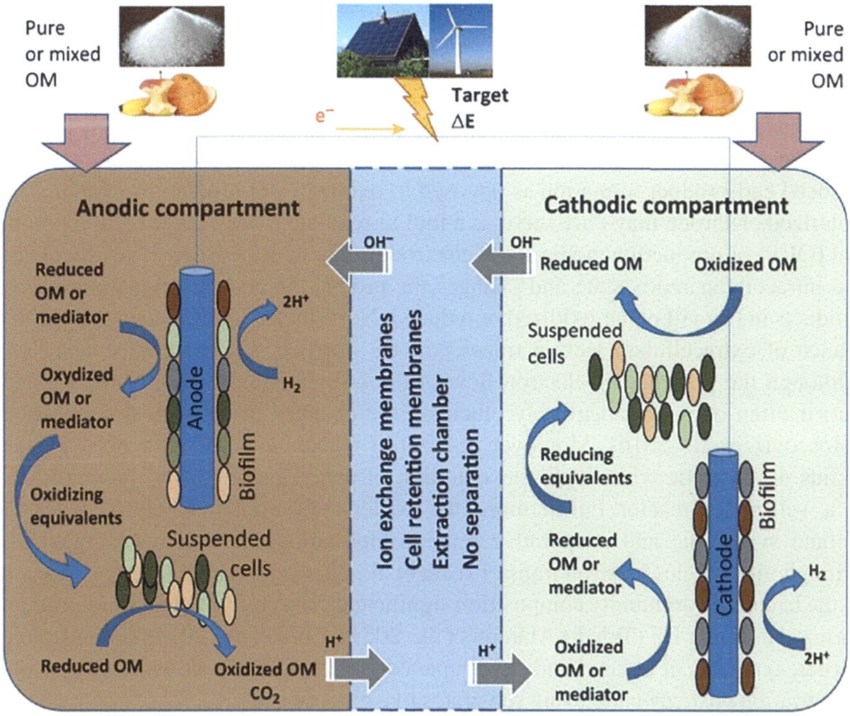

Fig. 1 Schematic representation of principle of electro-fermentation technology (Schievano et al., 2016)

by providing a controlled environment for electron transfer, ultimately leading to more effective conversion of substrates into desired products (Virdis et al., 2022).

In case of AEF, the final products undergo higher oxidation than the substrates. Due to this, the amount of intracellular NADH decreases, depriving the cell of electrons needed to produce a reduced molecule. The use of an anode as a strategic substitute for the oxygen needed in conventional aerobic respiration processes has been suggested. AEF may be advantageous when the reduced compound is an undesirable result. Energy and financial savings are two major benefits of the potential to convert aerobic processes to anaerobic ones (Vassilev et al., 2021). In summary, AEF is the process where working electrode is a polarized anodic electrode serves as a powerful equipment in EF. In general, bioengineered bacterial strains, such as *Escherichia coli*, *Clostridium cellobioparum*, *Shewanella oneidensis*, and *Geobacter sulfurreducens*, are widely used in AEF (Logan et al., 2008; Moscoviz et al., 2016). To shed light on the facts supporting the viability and efficacy of this approach, more investigation is nonetheless required.

2.2 Cathodic Electro-fermentation

In case of cathodic electro-fermentation (CEF), the electrode donates electrons to microbial cells besides electrons gained from substrate oxidation which in turn enhance reduction of intracellular NAD^+. This could eventually result in more reduced end-product formation as opposed to oxidized end-product formation. The polarized electrode may only serve as a tool to regulate oxidation–reduction potential (ORP) of production media at the electrode edge, which consecutively regulates the intracellular redox state and changes the production and dissemination of end products in form of either oxidized or reduced. Nevertheless, mechanism and significance of extracellular electron transfer in EF practice still need to be revealed. Although the presence of electron flow in the fermentation broth cannot be ruled out, it often does not adequately elucidate the changes in product dissemination (Moscoviz et al., 2016). Moreover, the entire reactor may be influenced by the radius of electrode when high concentration of redox intermediates are employed in a well-mixed reactor. Furthermore, the existence of a polarized electrode might initiate syntrophic and cooperative relationships among various microorganisms through interspecies electron transfer (Chu et al., 2020). Remarkably, any alteration in the bacterial community composition significantly interferes with the end-product formation during EF (Toledo-Alarcón et al., 2021). Choi et al. (2014) demonstrated earlier, synthesis of butanol and 1,3-Propanediol were carried out under CEF using *Clostridium pasteurianum* using substrates like glucose and glycerol respectively. In another study, *Clostridium beijerinckii* was utilized to facilitate the acetone-butanol-ethanol fermentation for simultaneous production of H_2 and butanol in order to investigate the efficacy of neutral red-assisted CEF (Zhang et al., 2021a). Compared to the open-circuit system, production of H_2 and butanol were enhanced by 172.7% and 71.4%, respectively, while adjusting cathodic potential at −0.75 V (vs. Ag/AgCl). (Zhang et al., 2021a). Further, impact of neutral red in butyric acid production was conducted at a cathode potential of −0.8 V (vs. Ag/AgCl) in a MFC-CEF system mixed microbial cultures (Zhang et al., 2021b). No significant improvement in butyric acid production was observed with further enhancement of neutral red since contribution of electric current to end product was minimal.

3 Fundamentals of Microbial Electrochemistry and Bio-electrochemical System

The EF process depends on microbial bio-electrochemical system (BES) which allows bidirectional flow of electrons between microbes and electrode for conversion of substrate to value added bioproducts (Hassan et al., 2021). In this process, electrons are harvested from organic substances mediated via biocatalytic activity of electrochemically active microorganism, thereby facilitated bio-electrochemical interaction (Schröder et al., 2015). In general, two types of interaction occurs between physical electrodes and microorganisms such as capacitive and Faraday (Fig. 2). In case of capacitive interaction, double layer capacity of electrodes can be achieved by either attachment or detachment of microbial cell to the electrode. Microbes adhere to an electrode, resulting in displacement of ions and water molecules from the double layer which in turn reduces the electrochemical capacity of the electrode. This allows a charge-balancing (capacitive) electric current to flow (Wang et al., 2020a, 2020b). Conversely, any redox reactions concerning microbial cells and any molecular species facilitating microbial extracellular electron transport are considered to be part of microbial Faraday processes. Microbial cells and biofilms during pseudo-capacitive interaction experience oxidative or reductive charging and discharging (acting as a microbial supercapacitor), and microbial electrocatalysis are prominent instances of these microbial Faraday reactions (Schröder et al., 2015; Zhang et al, 2019).

In a typical BES, electrodes functions synergistic approach with microorganisms to exchange electrons in various configurations. This process can be used for bioenergy production, wastewater treatment, bioremediation, and synthesis of high-value added biochemical and bioactive compounds (Ghangrekar & Neethu, 2020; Jin & Fallgren, 2022; Kondaveeti et al., 2023; Wilberforce et al., 2021). Microorganisms universally require electron mediators for executing all normal physiological functions. Both organic and inorganic materials in the environment serve as donors and acceptors (Rojas et al., 2021). Lithotrophic, lithoheterotrophic and lithoautotrophic microorganisms have the ability to utilize inorganic substances as electron donors to obtain energy while using CO_2 and organic substances, i.e., sole source of carbon, as electron acceptor (Rabaey et al., 2011).

In traditional microbial electrochemical technology, solid electroconductive materials as electrode can facilitate electron transfer from microbes and vice versa through direct electrochemical stimulation (Zheng et al., 2020). Lithotrophic organisms utilizes solid conductors as electron donors or acceptors which demonstrate electron transfer via nanowires, cytochromes, NAD, and other membrane-bound enzymes. The primary mechanisms for this electron transfer are extracellular electron transfer (EET) and transmembrane transport (Kracke et al., 2015). Additionally, soluble redox mediators such as riboflavin and neutral red also play a crucial role in mediating the EET and transmembrane trafficking to the cytoplasm (Roy et al., 2022). Redox mediators, for example reduced neutral red, were used for the metabolic activation in a pure and mixed culture of hydrogen-utilizing microorganisms which enlighten the involvement of menaquinone in the fumarate reductase complex. Furthermore, *Actinobacillus succiongenes* used carbon dioxide to ferment hydrogen and fumarate, oxidizing neutral red in the process to produce succinate.

Electrons from the electrode can be directly accepted in a variety of species which involves the movement of the electrons from the cathode to the attached biofilm; some of these bacteria include *Moorella thermoacetica*, *Clostridium* sp., and *Sporomusa* sp., which convert carbon dioxide to acetate. Thus, the cathodic interaction aids these acetogenic bacteria to thrive in autotropic environments (Lovley, 2017). To activate acetyl-CoA pathway, the metabolic process requires a terminal electron acceptor apart from oxygen, such as CO_2. The acetyl-CoA pathway is used by non-acetogenic

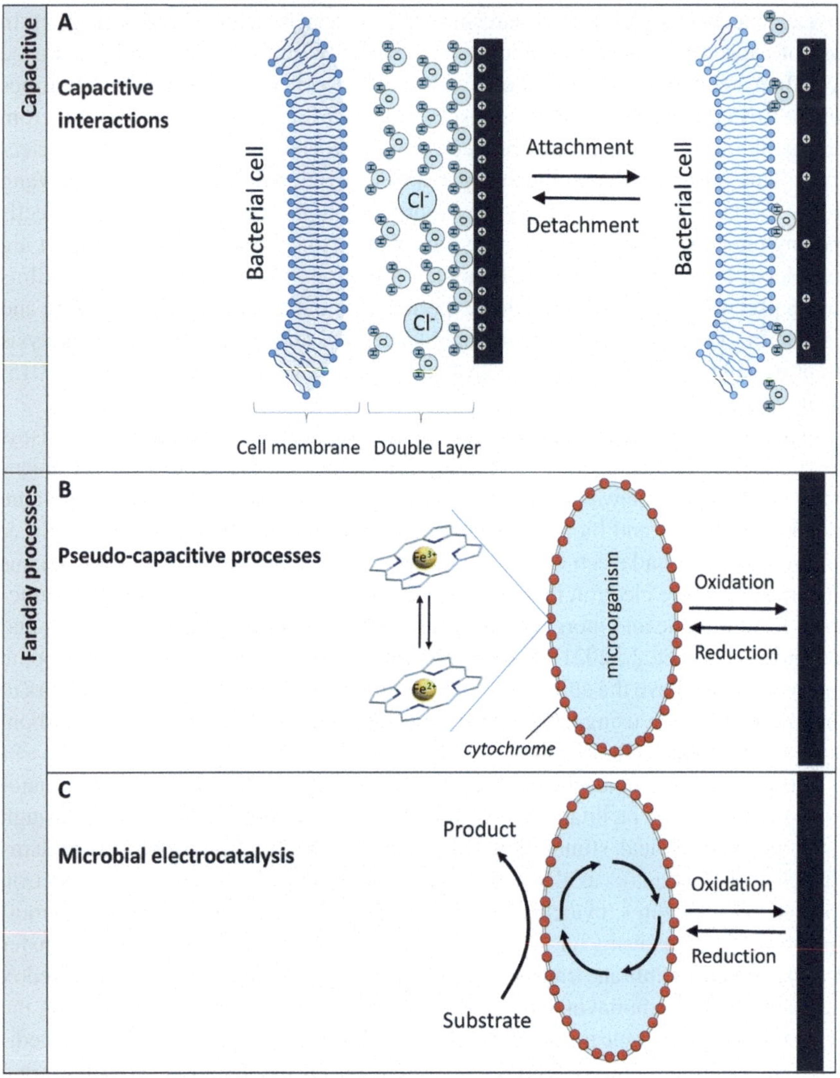

Fig. 2 Depiction of major electrochemical interaction between microorganism and electrodes in bioelectrochemical system (Schröder et al., 2015)

bacteria such as sulfur-reducing bacteria and other archaebacteria to neutralize CO_2. This phenomenon also coincides with electron transport in the Archaea domain. The methanogens belonged to the Archaea domain need hydrogen and carbon dioxide in order to produce methane. For instance, electro-methanogenesis process can be demonstrated in *Methanobacterium palustre* by delivering the electron directly from the cathode to the Archaea biofilm. It was discovered that thermophilic bacteria, such as *Methanococcus maripaludis*, can produce methane even without an external hydrogen source. It was shown that, in the absence of hydrogenase, the bacteria utilized electrons from the cathode (Pawar et al., 2022).

3.1 Microbial Electrolysis

Microbial electrocatalysis can be defined as facilitating electrochemical reactions through electrochemically active microorganisms through extracellular electron transfer. Reducing the overpotential needed to reach a specific reaction rate or raising the rate constant and current flow at a specific potential are the two methods involved (Liu et al., 2018). While "catalysis" generally refers to passive facilitation, microbial cell actively engage in microbial systems by harvesting energy from the reaction. Bioelectrochemical systems (BESs) in biotechnology can be expanded to include synthesis by microbial electrocatalysis, or it can be utilized for substrate degradation, with an emphasis on environmental applications such as microbial fuel cells (MFCs) for wastewater treatment and energy recovery. Microbial electrocatalysis has shown enormous promise in recent years, both for the electrochemical synthesis of chemicals for environmentally friendly and sustainable companies and for the biodegradation of pollutants for environmental applications (Gildemyn et al., 2017; Rosenbaum & Henrich, 2014).

Utilizing exoelectrogenic microorganisms, the MEC system transforms organic and biodegradable materials found in the waste stream into protons and electric current (Ahmad et al., 2022; Jain & He, 2018). At the anode, microorganisms work with the biodegradable waste to produce protons and power. Afterwards, the electrons are transported to the cathode, which facilitate reduction of the protons to produce H_2 (Adesra et al., 2021; Huang et al., 2022). Depending on the substrate, different microorganisms are employed in the MEC. For instance, acetate is utilized as substrate for *Aeromonas hydrophila*, *Thermincola* sp., *Geothrix fermentans*, and *Gluconobacter oxydans*; acetate lactate is used for *S. putrefaciens* and *S. oneidensis*; and acetate glucose serves as the substrate for *K. pneumoniae*, *Rhodoferax ferrireducens*, and *E. cloacae* (Kadier et al., 2020).

Being endothermic process, microbial electrolysis requires a small voltage between two electrodes to trigger electricity generation and produce H_2 at the cathode. However, small solar panels or low-grade microbial fuel cells can provide the applied voltage, which is typically 0.2–0.8 V (Rousseau et al., 2020). Microbial electrolysis cells (MECs) for the production of H_2 is highly regarded for commercialization because of enhanced production of H_2 as compared to conventional biorefineries or gas-fermentation. Additionally, MECs demonstrate higher performance efficiency than other microbial BES, such as MFCs (Koul et al., 2022; Rousseau et al., 2020).

A typical MEC system comprised of three major components such as ion exchange membrane, a cathode, and an anode, as shown in Fig. 3 (Saravanan et al., 2020). Organic substances in the anolyte and on the anode surface are oxidized by microbial cell, subsequently, releasing proton into the fermentation media which cross the separator to reach cathode. The electron (using an external circuit that is transmitted from anode to cathode) converts them to molecular hydrogen. Like other biological wastewater treatment, the application of MEC technology includes a diverse range of bacteria present in the mixed culture and a number of lateral reactions that stem from the complex composition of the wastewater. The entire MEC process is demonstrated using acetate, the frequently utilized carbon source for bioelectrochemical system (BES) bioanodes in laboratory investigations. (Cardeña et al., 2019). The chemical reactions for generation of hydrogen have been illustrated in the following equations (Eqs. (1), (2), (3)).

Anode:

$$CH_3COO^-_{(ac)} + 4H_2O_{(l)} \rightarrow 2HCO^-_{3(ac)} + 9H^+_{(ac)} + 8e^- \tag{1}$$

Cathode:

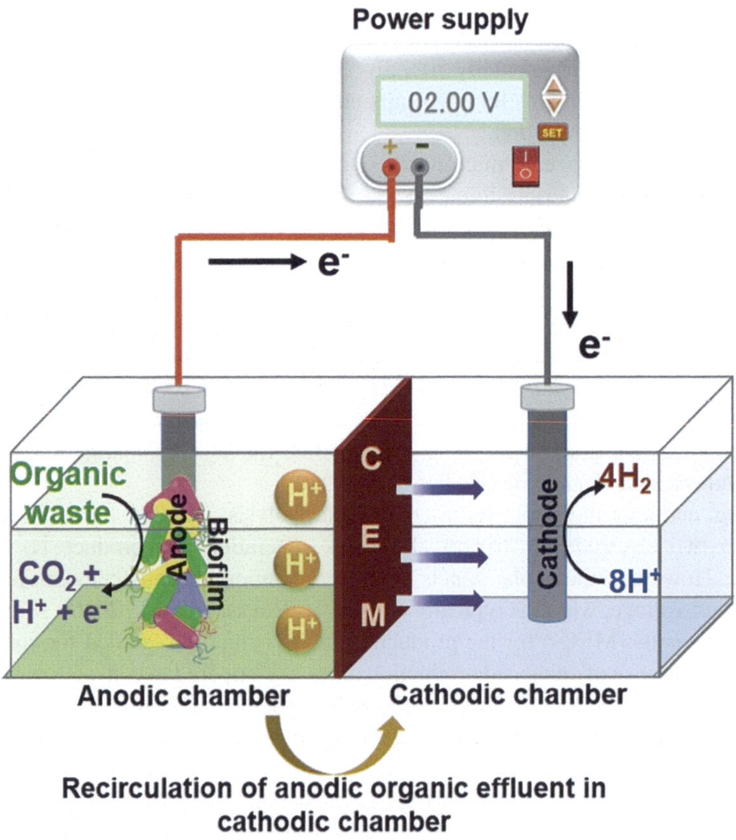

Fig. 3 Schematic of a microbial electrolysis cell used for hydrogen production

$$8H^+_{(ac)} + 8e^- \rightarrow 4H_{2(g)} \tag{2}$$

Overall Reaction:

$$CH_3COO^-_{(ac)} + 4H_2O_{(l)} \rightarrow 2HCO^{-\bullet}_{3(ac)} + H^+_{(ac)} + 4H_{2(g)} \tag{3}$$

The MEC technology offers a number of benefits over conventional treatment techniques and has shown to be a very effective way for recovering resources from several industrial wastewater effluents which includes, exploration of various substrates, yielding significantly higher H_2 than fermentation, for example yielding 90% hydrogen from acetate-based MECs, improvement of methane production from 60 to 98% by integrating traditional anaerobic digesters with MECs, integration with biorefineries to produce hydrogen and organic chemicals, increasing overall fermentation efficiency, reducing fossil fuel use, lowering greenhouse gas emissions, promoting renewable substrates for energy generation, processing organic wastes, and supporting ecological recycling (Koul et al., 2022; Segundo-Aguilar et al., 2021; Varanasi et al., 2019). However, MEC technology also has certain drawback such as reduction of H_2 yield over time due to undesired electron sinks in various metabolisms, complicated MEC reactor set up, struggle to meet effluent discharge standards, necessitating process refinement for maximum efficiency and require extensive knowledge to understand microbial metabolic pathway (Kadier et al., 2018; Leicester et al., 2020; Miller et al., 2019).

3.2 Microbial Electrosynthesis

Microbial electrosynthesis (MES) is considered as cutting-edge technology for capturing and using carbon dioxide (CO_2) for the bioelectrochemical conversion of CO_2 into value-added bioproducts (Nevin et al., 2010). This revolutionary biocathode-driven technology explores pure and mixed microbial culture to manufacture biofuels or biochemicals by either absorbing electrons from physical electrodes or utilizing electron carriers (e.g., H_2) at cathodic terminal end (Arends et al., 2017; Sciarria et al., 2018; Vassilev et al., 2019). The idea of electrically driven bioconversion of CO_2 was first presented in 2009 with methane production in a BES equipped with an abiotic anode and a biocathode employing *Methanobacterium palustre* as a catalyst (Cheng et al., 2009). In another study, acetic acid and trace amounts of 2-oxobutyrate production from CO_2 was achieved by using graphite electrode as a cathode and *Sporomusa ovata* as the biocatalyst (Nevin et al., 2010). Since then, MES has been thoroughly investigated for its ability to fix CO_2 and create basic organic compounds like ethanol, butyrate, acetate, and methane (Bian et al., 2018; Wang et al., 2021). If solar energy is used as the external power source, the MES process resembles the natural process of photosynthesis (Majumdar et al., 2017). The acetogens have typically been employed as biocatalysts at the cathode of MES systems because they fix CO_2 via the Wood-Ljungdahl (WL) route. Therefore, it has generally been reported that acetate is the primary result of MES. Still, it has been claimed that MES can synthesize other higher carbon organic compounds such as propionate, butyrate, ethanol, isopropanol, caproate, and caprylate from CO_2 (Das & Ghangrekar, 2018; Das et al., 2020).

The abiotic anodic chamber and the biotic cathodic chamber are the two chambers that make up a standard MES configuration (Fig. 4). Proton exchange membrane (PEM) facilitates movement of protons from the anodic to the cathodic chamber. Gaseous oxygen, protons, and electrons are produced when water molecules split at the anode. Protons migrate from the anodic chamber to the cathodic chamber across the PEM, oxygen leaves the anodic chamber, and electrons are pulled to the cathode by an external circuit. Biocatalysts in the cathodic chamber combine protons and electrons or energy carriers like H_2 and CO_2 to generate primarily volatile fatty acids such as formate, acetate, butyrate, etc. either via H_2 mediated electron transfer or direct electron transfer (DET), i.e., H_2 independent route (Pant et al., 2012). According to Karthikeyan et al. (2019), there are two types of extracellular electron transfer (EET) mechanisms occurred at cathodic end: direct electron transfer (DET) and indirect electron transfer (IDET). The IDET needs electron transfer mediators to deliver cathode generated electron to suspended biomass rather than a biofilm forming on the surface of cathode, as required by the DET. The current density and MES production rate are determined by the EET rate. Thus, two of the main MES research focus areas have been improving the EET rate and comprehending the EET mechanisms (Choi & Sang, 2016).

MES technology provides two major benefits which includes: (i) utilizing anaerobic electrotrophic bacteria as biocatalysts to produce multi-carbon organic compounds, or electro-commodities from CO_2 sequestration, (ii) serving as repository surplus electrical energy in chemical bonds (Das et al., 2019; Rabaey & Rozendal, 2010). Activation and subsequent conversion reactions of CO_2 require the supply of external energy due to its thermodynamic stability. MES technology offers several advantages, especially in countries with limited arable land, as it does not require cultivable land for biofuel production. Its efficiency, particularly when combined with photovoltaic systems, surpasses that of biological photosynthetic pathways, enabling the synthesis of high-value products with less energy. MES also reduces operational costs by not requiring large amounts of nutrients or high-quality water, and it minimizes environmental impact by avoiding excess nutrient and pollutant release. Additionally, MES simplifies the production process by reducing the need for extensive downstream processing, which is often necessary for extracting biofuels from biomass and can generate additional waste and costs (Liu et al., 2023; Rojas et al., 2021). However, MES technology also demonstrated certain limitations which includes difficulties in purification and separation of products which further decreasing the product yield, high operation cost affecting scale up process, and slow rate of conversion with long chain carbon molecules (Das et al., 2019).

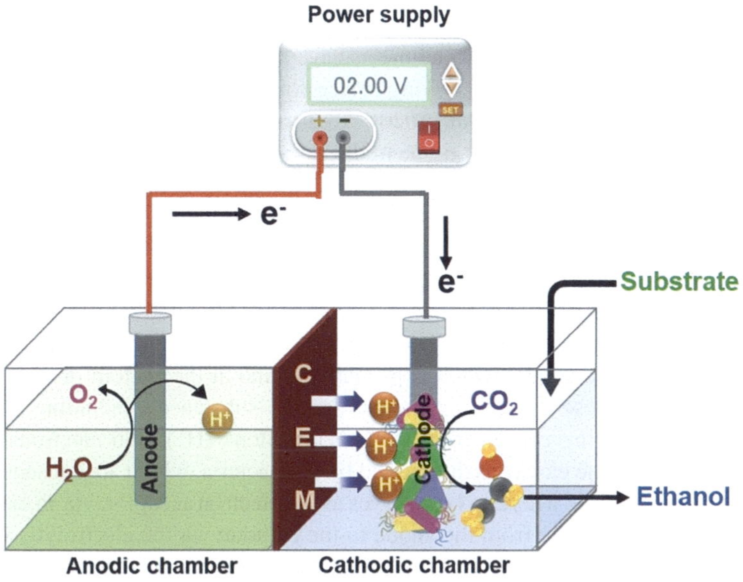

Fig. 4 Depiction of a typical microbial electrosynthesis system for production of value added bioproducts from CO_2

4 Exploring Electrochemically Active Microorganisms in Electro-fermentation

Microorganism belonged to the domains Bacteria and Archaea are predominant electrochemically active bacteria (EAB), and they can thrive in various environmental niches, such as soil, sediments, aquatic systems, and harsh environments. These microbes have the unique capacity of establishing electrical interactions with their extracellular environment as well as with other bacteria. Due to their ability to produce electricity within BES, EAB have recently gained an extensive attention for EF technology (Garbini et al., 2023). For instance, genus *Thiomicrorhabdus* was isolated by Yamamoto et al. (2023) from a heat vent which possesses unique multiheme cytochrome-c proteins involved in extracellular electron transfer. Remarkably, Ren et al. (2019) also discovered a number of genera of electroactive bacteria on varnish rock, a black, Fe/Mn-rich coating that forms on rock surfaces and is thought to be a harsh habitat. Recently, there are several reports of electroactive bacteria thriving in freshwater environments, including the sediments of lakes or rivers (Lu et al., 2021; Shi et al., 2021; Zhao et al., 2018). Furthermore, EAB was also found in brackish ecosystems, mudflat marine sediment, and salt marshes (Chabert et al., 2015; Rimboud et al., 2021; Yang & Chen, 2021). These bacteria utilizes electron acceptors like Mn^{4+} and Fe^{3+} to facilitate complete oxidation of organic matter to CO_2 in a mangrove silt, typically characterized by enriched organic matter and anoxic conditions (Kristensen et al., 2008; Szydlowski et al., 2020). Even while freshwater and marine sediments have been studied for EAB, soil ecosystems, which have a higher biodiversity range than water and sediment, have not been thoroughly investigated till date. Microbial communities thriving in rice soils have been extensively studied (Gustave et al., 2019; Kamaraj et al., 2020). Additionally, electrogenic bacteria were discovered as endophytes in plant tissue in the stems of *Angelica sinensis* and sweet potato roots (*Dioscorea esculenta*) (Ling et al., 2022). It has been also shown that exoelectrogenic bacteria thrive in the human stomach. Five gut bacteria that are capable of extracellular electron transfer have been found by Tahernia et al. (2020).

BES is a promising green technology to generate renewable bioelectricity from organic carbon-rich wastewater due to the adaptability of electrochemically active bacteria to various carbon sources. Under anaerobic conditions, electrogenic bacteria form a biofilm on the anode, which is primarily found at the bottom of a modular fuel cell (MFC). These bacteria catabolize organic substances, including various pollutants, by releasing carbon dioxide (CO_2), protons (H^+), and electrons (e^-). The anode receives the electrons discharged by the bacteria, which are subsequently migrated to the cathode, where oxygen serves as the electron acceptor, via an external connection. Protons move from the anode to the cathode via the electrolyte, which is made up of soil, water, or silt and offers organic matter that the bacterial colonies require. (Borello et al., 2021). Bacterial culture can be utilized in EF process either in pure or consortia form which are described below.

4.1 Pure Culture

Metabolic pathway engineering targets specific industrial biosynthesis of metabolites through fermentation using pure designated microbial strains, besides utilizing native microbiomes (Mariano et al., 2015). The primary goal of pure-culture fermentations is maximum product formation, requiring a pure, specific, and single substrate to ensure product selectivity and culture purity. The manufacturing of pure and sterile carbon sources, like glucose, glycerol, cornstarch, sugarcane hydrolysate, and vanillin, constitutes a significant cost (Lee et al., 2011). Sustainability in substrate production necessitates allocated land usage and agro-industrial revolutions, especially for upscaling biochemical production. For instance, approximately half of the total greenhouse gas emissions and energy expenditure are attributed to the manufacturing of dextrose from corn. (Mathew et al., 2015). In spite of air filtering and sterilizing technologies, pure culture stability and reactor sterility are also issues when significant volumes of ambient air are sparged (Deval et al., 2017). Gas sparging in fermentation broths frequently leads to foam formation, necessitating the constant use of antifoaming agents and foam-monitoring systems. In optimally balanced fermentation, all electron equivalents are directed into a single product, simplifying downstream separation and purification. Even in pure-culture fermentations of a single microbial species, multiple metabolic pathways can generate distinct metabolites, resulting in diverse end products while the cell maintains homeostasis (Bhagchandanii et al., 2020; Schievano et al., 2016).

The integration of bioelectrochemistry and fermentation can lead to several benefits, including increased productivity and yields, enhanced process kinetics, modified and reduced fermentation product spectrum, more efficient in situ supply of O_2, H_2, or electrons compared to sparged gases, improved process control through electron flux, and stimulation of thermodynamically unfavorable reactions (Rago et al., 2019). Early work in microbial electrochemistry introduced electrodes as electron donors for L-glutamic acid formation. Later, a shift toward more reduced products, such as butanol and ethanol, was demonstrated during glucose fermentation using *Clostridium acetobutylicum* (Kim & Kim, 1988; Rago et al., 2019). An electrode inoculated with a genetically modified strain of *Shewanella oneidensis* was used to perform the stoichiometric transformation of glycerol to ethanol (Zhou et al., 2015). In more recent work, the ability of *Pseudomonas putida* KT2440 to manufacture its own mediators, like phenazines, was utilized in fermentation using an engineered strain of the bacteria, enabling the production of biochemicals otherwise requiring oxygen sparging (Schmitz et al., 2015).

Kracke and Krömer (2014) established the fundamental networks of metabolic (anaerobic) carbon pathways in *Escherichia coli*, demonstrating significant efficiency to increase the rates of anaerobic fermentations of glycerol and sugar, improving the yields of 18 of the 20 products by 7–84%. Pandit and Mahadevan (2011) presented a step-by-step method for increasing productivity. Using no electrical augmentation, they first maximized growth rate and then used the same strategy to increase product yield. Applying 15 V of static potential to yeast culture of *Saccharomyces cerevisiae* significantly increased both ethanol yield and productivity without any resulting current (Mathew et al., 2015; Zhou et al., 2018). Similar results were reported for butanol and glutamate production using pure culture of *Clostridia* sp. and *Brevibacterium* sp., respectively, (Rago et al., 2019). The use of highly active catalytic cathodes, i.e., electrochemically deposited nickel phosphide (Ni–P) catalyst on cathode significantly improved the production of acetate and butyrate using engineered *Clostridium ljungdahlii* strain (Wang et al., 2020a, 2020b). Recently, Mogollón et al. (2023) reported enhanced production of butanol and acetone using electrochemically active *Clostridium saccharoperbutylacetonicum* N1-4 by tuning the extracellular redox potential which was accomplished by hovering the cathode potential at −600 mV versus Ag/AgCl or by introducing NADH, a redox agent, to the microbial broth.

4.2 Microbial Consortia

Enriched mixed microbial cultures often represent the core of many conventional microbial fermentation techniques used in food and beverage industries. Depending on the purpose of the fermentation, using cocultures in traditional fermentation instead of pure cultures can have a number of benefits, including optimizing the process with affordable and minimum requirement to prevent contamination and maintain axenic conditions, and improving its economics. During co-culture, syntrophy occurs in which one microorganism produces growth factors, eliminates inhibitors, or ensures ideal environmental circumstances to promote the growth of a second microorganism. For complete product synthesis, several cascade routes necessitate the presence of at least two different species. Numerous investigations have shown that under cocultivation conditions, new compounds can only be created by the synthesis of primary or secondary metabolites (Rago et al., 2019; Thuan et al., 2022). For instance, Zuroff et al. (2013) developed a symbiotic co-culture of the cellodextrin-fermenting yeast, *Saccharomyces cerevisiae* cdt-1, and the cellulolytic mesophile, *Clostridium phytofermentans,* to enhance production of ethanol from α-cellulose through solid-state fermentation. Industrial biotechnology can benefit from microbial consortia that has the ability to utilize mixed, impure, and unrefined substrates, including waste biomass. This is particularly advantageous when complex substrates, such as waste or mixed organic materials (e.g., lignocellulosic biomass), are used to produce chemicals or fuels (Schmidt, 2021).

EF approach offers significant advantages for enhancing the metabolisms of mixed microbial cultures, particularly in industrial biotechnology. One of the key challenges in using mixed cultures is managing the stability of product formation, which can be compromised by substrate impurities. EF addresses this issue by using electrodes to regulate and stabilize metabolic processes. By applying electrodes, EF can control the electron flow within microbial communities, directing metabolic pathways toward the desired products and minimizing the formation of undesired by-products (Jiang et al., 2020). This precise control helps maintain consistent product quality, even when using impure or unrefined substrates. Furthermore, EF can enhance overall metabolic activity, boosting the efficiency of biochemical reactions and increasing yield. This makes the EF approach crucial for improving industrial biotechnology processes, particularly when dealing with complex substrates such as waste biomass or lignocellulosic materials. By integrating EF, industries can achieve more stable and efficient production of chemicals and fuels, thereby optimizing resource use and reducing waste (Modestra et al., 2015).

To manipulate intricate microbiomes and correlate their organization, arrangement, and role to reactor efficiency, reliability, and product generation, electrochemical techniques were demonstrated to be novel tools for more conventional approaches like anaerobic digestion (Sen et al., 2016). Zhao et al. (2015) recently introduced electrodes into an anaerobic digester, achieving enrichment of bioelectrochemical system by *Geobacter* species in the anodic biofilm and suspended sludge. Although the reactor showed enhanced methanogenesis, 50% of which is produced through unfamiliar metabolic route. The authors postulated direct interspecies electron transfer between *Geobacter* and *Methanosaeta.* Other studies have demonstrated the potential to regulate metabolic pathways in specific microbiome-driven fermentations through extracellular electron transfer (EET) to/from electrodes, optimizing the conversion of substrates e.g., agro-industrial waste to desired product.

Flynn et al. (2010) genetically modified *Shewanella oneidensis* for fermentation of glycerol to produce ethanol and carbon dioxide, addressing unbalanced redox reactions suggesting implication of this practice to more reduced substrate. In a patented method (Hafez, 2019), electrically driven dark fermentation with a microbiome in a single-chambered bioreactor was used to produce H_2 from several organic substrates. Chemicals such as acetone, butanol, ethanol, acetic acid, and butyric acid could also be produced through similar approach with identical system. Toledo-Alarcón et al. (2019) investigated manipulation of microbial community structure and metabolic pathways through introduction of polarized electrodes to enhance production of H_2 in dark fermentation.

Various EF approaches have aimed to advance the conversion rates of glycerol and product yield. Selembo et al. (2009) reported that electrochemical approach significantly upsurge H_2 and biofuel production from pure glycerol. Subsequently, another study showed that glycerol fermentation when electrically stimulated altered metabolite production, leading to the formation of alcohols and medium-chain fatty acids, although 1,3-propanediol (1,3-PDO) production remained unchanged (Dennis et al., 2013). The 1,3-PDO production from glycerol fermentation can be stirred by applying potential of 20.9 V or varied electricity from 1 to 10 A/m^2 at cathodic end (Zhou et al., 2013, 2015). Subsequent research described that 42 g/L of 1,3-PDO was produced by glycerol EF using biocathodes polarized at 21.1 V (Xafenias et al., 2015). Introduction of polarized cathode (−700 mV vs. SHE, Standard Hydrogen Electrode) significantly enhanced the butyric acid production consuming glucose as substrate by mixed culture (Paiano et al., 2019). Dhungana et al. (2023) reported that simultaneous saccharification and electro-fermentation greatly improved ethanol production by 12–18% by using fungal consortia comprised of a cellulolytic isolate F1 (*Aspergillus niger*) and/or ligninocellulolytic isolate F2 (*Ganoderma sessile*), *Saccharomyces cerevisiae* (CDBT2).

5 Mechanism of Electro-fermentation

Electrodes serve as additional electron sources and sinks, enabling complete control over microbial metabolism in EF (Rago et al., 2015; Virdis et al., 2022). Depending on the reduction level of the target metabolite, microbes can exploit the electrodes as electron mediators to their benefit. The fermentation process can gain advantages from either supply or removal of electron, which is crucial for optimizing product-driven EF (Choi & Sang, 2016). A study on a rudimentary EF model for glucose and glycerol forecast harvesting amount of metabolic end products. Efficient chain elongation in organic metabolites can occur during an abundance of redox cofactors, e.g., NADH and NADPH, stirred by the cathode to facilitate specific reduction process (Sharma et al., 2013; Xiao et al., 2018). Conversely, the anode, i.e., electron sink, facilitating advanced biomass yield as well as improved power consumption as compared to conventional anaerobic catabolism. This drives to surplus ATP synthesis and potentially increased biomass output (Tiquia-Arashiro & Pant, 2020).

The mechanism of electron transfer to the anodic chamber involving exoelectrogenic bacteria have been extensively instigated over the years. However, recent research has shifted focus to the phenomenon of endoelectrogenesis, exploring how electron flow from the cathode can significantly influence microbial metabolic activities (Lee et al., 2022; Thulasinathan et al., 2022). Various electron-transfer mechanisms have been identified, including cytochrome dehydrogenase conglomerates, periplasmic hydrogenase-mediated ET, and DET via c-type cytochromes, as demonstrated in exoelectrogenic bacteria like *Geobacter* and *Shewanella*. These mechanisms highlight that electron transfer can preserve energy through both coupled and uncoupled reactions. Key electron carriers, such as cytochromes, ferredoxin, quinones, and flavins, play crucial roles in electrode–microbe interactions and have been examined for their catalytic potential. The electrons obtained from electrodes facilitate the catalytic generation of essential energy carriers like NADH and ATP (Conners et al., 2023; Gemünde et al., 2022; Zhao et al., 2021). These insights into electron transfer pathways not only deepen our understanding of microbial metabolism but also pave the way for optimizing bioelectrochemical systems for enhanced energy production and biochemical synthesis.

In electro-fermentation, applying a potential to the surface of electrode aids interaction between microbial catalysts and electrodes. This process not only doubles the electrocatalytic rate but also enables control over bacterial metabolic pathways, crucial for enhancing product generation (Chandrasekhar et al., 2021). The movement of hydrogen ions and electrons between the electrode edge and the microbe is directly regulated by the applied potential. The presence of conductive materials within the microbiome, coupled with external stimuli, impacts the rate of extracellular electron transfer (EET), thereby enhancing process specificity and plasticity in bio-electrochemical processes (Zheng et al., 2020).

Manufacturing electrodes to enhance redox kinetics at the cathodic end significantly boosts reduction equivalents over product generation. Carbon-based electrodes like graphite flakes, carbon cloth, carbon mesh, and graphite are favored for their economic viability (Lekshmi et al., 2023). These electrodes offer higher sensitivity, selectivity, and economic value compared to native electrodes, especially when subjected to surface modifications (Thapa et al., 2022; Umar et al., 2020). Two-dimensional (2D) and three-dimensional (3D) electrode structures are designed to provide macroporous configurations with increased surface areas for bacterial adhesion and electrostatic interactions, unlike older electrode designs. The enhanced conductivity, larger surface area, and other favorable surface properties of 3D electrodes directly facilitate better attachment of bacteria and higher rates of electron transfer (Li et al., 2020; Wangpraseurt et al., 2022). Additionally, 3D electrodes promote efficient extracellular electron transfer (EET) between microorganisms and the electrode interface. Compared to conventional 2D electrodes, amending and coating macroporous electrodes are significantly more advantageous in the realm of electro-fermentation (EF) (Zou et al., 2022).

5.1 Electron Transfer Mechanism

Utilizing the metabolic processes of microorganisms, microbial electrolysis cells produce energy by exchanging electrons across membranes. Protons, electrons, and carbon dioxide are released when the microorganisms in the anodic chamber oxidize the substrate. The mechanism of extracellular electron transfer (EET) involves transporting electrons into and out of the cell (Fig. 5). Three processes are involved in the flow of electrons: oxidation of the substrate, the electron transfer to the relevant proteins, and the last stage of EET (Croese et al., 2011). Typically, electrons flow from a very low redox potential electron donor to a relatively high redox potential electron acceptor. The multiheme c-type cytochrome seen in gram-negative bacteria is the primary component involved in electron transport. The filamentous conductive pill, which is present in bacteria, helps transfer electrons in addition to cytochrome. Other vital proteins that help with electron transfer are hydrogenase, formate dehydrogenase, ferredoxin, rubredoxin, and cytochrome. During the electron transfer cycle, several bioelectrochemical reactions can be catalyzed at the same time. This kind of reaction often arises due to moving of electrons from the cathode to the corresponding bacterial cell. Diffusible intracellular electron carriers, e.g., $NAD^+/NADH$, initially transport the electrons to proteins that are present in the outer membrane (Thapa et al., 2022).

It is widely known that three distinct routes have been established for the transfer of electrons from the cathode to the donor and vice versa, such as indirect transfer, direct transfer, and transfer via biological projections. One of the best methods for transporting electrons is thought to be direct transfer. It happens from the cathode to the attached microorganisms and is widely comprehended in biocathodes of microbial communities, especially *betaproteobacteria* (Perona-Vico et al., 2020). It is noteworthy that there is no requirement for a mediator to facilitate this procedure. Redox enzymes are assumed for assisting electrochemical reduction, though the precision of this mechanism has not yet been established. Important contenders for the extracellular transport of electrons include proteins like ferredoxin and other redox proteins. Two potential gram-negative and metal-reducing bacterial strains, *Shewanella* and *Geobacter* species have been extensively investigated to understand the detailed mechanism of EET. The reduction of metal is aided by the electron transfer process, which specifies the metal as an electron acceptor to facilitate the reduction reaction. Multi-heme C cytochrome is one such protein that is utilized for this phenomena particularly when the metal oxides or hydroxides that are employed as electron acceptors with poor water solubility (Ranawat & Rawat, 2018).

It has been noted that highly poisonous hexavalent chromium can be reduced at cathode by *Geobacter* and *Shewanella* species, reported as iron-reducing bacteria. This study suggested that in the electroactive bacteria, electron transfer may occur in both directions (that is, from microbe to cathode and from cathode to microbe). Indirect electron transfer is the process of transferring electrons through soluble electron mediators. This route has a great potential to improve EET competence in *Shewanella oneidensis* MR-1 (Harris et al., 2012). Earlier studies demonstrated that

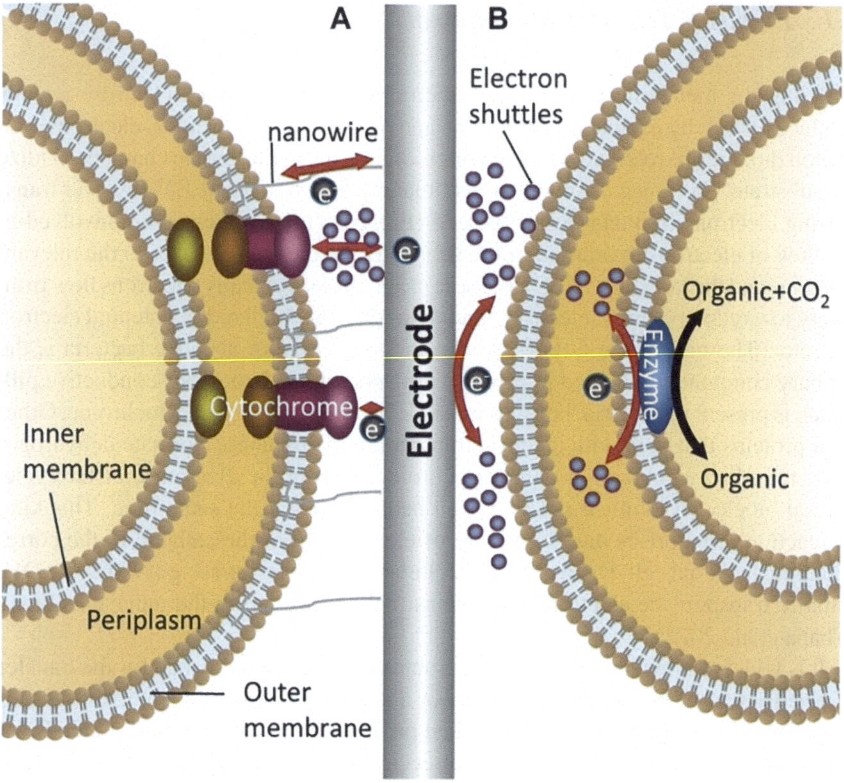

Fig. 5 Schematic representation of bidirectional electron transfer mechanism between microbial cell and electrodes: **a** Two direct electron transfer mechanisms: one mediated by nanowires and the other by outer membrane cytochromes, with or without the involvement of electron shuttles. **b** Indirect electron transfer, which is mediated by electron shuttles. Adapted with the permission of Zheng et al. (2020)

EET could occur via either direct contact or mediators (indirect pathway) (Liu et al., 2018). However, recent research suggests that these mechanisms are not efficient to buildup high current density owing to certain restrictions such as insufficient surface area and diffusion issues with mediators. Instead, microbes carry out conduction-based EET for electron exchange and produce high current density. This conduction-based method utilizes biological projections like pili. Utilizing such research concept, additional pathways proposed including redox conduction (electron hopping) and conduction through a biofilm matrix (Zacharoff & El-Naggar, 2017).

The process of "transfer via biological projections" occurs when electrons are transferred through nanowires or pili. This particular strategy is unique because it involves the use of two different projections: the biological projections that carry the electrons from the microorganisms to the electron carrier protein. The high current density generated by the nanowires facilitates the transfer of electrons. The pili that are utilized in this technique frequently cause cells to accumulate on the biofilm; they exhibit metal-like behavior that raises the temperature and lowers conductivity, both of which can be detrimental to the (MES). The multiheme c-type cytochrome is essential elements that support the flow of electrons (Kadier et al., 2016). The polypeptide chain is formed by the multiheme c-type cytochromes binding to one another through covalent connections. The two multiheme c-type cytochrome are separated by about 1–2 nm. The EET is made easier by the distinct structure of multiheme c-type cytochrome. Microbes are able to facilitate the extracellular transport of electrons through these three processes. However, a lot of research still needs to be undertaken to investigate these conductive EET pathways.

5.1.1 Cathode-Mediated Electron Transfer

Electricity is produced by reactions that occur in the anodic chamber which subsequently gets utilized in the cathodic chamber in order to produce useful biological products. Adding charged species, like chitosan, and expanding surface area with the use of a three-dimensional scaffold electrode support resulted in cathodic changes that increased current consumption. Many bio-based electrochemical reactions can be catalyzed by the electron-transfer event (EET) that occurs from the microbe to the cathodic chamber. At the anodic terminal, methanogenesis proceeds in a stepwise manner, involving the transfer of electrons to protons, hydrogen generation, and finally carbon dioxide reduction, where the cathode serves as a H_2 sink (Bora et al., 2022). At the cathodic chamber, a substance is needed to reduce in order to sustain the non-spontaneous hydrogen reaction. Due to their high methane generation productivity and documented ability to enhance carbon capture efficiency, carbon stick and graphite felt are among the most often used materials for cathode manufacture (Yaqoob et al., 2021). Increasing the porosity in carbon materials through the use of carbon nanostructures, such as MWCNT, doped with catalytic materials, e.g., cobalt tetramethyl phenyl porphyrin, manganese oxide, etc., is an alternate strategy to improve methanogenesis.

Both electrode terminals can be the site of the electrochemical pathways to produce methane by CO_2 using the DET method. It is possible to catalyze the process and so lower the overpotential by using valuable metals like platinum. Nevertheless, this procedure is quite costly and has negative environmental implications (LaBelle et al., 2019). Hence, stain steel has the alternative potential to be used in the construction of anaerobic reactors in conjunction with single-chambered MEC to improve the methanogenesis process. Additional alloy-based compounds, such as iron–graphite and Ti/Ru, can accelerate the digestion of sewage sludge via methanogenesis (Buitrón et al., 2019; Pawar et al., 2022). The cathode can alternatively be produced using carbon-based materials, such as graphite and reticular vitreous carbon, which are frequently used to produce the anode. Simple graphite achieved a higher coulombic efficiency (CE) of 100% compared to non-precious carbon-based materials and specific precious metals (LaBelle et al., 2019). These results were comparable to those of platinum-based cathodes, which produced an average (CE) of roughly 96% (Mier et al., 2021). Consequently, it is believed that plain graphite could be an effective substitute for cathodes based on platinum.

The mechanism of electron transfer from microorganisms to the cathodic terminal is being improved by a number of measures, including enhancing interfacial contacts and interfacial area. It has been attempted to attach nanoparticles to the cathode using carbon nanotubes, conjugated oligoelectrolytes (COEs), carbon nanotubes on reticulated vitreous carbon (NanoWeb-RVC), and nano-nickel (Bajracharya et al., 2022; Yan et al., 2015). Moreover, it has been noted that in a MES system, graphene-modified biocathodes sustained and improved the bioelectrochemical hydrogen production. An additional attempt was made to modify the surface with positive charges. The ability of cathode to transport electrons to microorganisms was enhanced by adding functional groups that were positively charged. However, the negatively charged bacteria *Sporamusa ovata* improved electron uptake at cathode end and preferred cathode attachment, reducing it to acetate and carbon dioxide due to attachment of bacterial cell wall to cathode (Perona-Vico et al., 2020).

Microbial biofilm was found to be enriched at positively charged anode end. However, because of the negatively charged alterations made to it, a repelling interaction took place at the cathodic end. As a result, attaching bacteria to a cathode creates a charge in the barrier, which has been shown to enhance the zeta-potential and hydrophobicity of cells in a biofilm that consumes electricity (Liang et al., 2019; Philips et al., 2016). Different alterations can be tried in contrast to the anode, depending on the changes in the cell surface and properties of cathode. In terms of the mechanism, gram-negative, metal-reducing bacteria such as *Shewanella* and *Geobacter* species are the two most thoroughly researched bacteria for extracellular electron transfer. Microorganisms use extracellular electron transfer to reduce metal, and the metal serves as an electron acceptor in this process. The multiheme c-type cytochromes present in *Geobacter* and *Shewanella* facilitate EET using insoluble metal (hydr)oxides as electron acceptors (Paquete et al., 2022). The extracellular electron transfer mechanism of *Shewanella* is represented in Fig. 6.

5.1.2 Anode-Mediated Electron Transfer

A more efficient anodic chamber can be built with a number of key qualities, including ease of construction, high conduction rate, non-corrosiveness, large surface area, high porosity, fouling resistance, scalability, and low construction costs. These characteristics are crucial for processes like adhesion and electron delivery that exo-electrogenic bacteria produce (Prathiba et al., 2022). Carbon is one of the most common materials utilized to create the anodic chamber because of its adjustable shape and compatibility with electrochemically active biofilm. Moreover, it is less expensive, has a much lower potential, and maintains its chemical stability in anaerobic environments (Li & Cheng, 2019). On the other hand, due to their large surface area, carbon-based materials such graphite fiber brushes and granules are some viable options for anode production. Additional materials utilized in the creation of the anodic assembly include carbon felt, vitreous carbon, and carbon cloth coated with Pt/C

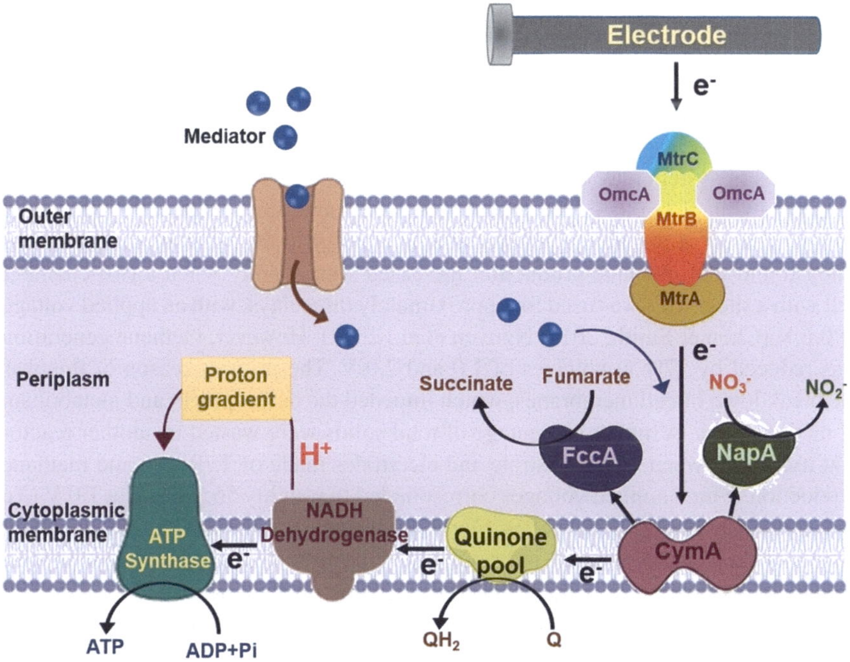

Fig. 6 Diagram illustrating extracellular electron transfer (EET) mechanisms in *Shewanella* spp., where electrons from the inner membrane transfer via c-type cytochrome *CymA* to metal reduction (MTR) proteins at the outer membrane, reaching electron acceptors for reduction at the cathode via mediators

catalyst (Banerjee et al., 2022). Hydrogen and methane might be produced in a MEC using the same materials used to make anodes in an MFC. A modified graphite fiber cloth with a nickel-supported multiwall carbon nanotube was found to be a viable substitute for creating the anodic chamber. As a result, methane production rises by 52% (Blasco-Gómez et al., 2017). In addition to carbon-based materials, alternative options for reducing operating costs in methane production include transition metals such as iron (Fe) and alloys like Ti/Ru and Ti/RuO_2 (Bhagchandanii et al., 2020).

5.2 *Source of Electrical Power*

The applied voltage is a crucial parameter that significantly influences the performance of BESs in methanogenesis. It is important to consider the specific range of applied electrical potentials, as inappropriate levels can be harmful to microbes (Lai et al., 2016). A study evaluated the impact of applied voltage on hydrogenotrophic methanogens. The results demonstrated the luxuriant growth of hydrogenotrophic methanogens at an applied voltage of 0.3 V, positively affecting methane production during the anaerobic digestion (AD) of activated sludge (Feng et al., 2023). Methanogenesis was shown to rise by around 22.4% when utilizing electrodes made of iron–graphite materials at a voltage of 0.3 V, in contrast to the standard anaerobic digestion method (ElMekawy et al., 2017). It was also noted, that the alkaline pH conditions of system may cause these bacteria to be inhibited by a small adjustment in applied voltage to 0.6 V. The production of methane was significantly reduced by 43% as a result of this abrupt pH increase. In contrast to the previous study, another study found that methane production increased significantly when a two-chamber cell with a short HRT was used for approximately three days, with an applied voltage of 0.6 V (Chen & Smith, 2018; Nguyen et al., 2022). However, methane generation was reduced by 37% at voltages of 1.0 and 2.0 V. The primary reason of this was the breakdown of cell membranes, which impeded the development and metabolism of methanogens. A higher percentage of total solids were wasted in another reactor that used waterweed as its substrate and electrodes made of Ti/RuO_2, and methane production at high applied voltages corresponded to roughly 56.3% and at 1.0 V (Yu et al., 2014; Liyanage et al., 2019; Flores-Rodriguez & Min, 2020). The divergent results of the several research indicated above point to the significance of applied voltage in an anaerobic BES and the various elements that affect it, including cell configuration, substrate type, and electrode material. The reliance of the applied voltage on these variables highlights the significance of doing voltage optimizations for each BES in order to improve and maximize the energy efficiency of system.

6 Types of Electro-fermenters

EF has been utilized to overcome the thermodynamic limitations of conventional microbial fermentation while fine-tuning metabolism to produce targeted biobased products. An EF system consists of electrically connected anodic and cathodic chambers, which can be operated independently for anodic EF or cathodic EF, depending on the desired metabolic modulation (Salar-García et al., 2024). The fermentation environment can be regulate and optimized to achieve greater purity of particular products. Anodic and cathodic charge can direct particular microbial metabolic feedbacks via different mechanisms. At anode, the solid conductor accepts electrons from oxidative reactions, acting as an electron sink. Maximizing electron subtraction by oxidizing transitional metabolic electron acceptors can enhance NADH intake and generate proton gradients adequate for ATP production. At the cathode, the electron imparts from the solid conductor to the cells facilitates the reduction of metabolites via enhanced reduced NADH pool and additional ATP production (Schievano et al., 2016).

EF uses a modest electrical current density to start and improve fermentation by regulating both extracellular and intracellular oxidation–reduction potentials (ORP) under imbalanced conditions (Moscoviz et al., 2016). The most crucial elements to enhance the device performance are the electrode composition and the reactor

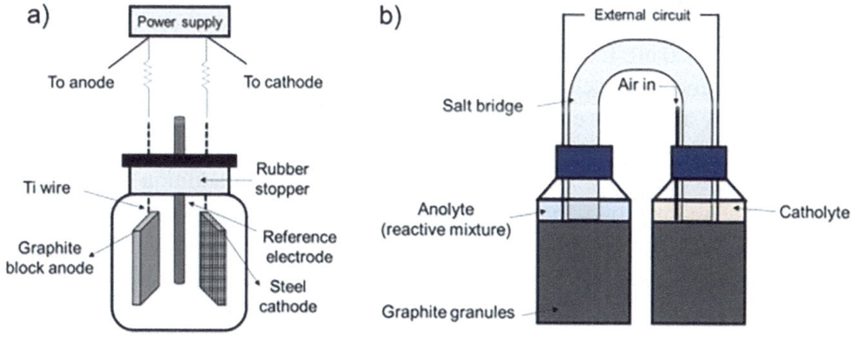

Fig. 7 Schematic diagram of single-chambered and double-chambered electro-fermentation. Adapted with the permission of Singh et al. (2020)

design. The major categories of EF reactors include (i) single-chamber and (ii) double-chamber EF reactor (Fig. 7). Irrespective of reactor configuration, for the simultaneous production of simplest and most commonly used fuels such as biogas (H_2 and/or CH_4) and organic acids through electro-fermentation, both pure (like glucose, lactate, and acetate) and complex compounds (like synthetic waste, waste-activated sludge, and food waste) are utilized as substrates. The materials based on carbon are the common materials used for electrodes preparation because of their inexpensive cost, good biocompatibility, high surface area, conductivity, and porosity that favor redox reactions in comparison with the electrodes based on metals (Sharma et al., 2019). The most common EF reactors configurations are given in the section described below.

6.1 Single-Chamber EF Reactor

In single-chamber reactor, the anodic and cathodic reactions take place in a similar chamber. Single-chamber EF reactors are a more accessible approach in comparison with double-chamber EF reactor, nevertheless, the metabolites produced in them demand post-treatment process. It has been demonstrated as an appropriate configuration for transforming a variety of raw materials into methane, hydrogen, bioethanol, etc. Later on, it has been shown that conductive particles stimulate anaerobic digestion. A research group previously used a 0.6 V voltage to produce methane. They submerged a cathode based on graphite-rod and a anode based on graphite-brush in a cylindrical reactor filled with activated sludge. Over 33 days, the electro-fermenter provided 2998.4 mL of built up methane, in contrast to 904.5 mL from traditional fermentation. Their findings indicate that using polarized electrodes during the fermentation process considerably increases methane generation (Zhao et al., 2015). Later on, they investigated the effect of different voltages on production of methane and the *E. densa* fermentation process by using single-chamber reactor with a submerged cathode. The largest volume of methane generated at 1.0 V (248.2 ± 21.0 mL L d^{-1}). Their studies demonstrated that fermenting bacteria as well as electroactive bacteria interact syntrophically and positively, resulting in process stability and better energy recovery (Zhen et al., 2016). Another study also confirmed that conductive graphite improved methane productivity by 54.3% and highest methane output by 72.2% in EF methanogenesis. In addition to increased active sites and redox groups that facilitated electron transport on the biofilm because of the presence of graphite particles, it also accelerated the formation of methanogens and phylum proteobacteria, subsequently improved EF methanogenesis (Ren et al., 2018). Besides producing various kinds of biogas, single-chamber electro-fermenters have also been used for producing various fatty acids. A previous study showed that anodic EFs can maximize titers as well as productivity by more than three times in a single-chamber microbial electrolysis cell (SCMEC) using cellulolytic bacterium *Cellulomonas* uda. The efficiency of SCMEC was relied on better anode biofilms from *Geobacter sulfurreducens* that inhibited cathodic hydrogen oxidation and enhanced lactate oxidation (Awate et al., 2017). The transformation of carbon dioxide into short-chain fatty acids/volatile fatty acids and methane was examined under four investigational settings (R1–R4). The setup for the study included an anaerobic fermenter as the control (R1), an anaerobic fermenter with electrodes in an open circuit (R2), an anaerobic fermenter with electrodes in a closed circuit with 100 Ω load (R3), and an EF reactor with a cathodic potential of −0.8 V versus Ag/AgCl (R4). The findings were evaluated with reference to methane, SCFA generation, current density, as well as inorganic carbon. The EF (R4) arrangement produced the highest volumes of SCFA (2050 mg/L) and methane (41.2 mL/day) in comparison with the other reaction setups. R3 recorded 1800 mg/L and 24 mL/day, R2 provided 1560 mg/L and 15 mL/day, whereas R1 exhibited 1430 mg/L and 10 mL/day of methane and SCFA, respectively. The investigation revealed that EF might efficiently expedite biochemical reactions and improve the transformation of carbon dioxide to organic molecules in a renewable way (Velvizhi et al., 2022). In a recent study, the rumen microbiota has been utilized for the production volatile fatty acids (VFAs) through EF by using solid-state electrodes as electron sources. The significant modifications such as increased microbial biomass (12%), dry matter digestibility (58%), VFA level (~84%), and reduction in NAD/NADH ratio (~50%) have been observed in RF at 0.9 V electric potential for 24 h in contrast to the control condition (0 V) (López-Hernández et al., 2023).

6.2 Double-Chamber EF Reactor

In double-chamber reactor, the reactions occur in a chamber partitioned through a membrane. A proton exchange membrane has been introduced into EF reactors for the electrical extraction of organic acids from the solution to avoid product inhibition. In double-chamber electro-fermenters the higher internal resistance (1–10 Ω cm^{-2}) leads to increase in voltage losses from +0.26 to +0.38 V in contrast to single-chamber electro-fermenters. Moreover, a pH gradient is also produced by the transfer of cations through proton exchange membrane, and this produces a voltage loss of about 0.06 V for every pH unit (Sriram et al., 2022). Earlier work on double-chamber Electro-fermenters involves the investigation of effects of cathodic reaction on the synthesis of methane, VFA and COD elimination in contrast to the results found in a non-bio-electrochemical reactor and a stirred tank reactor (Sasaki et al., 2013). In all instances, the same reactor configuration was employed, consisting of a cylindrical glass vessel with two concentric chambers separated by a Nafion® 117 membrane (Dupont, DE, USA). The major difference between the three methanogenic reactors under investigation was that the electrode side facing the counter chamber was wrapped with carbon fiber fabric in the bioelectrochemical and non-bioelectrochemical systems in contrast to the stirred tank reactor. Using a carbon bar as a counter electrode, 170 mL of 100 mM NaCl was used, which is 13.3 times the volume of the anodic chamber and the cathodic chamber. At a hydraulic retention time (HRT) of 4.0 days, the bioelectrochemical system generated 3.57 L L^{-1} day^{-1}, which was more than 3.5 times higher than the output rate of the non-bioelectrochemical reactor at the same HTR. This illustrates the advantageous impact of the applied potential (0.8 V vs. Ag/AgCl) on gas production. The stirred tank reactor produced the least amount of methane (1.37 L L^{-1} day^{-1} at a 12-day HRT) and showed system degradation, highlighting the advantages of utilizing carbon fiber fabric. On the other hand, biobutanol was produced in an electro-fermenter using a pure culture of C. pasteurianum fed minimum media enriched with glucose. The system was made up of a Nafion® 117 selective separator and an H-type reactor with graphite felt electrodes. Additionally, a comparison was made between the outcomes and the electrodes made of stainless steel. The findings of the various applied voltages demonstrated that the amount of butanol produced by both types of electrode materials is reduced by an applied voltage more than 1.5 V. These findings could be attributed to a surge in H_2 production as the voltage raised because of concurrence between the two pathways and the unpredictability of the biofilm (Mostafazadeh et al., 2016). Regarding electrode material of graphite felt instead of stainless steel as electrodes yielded higher butanol formation for all voltages applied, confirming the advantages of using materials based on carbon. As per central composite design, up to 13.31 g/L of butanol at a voltage of 1.32 V was produced. A similar arrangement incorporating a double-chamber EF and a Nafion® 117 membrane was employed for propionate production through lactate EF. In this instance, plated titanium mesh was used to create the anode and carbon cloth to create the cathode. The catholyte in the system was either butyrate or D-lactate, or both. The Ag/AgCl reference electrode was inoculated with anaerobic waste from a dairy processing factory. The ideal propionate yield of 0.96 g L^{-1} d^{-1} was attained at an initial lactate level of 150 mM with a minimum of one kWh g^{-1} of propionate of energy used up. At a pH of 5 and the original lactate level of 20 mM, an applied voltage of −1 V (vs. Ag/AgCl) increased propionate output to 0.44 g L^{-1} d^{-1}. These results could be explained by the presence of *Propionibacterium* sp. and *Tyzzerella* sp., both of which produce propionate (Isipato et al., 2020). Electro-fermenters have also used cation exchange membranes, including CMI7000 (Membranes International Inc., Ringwood, NJ, USA), in place of Nafion®. This type of separator was employed in a double-chamber reactor with an Ag/AgCl reference electrode (+0.2 V vs. SHE) and graphite felt for the working and counter electrodes. Synthetic effluent rich in glucose was introduced into the system and anaerobic sludge was used as an inoculant. The study discovered that at neutral pH levels, the synthesis of methane, hydrogen, and acetic acid increases with a greater negative working potential, but the production of methane and acetic acid decreases significantly with an increase in the set potential. However, by slightly acidifying the media (pH 6.2), the effects of changing the applied voltage on product distribution are lessened. Methane production under neutral conditions ranged from 4.3 to 6.7 mL at a fixed potential of −1.0 V. On the other hand, when the pH dropped by 6.2, output went up. These results demonstrate that the metabolite distribution may be fine-tuned, enabling improved control of the entire process, by varying the applied voltage and media pH (Jiang et al., 2018).

6.3 Hybrid Configurations of EF Reactor

MFCs are BESs that convert chemical energy stored in a range of substrates into electrical energy through microbial metabolism. This technology may supply the energy required to complete the electro-fermentation process, enabling researchers to create a self-sufficient system for converting carbon-rich raw materials into a variety of valuable products, including methane, H_2, and/or VFA, among others. The researchers generated a self-operating, membrane-less, air-cathode single-chamber electro-fermenter that yields significant amounts of bioelectricity, biohydrogen, and bioethanol from food waste. After nine days of operation, the system generated 162.4 mW m^{-2}, 21.9 mL h^{-1}, and 4.85% w/v of bioethanol. The results suggested that the fermentation of mixed acetate and butyrate might be the source of hydrogen production, but the presence of other metabolites like formate or lactate might trigger other metabolic pathways that restrict hydrogen production. In another study, MFC technology and acidogenic fermentation was used in a single-chamber biocatalized electro-fermenter. The hybrid system consisted of two pairs of graphite electrodes in a single reactor chamber. The anodes were submerged, while the cathodes were air-faced. The MFC-EF prototype achieved maximum power output of 72 mW m^{-2} and produced 343 mL of biohydrogen from synthetic wastewater with an organic load of 5000 mg L^{-1} in a closed circuit with an external resistance of 300 Ω. Submersible MECs are one option for bypassing the power source which combine MEC technology with anaerobic fermentation. This arrangement involves connecting two cathodic chambers to current anaerobic digester. A few of the cathodic sections is mostly utilized to generate electricity, while the other produces hydrogen. In a study, platinum-coated titanium mesh anodes and graphite felt cathodes partitioned by a cation exchange membrane has been used to produce hydrogen at rates ranging from 5 to 487 mL/d. From another perspective, EF could be employed as a H_2 source for pyrolysis (Van Eerten-Jansen et al., 2013). Presently, the hydrogen necessary in the final step of biomass pyrolysis is often derived from fossil fuels, which emit greenhouse gases into the environment and contribute to global warming. Electro-fermentation provides a possibility to convert the pyrolysis process into a more sustainable one (Bhagchandanii et al., 2020; Chandrasekhar et al., 2021). Alternatively, the benefits can be harnessed in reverse, where electro-fermentation could potentially reutilize the aqueous waste stream from the pyrolysis procedure, known as bio-oil aqueous phase (BOAP). In such instances, EF presents a viable choice to convert BOAP into H_2 or other valuable products (Borole et al., 2018).

7 Value-Added Biochemicals Through Electro-fermentation

The production of biochemicals and biofuels depends on a number of factors, including the type of bioelectrode, reduction potential, electrochemically redox mediators, and biocatalyst. This EF includes the manufacture of various organic acids and alcohols etc., which are detailed in the following sections.

7.1 Lactic Acid

An attractive substitute for increasing the yields of lactic acid (LA) produced by lactic acid bacteria (LAB) is electro-fermentation (EF). In addition, one of the key difficulties faced by LA-based biorefineries is preserving high production while controlling the media's ongoing acidification. One potential remedy for this problem is to increase the pH of the medium by the use of cathodic electro-fermentation techniques. In order to maximize its potential, EF has to be tested for extended periods of time in industrial-scale media, paired with synthetic biology techniques, and evaluated in single-electrode reactors to maximize its use in LA fermentations. Prior to being used, EF must first be tested in pure LAB cultures to determine its viability from a biological and financial standpoint and to pinpoint possible metabolic targets. Although effect of EF on lactic acid production has been seen in a number of mixed-culture settings, including wastewater extractions, it has not been evaluated in pure cultures of LAB. Research has demonstrated that cathodic EF treatments can increase the generation of lactic acid (Xue et al., 2018), but it has proven difficult to discern between the effects of community selection and biochemical causes. Conversely, anodic EF experiments usually focus on aerobic metabolism and try to supply a terminal electron acceptor that is renewable and does not require oxygen. Anaerobic metabolism, on the other hand, has the capacity to produce more oxidized products (Vassilev et al., 2018).

Schuppert and colleagues have used the microbe *Propionibacterium acidipropionici* to perform electrofermentation (EF) of lactate, and they used cobalt sepulchrate as an electron donor at −0.47 V versus SHE. They stated that in batch and continuous operations, propionate was the individual fermentation product (Schuppert et al., 1992). In the same way, Xu and their co-workers demonstrated how cathodic EF can improve lactic acid productivity and optical activity. They maintained a steady potential of +0.1 V versus SHE with a mixed microbial population made from food waste and activated sludge. According to their research, the in situ electron supply from cathode considerably elevated optical activity for the manufacture of L-lactic acid from 3.6 to 42.3% and greatly increased productivity by 4.73 times (Xue et al., 2018).

7.2 Alcohols

In electro-fermenters, electrons generated in the anodic chamber can be used to convert organic acids into bio-alcohols. The efficiency of these reactions and the overall electro-metabolic control strongly rely on the biocathode (Kucek et al., 2016). Biofuels such as ethanol and butanol can be synthesized in electro-fermenter. Awate and their team investigated the anodic EF of cellobiose using *Cellulomonas* uda for ethanol production, along with a genetically modified consortium of *Geobacter sulfurreducens* strains to eliminate non-ethanol fermentation by-products. They operated at +0.24 V versus Ag/AgCl in a single-chamber setup, achieving significant enhancements in yield (2.7-fold) and productivity (3.7-fold) compared to conventional fermentation methods (Awate et al., 2017). Similarly, Choi and their team conducted cathodic electro-fermentation of glucose with *Clostridium pasteurianum* DSM 525 at +0.045 V versus SHE. Their findings showed that the microbe and the cathode may transmit electrons directly to one another, creating a novel metabolic route during electricity-driven fermentation that might boost butanol output by as much as 150% (Choi et al., 2014). Building on this work, Khosravanipour and their team further investigated the influence of electrode materials, glucose concentration, temperature, and applied potential on optimizing butanol production by *Clostridium pasteurianum.* They achieved a maximum butanol concentration of 13.31 g/L when fermenting an initial glucose concentration of 120 g/L at 33.51 °C, using a fixed potential of +1.32 V versus Ag/AgCl (Khosravanipour Mostafazadeh et al., 2016).

A subsequent research demonstrated that applying electrical current during glycerol fermentation led to changes in metabolite production: alcohols and medium-chain fatty acids were produced, while the production of 1,3-propanediol (1,3-PDO) used in polymer synthesis, solvents, and antifreeze remained unchanged (Dennis et al., 2013). To enhance the conversion of glycerol to 1,3-PDO, researchers have explored imposing a cathodic potential (up to 20.9 V) (Zhou et al., 2013) or cathodic current (ranging from 1 to 10 A/m^2) (Zhou et al., 2015). In a glycerol electro-fermentation study, biocathodes polarized at 21.1 V achieved the highest concentrations of 1,3-PDO, reaching 42 g/L (Xafenias et al., 2015).

7.3 Hydrogen

Hydrogen has garnered significant global interest in recent decades; however, its large-scale production via biological methods is restricted by costly feed materials and process expenses. Theoretical hydrogen yields of 4 mol/mol can be attained through dark fermentation, but scaling up this process to larger scales poses limitations (Liu et al., 2005). However, integrating this process with anaerobic microbial electrolysis cells (MEC) has been shown to increase hydrogen yields by up to fourfold. The typical metabolic pathway during dark fermentation involves the production of acetate or butyrate through the following reactions:

$$C_6H_{12}O_6 + 2H_2O \rightarrow 4H_2 + 2CO_2 + 2C_2H_4O_2, \quad (4)$$

$$C_6H_{12}O_6 \rightarrow 2H_2 + 2CO_2 + C_4H_8O_2. \quad (5)$$

This restricts the achievable hydrogen yield to 2 mol/mol with butyrate as an end product and 4 mol/mol with acetate (Liu et al., 2005). Integrating this process with a microbial electrolysis cell (MEC) utilizing microbes such as *Geobacter, Shewanella, and Pseudomonas* allows the production of hydrogen from protons and electrons generated by the microbes, as depicted below:

$$\text{Anode:} C_2H_4O_2 + 2H_2O \rightarrow 2CO_2 + 8e^- + 8H^+, \quad (6)$$

$$\text{Cathode:} 8H^+ + 8e^- \rightarrow 4H_2. \quad (7)$$

Cardena and their team have evaluated H_2 productivity from cheese whey using EF in reactors without membrane in comparison with conventional dark fermentation. In their study, EF reactors utilized S.S. electrodes, with applied potentials (EW) on the working electrodes set at 0.5, 0.2, +0.1, and +0.4 V versus NHE. EF enhanced H_2 productivity across all EW values, with the highest performance observed at EW = 0.5 V (4.0 ± 0.7 L H_2/L-d and 2.5 ± 0.3 mol H_2/mol glucose). This method enhanced H_2 productivity through ethanol-type fermentation, surpassing conventional dark fermentation. Negative potentials reduced biomass production but enhanced H_2 productivity and medium-chain fatty acids, while positive potentials increased ethanol yield, influenced by microorganisms like *Enterococcus* sp. *Clostridium butyricum*, and *Clostridium tertium*, present under all tested conditions but with varying relative abundances. Implementing EW allowed for precise regulation of fermentation toward particular valuable products (Cardeña et al., 2024). Gu and their team has explored electrode potentials in electro-fermentation system (EFSs) influence metabolic products and global transcriptome expression in *Ethanoligenens harbinense*. Anodic EFS (AEFS) inoculated with *E. harbinense* at a poised potential of 0 V, as well as cathodic EFS (CEFS) with the same potential (vs. Ag/

AgCl reference electrode), achieved highest H_2 productivity of 1888–1986 mL/L-medium, marking a 23–26% increase compared to open-circuit EFSs (OC-EFSs). The highest H_2 yield of 1.190 ± 0.009–1.197 ± 0.001 mol H_2/mol glucose was observed in AEFS0 and OC-EFS. Ethanol productivity in AEFS-0.4 enhanced by 30.7 ± 13.3% in contrast to OC-EFSs. Furthermore, glucose uptake and cell growth were enhanced in EFS, bolstering cellular energy supply (Gu et al., 2022).

7.4 Polyhydroxybutyrate (PHB)

Polyhydroxyalkanoates are biodegradable and biocompatible polymers synthesized in intracellular granules by various microorganisms under growth-suppressed conditions. A study via Nishio and their team members demonstrated that PHB productivity in *Ralstonia eutropha* can be faster using extracellular electron transfer (EET) facilitated by a biocompatible mediator in an electrochemical system. When the electrode potential was kept at 0.6 V versus the standard H_2 electrode, forming the EET pathway, the rate of PHB output rose by 60% contrasted to conditions without EET. This enhancement suggests that the extracellular anode acted as an additional electron acceptor for microbial metabolism, thereby accelerating glycolysis and promoting PHB synthesis (Nishio et al., 2013). The introduction of additional electrons through the redox mediator accelerated the glycolytic pathway and facilitated $NADH/NAD^+$ redox cycling, leading to a spontaneous increase in ATP production and promoting PHB biosynthesis. Integrated hydrogen fuel cell with an EFS also has introduced a novel method to enhance PHB accumulation in *Ralstonia eutropha* H16, leveraging a sustainable energy source. EF significantly influenced cell growth, biomass production, substrate utilization, and PHB accumulation outcomes. The addition of a 10 mA electric current notably improved final residual biomass concentrations and maximum specific growth rates. Moreover, redox-mediated EF with a 10 mA electric current, facilitated by a redox mediator, achieved a substantial increase in PHB content (30% higher than the control). Implementing a two-stage cultivation strategy mitigated growth suppression associated with redox-mediated EF and enhanced the maximum PHB productivity of the system as shown in Fig. 8 (Lai & Lan, 2021). Li and their co-workers investigated electrode-driven PHB production from CO_2 using *Rhodobacter sphaeroides*. Electrode-associated cells of *R. sphaeroides* achieved PHB concentrations up to 23.50 ± 2.8% of the dry cell weight (DCW), whereas suspended cells exhibited faster growth but lower PHB content per cell. Gene expression analysis revealed up-regulation of phaA in electrode-associated cells, while phaB expression was down-regulated in suspended cells. Additionally, electrode-associated cells expressed unconventional CO_2 fixation enzymes like isocitrate dehydrogenase and formate dehydrogenase, which correlated with increased PHB synthesis. These findings highlight the potential for CO_2 conversion into polymeric substances through MES and offer new insights into the genetic regulation of electrode-associated cells in this process (Li et al., 2023).

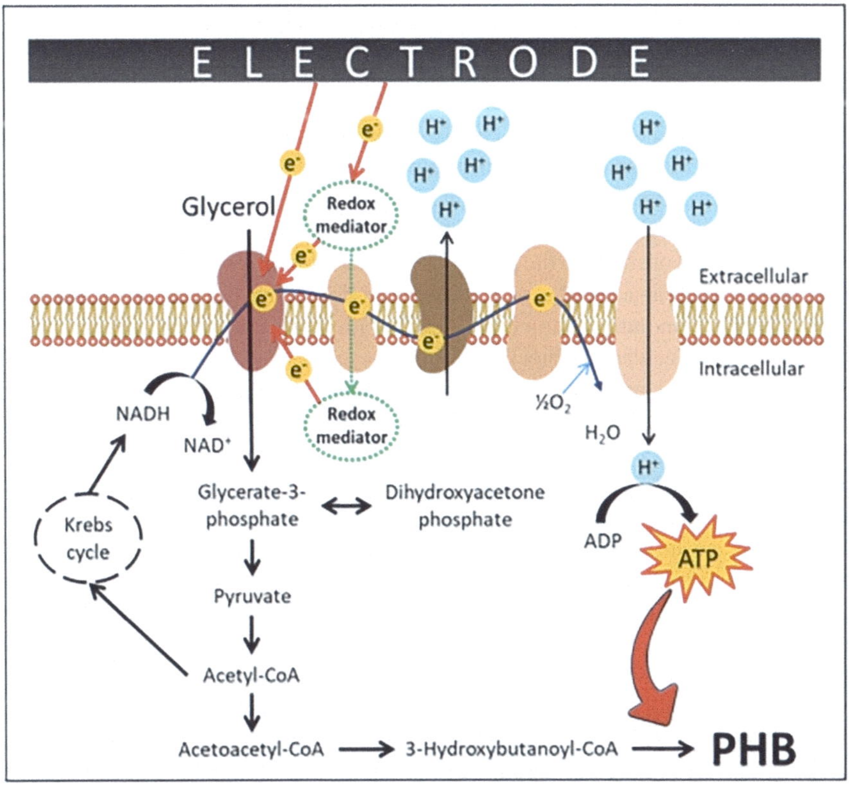

Fig. 8 Proposed model of biosynthesis pathway of PHB from *Ralstonia eutropha* through electro-fermentation technology. Adapted with the permission of Lai and Lan (2021)

7.5 *Mevalonate*

While EF represents a promising method for enhancing microbial production, its practical application has been limited. Matsumoto and their team integrated mevalonate pathway into synthetic electroactive *Escherichia coli*, equipped with an *Mtr*-based extracellular electron transfer pathway. The goal was to enhance the mevalonate pathway by regenerating intracellular cofactors through electrode-supplied electrons. Additionally, the adhE gene, responsible for acetyl-CoA accumulation—the precursor for mevalonate production—was deleted and combined with EF. This approach successfully boosted mevalonate production following 48 h of cultivation under a constant potential. These findings offer a novel perspective on enhancing bioproduction capabilities (Matsumoto et al., 2023).

8 Summary of Electro-fermentation: Challenges and Future Prospect

EF is a very promising, as of yet largely unexplored, biotechnological method to increase control over anaerobic fermentation processes, especially those involving mixed microbial cultures. It is evident that a number of distinct scientific and technical issues remain poorly understood, despite the encouraging outcomes to date. This poses a significant obstacle to the product's successful translation from the lab to the marketplace. The capacity of a polarized electrode to alter the rate, yield, titer, or product distribution during a microbial fermentation process is primarily unknown in terms of its molecular and biochemical basis.

A portion of the mechanistic models that were developed as a result of pure culture studies emphasized the critical role that the redox status of cells (such as the $NADH/NAD^+$ ratio) plays as a major factor influencing EF performance. It has not yet been established, therefore, whether these models apply to different microbial ecosystems or mixed microbial communities. Usually, this resulted in a weak ability to anticipate and model EF performance. Additionally, it was difficult to establish solid design criteria with a wide range of application scenarios that might support technology upscaling and spark potential stakeholders' financial interest. Because of this apprehension, the majority of EF processes are still primarily created and refined using just empirical trial-and-error methods, which usually have a very limited validity and a small range of application. More research is urgently required to help address this significant issue. This research may use labelled substrates (e.g., ^{13}C) and combine a variety of chemical, analytical, and biomolecular techniques to follow in vivo a number of parameters related to the EF process, like the intracellular concentration of redox carriers and other important metabolites and the real-time redox state of cells.

Developing efficient design requirements for EF reactors is another essential subject that deserves more attention. In fact, until now, EF has been carried out using bioelectrochemical reactors that have been specifically created and optimized for different processes and uses, such as microbial fuel cells for generating energy or microbial electrolysis cells for the production of hydrogen or methane. Achieving a high current density in EF is not often the final design criterion to satisfy process environmental sustainability and financial viability, which is a major distinction from the previously listed processes. In fact, the electron flow in EF is usually marginal in relation to the carbon flow and productivity of the target metabolite, regardless of whether the EF is cathodic, anodic, or based on a modified ORP of the reaction medium. This result suggests that, in theory, EF should not experience the same scalability issues that generally plague all (bio)electrochemical systems. It also suggests that conventional anaerobic bioreactors outfitted with electrodes of comparatively small surface area could likely be used for this purpose; put another way, the process of upscaling EF reactors could begin with the retrofitting of conventional fermenters. Despite these advantages, it should be highlighted that no efforts have been made to determine how the electrode's surface area, morphology, topography, geometry, and material affect the effectiveness of EF. All of these problems obviously stem from the need for a more thorough comprehension of the fundamental microbe-electrode interactions, which are crucial for the use of this innovative technology.

References

Adesra, A., Srivastava, V. K., & Varjani, S. (2021). Valorization of dairy wastes: integrative approaches for value added products. *Indian Journal of Microbiology, 61*(3), 270–278.

Ahmad, A., Chowdhary, P., Khan, N., Chaurasia, D., Varjani, S., Pandey, A., & Chaturvedi, P. (2022). Effect of sewage sludge biochar on the soil nutrient, microbial abundance, and plant biomass: A sustainable approach towards mitigation of solid waste. *Chemosphere, 287*, Article 132112.

Albuquerque, M. G. E., Eiroa, M., Torres, C., Nunes, B. R., & Reis, M. A. M. (2007). Strategies for the development of a side stream process for polyhydroxyalkanoate (PHA) production from sugar cane molasses. *Journal of Biotechnology, 130*(4), 411–421.

Amaro-Reyes, A., Marcial-Ramírez, D., Vázquez-Landaverde, P. A., Utrilla, J., Escamilla-García, M., Regalado, C., & Favela-Camacho, S. E. (2024). Electrostatic fermentation: Molecular response insights for tailored beer production. *Foods, 13*(4), 600.

Arends, J. B., Patil, S. A., Roume, H., & Rabaey, K. (2017). Continuous long-term electricity-driven bioproduction of carboxylates and isopropanol from CO_2 with a mixed microbial community. *Journal of CO_2 Utilization, 20*, 141–149.

Awate, B., Steidl, R. J., Hamlischer, T., & Reguera, G. (2017). Stimulation of electro-fermentation in single-chamber microbial electrolysis cells driven by genetically engineered anode biofilms. *Journal of Power Sources, 356*, 510–518. https://doi.org/10.1016/j.jpowsour.2017.02.053

Bajracharya, S., Krige, A., Matsakas, L., Rova, U., & Christakopoulos, P. (2022). Advances in cathode designs and reactor configurations of microbial electrosynthesis systems to facilitate gas electro-fermentation. *Bioresource Technology, 354*, Article 127178.

Banerjee, A., Calay, R. K., & Mustafa, M. (2022). Review on material and design of anode for microbial fuel cell. *Energies, 15*(6), 2283.

Bhagchandanii, D. D., Babu, R. P., Sonawane, J. M., Khanna, N., Pandit, S., Jadhav, D. A., Khilari, S., & Prasad, R. (2020). A comprehensive understanding of electro-fermentation. *Fermentation, 6*(3), 92.

Bian, B., Alqahtani, M. F., Katuri, K. P., Liu, D., Bajracharya, S., Lai, Z., Rabaey, K., & Saikaly, P. E. (2018). Porous nickel hollow fiber cathodes coated with CNTs for efficient microbial electrosynthesis of acetate from CO_2 using *Sporomusa ovata*. *Journal of Materials Chemistry A, 6*(35), 17201–17211.

Blasco-Gómez, R., Batlle-Vilanova, P., Villano, M., Balaguer, M. D., Colprim, J., & Puig, S. (2017). On the edge of research and technological application: A critical review of electromethanogenesis. *International Journal of Molecular Sciences, 18*(4), 874.

Bora, A., Mohanrasu, K., Swetha, T. A., Ananthi, V., Sindhu, R., Chi, N. T. L., Pugazhendhi, A., Arun, A., & Mathimani, T. (2022). Microbial electrolysis cell (MEC): Reactor configurations, recent advances and strategies in biohydrogen production. *Fuel, 328*, Article 125269.

Borello, D., Gagliardi, G., Aimola, G., Ancona, V., Grenni, P., Bagnuolo, G., Garbini, G. L., Rolando, L., & Caracciolo, A. B. (2021). Use of microbial fuel cells for soil remediation: A preliminary study on DDE. *International Journal of Hydrogen Energy, 46*(16), 10131–10142.

Borole, A. P., Tsouris, C., Pavlostathis, S. G., Yiacoumi, S., Lewis, A. J., Zeng, X., & Park, L. (2018). Efficient conversion of aqueous-waste-carbon compounds into electrons, hydrogen, and chemicals via separations and microbial electrocatalysis. *Frontiers in Energy Research, 6*(SEP), 1–21. https://doi.org/10.3389/fenrg.2018.00094

Boucher, D. G., Carroll, E., Nguyen, Z. A., Jadhav, R. G., Simoska, O., Beaver, K., & Minteer, S. D. (2023). Bioelectrocatalytic synthesis: Concepts and applications. *Angewandte Chemie, 135*(46), Article e202307780.

Buitrón, G., Cardeña, R., & Arcila, J. S. (2019). Bioelectrosynthesis of methane integrated with anaerobic digestion. In *Microbial electrochemical technology* (pp. 899–919). Elsevier.

Cardeña, R., Cercado, B., & Buitrón, G. (2019). Microbial electrolysis cell for biohydrogen production. In *Biohydrogen* (pp. 159–185). Elsevier.

Cardeña, R., Valencia-Ojeda, C., Chazaro-Ruiz, L. F., & Razo-Flores, E. (2024). Regulation of the dark fermentation products by electro-fermentation in reactors without membrane. *International Journal of Hydrogen Energy, 49*, 107–116.

Chabert, N., Ali, O. A., & Achouak, W. (2015). All ecosystems potentially host electrogenic bacteria. *Bioelectrochemistry, 106*, 88–96.

Chandrasekhar, K., Kumar, A. N., Kumar, G., Kim, D. H., Song, Y. C., & Kim, S. H. (2021). Electro-fermentation for biofuels and biochemicals production: Current status and future directions. *Bioresource Technology, 323*, Article 124598.

Cheah, Y. K., Vidal-Antich, C., Dosta, J., & Mata-Álvarez, J. (2019). Volatile fatty acid production from mesophilic acidogenic fermentation of organic fraction of municipal solid waste and food waste under acidic and alkaline pH. *Environmental Science and Pollution Research, 26*, 35509–35522.

Chen, S., & Smith, A. L. (2018). Methane-driven microbial fuel cells recover energy and mitigate dissolved methane emissions from anaerobic effluents. *Environmental Science: Water Research & Technology, 4*(1), 67–79.

Cheng, S., Xing, D., Call, D. F., & Logan, B. E. (2009). Direct biological conversion of electrical current into methane by electromethanogenesis. *Environmental Science & Technology, 43*(10), 3953–3958.

Choi, O., & Sang, B. I. (2016). Extracellular electron transfer from cathode to microbes: Application for biofuel production. *Biotechnology for Biofuels, 9*, 1–14.

Choi, O., Kim, T., Woo, H. M., & Um, Y. (2014). Electricity-driven metabolic shift through direct electron uptake by electroactive heterotroph *Clostridium pasteurianum*. *Scientific Reports, 4*(1), 6961.

Chu, N., Liang, Q., Jiang, Y., & Zeng, R. J. (2020). Microbial electrochemical platform for the production of renewable fuels and chemicals. *Biosensors and Bioelectronics, 150*, Article 111922.

Conners, E. M., Davenport, E. J., Gallagher III, B. M., & Bose, A. (2023). Electron transfer processes between microbes and electrodes in bioelectrochemical reactors. In *Material-microbes interactions* (pp. 59–81). Academic Press.

Croese, E., Pereira, M. A., Euverink, G. J. W., Stams, A. J., & Geelhoed, J. S. (2011). Analysis of the microbial community of the biocathode of a hydrogen-producing microbial electrolysis cell. *Applied Microbiology and Biotechnology, 92*, 1083–1093.

Das, S., & Ghangrekar, M. M. (2018). Value added product recovery and carbon dioxide sequestration from biogas using microbial electrosynthesis. *Indian Journal of Experimental Biology, 56*, 470.

Das, S., Das, S., Das, I., & Ghangrekar, M. M. (2019). Application of bioelectrochemical systems for carbon dioxide sequestration and concomitant valuable recovery: A review. *Materials Science for Energy Technologies, 2*(3), 687–696.

Das, S., Diels, L., Pant, D., Patil, S. A., & Ghangrekar, M. M. (2020). Microbial electrosynthesis: A way towards the production of electro-commodities through carbon sequestration with microbes as biocatalysts. *Journal of the Electrochemical Society, 167*(15), Article 155510.

Dennis, P. G., Harnisch, F., Yeoh, Y. K., Tyson, G. W., & Rabaey, K. (2013). Dynamics of cathode-associated microbial communities and metabolite profiles in a glycerol-fed bioelectrochemical system. *Applied and Environmental Microbiology, 79*(13), 4008–4014.

Deval, A. S., Parikh, H. A., Kadier, A., Chandrasekhar, K., Bhagwat, A. M., & Dikshit, A. K. (2017). Sequential microbial activities mediated bioelectricity production from distillery wastewater using bio-electrochemical system with simultaneous waste remediation. *International Journal of Hydrogen Energy, 42*(2), 1130–1141.

Dhungana, P., Prajapati, B., Bhatt, P., Regmi, D., Yadav, M., Maharjan, S., Lamsal, U., Kathariya, S., Chaudhary, P., & Joshi, J. (2023). Production of bioethanol from Saccharum spontaneum by simultaneous saccharification and electro-fermentation using mixed culture of microbes. *Biofuels, 14*(2), 191–199.

d'Ippolito, G., Dipasquale, L., Vella, F. M., Romano, I., Gambacorta, A., Cutignano, A., & Fontana, A. (2010). Hydrogen metabolism in the extreme thermophile Thermotoga neapolitana. *International Journal of Hydrogen Energy, 35*(6), 2290–2295.

ElMekawy, A., Hegab, H. M., Losic, D., Saint, C. P., & Pant, D. (2017). Applications of graphene in microbial fuel cells: The gap between promise and reality. *Renewable and Sustainable Energy Reviews, 72*, 1389–1403.

Feng, D., Xia, A., Huang, Y., Zhu, X., Zhu, X., Show, P. L., & Liao, Q. (2023). Continuous electromethanogenesis of propionate wastewater: Effect of external voltage and hydraulic retention time. *Chemical Engineering Journal, 454*, Article 140267.

Flores-Rodriguez, C., & Min, B. (2020). Enrichment of specific microbial communities by optimum applied voltages for enhanced methane production by microbial electrosynthesis in anaerobic digestion. *Bioresource Technology, 300*, 122624.

Flynn, J. M., Ross, D. E., Hunt, K. A., Bond, D. R., & Gralnick, J. A. (2010). Enabling unbalanced fermentations by using engineered electrode-interfaced bacteria. *MBio, 1*(5), 10–1128.

Garbini, G. L., Barra Caracciolo, A., & Grenni, P. (2023). Electroactive bacteria in natural ecosystems and their applications in microbial fuel cells for bioremediation: A review. *Microorganisms, 11*(5), 1255.

Gemünde, A., Lai, B., Pause, L., Krömer, J., & Holtmann, D. (2022). Redox mediators in microbial electrochemical systems. *ChemElectroChem, 9*(13), Article e202200216.

Ghangrekar, M. M., & Neethu, B. (2020). Bioelectrochemical system for bioremediation and energy generation. In *Microbial bioremediation & biodegradation* (pp. 365–391).

Gildemyn, S., Rozendal, R. A., & Rabaey, K. (2017). A Gibbs free energy-based assessment of microbial electrocatalysis. *Trends in Biotechnology, 35*(5), 393–406.

Gu, J., Li, Z., Xie, G., Yang, Y., Liu, B., Ren, Z. J., & Xing, D. (2022). Electro-fermentation enhances H_2 and ethanol co-production by regulating electron transfer and substrate transmembrane transport. *Chemical Engineering Journal, 429*(May 2021), 132223. https://doi.org/10.1016/j.cej.2021.132223

Gustave, W., Yuan, Z. F., Sekar, R., Toppin, V., Liu, J. Y., Ren, Y. X., Zhang, J., & Chen, Z. (2019). Relic DNA does not obscure the microbial community of paddy soil microbial fuel cells. *Research in Microbiology, 170*(2), 97–104.

Hafez, H. M. (2019). *U.S. Patent No. 10,351,879*. U.S. Patent and Trademark Office.

Harris, H. W., El-Naggar, M. Y., & Nealson, K. H. (2012). Shewanella oneidensis MR-1 chemotaxis proteins and electron-transport chain components essential for congregation near insoluble electron acceptors.

Hassan, R. Y., Febbraio, F., & Andreescu, S. (2021). Microbial electrochemical systems: Principles, construction and biosensing applications. *Sensors, 21*(4), 1279.

Huang, Q., Liu, Y., & Dhar, B. R. (2022). A critical review of microbial electrolysis cells coupled with anaerobic digester for enhanced biomethane recovery from high-strength feedstocks. *Critical Reviews in Environmental Science and Technology, 52*(1), 50–89.

Isipato, M., Dessì, P., Sánchez, C., Mills, S., Ijaz, U. Z., Asunis, F., Spiga, D., De Gioannis, G., Mascia, M., Collins, G., Muntoni, A., & Lens, P. N. L. (2020). Propionate production by bioelectrochemically-assisted lactate fermentation and simultaneous CO_2 recycling. *Frontiers in Microbiology, 11*(December), 1–16. https://doi.org/10.3389/fmicb.2020.599438

Jain, A., & He, Z. (2018). "NEW" resource recovery from wastewater using bioelectrochemical systems: Moving forward with functions. *Frontiers of Environmental Science & Engineering, 12*, 1–13.

Jiang, Y., Chu, N., Zhang, W., Zhang, L., & Zeng, R. J. (2020). Electro-fermentation regulates mixed culture chain elongation with fresh and acclimated cathode. *Energy Conversion and Management, 204*, Article 112285.

Jiang, Y., Lu, L., Wang, H., Shen, R., Ge, Z., Hou, D., Chen, X., Liang, P., Huang, X., & Ren, Z. J. (2018). Electrochemical control of redox potential arrests methanogenesis and regulates products in mixed culture electro-fermentation. *ACS Sustainable Chemistry and Engineering, 6*(7), 8650–8658. https://doi.org/10.1021/acssuschemeng.8b00948

Jin, S., & Fallgren, P. H. (2022). Feasibility of using bioelectrochemical systems for bioremediation. In *Microbial biodegradation and bioremediation* (pp. 493–507). Elsevier.

Kadier, A., Jain, P., Lai, B., Kalil, M. S., Kondaveeti, S., Fahad Saud Alabbosh, K., Abu-Reesh, I. M. & Mohanakrishna, G. (2020). Biorefinery perspectives of microbial electrolysis cells (MECs) for hydrogen and valuable chemicals production through wastewater treatment.

Kadier, A., Kalil, M. S., Chandrasekhar, K., Mohanakrishna, G., Saratale, G. D., Saratale, R. G., Kumar, G., Pugazhendhi, A., & Sivagurunathan, P. (2018). Surpassing the current limitations of high purity H_2 production in microbial electrolysis cell (MECs): Strategies for inhibiting growth of methanogens. *Bioelectrochemistry, 119*, 211–219.

Kadier, A., Simayi, Y., Abdeshahian, P., Azman, N. F., Chandrasekhar, K., & Kalil, M. S. (2016). A comprehensive review of microbial electrolysis cells (MEC) reactor designs and configurations for sustainable hydrogen gas production. *Alexandria Engineering Journal, 55*(1), 427–443.

Kamaraj, Y., Punamalai, G., Kandasamy, S., & Kasinathan, K. (2020). Influence of long-term organic and conventional fertilization on bacterial communities involved in bioelectricity production from paddy field-microbial fuel cells. *Archives of Microbiology, 202*(8), 2279–2289.

Karthikeyan, R., Singh, R., & Bose, A. (2019). Microbial electron uptake in microbial electrosynthesis: A mini-review. *Journal of Industrial Microbiology and Biotechnology, 46*(9–10), 1419–1426.

Khosravanipour Mostafazadeh, A., Drogui, P., Brar, S. K., Tyagi, R. D., Le Bihan, Y., Buelna, G., & Rasolomanana, S. D. (2016). Enhancement of biobutanol production by electromicrobial glucose conversion in a dual chamber fermentation cell using C. pasteurianum. *Energy Conversion and Management, 130*, 165–175. https://doi.org/10.1016/j.enconman.2016.10.050

Kim, D. H., Han, S. K., Kim, S. H., & Shin, H. S. (2006). Effect of gas sparging on continuous fermentative hydrogen production. *International Journal of Hydrogen Energy, 31*(15), 2158–2169.

Kim, T. S., & Kim, B. H. (1988). Electron flow shift in Clostridium acetobutylicum fermentation by electrochemically introduced reducing equivalent. *Biotechnology letters, 10*, 123–128.

Kondaveeti, S., Govindarajan, D., Mohanakrishna, G., Thatikayala, D., Abu-Reesh, I. M., Min, B., Nambi, I. M., Al-Raoush, R. I., & Aminabhavi, T. M. (2023). Sustainable bioelectrochemical systems for bioenergy generation via waste treatment from petroleum industries. *Fuel, 331*, Article 125632.

Koul, Y., Devda, V., Varjani, S., Guo, W., Ngo, H. H., Taherzadeh, M. J., Chang, J. S., Wong, J. W., Bilal, M., Kim, S. H., Bui, X. T., & Parra-Saldívar, R. (2022). Microbial electrolysis: A promising approach for treatment and resource recovery from industrial wastewater. *Bioengineered, 13*(4), 8115–8134.

Kracke, F., & Krömer, J. O. (2014). Identifying target processes for microbial electrosynthesis by elementary mode analysis. *BMC Bioinformatics, 15*, 1–14.

Kracke, F., Vassilev, I., & Krömer, J. O. (2015). Microbial electron transport and energy conservation—The foundation for optimizing bioelectrochemical systems. *Frontiers in Microbiology, 6*, 575.

Kristensen, E., Bouillon, S., Dittmar, T., & Marchand, C. (2008). Organic carbon dynamics in mangrove ecosystems: A review. *Aquatic Botany, 89*(2), 201–219.

Kucek, L. A., Xu, J., Nguyen, M., & Angenent, L. T. (2016). Waste conversion into n-caprylate and n-caproate: Resource recovery from wine lees using anaerobic reactor microbiomes and in-line extraction. *Frontiers in Microbiology, 7*(NOV), 1–14. https://doi.org/10.3389/fmicb.2016.01892

LaBelle, E. V., Marshall, C. W., & May, H. D. (2019). Microbiome for the electrosynthesis of chemicals from carbon dioxide. *Accounts of Chemical Research, 53*(1), 62–71.

Lai, B., Yu, S., Bernhardt, P. V., Rabaey, K., Virdis, B., & Krömer, J. O. (2016). Anoxic metabolism and biochemical production in Pseudomonas putida F1 driven by a bioelectrochemical system. *Biotechnology for Biofuels, 9*, 1–13.

Lai, Y. H., & Lan, J. C. W. (2021). Enhanced polyhydroxybutyrate production through incorporation of a hydrogen fuel cell and electro-fermentation system. *International Journal of Hydrogen Energy, 46*(31), 16787–16800. https://doi.org/10.1016/j.ijhydene.2020.08.117

Lee, J. W., Kim, H. U., Choi, S., Yi, J., & Lee, S. Y. (2011). Microbial production of building block chemicals and polymers. *Current Opinion in Biotechnology, 22*(6), 758–767.

Lee, S. H., Kang, H. J., Kim, Y., Kim, N. K., & Park, H. D. (2022). Different contribution of exoelectrogens in methanogenesis via direct interspecies electron transfer (DIET) by the different substrate in continuous anaerobic bioreactor. *Bioresource Technology, 364*, Article 128115.

Leicester, D., Amezaga, J., & Heidrich, E. (2020). Is bioelectrochemical energy production from wastewater a reality? Identifying and standardising the progress made in scaling up microbial electrolysis cells. *Renewable and Sustainable Energy Reviews, 133*, Article 110279.

Lekshmi, G. S., Bazaka, K., Ramakrishna, S., & Kumaravel, V. (2023). Microbial electrosynthesis: Carbonaceous electrode materials for CO_2 conversion. *Materials Horizons, 10*(2), 292–312.

Li, C., & Cheng, S. (2019). Functional group surface modifications for enhancing the formation and performance of exoelectrogenic biofilms on the anode of a bioelectrochemical system. *Critical Reviews in Biotechnology, 39*(8), 1015–1030.

Li, S., Kim, M., Min, K., Wu, G., Cui, M., Kim, C., Oh, Y. K., Kim, S., Lee, S. Y., Kang, S. G., Nygard, Y., & Kim, J. R. (2023). Electron uptake from solid electrodes promotes the more efficient conversion of CO_2 to polyhydroxybutyrate by using *Rhodobacter sphaeroides*. *Chemical Engineering Journal, 469*, 143785.

Li, X., Wang, S., Li, L., Zu, X., Sun, Y., & Xie, Y. (2020). Opportunity of atomically thin two-dimensional catalysts for promoting CO_2 electroreduction. *Accounts of Chemical Research, 53*(12), 2964–2974.

Liang, D., He, W., Li, C., Yu, Y., Zhang, Z., Ren, N., & Feng, Y. (2019). Bidirectional electron transfer biofilm assisted complete bioelectrochemical denitrification process. *Chemical Engineering Journal, 375*, Article 121960.

Ling, L., Yang, C., Li, Z., Luo, H., Feng, S., Zhao, Y., & Lu, L. (2022). Plant endophytic bacteria: A potential resource pool of electroactive micro-organisms. *Journal of Applied Microbiology, 132*(3), 2054–2066.

Liu, H., Grot, S., & Logan, B. E. (2005). Electrochemically assisted microbial production of hydrogen from acetate. *Environmental Science and Technology, 39*(11), 4317–4320. https://doi.org/10.1021/es050244p

Liu, X., Shi, L., & Gu, J. D. (2018). Microbial electrocatalysis: Redox mediators responsible for extracellular electron transfer. *Biotechnology Advances, 36*(7), 1815–1827.

Liu, Z., Xue, X., Cai, W., Cui, K., Patil, S. A., & Guo, K. (2023). Recent progress on microbial electrosynthesis reactor designs and strategies to enhance the reactor performance. *Biochemical Engineering Journal, 190*, Article 108745.

Liyanage, H., Udayanga, T., & Babel, S. (2019). *Use of different pretreatment methods to enhance the methane production by anaerobic digestion of food waste*. Doctoral dissertation, Thammasat University.

Logan, B. E., Call, D., Cheng, S., Hamelers, H. V., Sleutels, T. H., Jeremiasse, A. W., & Rozendal, R. A. (2008). Microbial electrolysis cells for high yield hydrogen gas production from organic matter. *Environmental Science & Technology, 42*(23), 8630–8640.

López-Hernández, R., Cercado-Quezada, B., Gómez-Velázquez, H. D. J., Robles-Rodríguez, C., González-Dávalos, L., Varela-Echavarría, A., Shimada, A., & Mora, O. (2023). Single-chamber electrofermentation of rumen fluid increases microbial biomass and volatile fatty acid production without major changes in diversity. *Fermentation, 9*(6). https://doi.org/10.3390/fermentation9060502

Lovley, D. R. (2017). Syntrophy goes electric: Direct interspecies electron transfer. *Annual Review of Microbiology, 71*(1), 643–664.

Lu, S., Leininger, A., Jassby, D., Hoek, E. M., May, H. D., & Ren, Z. J. (2024). Temperature dependence of acids recovery from wastewater electro-fermentation. *Resources, Conservation and Recycling, 203*, Article 107407.

Lu, X., von Haxthausen, K. A., Brock, A. L., & Trapp, S. (2021). Turnover of lake sediments treated with sediment microbial fuel cells: A long-term study in a eutrophic lake. *Science of the Total Environment, 796*, Article 148880.

Lyu, X., Araujo, L. D., Quek, S. Y., Du Toit, W. J., & Kilmartin, P. A. (2021). Effect of oxygen and nitrogen sparging during grape fermentation on volatile sulphur compounds. *South African Journal of Enology and Viticulture, 42*(2), 154–164.

Majumdar, P., Pant, D., & Patra, S. (2017). Integrated photobioelectrochemical systems: A paradigm shift in artificial photosynthesis. *Trends in Biotechnology, 35*(4), 285–287.

Manzini, E., Scaglia, B., Schievano, A., & Adani, F. (2015). Dark fermentation effectiveness as a key step for waste biomass refineries: Influence of organic matter macromolecular composition and bioavailability. *International Journal of Energy Research, 39*(11), 1519–1527.

Mariano, A. P., Ezeji, T. C., & Qureshi, N. (2015). Butanol production by fermentation: Efficient bioreactors. In S. W. Snyder (Ed.), *Commercializing biobased products: Opportunities, challenges, benefits, and risks* (pp. 48–70). Royal Society of Chemistry.

Mathew, A. S., Wang, J., Luo, J., & Yau, S. T. (2015). Enhanced ethanol production via electrostatically accelerated fermentation of glucose using Saccharomyces cerevisiae. *Scientific Reports, 5*(1), 15713.

Matsumoto, T., Higuma, K., Yamada, R., & Ogino, H. (2023). Mevalonate production by electro-fermentation in *Escherichia coli* via Mtr-based electron transfer system. *Biochemical Engineering Journal, 191*(September 2022), 108772. https://doi.org/10.1016/j.bej.2022.108772

Mier, A. A., Olvera-Vargas, H., Mejía-López, M., Longoria, A., Verea, L., Sebastian, P. J., & Arias, D. M. (2021). A review of recent advances in electrode materials for emerging bioelectrochemical systems: From biofilm-bearing anodes to specialized cathodes. *Chemosphere, 283*, Article 131138.

Miller, A., Singh, L., Wang, L., & Liu, H. (2019). Linking internal resistance with design and operation decisions in microbial electrolysis cells. *Environment International, 126*, 611–618.

Modestra, J. A., Babu, M. L., & Mohan, S. V. (2015). Electro-fermentation of real-field acidogenic spent wash effluents for additional biohydrogen production with simultaneous treatment in a microbial electrolysis cell. *Separation and Purification Technology, 150*, 308–315.

Mogollón, C. A. G., Díaz, J. C. Q., & Posada, J. O. G. (2023). Production of acetone, butanol, and ethanol by electro-fermentation with Clostridium saccharoperbutylacetonicum N1-4. *Bioelectrochemistry, 152*, Article 108414.

Moscoviz, R., Toledo-Alarcón, J., Trably, E., & Bernet, N. (2016). Electro-fermentation: How to drive fermentation using electrochemical systems. *Trends in Biotechnology, 34*(11), 856–865.

Mostafazadeh, A. K., Drogui, P., Brar, S. K., Tyagi, R. D., Le Bihan, Y., Buelna, G., & Rasolomanana, S. D. (2016). Enhancement of biobutanol production by electromicrobial glucose conversion in a dual chamber fermentation cell using *C. pasteurianum. Energy Conversion and Management, 130*, 165–175.

Nevin, K. P., Woodard, T. L., Franks, A. E., Summers, Z. M., & Lovley, D. R. (2010). Microbial electrosynthesis: Feeding microbes electricity to convert carbon dioxide and water to multicarbon extracellular organic compounds. *MBio, 1*(2), 10–1128.

Nguyen, T. T., Ta, D. T., Lin, C. Y., & Chu, C. Y. (2022). Biohythane production from swine manure and pineapple waste in a single-stage two-chamber digester using gel-entrapped anaerobic microorganisms. *International Journal of Hydrogen Energy, 47*(60), 25245–25255.

Nishio, K., Kimoto, Y., Song, J., Konno, T., Ishihara, K., Kato, S., Hashimoto, K., & Nakanishi, S. (2013). Extracellular electron transfer enhances polyhydroxybutyrate productivity in Ralstonia eutropha. *Environmental Science and Technology Letters, 1*(1), 40–43. https://doi.org/10.1021/ez400085b

Paiano, P., Menini, M., Zeppilli, M., Majone, M., & Villano, M. (2019). Electro-fermentation and redox mediators enhance glucose conversion into butyric acid with mixed microbial cultures. *Bioelectrochemistry, 130*, Article 107333.

Pandit, A. V., & Mahadevan, R. (2011). In silico characterization of microbial electrosynthesis for metabolic engineering of biochemicals. *Microbial Cell Factories, 10*, 1–14.

Pant, D., Singh, A., Van Bogaert, G., Olsen, S. I., Nigam, P. S., Diels, L., & Vanbroekhoven, K. (2012). Bioelectrochemical systems (BES) for sustainable energy production and product recovery from organic wastes and industrial wastewaters. *RSC Advances, 2*(4), 1248–1263.

Paquete, C. M., Morgado, L., Salgueiro, C. A., & Louro, R. O. (2022). Molecular mechanisms of microbial extracellular electron transfer: The importance of multiheme cytochromes. *Frontiers in Bioscience-Landmark, 27*(6), 174.

Pawar, A. A., Karthic, A., Lee, S., Pandit, S., & Jung, S. P. (2022). Microbial electrolysis cells for electromethanogenesis: Materials, configurations and operations. *Environmental Engineering Research, 27*(1).

Perona-Vico, E., Feliu-Paradeda, L., Puig, S., & Bañeras, L. (2020). Bacteria coated cathodes as an in-situ hydrogen evolving platform for microbial electrosynthesis. *Scientific Reports, 10*(1), 19852.

Philips, J., Verbeeck, K., Rabaey, K., & Arends, J. B. (2016). Electron transfer mechanisms in biofilms. In *Microbial electrochemical and fuel cells* (pp. 67–113). Woodhead Publishing.

Prathiba, S., Kumar, P. S., & Vo, D. V. N. (2022). Recent advancements in microbial fuel cells: A review on its electron transfer mechanisms, microbial community, types of substrates and design for bio-electrochemical treatment. *Chemosphere, 286*, Article 131856.

Rabaey, K., & Rozendal, R. A. (2010). Microbial electrosynthesis—Revisiting the electrical route for microbial production. *Nature Reviews Microbiology, 8*(10), 706–716.

Rabaey, K., Girguis, P., & Nielsen, L. K. (2011). Metabolic and practical considerations on microbial electrosynthesis. *Current Opinion in Biotechnology, 22*(3), 371–377.

Raghavendran, V., Webb, J. P., Cartron, M. L., Springthorpe, V., Larson, T. R., Hines, M., Mohammed, H., Zimmerman, W. B., Poole, R. K., & Green, J. (2020). A microbubble-sparged yeast propagation–fermentation process for bioethanol production. *Biotechnology for Biofuels, 13*, 1–16.

Rago, L., Pant, D., & Schievano, A. (2019). Electro-fermentation—Microbial electrochemistry as new frontier in biomass refineries and industrial fermentations. In *Advanced bioprocessing for alternative fuels, biobased chemicals, and bioproducts* (pp. 265–287). Woodhead Publishing.

Rago, L., Ruiz, Y., Baeza, J. A., Guisasola, A., & Cortés, P. (2015). Microbial community analysis in a long-term membrane-less microbial electrolysis cell with hydrogen and methane production. *Bioelectrochemistry, 106*, 359–368.

Ranawat, P., & Rawat, S. (2018). Metal-tolerant thermophiles: Metals as electron donors and acceptors, toxicity, tolerance and industrial applications. *Environmental Science and Pollution Research, 25*, 4105–4133.

Ren, G., Hu, A., Huang, S., Ye, J., Tang, J., & Zhou, S. (2018). Graphite-assisted electro-fermentation methanogenesis: Spectroelectrochemical and microbial community analyses of cathode biofilms. *Bioresource Technology, 269*(July), 74–80. https://doi.org/10.1016/j.biortech.2018.08.078

Ren, G., Yan, Y., Nie, Y., Lu, A., Wu, X., Li, Y., Wang, C., & Ding, H. (2019). Natural extracellular electron transfer between semiconducting minerals and electroactive bacterial communities occurred on the rock varnish. *Frontiers in Microbiology, 10*, 293.

Rimboud, M., Etcheverry, L., Barakat, M., Achouak, W., Bergel, A., & Délia, M. L. (2021). Hypersaline microbial fuel cell equipped with an oxygen-reducing microbial cathode. *Bioresource Technology, 337*, Article 125448.

Rojas, C. A., De Santiago Torio, A., Park, S., Bosak, T., & Klepac-Ceraj, V. (2021). Organic electron donors and terminal electron acceptors structure anaerobic microbial communities and interactions in a permanently stratified sulfidic lake. *Frontiers in Microbiology, 12*, Article 620424.

Rosenbaum, M. A., & Henrich, A. W. (2014). Engineering microbial electrocatalysis for chemical and fuel production. *Current Opinion in Biotechnology, 29*, 93–98.

Rousseau, R., Etcheverry, L., Roubaud, E., Basséguy, R., Délia, M. L., & Bergel, A. (2020). Microbial electrolysis cell (MEC): Strengths, weaknesses and research needs from electrochemical engineering standpoint. *Applied Energy, 257*, Article 113938.

Roy, A. S., Sharma, A., Thapa, B. S., Pandit, S., Lahiri, D., Nag, M., Sarkar, T., Pati, S., Ray, R. R., Shariati, M. A., Wilairatana, P., & Mubarak, M. S. (2022). Microbiomics for enhancing electron transfer in an electrochemical system. *Frontiers in Microbiology, 13*, Article 868220.

Roy, S., Schievano, A., & Pant, D. (2016). Electro-stimulated microbial factory for value added product synthesis. *Bioresource Technology, 213*, 129–139.

Salar-García, M. J., Ortiz-Martínez, V. M., Sánchez-Segado, S., Valero Sánchez, R., Sáez López, A., Lozano Blanco, L. J., & Godínez-Seoane, C. (2024). Sustainable production of biofuels and biochemicals via electro-fermentation technology. *Molecules, 29*(4), 834.

Saravanan, A., Karishma, S., Kumar, P. S., Yaashikaa, P. R., Jeevanantham, S., & Gayathri, B. (2020). Microbial electrolysis cells and microbial fuel cells for biohydrogen production: Current advances and emerging challenges. In *Biomass conversion and biorefinery* (pp. 1–21).

Sasaki, D., Sasaki, K., Watanabe, A., Morita, M., Matsumoto, N., Igarashi, Y., & Ohmura, N. (2013). Operation of a cylindrical bioelectrochemical reactor containing carbon fiber fabric for efficient methane fermentation from thickened sewage sludge. *Bioresource Technology, 129*, 366–373. https://doi.org/10.1016/j.biortech.2012.11.048

Schievano, A., Pepé Sciarria, T., Vanbroekhoven, K., De Wever, H., Puig, S., Andersen, S. J., Rabaey, K., & Pant, D. (2016). Electro-fermentation—Merging electrochemistry with fermentation in industrial applications. *Trends in Biotechnology, 34*(11), 866–878. https://doi.org/10.1016/j.tibtech.2016.04.007

Schmidt, S. (2021). Applications of mixed microbial cultures in industrial biotechnology. In *The autotrophic biorefinery: Raw materials from biotechnology. De Gruyter* (pp. 353–384).

Schmitz, S., Nies, S., Wierckx, N., Blank, L. M., & Rosenbaum, M. A. (2015). Engineering mediator-based electroactivity in the obligate aerobic bacterium Pseudomonas putida KT2440. *Frontiers in Microbiology, 6*, 284.

Schröder, U., Harnisch, F., & Angenent, L. T. (2015). Microbial electrochemistry and technology: Terminology and classification. *Energy & Environmental Science, 8*(2), 513–519.

Schuppert, B., Schink, B., & Trösch, W. (1992). Batch and continuous production of propionic acid from whey permeate by Propionibacterium acidi-propionici in a three-electrode amperometric culture system. *Applied Microbiology and Biotechnology, 37*, 549–553.

Sciarria, T. P., Batlle-Vilanova, P., Colombo, B., Scaglia, B., Balaguer, M. D., Colprim, J., Puig, S., & Adani, F. (2018). Bio-electrorecycling of carbon dioxide into bioplastics. *Green Chemistry, 20*(17), 4058–4066.

Segundo-Aguilar, A., González-Gutiérrez, L. V., Payá, V. C., Feliu, J., Buitrón, G., & Cercado, B. (2021). Energy and economic advantages of simultaneous hydrogen and biogas production in microbial electrolysis cells as a function of the applied voltage and biomass content. *Sustainable Energy & Fuels, 5*(7), 2003–2017.

Selembo, P. A., Perez, J. M., Lloyd, W. A., & Logan, B. E. (2009). High hydrogen production from glycerol or glucose by electrohydrogenesis using microbial electrolysis cells. *International Journal of Hydrogen Energy, 34*(13), 5373–5381.

Sen, B., Aravind, J., Kanmani, P., & Lay, C. H. (2016). State of the art and future concept of food waste fermentation to bioenergy. *Renewable and Sustainable Energy Reviews, 53*, 547–557.

Sharma, M., Alvarez-Gallego, Y., Achouak, W., Pant, D., Sarma, P. M., & Dominguez-Benetton, X. (2019). Electrode material properties for designing effective microbial electrosynthesis systems. *Journal of Materials Chemistry A, 7*(42), 24420–24436. https://doi.org/10.1039/c9ta04886c

Sharma, M., Aryal, N., Sarma, P. M., Vanbroekhoven, K., Lal, B., Benetton, X. D., & Pant, D. (2013). Bioelectrocatalyzed reduction of acetic and butyric acids via direct electron transfer using a mixed culture of sulfate-reducers drives electrosynthesis of alcohols and acetone. *Chemical Communications, 49*(58), 6495–6497.

Shi, K., Liang, B., Guo, Q., Zhao, Y., Sharif, H. M. A., Li, Z., Chen, E., & Wang, A. (2021). Accelerated bioremediation of a complexly contaminated river sediment through ZVI-electrode combined stimulation. *Journal of Hazardous Materials, 413*, Article 125392.

Singh, V., Haque, S., Niwas, R., Srivastava, A., Pasupuleti, M., & Tripathi, C. (2017). Strategies for fermentation medium optimization: An in-depth review. *Frontiers in Microbiology, 7*, 2087.

Singh, M., Joshi, G., & Ali, M. A. (2020). Electro-fermentation technology: Emerging platform. In *Bioelectrochemical systems: Vol. 1 Principles and processes* (pp. 71–79).

Sravan, J. S., Butti, S. K., Sarkar, O., & Mohan, S. V. (2019). Electrofermentation: Chemicals and fuels. In *Microbial electrochemical technology* (pp. 723–737). Elsevier.

Sriram, S., Wong, J. W. C., & Pradhan, N. (2022). Recent advances in electro-fermentation technology: A novel approach towards balanced fermentation. *Bioresource Technology, 360*(July), Article 127637. https://doi.org/10.1016/j.biortech.2022.127637

Szydlowski, L., Lan, T. C., Shibata, N., & Goryanin, I. (2020). Metabolic engineering of a novel strain of electrogenic bacterium *Arcobacter butzleri* to create a platform for single analyte detection using a microbial fuel cell. *Enzyme and Microbial Technology, 139*, Article 109564.

Tahernia, M., Plotkin-Kaye, E., Mohammadifar, M., Gao, Y., Oefelein, M. R., Cook, L. C., & Choi, S. (2020). Characterization of electrogenic gut bacteria. *ACS Omega, 5*(45), 29439–29446.

Thapa, B. S., Kim, T., Pandit, S., Song, Y. E., Afsharian, Y. P., Rahimnejad, M., Kim, J. R., & Oh, S. E. (2022). Overview of electroactive microorganisms and electron transfer mechanisms in microbial electrochemistry. *Bioresource Technology, 347*, Article 126579.

Thuan, N. H., Tatipamula, V. B., Canh, N. X., & Van Giang, N. (2022). Recent advances in microbial co-culture for production of value-added compounds. *3 Biotech, 12*(5), 115.

Thulasinathan, B., Jayabalan, T., Arumugam, N., Kulanthaisamy, M. R., Kim, W., Kumar, P., Govarthanan, M., & Alagarsamy, A. (2022). Wastewater substrates in microbial fuel cell systems for carbon-neutral bioelectricity generation: An overview. *Fuel, 317*, Article 123369.

Tiquia-Arashiro, S. M., & Pant, D. (Eds.). (2020). *Microbial electrochemical technologies*. CRC Press, Taylor & Francis Group.

Toledo-Alarcón, J., Fuentes, L., Etchebehere, C., Bernet, N., & Trably, E. (2021). Glucose electro-fermentation with mixed cultures: A key role of the Clostridiaceae family. *International Journal of Hydrogen Energy, 46*(2), 1694–1704.

Toledo-Alarcón, J., Moscoviz, R., Trably, E., & Bernet, N. (2019). Glucose electro-fermentation as main driver for efficient H2-producing bacteria selection in mixed cultures. *International Journal of Hydrogen Energy, 44*(4), 2230–2238.

Umar, M. F., Abbas, S. Z., Mohamad Ibrahim, M. N., Ismail, N., & Rafatullah, M. (2020). Insights into advancements and electrons transfer mechanisms of electrogens in benthic microbial fuel cells. *Membranes, 10*(9), 205.

Van Eerten-Jansen, M. C. A. A., Ter Heijne, A., Grootscholten, T. I. M., Steinbusch, K. J. J., Sleutels, T. H. J. A., Hamelers, H. V. M., & Buisman, C. J. N. (2013). Correction to bioelectrochemical production of caproate and caprylate from acetate by mixed cultures. *ACS Sustainable Chemistry & Engineering, 1*(8), 1069–1069. https://doi.org/10.1021/sc400143v

Varanasi, J. L., Veerubhotla, R., Pandit, S., & Das, D. (2019). Biohydrogen production using microbial electrolysis cell: Recent advances and future prospects. In *Microbial electrochemical technology* (pp. 843–869).

Vassilev, I., Averesch, N. J., Ledezma, P., & Kokko, M. (2021). Anodic electro-fermentation: Empowering anaerobic production processes via anodic respiration. *Biotechnology Advances, 48*, Article 107728.

Vassilev, I., Gießelmann, G., Schwechheimer, S. K., Wittmann, C., Virdis, B., & Krömer, J. O. (2018). Anodic electro-fermentation: Anaerobic production of L-Lysine by recombinant Corynebacterium glutamicum. *Biotechnology and Bioengineering, 115*(6), 1499–1508. https://doi.org/10.1002/bit.26562

Vassilev, I., Kracke, F., Freguia, S., Keller, J., Krömer, J. O., Ledezma, P., & Virdis, B. (2019). Microbial electrosynthesis system with dual biocathode arrangement for simultaneous acetogenesis, solventogenesis and carbon chain elongation. *Chemical Communications, 55*(30), 4351–4354.

Velvizhi, G., Sarkar, O., Rovira-Alsina, L., Puig, S., & Mohan, S. V. (2022). Conversion of carbon dioxide to value added products through anaerobic fermentation and electro fermentation: A comparative approach. *International Journal of Hydrogen Energy, 47*(34), 15442–15455.
Virdis, B., Hoelzle, R. D., Marchetti, A., Boto, S. T., Rosenbaum, M. A., Blasco-Gómez, R., Puig, S., Freguia, S., & Villano, M. (2022). Electro-fermentation: Sustainable bioproductions steered by electricity. *Biotechnology Advances, 59*, Article 107950.
Wang, G., Huang, Q., Song, T. S., & Xie, J. (2020a). Enhancing microbial electrosynthesis of acetate and butyrate from CO2 reduction involving engineered Clostridium ljungdahlii with a nickel-phosphide-modified electrode. *Energy & Fuels, 34*(7), 8666–8675.
Wang, Y., Zheng, H., Chen, Y., Wen, Q., & Wu, J. (2020b). Macroporous composite capacitive bioanode applied in microbial fuel cells. *Chinese Chemical Letters, 31*(1), 205–209.
Wang, H., Liu, Y., Du, H., Zhu, J., Peng, L., Yang, C., & Luo, F. (2021). Exploring the effect of voltage on biogas production performance and the methanogenic pathway of microbial electrosynthesis. *Biochemical Engineering Journal, 171*, Article 108028.
Wangpraseurt, D., You, S., Sun, Y., & Chen, S. (2022). Biomimetic 3D living materials powered by microorganisms. *Trends in Biotechnology, 40*(7), 843–857.
Wilberforce, T., Sayed, E. T., Abdelkareem, M. A., Elsaid, K., & Olabi, A. G. (2021). Value added products from wastewater using bioelectrochemical systems: Current trends and perspectives. *Journal of Water Process Engineering, 39*, Article 101737.
Xafenias, N., Anunobi, M. O., & Mapelli, V. (2015). Electrochemical startup increases 1, 3-propanediol titers in mixed-culture glycerol fermentations. *Process Biochemistry, 50*(10), 1499–1508.
Xiao, W., Wang, R. S., Handy, D. E., & Loscalzo, J. (2018). NAD (H) and NADP (H) redox couples and cellular energy metabolism. *Antioxidants & Redox Signaling, 28*(3), 251–272.
Xue, G., Lai, S., Li, X., Zhang, W., You, J., Chen, H., Qian, Y., Gao, P., Liu, Z., & Liu, Y. (2018). Efficient bioconversion of organic wastes to high optical activity of L-lactic acid stimulated by cathode in mixed microbial consortium. *Water Research, 131*, 1–10. https://doi.org/10.1016/j.watres.2017.12.024
Yamamoto, M., Takaki, Y., Kashima, H., Tsuda, M., Tanizaki, A., Nakamura, R., & Takai, K. (2023). In situ electrosynthetic bacterial growth using electricity generated by a deep-sea hydrothermal vent. *The ISME Journal, 17*(1), 12–20.
Yan, H., Catania, C., & Bazan, G. C. (2015). Membrane-intercalating conjugated oligoelectrolytes: Impact on bioelectrochemical systems. *Advanced Materials, 27*(19), 2958–2973.
Yang, X., & Chen, S. (2021). Microorganisms in sediment microbial fuel cells: Ecological niche, microbial response, and environmental function. *Science of the Total Environment, 756*, Article 144145.
Yaqoob, A. A., Ibrahim, M. N. M., & Umar, K. (2021). Electrode material as anode for improving the electrochemical performance of microbial fuel cells. In *Energy storage battery systems—Fundamentals and applications*. IntechOpen.
Yu, B., Xu, J., Yuan, H., Lou, Z., Lin, J., & Zhu, N. (2014). Enhancement of anaerobic digestion of waste activated sludge by electrochemical pretreatment. *Fuel, 130*, 279–285.
Zacharoff, L. A., & El-Naggar, M. Y. (2017). Redox conduction in biofilms: From respiration to living electronics. *Current Opinion in Electrochemistry, 4*(1), 182–189.
Zhang, C., He, D., Ma, J., Tang, W., & Waite, T. D. (2019). Comparison of faradaic reactions in flow-through and flow-by capacitive deionization (CDI) systems. *Electrochimica Acta, 299*, 727–735.
Zhang, J., Li, F., Liu, D., Liu, Q., & Song, H. (2024). Engineering extracellular electron transfer pathways of electroactive microorganisms by synthetic biology for energy and chemicals production. *Chemical Society Reviews*.
Zhang, Y., Li, J., Meng, J., & Wang, X. (2021a). A cathodic electro-fermentation system for enhancing butyric acid production from rice straw with a mixed culture. *Science of the Total Environment, 767*, Article 145011.

Zhang, Y., Li, J., Meng, J., Sun, K., & Yan, H. (2021b). A neutral red mediated electro-fermentation system of *Clostridium beijerinckii* for effective co-production of butanol and hydrogen. *Bioresource Technology, 332*, Article 125097.
Zhao, J., Li, F., Cao, Y., Zhang, X., Chen, T., Song, H., & Wang, Z. (2021). Microbial extracellular electron transfer and strategies for engineering electroactive microorganisms. *Biotechnology Advances, 53*, Article 107682.
Zhao, Y., Li, Z., Ma, J., Yun, H., Qi, M., Ma, X., Wang, H., Wang, A., & Liang, B. (2018). Enhanced bioelectroremediation of a complexly contaminated river sediment through stimulating electroactive degraders with methanol supply. *Journal of Hazardous Materials, 349*, 168–176.
Zhao, Z., Zhang, Y., Wang, L., & Quan, X. (2015). Potential for direct interspecies electron transfer in an electric-anaerobic system to increase methane production from sludge digestion. *Scientific Reports, 5*(1), 11094.
Zhen, G., Kobayashi, T., Lu, X., Kumar, G., & Xu, K. (2016). Biomethane recovery from Egeria densa in a microbial electrolysis cell-assisted anaerobic system: Performance and stability assessment. *Chemosphere, 149*, 121–129.
Zheng, T., Li, J., Ji, Y., Zhang, W., Fang, Y., Xin, F., Dong, W., Wei, P., Ma, J., & Jiang, M. (2020). Progress and prospects of bioelectrochemical systems: Electron transfer and its applications in the microbial metabolism. *Frontiers in Bioengineering and Biotechnology, 8*, 10.
Zhou, M., Chen, J., Freguia, S., Rabaey, K., & Keller, J. (2013). Carbon and electron fluxes during the electricity driven 1,3-propanediol biosynthesis from glycerol. *Environmental Science & Technology, 47*(19), 11199–11205.
Zhou, M., Freguia, S., Dennis, P. G., Keller, J., & Rabaey, K. (2015). Development of bioelectrocatalytic activity stimulates mixed-culture reduction of glycerol in a bioelectrochemical system. *Microbial Biotechnology, 8*(3), 483–489.
Zhou, M., Yan, B., Wong, J. W., & Zhang, Y. (2018). Enhanced volatile fatty acids production from anaerobic fermentation of food waste: A mini-review focusing on acidogenic metabolic pathways. *Bioresource Technology, 248*, 68–78.
Zhou, M., Zhou, J., Tan, M., Du, J., Yan, B., Wong, J. W., & Zhang, Y. (2017). Enhanced carboxylic acids production by decreasing hydrogen partial pressure during acidogenic fermentation of glucose. *Bioresource Technology, 245*, 44–51.
Zou, L., Zhu, F., Chang, F. X., & Yong, Y. C. (2022). Extracellular electrons powered microbial CO_2 upgrading: Microbial electrosynthesis and artificial photosynthesis. *One-carbon feedstocks for sustainable bioproduction* (pp. 243–271). Springer International Publishing.
Zuroff, T. R., Xiques, S. B., & Curtis, W. R. (2013). Consortia-mediated bioprocessing of cellulose to ethanol with a symbiotic *Clostridium phytofermentans*/yeast co-culture. *Biotechnology for Biofuels, 6*, 1–12.

Chapter 11
Application of Dark Fermentation and Electro-Fermentation for CO_2 Conversion to Value-Added Products

Manikanta Manmadha Doki, Lakshmi Pathi Thulluru, Shamik Chowdhury, and M. M. Ghangrekar

1 Introduction

About 3.55×10^{14} kg of carbon dioxide (CO_2), 10^{12} m^3 of municipal wastewater, and 2.01×10^{12} kg of municipal solid waste are released into the environment from anthropogenic activities every year (Kaza et al., 2018; Mac Dowell et al., 2017; WWAP, 2017). This carbon-based waste (solid, liquid, and gas) presently encumbers society due to the exorbitantly high capital and operational expenses and substantial land requirements associated with traditional waste-valorization technologies. It has, therefore, encouraged us to reduce our reliance on fossil fuels and petroleum-derived synthetic products that will at least alleviate the anthropogenic CO_2 emissions into the atmosphere. Additionally, the volatility of energy prices sourced from fossil fuels and their ecological consequences has drawn colossal attention to carbon-neutral and renewable energy sources. Hence, a paradigm shift toward a circular economy can endorse a closed-loop biorefinery model that harnesses the energy within waste to produce marketable products like biomaterials, biochemicals, and biofuels, thereby closing the resource loop.

A possible solution is to use microbial electrochemical technologies (METs) that exploit electroactive microorganisms to catalyze the bioelectrochemical reactions to convert these carbon-based wastes into bioenergy and bioproducts. It is a system capable of carrying out the oxidation at the electron acceptor end (anode)

M. M. Doki · M. M. Ghangrekar (✉)
Department of Civil Engineering, Indian Institute of Technology Kharagpur, Kharagpur, India
e-mail: ghangrekar@civil.iitkgp.ac.in

L. P. Thulluru · S. Chowdhury · M. M. Ghangrekar
School of Environmental Science and Engineering, Indian Institute of Technology Kharagpur, Kharagpur, India

M. M. Ghangrekar
National Institute of Technology Puducherry, Karaikal, India

© The Author(s), under exclusive license to Springer Nature Singapore Pte Ltd. 2025

S. Sevda (ed.), *Electro-Fermentation for Synthesis of Chemicals, Biochemicals and Biofuels*, https://doi.org/10.1007/978-981-96-6358-3_11

and reduction at the electron donor end (cathode) separately, under ambient conditions. Initially, METs demonstrated methane (electromethanogenesis) and acetic acid biosynthesis at the biotic cathode using microorganisms such as *Methanobacterium palustre* and *Sporomusa ovata*, supported by an abiotic anode (Cheng et al., 2009; Nevin et al., 2011). These biosynthesizing systems were then termed as microbial electrosynthesis (MES), which performs bioreduction of CO_2 with the help of electrons donated at the cathode.

On the other hand, controlling the fermentation pathways aided by an externally imposed potential is defined as electro-fermentation (EF). The captured electrons, that help to valorize organic waste or CO_2 store the chemical energy in the covalent bond, are sourced from renewable energies like solar and wind energy, among others (Rabaey & Rozendal, 2010). Another process that valorizes organic matter is the acidogenesis process, which transforms it into a wide range of products, such as acetic acid, butyric acid, and hydrogen (H_2). This approach is called dark fermentation (DF). The ability to derive H_2 from DF has attracted immense attention, recognizing it as a valuable green fuel and sustainable alternative for waste management.

These systems, however, rely on the oxidation–reduction potential (ORP) of the environment around the electrode. This ORP refers to the ability of the electrolyte to accept or donate electrons during biochemical reactions. The polarized electrode often influences ORP and, in turn, it affects fermentation pathways in EF by altering the balance of reduced and oxidized nicotinamide-adenine dinucleotide (NAD), commonly represented as the $NADH/NAD^+$ ratio (Paiano et al., 2019). The $NADH/NAD^+$ ratio plays a vital role in cellular metabolism, maintaining the intracellular redox balance that drives the energy production for carrying out the EF. As an electron carrier and signaling molecule, it participates in a wide range of biological processes in living cells. A high $NADH/NAD^+$ ratio typically supports oxidative processes, encouraging the breakdown of molecules for energy production (Hu et al., 2021).

The present chapter elucidates the fundamental mechanisms and scientific aspects of organic waste and CO_2 valorization technologies, with an emphasis on DF and EF. Additionally, it summarizes the products synthesized in these technologies. Further, it elucidates the factors that influence the performance of DF and EF and the need for optimizing their performance to maximize the product yield. Finally, it identifies the challenges and potential solutions enabling its field-scale application in the foreseeable future.

2 Fundamentals of Dark Fermentation and Electrofermentation

2.1 Mechanism of Dark Fermentation

Dark fermentation typically entails three stages, namely hydrolysis, acidogenesis, and acetogenesis. The hydrolytic fermentative bacteria primarily break down complex polymeric substrates into monomers or simple compounds, which are utilized in subsequent stages (Fig. 1). The acidogenic microbes oxidize the monomers to produce volatile fatty acids (VFAs), H_2, and CO_2, simultaneously. Subsequently, the acetogenic microbes utilize these VFAs to synthesis acetic acid, H_2, and CO_2 (Srivastava et al., 2024).

Delving further into the fundamental mechanism, glucose, for instance, serves as a template substrate. Here, the H_2-producing bacteria converts glucose into pyruvate following the glycolytic pathway, generating adenosine triphosphate (ATP) from both adenosine diphosphate (ADP) and reducing the equivalent NADH. The hydrogenase and pyruvate ferredoxin oxidoreductase enzymes oxidize pyruvate to H_2, CO_2, and acetyl coenzyme A (acetyl-CoA). However, the pyruvate can also be oxidized to formate acetyl-CoA based on the microbes and ambient conditions involved in the

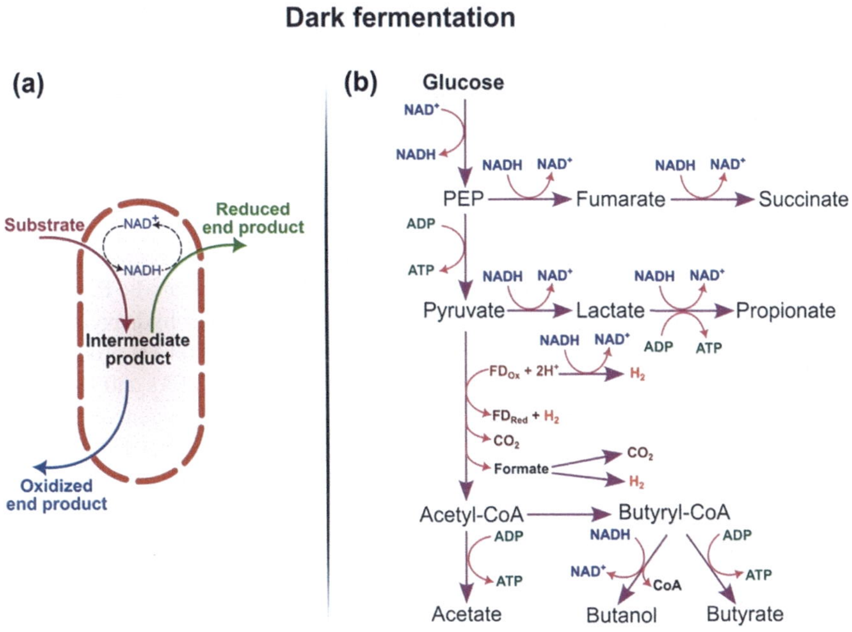

Fig. 1 **a** Principle and **b** mechanism of production of biochemicals in dark fermentation

bioreactors. Additionally, compounds like ethanol, butyrate, and acetate are synthesized during the subsequent oxidation of acetyl-CoA. On the other hand, the DF of a complex substrate yields several intermediates and auxiliary products contingent upon bioreactor operating conditions, including (i) substrate characteristics, (ii) organic loading rate (OLR), (iii) temperature, (iv) pH, (v) yield, (vi) redox potential, (v) solid retention time (SRT), (vi) hydraulic retention time (HRT), and (vii) headspace gaseous composition, along with other ambient conditions (Ghimire et al., 2015; Pandey et al., 2022).

2.2 *Mechanism of Electro-fermentation*

The EF incorporates electrochemistry to modulate microbial metabolism while regulating the fermentation pathways, aiding the synthesis of valuable products from organic wastes (Fig. 2). The electrodes proposed here serve as surplus electron donor (cathode) and acceptor (anode) that influence the fermentation pathways followed by the biocatalyst (mixed and pure cultures) (Schievano et al., 2016). The extracellular electron transport observed during the process can be further categorized as anodic electron transfer, which features an outward electron transfer toward the anodic electron sinks. Consequently, the cathodic electron flow involves inward electron transfer, where the electrons are utilized by the electrogens (Moscoviz et al., 2018). Additionally, this is maintained through the ORP and $NADH/NAD^+$ ratio. Further, the anodic electron sink in EF promotes ATP synthesis that creates a proton gradient contrary to the cathodic electron source that contributes to generating the NADH cofactor (Kracke & Krömer, 2014).

2.2.1 Anodic Electro-fermentation

The applied potential, substrate type, and inocula employed in the anodic EF govern the product generation such as H_2, ethanol, acetoin, lysine, poly-b-hydroxybutyrate (PHB), and 2-keto-gluconate (Jiang et al., 2019). The anodic EF for H_2 production, particularly couples DF that degrades organic matter to produce VFAs and H_2 in the anodic chamber (Fig. 1). At the same time, the microbial electrolysis generates H_2 in the cathodic chamber under an applied electric voltage (Nikhil et al., 2015). Further, the synthesis of H_2 at the anode is regulated by factors including applied voltage and the concentration of autocrine mediators; however, scenarios like reduction in pH resulting from the accumulation of VFAs can significantly impede H_2 production. Moreover, the utilization of H_2 by electroactive bacteria (*Geobacter sulfurreducens* and *Escherichia coli*) and methanogens can also impair the performance. However, practices like applying chemical inhibitors and implementing appropriate operational controls help to minimize the consumption of H_2 (Awate et al., 2017).

The anode of anodic EF acts as an electron acceptor by removing excess electrons generated during substrate degradation. However, engineered strains like *Shewanella*

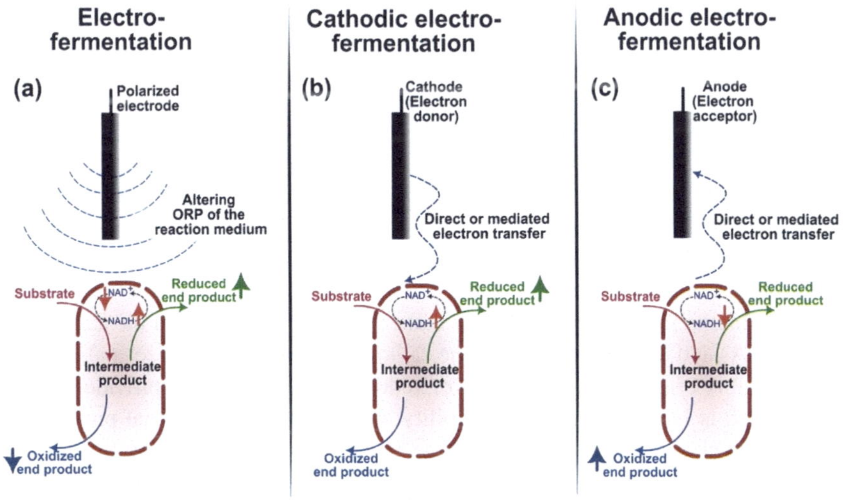

Fig. 2 Principle of bioelectrochemical production in **a** electro-fermentation (EF), specifically in **b** cathodic and **c** anodic EF

oneidensis and *Escherichia coli* have been employed for ethanol and acetic acid production using glycerol as a substrate in anodic EF (Flynn et al., 2010; Sturm-Richter et al., 2015). On the other hand, a synergistic relationship is observed in co-cultures like *Geobacter sulfurreducens* and *Clostridium cellobioparum* while synthesizing ethanol from glycerol. Here, *G. sulfurreducens* played a crucial role in eliminating fermentative byproducts (Speers et al., 2014).

Furthermore, research on substrates like lactate has demonstrated acetoin synthesis using engineered strains of *S. oneidensis* (Bursac et al., 2017). Moreover, electron mediators, such as methylene blue, can enhance acetoin production from glucose in the presence of *E. coli* mutants (Förster et al., 2017). As a result, anaerobic fermentation in anodic EF reduces substrate loss compared to aerobic fermentation. Additionally, the anodic EF allows for more significant biomass loading than the cathodic EF due to the low potential applied at the cathode (Vassilev et al., 2018).

2.2.2 Cathodic Electro-fermentation

In contrast to anodic EF, cathodic EF relies on the cathode as the sole electron source, which provides the reducing power, regulating both product selectivity and production rate (Xue et al., 2018). The cathodic EF represents a type of MES, where a carbon substrate instead of CO_2 is stoichiometrically bioreduced to the desired product (Fig. 2). This approach has demonstrated the ability to produce organics like butanol, acetate, lactate, butyrate, 1,3-propanediol, lysine, and polyhydroxyalkanoates (PHA) (Jiang et al., 2019). The electrons supplied by the electrode facilitate the NADH formation, which enables the synthesis of reduced products from carbon substrates like sucrose and glycerol (Choi et al., 2012).

However, the working potential significantly influences product selectivity, which shapes the microbial community on the biocathode over an extended operational period (Xafenias et al., 2015). Further, the PHA synthesis using mixed cultures is predominantly supported by a microaerophilic environment along with a continuous supply of electrons (Srikanth et al., 2012). In contrast, mediators such as anthraquinone-2,6-disulfonate and neutral red can enhance lactic acid and ethanol production in *Corynebacterium glutamicum* and *Klebsiella pneumoniae* (Harrington et al., 2015; Sasaki et al., 2014).

The reducing power supplied by the cathode lowers the ORP and the accumulation of VFAs and hydrogen sulfide in the cathodic chamber (De Vrieze et al., 2018). However, this differs from the usage of reducing agents like H_2 provided externally, which is limited by its low solubility along with the low partial pressure of H_2 (Omar et al., 2018). However, the reducing power of the electrode can be further enhanced by increasing its surface area, altering its structure, and augmenting microbial interaction with the electrode (Martins et al., 2018). Further, this interaction is also influenced by the redox potential and the application of conductive materials over the electrode, such as metal oxides (iron oxide minerals), granular activated carbon, or conductive biochar (Lovley et al., 2017). These materials play a vital role, as most microorganisms exhibit a weak extracellular electron transfer (EET) ability (Liu et al., 2017).

3 Factors Influencing Dark Fermentation and Electro-fermentation

The DF and EF generate a mixture of gases, organic acids, and solvents. The contributional percentage of these products depends on substrate characteristics, the type of inocula, and other operational conditions, which are briefly explained below.

3.1 *Microbial Strains and Electron Transfers*

Electroactive pure cultures such as species of *Geobacteraceae* and *Shewanellaceae* are significant due to their capacity for electron transfer, facilitating biofilm growth on the electrode (Thrash & Coates, 2008). Conversely, a few bacteria, like *Clostridium pasteurianum*, exhibit the characteristics of electroactive electrogen while still possessing the capability to metabolize complex carbon sources, including carbohydrates and alcohol (Arends & Verstraete, 2012; Choi et al., 2014). However, this is not the case with most pure cultures. This scenario has, thus, motivated investigations into co-culturing to incorporate the features of both electroactive and fermentative properties in EF. For instance, the co-culturing of *Clostridium cellobioparum* and *Geobacter sulfurreducens* has been shown to synthesize ethanol using glycerol (Gemma, 2014).

Usually, when fermentative bacteria do not exhibit electroactive behavior, adding solubilizing agents like redox mediators in the fermentation medium provides a conducive extracellular ORP (Thrash & Coates, 2008). These electron shuttles, such as neutral red and methyl viologen, can undergo continuous oxidation and reduction cycles between the electrode and fermentative bacteria (Sturm-Richter et al., 2015). Additionally, fermentation products like H_2, formate, and acetate could act as mediators for synthesizing other targeted products (Moscoviz et al., 2016). Moreover, efforts have been initiated to incorporate electroactive properties in metabolically engineered fermentative strains as well as engineer electroactive bacterial strains to assimilate diverse carbon substrates (Flynn et al., 2010).

The electrodes in EF behave as non-soluble electron acceptors (anode) and donors (cathode), which do not act as a limiting step but strengthen the electrochemical reactions. Electron transfer between electrodes and electrogens can occur via direct contact, nanowire-like extensions (conductive pili) of electroactive bacteria, redox mediators, or extracellular polymeric substances released during biofilm formation (Choi & Sang, 2016). These electron transfers promote an efficient supply of electrons during anodic EF and the conversion of substrates into desired reduced products during cathodic EF.

3.2 Syntropic Interactions

The microbial community in EF mimics microbial fuel cell (MFC), where electroactive bacteria conduct anodic respiration within the anodic biofilm, while fermentative bacteria convert substrates into usable metabolites for the electroactive bacteria (Kouzuma et al., 2015). This syntropic relationship establishes thermodynamically favorable conditions by removing fermentative byproducts (Kiely et al., 2011). These interactions occur through interspecies electron transfer within the biofilm formed on the electrode (Rotaru et al., 2012). The biofilm exhibits a spatial arrangement, with electroactive bacteria positioned nearest to the electrode and the fermenters present at the upper layers of the biofilm (Malvankar et al., 2012). Furthermore, biofilm thickness may serve as a limiting factor for these interactions, as an increased thickness reduces electron transfer, resulting in a gradient within the biofilm concerning redox mediators and pH (Beyenal & Babauta, 2012). However, the accumulation of metabolites such as VFAs and H_2 in the fermentative media and headspace inhibits further production of these metabolites and promotes the growth of fermentative bacteria (Xafenias et al., 2015). Electroactive bacteria like the *Geobacteraceae* sp. consume these metabolites via electron transfer mechanisms, thereby stimulating the fermentation process and enhancing the purity of the desired product by eliminating unwanted byproducts (Wang et al., 2015).

3.3 Electrical Current

The EF is often considered an autonomous fermentation process facilitated by an electrode that serves as a driving mechanism (Flynn et al., 2010). While MFC seeks to optimize electron recovery efficiency via electricity generation, EF focuses on utilizing the electrode to modify the metabolic pathways of organic substrate fermentation (Moscoviz et al., 2016). The EF coefficient' here is analogous to 'coulombic efficiency' in MFCs. It is the ratio of the charge supplied by the electrical circuit to the total charge accumulated in the synthesized product (Choi et al., 2014). However, the EF coefficient is typically close to zero, which signifies that a low current density can considerably influence the fermentation pattern. On the other hand, redox pair homeostasis like $NADH/NAD^+$ observed at the cellular level is crucial for cellular metabolism (Liu et al., 2013). The majority of metabolism regulating enzymes are responsive to intracellular and extracellular ORP and the electron transfer associated with the stability of the $NADH/NAD^+$ ratio (Green & Paget, 2004). Moreover, the $NADH/NAD^+$ ratio further depends on the electron transfers facilitated by the electrodes and other electron carriers (Choi et al., 2012).

3.4 Substrate Characteristics

Multiple carbon sources have been identified as feedstock for producing valuable products in both DF and EF. When selecting these substrates, it is crucial to ensure they are renewable, sustainable, and do not compete with primary resources intended for human consumption (Baumann & Westermann, 2016). Given this, cost-effective and readily available biomass-based waste, such as sewage sludge, wastewater, biodiesel residues, lignocellulosic feedstock, and residues from dairy industries are some of the lucrative options (Park et al., 2014).

Sewage sludge is an unavoidable organic byproduct released during wastewater treatment. It consists of multiple microorganisms and approximately 10% (w/w) carbohydrates, 50% proteins, 10% lipids, and 30% other constituents, including ribonucleic acid and fiber (Albert & Martens, 1997). In the recent years, sludge has been productively converted into biodiesel, biogas, bioplastics, biosurfactants, bio-flocculants, biopesticides, bioethanol, and other useful commodities. However, during DF and EF of sewage sludge, possible accumulation of VFAs leads to a reduction in pH that acts as an inhibitor by permeating through the cell membrane, lowering microbial activity, and shifting the metabolic pathway (Pind et al., 2003). Therefore, an alkaline pH (7.0–12.0) is regarded as a beneficial parameter facilitating the synthesis of short-chain fatty acids (SCFAs), notably acetic and propionic acids, in both primary and secondary sludge (Yang et al., 2014). In addition, investigations have indicated that the presence of saponin or isolated strains like *Citrobacter freundii* to augment the synthesis of SCFAs (Huang et al., 2016; Yang et al., 2015).

Wastewaters, particularly those derived from dairy sources, after acidogenic process results in abundant acetic, propionic, butyric, and valeric acids (Demirel & Yenigun, 2004). A shorter HRT and a higher OLR promote the proliferation of acid-forming bacteria, which in turn, augment VFA production (Pandey et al., 2017). Furthermore, surmounting rate-limiting factors such as pH regulation, low-temperature thermochemical treatment, enhancing hydrolysis rates, maintaining the substrate characteristics, and reactor health improve the VFA yield. Moreover, wastewater from other sources, including paper mills, the lumber sector, pharmaceuticals, and the sugar industry can also be exploited to generate VFAs (Pandey et al., 2022).

3.5 pH and Temperature

The enzymes involved in microbial activity are often sensitive to the pH. Therefore, regulating the pH during VFA-producing acidogenic fermentation is crucial (Li et al., 2013). The optimum growth of most acetogens occurs in the acidic pH range of 5.5–6.5, which suppresses the proliferation of methanogens that require a pH of 7.0 (Leung & Wang, 2016). Conversely, an alkaline pH facilitates the passage of undissociated VFAs across the cell membrane into the electrolyte. On the contrary, in acidic electrolyte conditions, this transport across the cell membrane demands more energy (Rodríguez et al., 2006).

Apart from the substrate type, the prevailing pH can also influence the content of VFAs produced during acidic fermentation. Investigations have indicated that a pH below 5 and an alkaline pH ranging from 8 to 11 are suitable for improving butyric acid in VFA (Begum et al., 2018; Wang et al., 2014). The optimal pH for increasing acetic acid is 5, while for propionic acid is 6.5 (Jiang et al., 2013; Kim et al., 2011).

Temperature significantly influences the functioning of facultative bacteria during the hydrolysis process. It impacts microbial growth, and different microbial species react differently to temperature. Usually, elevated temperatures promote the effective solubilization of organics, improving the hydrolysis phase in DF (Jiang et al., 2013). Nonetheless, microbial species exhibit exceptional performance despite temperature variations ranging from 10 to 60 °C (Ho, 2014). Additionally, multiple investigations have indicated that the optimal temperature for VFA production is between 20 and 40 °C, which falls within the mesophilic range. This ensures the cost-effectiveness of the entire process, as no external energy is necessary to elevate the temperature further (Zhou et al., 2018).

Furthermore, research has also shown the impact of thermophilic temperatures on the selectivity of a certain VFA. For instance, the temperature at which maximum acetic acid production takes place is 35 °C, while propionic and butyric acid are produced predominantly at 45 °C and 55 °C, respectively (He et al., 2012; Jiang et al., 2013). Nonetheless, DF conducted at thermophilic temperatures enhance the solubility of supporting hydrolysis. However, it is crucial to recognize that certain acidogenic bacteria are susceptible to these drastic temperature variations, impeding VFA production (Strazzera et al., 2018).

3.6 Hydraulic Retention Time and Organic Loading Rate

HRT is the average period for which the soluble component persists in the DF reactor. It determines the daily flow rate at which the influent is treated in the context of reactor operation. However, operating the reactors at elevated HRT values hinders the acidogenic fermentation in the DF reactor (Pandey et al., 2022). Conversely, reduced HRT may lead to the washout of slow-growing bacteria (Zhou et al., 2018). Consequently, determining an optimal HRT facilitates the proliferation of predominant microbial species in the DF reactor, allowing adequate time for substrate solubilization, which overcomes the rate-limiting processes and augments the VFA production process (Pandey et al., 2022). Furthermore, analyses of the VFA composition at different HRTs indicate that shorter HRTs predominantly yield acetic acid, whereas longer HRTs favor propionic acid production (Lim et al., 2008). Nonetheless, a notable alteration in VFA composition was not observed during the co-fermentation of activated sludge and fruit and vegetable waste at different HRTs (Zhou et al., 2018).

The OLR determines the accessibility of carbon substrate during DF. Generally, a greater OLR assists VFA production as it enhances substrate availability across the reactor (Pandey et al., 2022). However, investigations have revealed that the performance boost in terms of VFA generation only lasts up to a certain increase in OLR, beyond which it starts to decline (Tang et al., 2016). One possible explanation is that an increase in OLR may lead to a rise in the viscosity of the medium, which impacts the microbe and substrate interaction, impeding the hydrolysis rate (Lim et al., 2008). Also, during DF, the OLR can alter the dominance of a certain VFA in the product yield. Low values of OLR are associated with increased production of propionic and butyric acids, while high values of OLR promote the generation of acetic and valeric acids in DF (Jiang et al., 2013; Khan et al., 2016).

3.7 Redox Potential

The ORP of the chemical species is often called the redox potential and is a reflection of its ability to donate or accept electrons (Liu et al., 2013). The reduced equivalent pairs, i.e., $NADH/NAD^+$ and H^+/H_2 have an ORP of –320 mV and –420 mV, respectively. The former exhibits a relatively higher ORP, which indicates a greater propensity for electron acceptance than the latter, rendering it inappropriate for H_2 production. It is essential for strict anaerobes to effectively use NADH to convert hydrogen ions into H_2 gas to attain a maximum theoretical yield of 4 mol H_2/mol hexose (Cao et al., 2022). It is well-established that the strict anaerobes are oxygen intolerant due to their lack of superoxide dismutase activity, which is essential for scavenging oxygen radicals (ORP range, +100 to −250 mV). However, the facultative anaerobes can grow aerobically and produce H_2 in oxygen-deficient environments (ORP range, +300 mV to −100 mV) (Vongvichiankul et al., 2017). Interestingly, ORP can act as an effective tool during fermentation to determine the presence of oxygen and influence the growth of fermentative microbes. It also helps regulate metabolic processes and ultimately aids in identifying the type of fermentation, which leads to the synthesis of metabolites of different titers (Sim et al., 2023).

3.8 Headspace Gas Pressure

The effectiveness of the fermentation process can be influenced by the headspace pressure exerted on the fermentation medium. It mainly depends upon the H_2 concentration in the headspace gas mixture, which may alter the electron flow associated with several metabolic processes impacting VFA formation (Yuan et al., 2019). A reduction in headspace pressure typically promotes homoacidogens that utilize H_2 and CO_2 to produce acetic acid that effectively minimizes energy loss (Yin et al., 2014). The predominant microbial species in mesophilic environments comprises *Acetobacterium* sp. (27–30 °C), *Acetobacterium woodii* sp., *Acetoanaerobium* sp., *Acetohalobium* sp. (38–40 °C), *Eubacterium* sp. and *Butyribacterium* sp. (30–37 °C). In contrast, some thermophilic strains include *Acetogenium* (66 °C) and *thermoautotrophicum* (55–60 °C) (Saady, 2013). As a result, lower headspace pressures help improve the hydrolysis process, which converts them into VFAs (Chen et al., 2017). Additionally, methods for optimizing the DF that efficiently utilize this H_2 include combining it with a homoacetogenic reactor, which uses the H_2 and CO_2 to produce acetic acid and improve the hydrolysis process in DF (Yin et al., 2014).

4 Integrated Approach Toward Sustainable Utilization of CO_2 into Value-Added Products

The DF is often selected as a cost-effective and carbon-neutral technology for its unique characteristics in utilizing diverse organic feedstocks to generate different bio-based products, including biofuels (biohydrogen, biomethane), solvents (ethanol, acetone, butanol, proponal), and SCFAs (acetic, propionic and butyric acids) under ambient pressure and temperature. However, its thermodynamic limitation because of redox imbalances (electron distribution) in the fermentation reactions results in lower H_2 yields and partial conversion of organic substrates, which restricts the commercialization of DF (Sravan et al., 2018). As a standalone process, it cannot metabolize the theoretical yield of 4 mol H_2/mol hexose, with its maximum yield of 33% from sugars (Ghimire et al., 2015). The DF effluent primarily consists of VFAs, which need further downstream processing to recover the individual acids as precursors. It can be used as platform chemicals or substrates in different processes. These include MFCs, microbial electrolysis cells, EF, photo-fermentation, and microalgal cultivation (Kumar et al., 2018). Moreover, CO_2 is also produced in larger quantities during DF, which alters the further usage of H_2 as fuel. Therefore, these limitations must be surmounted to optimize the recovery of valuable products at an industrial scale through advanced hybrid strategies that enhance the process to maximize energy recovery and support the broader bioeconomy (Sravan et al., 2018).

Recently, longer-chain carbon compounds have been of greater interest because of their increased commercialized value. However, due to the scarcity of biocatalysts that can utilize the electrons and inorganic CO_2 to synthesize them. The EF can achieve this by surmounting thermodynamic instabilities by providing external electron flux while stimulating the metabolism for the synthesis of targeted bioproducts. This induced external flux has the ability to elongate the chain length (Jiang et al., 2019). Thus, augmenting DF with EF enables an integrated hybrid process, as illustrated in Fig. 3, which provides a platform to produce medium-chain carboxylic acids (C6–C12) by utilizing SCFAs generated from DF effluent.

An investigation by Dennis et al. (2013) reported that the cathodic electro-fermentation of glycerol in a bioelectrochemical system produced a variety of metabolites, including ethanol, 1,3-propanediol, propionate, and valerate (Dennis et al., 2013). Initially, the microbial community abundance transitioned from *Citrobacter* sp. to *Clostridium* sp. after the sixth week of the inoculation period,

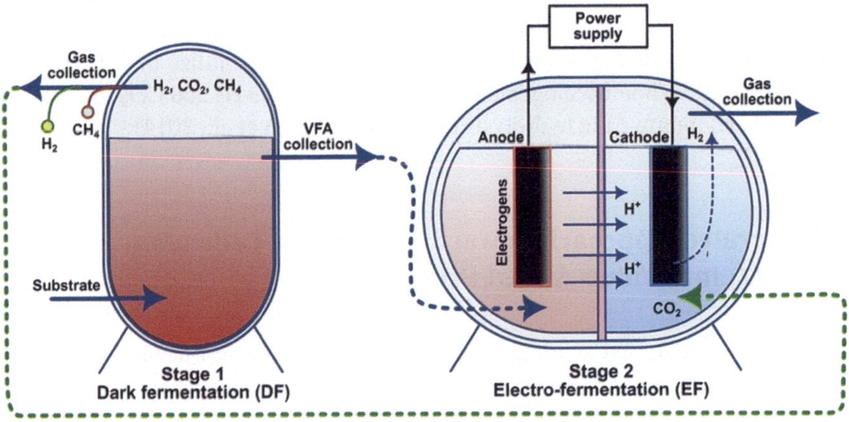

	DF	EF	DF + EF
H_2 yield	↓	↑	↑↑
Energy efficiency	↓	↑	↑↑
Substrate utilization	↑	↑↑	↑↑↑

Fig. 3 Integration of dark fermentation and electro-fermentation for enhanced resource recovery

indicating the occurrence of chain elongation within the cathode. The hypothesis was that electron flux maintained a reducing environment by enhancing the capacity of microorganisms to synthesize NADH. Another investigation conducted by Ma et al. (2022) demonstrated that the production of medium-chain fatty acids, including caproate, from VFAs and CO_2 through the chain elongation process in cathodic EF (Ma et al., 2022). It was postulated that ethanol-producing microorganisms, such as *Oscillibacter* species, utilize cathode as an electron donor to reduce acetate to ethanol in the presence of H_2. Subsequently, they utilize ethanol as an electron donor for chain elongation through indirect EET. The present hypothesis, therefore, proposes an efficient integration system to utilize the organic carbon in wastewater to synthesize industrially relevant chemical precursors.

5 Future Perspective and Conclusion

Platforms like EF, DF, and MES have shown substantial potential for producing value-added chemicals and biofuels. Critical factors, such as product and energy efficiencies in these systems, remain inadequately low. The significant bottlenecks include limited electron transfer efficiency, insufficient gene editing tools for engineering metabolic pathways, under-explored electrode materials, and reactor modifications in bioelectrochemical systems. Including this modular approach could decrease the intracellular $NADH/NAD^+$ ratio and enhance the production of exoelectrogens in the cathodic EF. Further, ideal electrode materials should possess a biocompatible surface that facilitates cell-electrode adhesion, an adequate specific surface area, robust chemical stability, superior mechanical strength, high conductivity, scalability, facile synthesis, and cost-effectiveness. Moreover, technical parameters such as electrical internal resistance in laboratory-size reactors, along with temperature, pH, and salinity of the operational environment, must be optimized to ensure a successful demonstration at an industrial scale.

References

Albert, D. B., & Martens, C. S. (1997). Determination of low-molecular-weight organic acid concentrations in seawater and pore-water samples via HPLC. *Marine Chemistry, 56*(1–2), 27–37.

Arends, J. B., & Verstraete, W. (2012). 100 years of microbial electricity production: Three concepts for the future. *Microbial Biotechnology,* 5(3), 333–346.

Awate, B., Steidl, R. J., Hamlischer, T., & Reguera, G. (2017). Stimulation of electro-fermentation in single-chamber microbial electrolysis cells driven by genetically engineered anode biofilms. *Journal of Power Sources, 356*, 510–518.

Baumann, I., & Westermann, P. (2016). Microbial production of short chain fatty acids from lignocellulosic biomass: Current processes and market. *BioMed Research International, 2016*(1), 8469357.

Begum, S., Anupoju, G. R., Sridhar, S., Bhargava, S. K., Jegatheesan, V., & Eshtiaghi, N. (2018). Evaluation of single and two stage anaerobic digestion of landfill leachate: Effect of pH and initial organic loading rate on volatile fatty acid (VFA) and biogas production. *Bioresource Technology, 251*, 364–373.

Beyenal, H., & Babauta, J. T. (2012). Microscale gradients and their role in electron-transfer mechanisms in biofilms. *Biochemical Society Transactions, 40*(6), 1315–1318.

Bursac, T., Gralnick, J. A., & Gescher, J. (2017). Acetoin production via unbalanced fermentation in Shewanella oneidensis. *Biotechnology and Bioengineering, 114*(6), 1283–1289.

Cao, Y., Liu, H., Liu, W., Guo, J., & Xian, M. (2022). Debottlenecking the biological hydrogen production pathway of dark fermentation: Insight into the impact of strain improvement. *Microbial Cell Factories, 21*(1), 166.

Chen, Y., Jiang, X., Xiao, K., Shen, N., Zeng, R. J., & Zhou, Y. (2017). Enhanced volatile fatty acids (VFAs) production in a thermophilic fermenter with stepwise pH increase—Investigation on dissolved organic matter transformation and microbial community shift. *Water Research, 112*, 261–268.

Cheng, S., Xing, D., Call, D. F., & Logan, B. E. (2009). Direct biological conversion of electrical current into methane by electromethanogenesis. *Environmental Science & Technology, 43*(10), 3953–3958.

Choi, O., Kim, T., Woo, H. M., & Um, Y. (2014). Electricity-driven metabolic shift through direct electron uptake by electroactive heterotroph Clostridium pasteurianum. *Scientific Reports, 4*(1), 6961.

Choi, O., & Sang, B. I. (2016). Extracellular electron transfer from cathode to microbes: Application for biofuel production. *Biotechnology for Biofuels, 9*, 1–14.

Choi, O., Um, Y., & Sang, B. I. (2012). Butyrate production enhancement by Clostridium tyrobutyricum using electron mediators and a cathodic electron donor. *Biotechnology and Bioengineering, 109*(10), 2494–2502.

De Vrieze, J., Arends, J. B., Verbeeck, K., Gildemyn, S., & Rabaey, K. (2018). Interfacing anaerobic digestion with (bio) electrochemical systems: Potentials and challenges. *Water Research, 146*, 244–255.

Demirel, B., & Yenigun, O. (2004). Anaerobic acidogenesis of dairy wastewater: The effects of variations in hydraulic retention time with no pH control. *Journal of Chemical Technology & Biotechnology: International Research in Process, Environmental & Clean Technology, 79*(7), 755–760.

Dennis, P. G., Harnisch, F., Yeoh, Y. K., Tyson, G. W., & Rabaey, K. (2013). Dynamics of cathode-associated microbial communities and metabolite profiles in a glycerol-fed bioelectrochemical system. *Applied and Environmental Microbiology, 79*(13), 4008–4014.

Flynn, J. M., Ross, D. E., Hunt, K. A., Bond, D. R., & Gralnick, J. A. (2010). Enabling unbalanced fermentations by using engineered electrode-interfaced bacteria. *Mbio, 1*(5), 10–1128.

Förster, A. H., Beblawy, S., Golitsch, F., & Gescher, J. (2017). Electrode-assisted acetoin production in a metabolically engineered Escherichia coli strain. *Biotechnology for Biofuels, 10*, 1–11.

Gemma, R. (2014). *Fermentation of glycerol into ethanol in a microbial electrolysis cell driven by a customized consortium.*

Ghimire, A., Frunzo, L., Pirozzi, F., Trably, E., Escudie, R., Lens, P. N., & Esposito, G. (2015). A review on dark fermentative biohydrogen production from organic biomass: Process parameters and use of byproducts. *Applied Energy, 144*, 73–95.

Green, J., & Paget, M. S. (2004). Bacterial redox sensors. *Nature Reviews Microbiology, 2*(12), 954–966.

Harrington, T. D., Mohamed, A., Tran, V. N., Biria, S., Gargouri, M., Park, J. J., & Beyenal, H. (2015). Neutral red-mediated microbial electrosynthesis by Escherichia coli, Klebsiella pneumoniae, and Zymomonas mobilis. *Bioresource Technology, 195*, 57–65.

He, M., Sun, Y., Zou, D., Yuan, H., Zhu, B., Li, X., & Pang, Y. (2012). Influence of temperature on hydrolysis acidification of food waste. *Procedia Environmental Sciences, 16*, 85–94.

Ho, D. (2014). *A high-rate high temperature anaerobic digestion system for sludge treatment.*

Hu, Q., Wu, D., Walker, M., Wang, P., Tian, R., & Wang, W. (2021). Genetically encoded biosensors for evaluating NAD^+/NADH ratio in cytosolic and mitochondrial compartments. *Cell Reports Methods, 1*(7).

Huang, X., Mu, T., Shen, C., Lu, L., & Liu, J. (2016). Alkaline fermentation of waste activated sludge stimulated by saponin: Volatile fatty acid production, mechanisms and pilot-scale application. *Water Science and Technology, 74*(12), 2860–2869.

Jiang, J., Zhang, Y., Li, K., Wang, Q., Gong, C., & Li, M. (2013). Volatile fatty acids production from food waste: Effects of pH, temperature, and organic loading rate. *Bioresource Technology, 143*, 525–530.

Jiang, Y., May, H. D., Lu, L., Liang, P., Huang, X., & Ren, Z. J. (2019). Carbon dioxide and organic waste valorization by microbial electrosynthesis and electro-fermentation. *Water Research, 149*, 42–55.

Kaza, S., Yao, L., Bhada-Tata, P., & Van Woerden, F. (2018). *What a waste 2.0: A global snapshot of solid waste management to 2050*. World Bank Publications.

Khan, M. A., Ngo, H. H., Guo, W. S., Liu, Y., Nghiem, L. D., Hai, F. I., & Wu, Y. (2016). Optimization of process parameters for production of volatile fatty acid, biohydrogen and methane from anaerobic digestion. *Bioresource Technology, 219*, 738–748.

Kiely, P. D., Regan, J. M., & Logan, B. E. (2011). The electric picnic: Synergistic requirements for exoelectrogenic microbial communities. *Current Opinion in Biotechnology, 22*(3), 378–385.

Kim, D. H., Kim, S. H., Jung, K. W., Kim, M. S., & Shin, H. S. (2011). Effect of initial pH independent of operational pH on hydrogen fermentation of food waste. *Bioresource Technology, 102*(18), 8646–8652.

Kouzuma, A., Kato, S., & Watanabe, K. (2015). Microbial interspecies interactions: Recent findings in syntrophic consortia. *Frontiers in Microbiology, 6*, 477.

Kracke, F., & Krömer, J. O. (2014). Identifying target processes for microbial electrosynthesis by elementary mode analysis. *BMC Bioinformatics, 15*, 410.

Kumar, G., Shobana, S., Nagarajan, D., Lee, D. J., Lee, K. S., Lin, C. Y., & Chang, J. S. (2018). Biomass based hydrogen production by dark fermentation—Recent trends and opportunities for greener processes. *Current Opinion in Biotechnology, 50*, 136–145.

Leung, D. Y., & Wang, J. (2016). An overview on biogas generation from anaerobic digestion of food waste. *International Journal of Green Energy, 13*(2), 119–131.

Li, X., Mu, H., Chen, Y., Zheng, X., Luo, J., & Zhao, S. (2013). Production of propionic acid-enriched volatile fatty acids from co-fermentation liquid of sewage sludge and food waste using Propionibacterium acidipropionici. *Water Science and Technology, 68*(9), 2061–2066.

Lim, S. J., Kim, B. J., Jeong, C. M., Ahn, Y. H., & Chang, H. N. (2008). Anaerobic organic acid production of food waste in once-a-day feeding and drawing-off bioreactor. *Bioresource Technology, 99*(16), 7866–7874.

Liu, C. G., Xue, C., Lin, Y. H., & Bai, F. W. (2013). Redox potential control and applications in microaerobic and anaerobic fermentations. *Biotechnology Advances, 31*(2), 257–265.

Liu, Y., He, P., Shao, L., Zhang, H., & Lü, F. (2017). Significant enhancement by biochar of caproate production via chain elongation. *Water Research, 119*, 150–159.

Lovley, D. R. (2017). Syntrophy goes electric: Direct interspecies electron transfer. *Annual Review of Microbiology, 71*(1), 643–664.

Ma, J., Wang, Z., Li, L., Shi, Z., Ke, S., & He, Q. (2022). Granular activated carbon stimulated caproate production through chain elongation in fluidized cathode electro-fermentation systems. *Journal of Cleaner Production, 365*, Article 132757.

Mac Dowell, N., Fennell, P. S., Shah, N., & Maitland, G. C. (2017). The role of CO_2 capture and utilization in mitigating climate change. *Nature Climate Change, 7*(4), 243–249.

Malvankar, N. S., Lau, J., Nevin, K. P., Franks, A. E., Tuominen, M. T., & Lovley, D. R. (2012). Electrical conductivity in a mixed-species biofilm. *Applied and Environmental Microbiology, 78*(16), 5967–5971.

Martins, G., Salvador, A. F., Pereira, L., & Alves, M. M. (2018). Methane production and conductive materials: A critical review. *Environmental Science & Technology, 52*(18), 10241–10253.

Moscoviz, R., Toledo-Alarcón, J., Trably, E., & Bernet, N. (2016). Electro-fermentation: How to drive fermentation using electrochemical systems. *Trends in Biotechnology, 34*(11), 856–865.

Moscoviz, R., Trably, E., & Bernet, N. (2018). Electro-fermentation triggering population selection in mixed-culture glycerol fermentation. *Microbial Biotechnology, 11*(1), 74–83.

Nevin, K. P., Hensley, S. A., Franks, A. E., Summers, Z. M., Ou, J., Woodard, T. L., & Lovley, D. R. (2011). Electrosynthesis of organic compounds from carbon dioxide is catalyzed by a diversity of acetogenic microorganisms. *Applied and Environmental Microbiology, 77*(9), 2882–2886.

Nikhil, G. N., Subhash, G. V., Yeruva, D. K., & Mohan, S. V. (2015). Synergistic yield of dual energy forms through biocatalyzed electrofermentation of waste: Stoichiometric analysis of electron and carbon distribution. *Energy, 88*, 281–291.

Omar, B., Abou-Shanab, R., El-Gammal, M., Fotidis, I. A., Kougias, P. G., Zhang, Y., & Angelidaki, I. (2018). Simultaneous biogas upgrading and biochemicals production using anaerobic bacterial mixed cultures. *Water Research, 142*, 86–95.

Paiano, P., Menini, M., Zeppilli, M., Majone, M., & Villano, M. (2019). Electro-fermentation and redox mediators enhance glucose conversion into butyric acid with mixed microbial cultures. *Bioelectrochemistry, 130*, 107333.

Pandey, A. K., Pilli, S., Bhunia, P., Tyagi, R. D., Surampalli, R. Y., Zhang, T. C., & Pandey, A. (2022). Dark fermentation: Production and utilization of volatile fatty acid from different wastes—A review. *Chemosphere, 288*, 132444.

Pandey, A. K., Singh, B., Upadhyay, S., Pandey, S., Mishra, V. K., & Jain, P. A. (2017). Solution for sustainable development for developing countries: Waste water treatment by use of membranes—A review. *International Journal of Current Microbiology and Applied Sciences, 6*(8), 1212–1228.

Park, G. W., Fei, Q., Jung, K., Chang, H. N., Kim, Y. C., Kim, N. J., & Cho, J. (2014). Volatile fatty acids derived from waste organics provide an economical carbon source for microbial lipids/biodiesel production. *Biotechnology Journal, 9*(12), 1536–1546.

Pind, P. F., Angelidaki, I., & Ahring, B. K. (2003). Dynamics of the anaerobic process: Effects of volatile fatty acids. *Biotechnology and Bioengineering, 82*(7), 791–801.

Rabaey, K., & Rozendal, R. A. (2010). Microbial electrosynthesis—Revisiting the electrical route for microbial production. *Nature Reviews Microbiology, 8*(10), 706–716.

Rodríguez, J., Kleerebezem, R., Lema, J. M., & Van Loosdrecht, M. C. (2006). Modeling product formation in anaerobic mixed culture fermentations. *Biotechnology and Bioengineering, 93*(3), 592–606.

Rotaru, A. E., Shrestha, P. M., Liu, F., Ueki, T., Nevin, K., Summers, Z. M., & Lovley, D. R. (2012). Interspecies electron transfer via hydrogen and formate rather than direct electrical connections in co-cultures of Pelobacter carbinolicus and Geobacter sulfurreducens. *Applied and Environmental Microbiology, 78*(21), 7645–7651.

Saady, N. M. C. (2013). Homoacetogenesis during hydrogen production by mixed cultures dark fermentation: Unresolved challenge. *International Journal of Hydrogen Energy, 38*(30), 13172–13191.

Sasaki, K., Tsuge, Y., Sasaki, D., & Kondo, A. (2014). Increase in lactate yield by growing Corynebacterium glutamicum in a bioelectrochemical reactor. *Journal of Bioscience and Bioengineering, 117*(5), 598–601.

Schievano, A., Sciarria, T. P., Vanbroekhoven, K., De Wever, H., Puig, S., Andersen, S. J., & Pant, D. (2016). Electro-fermentation–merging electrochemistry with fermentation in industrial applications. *Trends in Biotechnology, 34*(11), 866–878.

Sim, X. Y., Tan, J. P., He, N., Yeap, S. K., Hui, Y. W., Luthfi, A. A. I., & Jamali, N. S. (2023). Unraveling the effect of redox potential on dark fermentative hydrogen production. *Renewable and Sustainable Energy Reviews, 187*, 113755.

Speers, A. M., Young, J. M., & Reguera, G. (2014). Fermentation of glycerol into ethanol in a microbial electrolysis cell driven by a customized consortium. *Environmental Science & Technology, 48*(11), 6350–6358.

Sravan, J. S., Butti, S. K., Sarkar, O., Krishna, K. V., & Mohan, S. V. (2018). Electrofermentation of food waste—Regulating acidogenesis towards enhanced volatile fatty acids production. *Chemical Engineering Journal, 334*, 1709–1718.

Srikanth, S., Reddy, M. V., & Mohan, S. V. (2012). Microaerophilic microenvironment at biocathode enhances electrogenesis with simultaneous synthesis of polyhydroxyalkanoates (PHA) in bioelectrochemical system (BES). *Bioresource Technology, 125*, 291–299.

Srivastava, P., García-Quismondo, E., Palma, J., & González-Fernández, C. (2024). Coupling dark fermentation and microbial electrolysis cells for higher hydrogen yield: Technological competitiveness and challenges. *International Journal of Hydrogen Energy, 52*, 223–239.

Strazzera, G., Battista, F., Garcia, N. H., Frison, N., & Bolzonella, D. (2018). Volatile fatty acids production from food wastes for biorefinery platforms: A review. *Journal of Environmental Management, 226*, 278–288.

Sturm-Richter, K., Golitsch, F., Sturm, G., Kipf, E., Dittrich, A., Beblawy, S., & Gescher, J. (2015). Unbalanced fermentation of glycerol in Escherichia coli via heterologous production of an electron transport chain and electrode interaction in microbial electrochemical cells. *Bioresource Technology, 186*, 89–96.

Tang, J., Wang, X., Hu, Y., Zhang, Y., & Li, Y. (2016). Lactic acid fermentation from food waste with indigenous microbiota: Effects of pH, temperature and high OLR. *Waste Management, 52*, 278–285.

Thrash, J. C., & Coates, J. D. (2008). Direct and indirect electrical stimulation of microbial metabolism. *Environmental Science & Technology, 42*(11), 3921–3931.

Vassilev, I., Gießelmann, G., Schwechheimer, S. K., Wittmann, C., Virdis, B., & Krömer, J. O. (2018). Anodic electro-fermentation: Anaerobic production of L-lysine by recombinant *Corynebacterium glutamicum. Biotechnology and Bioengineering, 115*(6), 1499–1508.

Vongvichiankul, C., Deebao, J., & Khongnakorn, W. (2017). Relationship between pH, oxidation reduction potential (ORP) and biogas production in mesophilic screw anaerobic digester. *Energy Procedia, 138*, 877–882.

Wang, K., Yin, J., Shen, D., & Li, N. (2014). Anaerobic digestion of food waste for volatile fatty acids (VFAs) production with different types of inoculum: Effect of pH. *Bioresource Technology, 161*, 395–401.

Wang, V. B., Sivakumar, K., Yang, L., Zhang, Q., Kjelleberg, S., Loo, S. C. J., & Cao, B. (2015). Metabolite-enabled mutualistic interaction between Shewanella oneidensis and Escherichia coli in a co-culture using an electrode as electron acceptor. *Scientific Reports, 5*(1), 11222.

WWAP. (2017). The United Nations world water development report 2017. In *Wastewater: The untapped resource*. UNESCO.

Xafenias, N., Anunobi, M. O., & Mapelli, V. (2015). Electrochemical startup increases 1,3-propanediol titers in mixed-culture glycerol fermentations. *Process Biochemistry, 50*(10), 1499–1508.

Xue, G., Lai, S., Li, X., Zhang, W., You, J., Chen, H., & Liu, Y. (2018). Efficient bioconversion of organic wastes to high optical activity of L-lactic acid stimulated by cathode in mixed microbial consortium. *Water Research, 131*, 1–10.

Yang, X., Wan, C., Lee, D. J., Du, M., Pan, X., & Wan, F. (2014). Continuous volatile fatty acid production from waste activated sludge hydrolyzed at pH 12. *Bioresource Technology, 168*, 173–179.

Yang, X., Wen, L., Liu, X., Chen, S., Wang, Y., & Wan, C. (2015). Bio-augmentative volatile fatty acid production from waste activated sludge hydrolyzed at pH 12. *RSC Advances, 5*(62), 50033–50039.

Yin, J., Wang, K., Yang, Y., Shen, D., Wang, M., & Mo, H. (2014). Improving production of volatile fatty acids from food waste fermentation by hydrothermal pretreatment. *Bioresource Technology, 171*, 323–329.

Yuan, Y., Hu, X., Chen, H., Zhou, Y., Zhou, Y., & Wang, D. (2019). Advances in enhanced volatile fatty acid production from anaerobic fermentation of waste activated sludge. *Science of the Total Environment, 694*, 133741.

Zhou, M., Yan, B., Wong, J. W., & Zhang, Y. (2018). Enhanced volatile fatty acids production from anaerobic fermentation of food waste: A mini-review focusing on acidogenic metabolic pathways. *Bioresource Technology, 248*, 68–78.

Chapter 12
Application of Microbial Electrosynthesis Using Bioelectrochemical Systems

Monali Chhatbar, Parini Surti, Ambika Arkatkar, Suresh Kumar Kailasa, Alka Mungray, and Arvind Kumar Mungray

1 Introduction

The dearth of energy and increasing air pollution are two major issues that the human population is facing today. These problems are equally affecting the environment and decreasing the survival chances of all living beings on the planet. The world health organization, global health observatory data states that, 6.7 million deaths in the year 2019 was from exposure to air pollution (ambient and household both) and 99% of the world population lives in the places where air pollution level exceeds the World Health Organization guideline limits. The major culprit of air pollution are gases such as carbon dioxide (CO_2), carbon monoxide (CO), nitrogen oxides (NO_x), and sulfur oxides (SO_x). The possible solution proposed is decreasing the use of energy derived from nonrenewable resources and other means of air pollution. The market is already pushing toward the use of electric vehicles, hydrogen powered vehicles, promoting the use of public transport, etc. The production of electricity using renewable resources is also one of the solutions to curb air pollution and reduce environmental damage. The acceptance of these solutions is slow as compare to the increasing speed of energy consumption and air pollution.

It is important to reduce the generation of CO_2 but the disposal of already generated CO_2 is also an equally important point to be noted (Ibrahim et al., 2023). The microbial world is full of different wonders of the world. There are many microbes that can

M. Chhatbar · A. Mungray · A. K. Mungray (✉)
Department of Chemical Engineering, Sardar Vallabhbhai National Institute of Technology Surat, Surat, Gujarat, India
e-mail: akm@ched.svnit.ac.in

P. Surti · S. K. Kailasa
Department of Chemistry, Sardar Vallabhbhai National Institute of Technology Surat, Surat, Gujarat, India

A. Arkatkar
Department of Biotechnology, Veer Narmad South Gujarat University, Surat, Gujarat, India

© The Author(s), under exclusive license to Springer Nature Singapore Pte Ltd. 2025
S. Sevda (ed.), *Electro-Fermentation for Synthesis of Chemicals, Biochemicals and Biofuels*, https://doi.org/10.1007/978-981-96-6358-3_12

survive in extreme environments, few microbes called exoelectrogens can generate electricity, and microbes like methanogens can use biodegradable matter under anaerobic conditions and convert it into methane (Harnisch et al., 2024). The bioprocesses inside the microbial cell are very sensitive as well as adaptable. These bioprocesses should be supported with required growth conditions and can be successfully used to produce fuel and other value-added products.

The term bio-electro-synthesis symbolize a synthesis process for which biocatalysts is used and the process is driven by electrical energy. Microbes are known to have a good efficiency of electrosynthesis (Rabaey & Rozendal, 2010). The products of this process can be fuel like hydrogen and other chemicals as well. The concept of microbial electrosynthesis (MES) was introduced in the year 2010, where in the capacity of acetogens was explored. The acetogens could use electrons at the cathode electrodes that can reduce CO_2 and further lead to the formation of acetate as an important product (Lekshmi et al., 2023). The capacity of MES to reduce CO_2 has been the base line of many research studies that discussed the production of methane, butyrate, n-caproate, isobutyric acid, propionic acid, acetate, etc., using mixed or pure microbial cultures (Dessì et al., 2021). The studies in the field are supported by major pillars—microbes, electrodes, and their bioprocess (Jourdin & Burdyny, 2021). Thus, researchers are working to understand and enhance the microbe electrode interaction that is governed by many factors like electrode material, microbial species, reactor design, reactor conditions, etc. The second area of interest is the bioprocess that should be understood for technical advancement in MES. The top priority area for application of MES is the industrial sector (Zhang & Tremblay, 2019).

2 Bioelectrochemical Systems (Microbes as a Driving Force)

The bioelectrochemical systems (BES) are based on fundamental concept of the interaction between the electrodes and microorganisms (Santoro et al., 2017). The BES were developed based on the discovery of electrogenic bacteria by Potter in 1991 and recently BES have been used in the real field applications (Pant et al., 2010; Qian & Morse, 2011). The electrochemically active bacteria are the driving force behind the mechanism of BES, and these bacteria catalyze the organic matter degradation at the anode to produce electrons and protons as represented in Fig. 1. The protons migrate through the cation exchange membrane and electrons travel to the cathode through the external circuit. The protons and electrons are combined at the cathode in the presence of an electron acceptor for the reduction reaction (Escapa et al., 2016; Khan et al., 2017).

The BES have emerged as a sustainable tool in the recent years which can be categorized as shown in Fig. 2. The inoculum source for the BES systems can be domestic wastewater, anaerobic sludge, activated sludge, and marine sediments (Erable et al.,

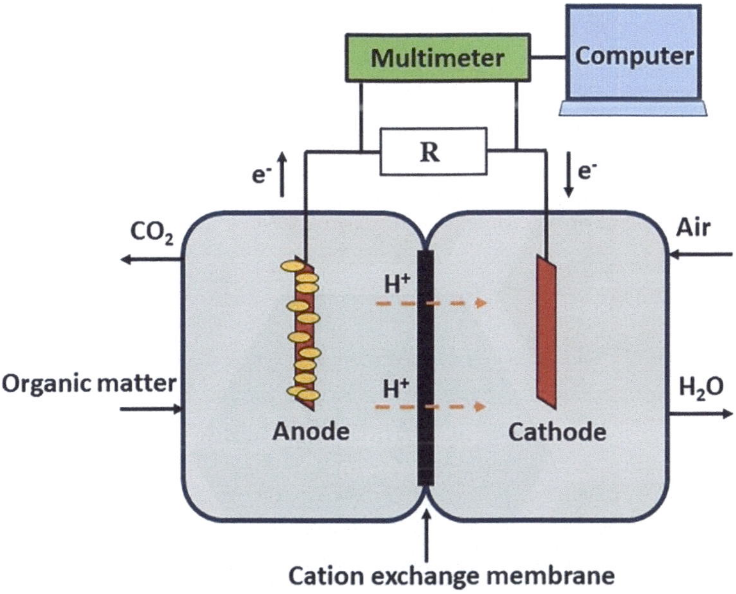

Fig. 1 General schematic diagram of bioelectrochemical system

2009; Liu & Logan, 2004; Lobato et al., 2012; Rezaei et al., 2007). The microbial community at the anode can be regulated and changed based on the source of inoculum used as well as the substrate type (Ieropoulos et al., 2010; Pant et al., 2010).

The type of substrate used as a carbon source can evidently influence the metabolic activity of the electrochemically active microbes and inevitably the bio-energy generation (Khater et al., 2017; Zhang et al., 2017a, 2017b). The substrate can be simple molecules like alcohols, acetate, glucose, and carboxylic acids as well as highly complex wastewaters such as brewery wastewater, meat processing wastewater, urine, domestic wastewater, or any type of industrial wastewater having organic content (Feng et al., 2008; Heilmann & Logan, 2006; Ieropoulos et al., 2013; Song et al., 2009; Zhang et al., 2017a, 2017b). The development of microbial consortia rich in electrochemically active bacteria clearly depends on the substrate biodegradability as electrogenic bacteria can easily oxidize simple carbon sources (D'Angelo et al., 2017). The electrogenic bacteria may need coordination from fermentative and hydrolytic microbes while oxidizing complex substrate to produce short chain fatty acids and fermentation products that can be readily oxidized by electrogenic bacteria (Lovley, 2006; Mateo et al., 2018).

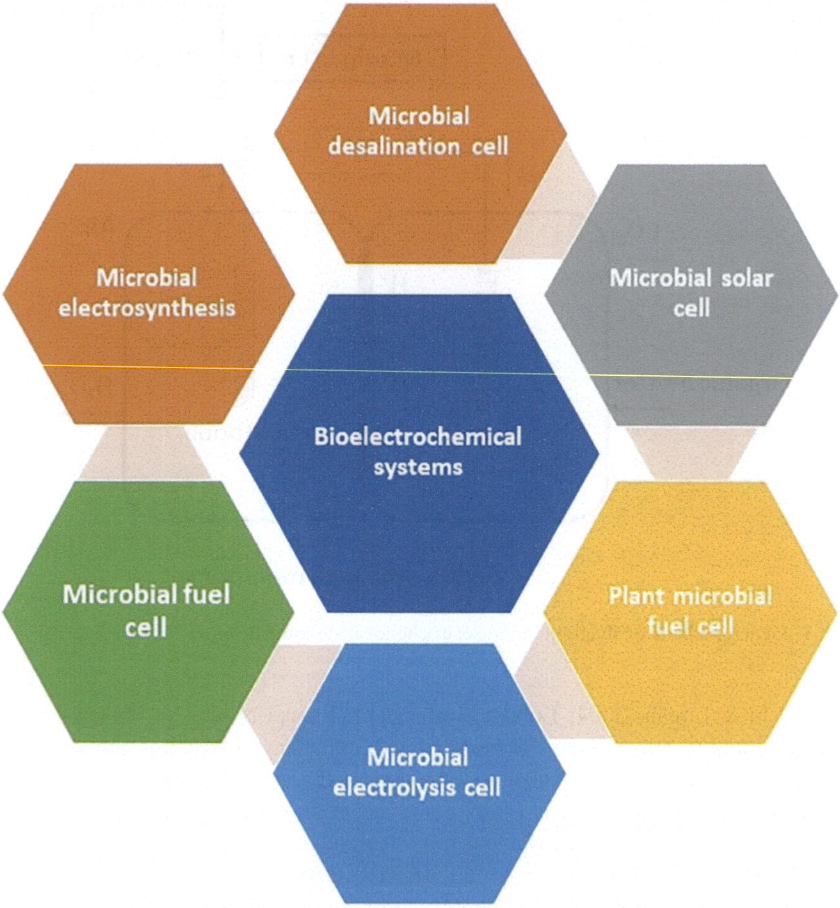

Fig. 2 Types of bioelectrochemical systems

2.1 Function of Microbes in BES

The bio-energy generation by bacteria in BES is a plausible solution to treat wastewater as bacteria can self-replicate and are biocatalysts for the oxidation of organic matter. The bacterial reaction can be carried out based on the tolerance of bacteria at various temperatures from room temperature (15–35 °C) to high temperature (50–60 °C) and at low temperatures (<15 °C) (Madigan et al., 1997). The chemical energy from various types of wastes can be converted to electrical energy through catalytic activity of various electrogenic microorganisms. The microbial community in the BES is at the core of the working mechanism and microbes play key role in carrying out various functions as shown in Fig. 3. The biofilm generation dynamics has the central importance to efficiently recover bio-energy in BES and has been confirmed

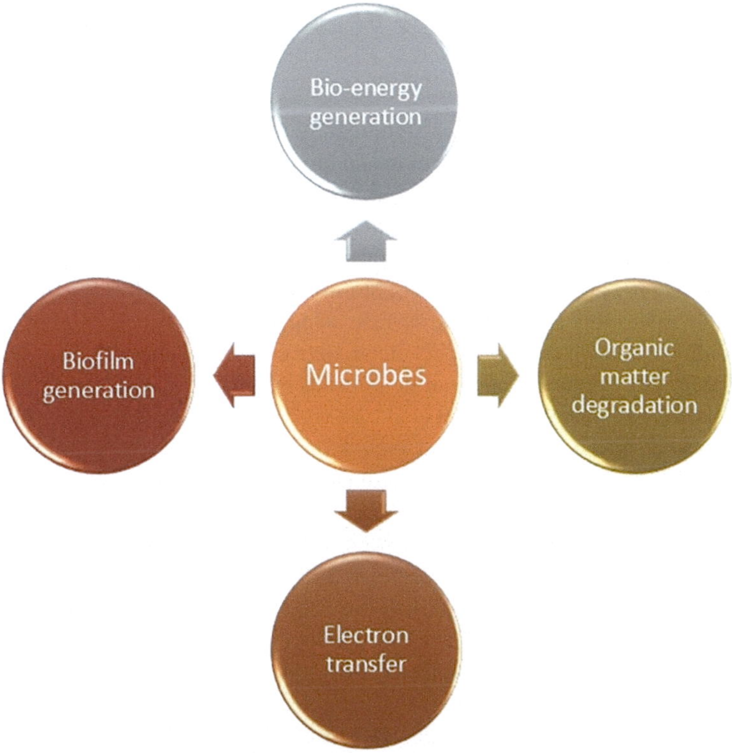

Fig. 3 Function of microbes in bioelectrochemical systems

by various laboratory-scale studies of bio-energy generation specifically targeting biofilm behavior and metabolism (Nath & Ghangrekar, 2020).

In BES, the electrons released due to organic matter oxidation are captured by anode and then transferred to cathode through external circuit resulting in bio-energy generation (Santoro et al., 2016). The primary function of the microbes is to generate biofilm on the anode surface and catalyze the electron transfer to anode generated during the degradation of organic content in wastewater. The electrons from microorganisms to anode are transferred through three mechanisms: (i) nanowire mediated direct electron transfer, (ii) direct electron transfer by outer membrane cytochromes with or without shuttle, and (iii) indirect electron transfer mediated by electron shuttles (Choi & Sang, 2016).

2.2 Structural Compatibility of BES for Efficient Microbial Function

The electroactive biofilm generation significantly impacts the performance of BES; however, the effective microbial function and biofilm generation depends on various structural aspects of BES (Fig. 4). The electrodes must be suitable as biofilm carrier apart from being a conductor, and the biofilm dynamics can be altered by their properties such as biocompatibility, surface roughness, and electron conductivity (Shabani et al., 2021). The modulation of biofilm to improve the BES function can be achieved by modifying the electrode surface using chemical and physical treatment, use of composite electrode, and highly electroactive coating (Chen et al., 2020; Jiang et al., 2021). The anode biofilm can be affected by the oxygen diffusion from the aerobic cathode via membrane (Xin et al., 2020). The membrane material should be chosen based on its oxygen mass transfer and ion exchange capacity; however, it can be improved by using chemical modification to enhance the electroactive biofilm on the anode surface (Roshanravan et al., 2021).

The low-cost sources of electroactive bacteria such as activated sludge, sediments, and paddy soils are viable to use as anolytes in BES (Birjandi et al., 2020; Salar-García et al., 2020; Tucci et al., 2021). This low-cost organic matter loading is valuable in maintaining microbial diversity and highly conductive environment for enhancement of carbon degradation and electron flux generation. The dominance of electroactive biofilm can be also mediated by addition of metabolites, low temperature, and reduced culture time (Jadhav et al., 2019; Xing et al., 2021). The addition of metabolites can improve the electron flux in BES. The incorporation of phosphate in soil MFCs can enhance the bio-energy generation by 52% and tetracycline degradation by 67% (Zhao et al., 2021). The nutrient delivery, biofilm dynamics, removal of metabolic waste, extracellular polymeric substances, and oxidation–reduction potential can be improved by increasing the mass flux and fluid flow as the flow contributes in

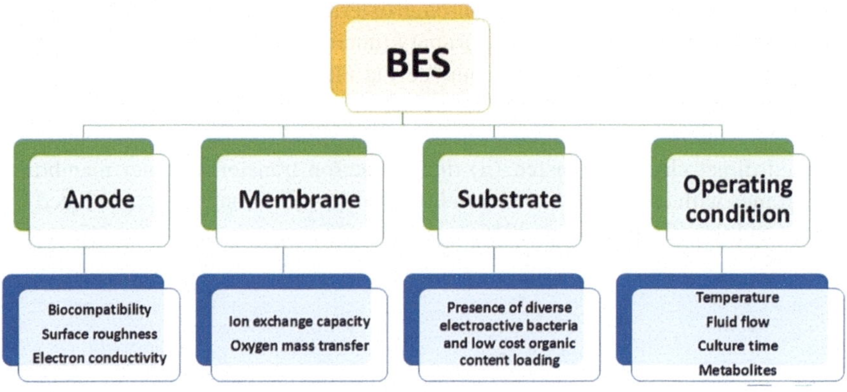

Fig. 4 BES design parameters influencing microbial function

maintaining the pH in buffer-less condition in anode and stimulating high current microbes (Roveto et al., 2021; Xing et al., 2021). The hydrogen recovery can be improved in a microbial electrolysis cell (MEC) by combining high flow rate with recycle process (Lewis & Borole, 2016).

3 Microbial Electrosynthesis

3.1 Working Principle

MES is a variant of BES where microbe/electrode interactions are exploited for the cathodic reduction of substrates to valuable products using an external power source (Surti et al., 2021). At least one of the electrodes (anode/cathode) in MES is biotic (Fig. 5). The electrons may be derived from the microbial metabolism of electrogens at the biotic anode. Alternatively, electrotrophic microbes can integrate electrons from the cathode for biocatalytic transformations (biotic cathode). Enzymes may also be utilized at the cathode for the biotransformations of the substrate (Zhang et al., 2017a, 2017b). Electron exchanges at the microbe/electrode interface may occur directly through microbial nanowires and outer membrane cytochromes or indirectly through redox mediators (Surti et al., 2022). The basic principle of MES is essentially based on converting externally supplied electrical energy to chemical energy using microbial catalysts.

The carbon source at the biotic anode is oxidized to generate electrons which are fed to the external circuit. Substrate-to-product conversion is achieved at the cathode with the additional electrical input from an external power source. Wastewater treatment can be achieved at a biotic anode and chemical oxygen demand of the wastewater can serve as a C source for electrogenic microbes (Kong et al., 2020). Other organic substrates such as acetate, glucose, butyrate and complex substrates like starch, cellulose, and sludge may also be used as a C source. It is reasonable to say that the C source at the biotic anode should be of lower value than the product synthesized at the cathode. In the case of abiotic anode, water splitting reaction (Rabaey & Rozendal, 2010), sulfide oxidation (Gong et al., 2013), glycerol/glucose oxidation (Jourdin & Burdyny, 2021), and ferrocyanide (Yang et al., 2021) can serve as a source of electrons. Oxygen evolved from water splitting may cross over toward the cathode and poison the anaerobic electrotrophs compromising the productivity. The choice of anode oxidation reaction should be energetically favorable and economically feasible. Anode material also significantly affects the overpotentials required to achieve this oxidation reaction. C-based materials are routinely used in the case of biotic anodes; however, abiotic anodes comprise of graphite, platinum, titanium, metal, metal oxides, and mixed metal oxides (Bian et al., 2020).

The MES cathode receives electrons from the external circuit that are used to reduce a substrate to a product. Certain products such as H_2, H_2O_2, and C compounds like formate, acetate, and ethanol can be generated at the abiotic cathode depending

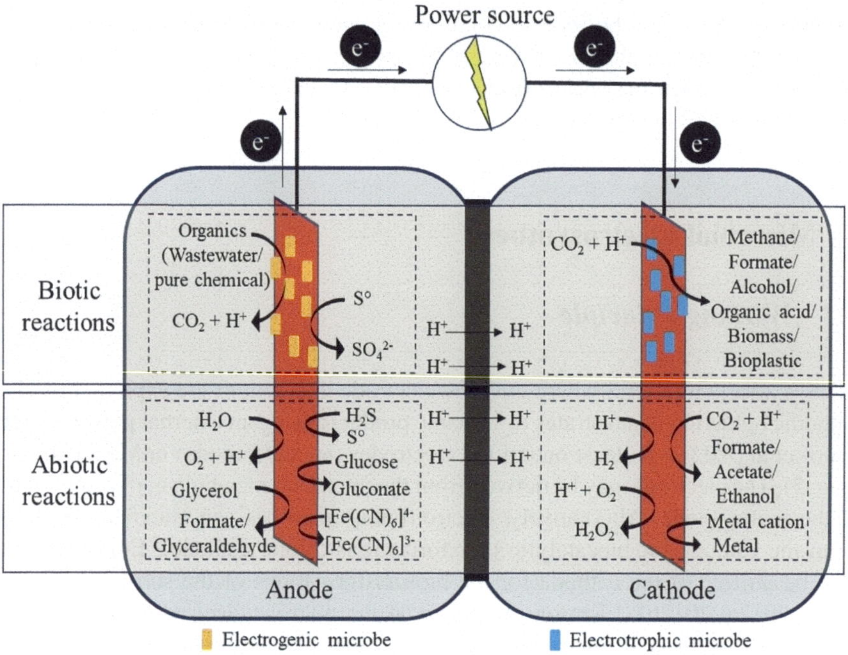

Fig. 5 Schematic of microbial electrosynthesis

on the electrode material and applied potentials (Kong et al., 2020; Prévoteau et al., 2020; Rabaey & Rozendal, 2010). However, reducing CO_2 abiotically requires large activation energy and hence cathodic overpotentials. The abiotically generated H_2 and formate can serve as intermediates for the biotic production of methane and multi-carbon compounds. Metal ion reductions can be achieved at the abiotic cathode for bioremediation of contaminated soil and water. Microbial communities at the biotic cathode are used to produce a range of C compounds from inorganic C sources. Gas electrofermentations fix inorganic CO and CO_2 gases to produce valuable chemicals or biofuels by direct microbial reduction or by in situ H_2 production (Bajracharya et al., 2022). These C sources enable MES to be used as carbon capture technology for value-added chemical production. Other C sources like carbonates and simple organic compounds such as formate, acetate, and methanol may also be used.

Cathode properties significantly impact the MES efficiency as it houses the microbial catalyst and feeds electrons into microbial metabolism. Biocompatibility, high surface area, porosity, chemical stability, low charge transfer resistance, and economical are properties desired in MES cathode (Bajracharya et al., 2022). Platinum and nickel cathodes are used for their catalytic efficiency in H_2 evolution reactions. Copper electrodes are suitable for CO_2 reduction but are prone to corrosion. Routinely used biotic cathode involves carbon-based materials owing to their biocompatibility, stainless steel for their conductivity while carbon-metal composites are used for their high conductivity, surface area, and catalytic activity. Besides, carbon electrode modification with gold, palladium, and Ni nanoparticles enhance productivity in MES (Aryal et al., 2017). Cathode architecture (2D and 3D) impacts the electron supply and mass transfer to electrotrophs that governs the efficiency and productivity of the MES (Tian et al., 2020). Although 3D cathodes have a high surface area, enhanced microbial colonization and higher electron transfer than the 2D cathode, their cost and biofouling are limitations to scale-up and long-term operations in MES.

3.2 Microbial Capacity

The microbial community at the biotic cathode accepts the electrons from the cathode and integrates them into their metabolism. Metabolic capacity of the microbial catalyst and the applied cathode potential significantly influence the product synthesis in MES. These microbes have specialized cell membrane structures and proteins for electron uptake or they may utilize secreted or supplemented redox moieties for electron transfers (Deutzmann et al., 2015; Liu et al., 2020; Marsili et al., 2008; Reguera et al., 2005). Outer membrane cytochromes, nanopili, secreted vesicles, extracellular enzymes, flavin, quinone, and phenazine electron shuttles are the major pathways known for electron transfer (Fig. 6). H_2 and formate generated from abiotic reactions at the cathode may also serve as electron carriers. Most of the information on the electron transfer pathway is derived from studies on *Geobacter* and *Shewanella.* However, there is a lack of information on electron uptake by several microbes used in MES.

Microbes that can assimilate inorganic C can be utilized to capture atmospheric carbon to form reduced organic compounds. Major types of obligate anaerobes

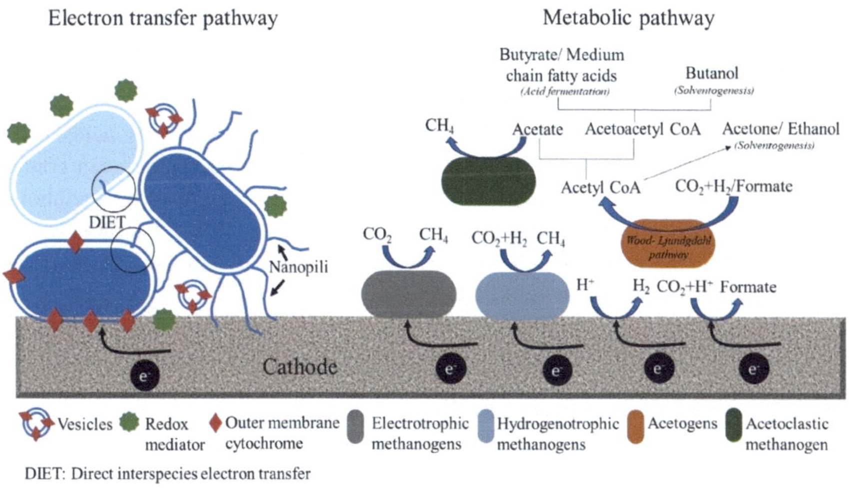

Fig. 6 Electron transfer and metabolic pathway at the biocathode

utilized in MES for product synthesis from inorganic C: (a) autotrophic acetogens that produce acetate and other organic compounds from CO_2 through the Wood–Ljungdahl pathway (Nevin et al., 2011). Acetyl CoA is the main intermediate synthesized through the Wood–Ljungdahl pathway. Various organic acids are synthesized through acid fermentation while alcohol and acetone through solventogenesis from acetyl CoA. *Sporomusa ovata*, *Clostridium ljungdahlii*, and *Acetobacterium woodii* are widely studied MES acetogens. Thermophilic acetogens like *Moorella thermoacetica* and *M. thermoautotrophica* have also been reported in MES; (b) methanogens produce methane through hydrogenotrophic methanogenesis (Cheng et al., 2009). *Methanobacterium palustre* and *Methanococcus maripaludis* have been reported for methane production in MES; (c) photoautotrophs that utilize light as an energy source to fix CO_2. *Rhodopseudomonas palustris* has been used to produce polyhydroxybutyrate (PHB) and *Rhodobacter sphaeroides* for carotenoids in MES (Lee et al., 2022; Rengasamy et al., 2020).

Knallgas bacteria are aerobic autotrophic hydrogen oxidizing bacteria which are an exception used in oxic MES (Eberhard Reiner et al., 2024). Examples are *Cupriavidus necator* and *Kyrpidia spormannii*. These microbes are utilized for biomass production rather than product synthesis. Assimilative C reducing Calvin cycle is the main metabolic pathway for these aerobic microbes. Oxic MES can be used for the treatment of gaseous waste stream consisting of CO_2 and O_2.

Most studies with MES favored mixed culture anaerobes mainly due to difficulties with handling the pure culture of obligate anaerobes. Besides, the productivity of mixed cultures is higher than pure culture due to the robustness, adaptivity, versatility, and oxygen tolerance of mixed consortium. There is limited knowledge of the metabolic pathways that operate in synergy to drive product synthesis in mixed consortia. The majority of mixed microbial culture studies report simpler product synthesis such as methane, acetate, and ethanol due to thermodynamic limitations of microbial catalysts. One major shortcoming of microbes in MES is byproduct synthesis which complicates the downstream product purification. On the contrary to favoring mixed microbial consortia, pure cultures with known metabolic pathways are desired for MES. Defined co-cultures of acetogens and chain-elongating bacteria have also been reported for the synthesis of long-chain fatty acids from simpler intermediate products (Herzog et al., 2023). Synthetic biology can confer electron uptake pathways to industrially exploited microbes or extend the metabolic pathways to electrotrophs (Surti et al., 2021). Fine-tuning of the metabolic pathways and achieving standardized MES reactor designs may expand the scope of electrobiosynthesis (Harnisch et al., 2024).

3.3 Product Derived from MES

A spectrum of products can be synthesized using MES ranging from simpler products like formate to multicarbon acids and alcohols (Table 1). For the production of simple organic C compounds, inorganic CO_2 is a preferred substrate. However, energetically

Table 1 Spectrum of products synthesized with MES

Product	Anode reaction/substrate	Microbial community at anode (A)/cathode (C)	Cathode reaction/substrate	Applied potential (V)	Productivity	Refs.
H_2	Organic oxidation from sugarcane bagasse	*Shewanella oneidensis* (A)	H_2 evolution	−0.8	0.25 m^3/m^3 day	Gautam et al. (2024)
H_2O_2	Acetate oxidation	Sewage sludge (A)	O_2 reduction	−0.6	10.82 mg/L h, 454.44 mg/L	Zou et al. (2021)
CH_4	O_2 evolution	Anaerobic sludge (C)	Carbonate solution	−0.2	0.84 L/L_c day	Nwanebu et al. (2024)
Formate	Acetate oxidation	*Escherichia coli* (A)	CO_2 reduction by *E. coli* formate dehydrogenase	−1.0	30 mg/L h	Singh et al. (2020)
Acetate	O_2 evolution	Anaerobic sludge (C)	CO_2	−0.89	12.5 g/ L	Bian et al. (2024)
Butyrate	O_2 evolution	Enriched culture from cow rumen (C)	CO_2 and methanol (co-substrate)	−0.95	0.36 g/L day, 8.6 g/L	Yao et al. (2024)
Caproate	O_2 evolution	Enriched culture from anaerobic sludge (C)	CO_2 and ethanol (co-substrate)	−1.0	1.51 g/L	Krishna Chaitanya et al. (2024)
Ethanol	–	Mixed culture enriched with *Clostridium ljungdahlii* (C)	CO_2	−0.8	10.95 g/m^2 day, 5.28 g/L	Romans-Casas et al. (2023)
Butanol	O_2 evolution	*Clostridium beijerinckii* (C)	Glucose	−0.8	1529 mg/L h	Molognoni et al. (2023)
PHB	O_2 evolution	*Cupriavidus necator* (C)	Flue gas containing CO_2	−(0.8–1.0)	333 mg/L	Langsdorf et al. (2024)
Biomass	–	*Krypidia spormannii* (C)	CO_2	−0.5	31.03 μm^3/μm^2 day	Rominger et al. (2023)
Metal	Organic oxidation from wastewater	Mixed culture (A/C)	Metal cations reduction	−0.9	73–98% removal of Cu, Pb, Zn and Cd	Shi et al. (2023)

stable CO_2 requires a poised cathode at low potentials to produce commodity chemicals (Karthikeyan et al., 2019). Significant external power input is necessary to achieve low potential. Cathode potential lower than −0.59 V produces H_2 and below −1.5 V produces formate electrochemically. These products can serve as intermediates for multicarbon product synthesis. Acetate and other organic acids from MES cathode have been upgraded to high-value products like PHB and antioxidants in the MES-integrated downstream process (Nguyen et al., 2024; Zhang et al., 2021). Theoretically, any industrial microbial product synthesis process can be adapted to MES. However, limitations to understanding microbial electron uptake, electrode biocompatibility, availability of efficient microbial factories and feasible MES reactor design need to be addressed for the industrial scale-up of this technology for sustainable and energy-efficient product synthesis.

4 Future Prospects

The increasing environmental air pollution has made many changes on the thought process of the stake holders, consumers, and governing authorities. The consumers are opting for technologies that can support environment sustainability. The norms of industrial waste management are more stringent and compliance to it has been made compulsory. The MES technology can serve as a better option in different fields like (a) Production of valuable chemicals and biofuel that can reduce the dependency on fossil fuel. (b) Carbon capture technology enhancement can be achieved that can be an important part to resist climate change. It can be a support technology to achieve net zero carbon emission. (c) Can be a part of wastewater treatment facility. (d) It can be employed as a bioremediation technology to clean up the contaminated site. It has the potential to convert harmful pollutants into less hazardous from.

Due to dynamic nature of microbial process, the large-scale application needs streamlined maintenance and monitoring of MES. The technology can be applicable for the production of chemicals, biofertilizers and biofuels. The technologies like solar and wind energy production requires efficient storage of energy at a large scale, MES can be used to store the energy by converting it to chemical form. As the system has the potential to be exploited on both the areas of pollutant minimization and production of value-added products, it has a promising applicability. Other BES systems have the capacity to use biodegradable matter but this technology is unique in capturing gaseous form of pollutant. The major challenges faced by the system are economical electrode material, efficient microbial strains, and an optimized reactor design. The future studies might open new avenues for MES. Meanwhile, policy support and awareness drive should be taken up for large-scale implementation of MES system.

References

Aryal, N., Ammam, F., Patil, S. A., & Pant, D. (2017). An overview of cathode materials for microbial electrosynthesis of chemicals from carbon dioxide. *Green Chemistry, 19*(24). https://doi.org/10.1039/C7GC01801K

Bajracharya, S., Krige, A., Matsakas, L., Rova, U., & Christakopoulos, P. (2022). Advances in cathode designs and reactor configurations of microbial electrosynthesis systems to facilitate gas electro-fermentation. *Bioresource Technology, 354*. https://doi.org/10.1016/j.biortech.2022.127178

Bian, B., Bajracharya, S., Xu, J., Pant, D., & Saikaly, P. E. (2020). Microbial electrosynthesis from CO_2: Challenges, opportunities and perspectives in the context of circular bioeconomy. *Bioresource Technology, 302*. https://doi.org/10.1016/j.biortech.2020.122863

Bian, Y., Leininger, A., May, H. D., & Ren, Z. J. (2024). H_2 mediated mixed culture microbial electrosynthesis for high titer acetate production from CO_2. *Environmental Science and Ecotechnology, 19*. https://doi.org/10.1016/j.ese.2023.100324

Birjandi, N., Younesi, H., Ghoreyshi, A. A., & Rahimnejad, M. (2020). Enhanced medicinal herbs wastewater treatment in continuous flow bio-electro-Fenton operations along with power generation. *Renewable Energy, 155*, 1079–1090. https://doi.org/10.1016/j.renene.2020.04.013

Chen, X., Li, Y., Yuan, X., Li, N., He, W., Feng, Y., & Liu, J. (2020). Surface modification by β-cyclodextrin/polyquaternium-11 composite for enhanced biofilm formation in microbial fuel cells. *Journal of Power Sources, 480*, 228789. https://doi.org/10.1016/j.jpowsour.2020.228789

Cheng, S., Xing, D., Call, D. F., & Logan, B. E. (2009). Direct biological conversion of electrical current into methane by electromethanogenesis. *Environmental Science & Technology, 43*(10). https://doi.org/10.1021/es803531g

Choi, O., & Sang, B. I. (2016). Extracellular electron transfer from cathode to microbes: Application for biofuel production. *Biotechnology for Biofuels, 9*, 1–14. https://doi.org/10.1186/s13068-016-0426-0

D'Angelo, A., Mateo, S., Scialdone, O., Cañizares, P., Fernandez-Morales, F. J., & Rodrigo, M. A. (2017). Optimization of the performance of an air-cathode MFC by changing solid retention time. *Journal of Chemical Technology & Biotechnology, 92*(7), 1746–1755. https://doi.org/10.1002/JCTB.5175

Dessì, P., Rovira-Alsina, L., Sánchez, C., Dinesh, G. K., Tong, W., Chatterjee, P., Tedesco, M., Farràs, P., Hubertus, M. V. H., & Puig, S. (2021). Microbial electrosynthesis: Towards sustainable biorefineries for production of green chemicals from CO_2 emissions. *Biotechnology Advances, 46*, 107675. https://doi.org/10.1016/j.biotechadv.2020.107675

Deutzmann, J. S., Sahin, M., & Spormann, A. M. (2015). Extracellular enzymes facilitate electron uptake in biocorrosion and bioelectrosynthesis. *Mbio, 6*(2), 1–8. https://doi.org/10.1128/mBio.00496-15

Eberhard Reiner, J., Korth, B., & Edel, M. (2024). Oxygen in the mix: Is oxic microbial electrosynthesis a potential alternative for biomass production? *ChemElectroChem*. https://doi.org/10.1002/celc.202400397

Erable, B., Etcheverry, L., & Bergel, A. (2009). Increased power from a two-chamber microbial fuel cell with a low-pH air-cathode compartment. *Electrochemistry Communications, 11*(3), 619–622. https://doi.org/10.1016/j.elecom.2008.12.058

Escapa, A., Mateos, R., Martínez, E. J., & Blanes, J. (2016). Microbial electrolysis cells: An emerging technology for wastewater treatment and energy recovery. From laboratory to pilot plant and beyond. *Renewable and Sustainable Energy Reviews, 55*, 942–956. https://doi.org/10.1016/J.RSER.2015.11.029

Feng, Y., Wang, X., Logan, B. E., & Lee, H. (2008). Brewery wastewater treatment using air-cathode microbial fuel cells. *Applied Microbiology and Biotechnology, 78*, 873–880. https://doi.org/10.1007/s00253-008-1360-2

Gautam, R., Ress, N. V., Wilckens, R. S., & Ghosh, U. K. (2024). Hydrogen production in microbial electrolysis cell and reactor digestate valorization for biochar—A noble attempt towards

circular economy. *International Journal of Hydrogen Energy, 52.* https://doi.org/10.1016/j.ijhydene.2023.07.190
Gong, Y., Ebrahim, A., Feist, A. M., Embree, M., Zhang, T., Lovley, D., & Zengler, K. (2013). Sulfide-driven microbial electrosynthesis. *Environmental Science & Technology, 47*(1). https://doi.org/10.1021/es303837j
Harnisch, F., Deutzmann, J. S., Boto, S. T., & Rosenbaum, M. A. (2024). Microbial electrosynthesis: Opportunities for microbial pure cultures. *Trends in Biotechnology.* https://doi.org/10.1016/j.tibtech.2024.02.004
Heilmann, J., & Logan, B. E. (2006). Production of electricity from proteins using a microbial fuel cell. *Water Environment Research, 78*(5), 531–537. https://doi.org/10.2175/106143005X73046
Herzog, J., Mook, A., Guhl, L., Bäumler, M., Beck, M. H., Weuster-Botz, D., Bengelsdorf, F. R., & Zeng, A. (2023). Novel synthetic co-culture of *Acetobacterium woodii* and *Clostridium drakei* using CO_2 and in situ generated H_2 for the production of caproic acid via lactic acid. *Engineering in Life Sciences, 23*(1). https://doi.org/10.1002/elsc.202100169
Ibrahim, I., Salehmin, M. N. I., Balachandran, K., Hil Me, M. F., Loh, K. S., Abu Bakar, M. H., Jong, B. C., & Lim, S. S. (2023). Role of microbial electrosynthesis system in CO_2 capture and conversion: A recent advancement toward cathode development. *Frontiers in Microbiology, 14*, 1192187. https://doi.org/10.3389/FMICB.2023.1192187/BIBTEX
Ieropoulos, I. A., Greenman, J., & Melhuish, C. (2013). Miniature microbial fuel cells and stacks for urine utilisation. *International Journal of Hydrogen Energy, 38*(1), 492–496. https://doi.org/10.1016/j.ijhydene.2012.09.062
Ieropoulos, I., Winfield, J., & Greenman, J. (2010). Effects of flow-rate, inoculum and time on the internal resistance of microbial fuel cells. *Bioresource Technology, 101*(10), 3520–3525. https://doi.org/10.1016/j.biortech.2009.12.108
Jadhav, D. A., Chendake, A. D., Schievano, A., & Pant, D. (2019). Suppressing methanogens and enriching electrogens in bioelectrochemical systems. *Bioresource Technology, 277*, 148–156. https://doi.org/10.1016/j.biortech.2018.12.098
Jiang, P. Y., Xiao, Z. H., Wang, Y. F., Li, N., & Liu, Z. Q. (2021). Enhanced performance of microbial fuel cells using Ag nanoparticles modified Co, N co-doped carbon nanosheets as bifunctional cathode catalyst. *Bioelectrochemistry, 138*, 107717. https://doi.org/10.1016/j.bioelechem.2020.107717
Jourdin, L., & Burdyny, T. (2021). Microbial electrosynthesis: Where do we go from here? *Trends in Biotechnology, 39*(4), 359–369. https://doi.org/10.1016/j.tibtech.2020.10.014
Karthikeyan, R., Singh, R., & Bose, A. (2019). Microbial electron uptake in microbial electrosynthesis: A mini-review. *Journal of Industrial Microbiology and Biotechnology, 46*(9–10), 1419–1426. https://doi.org/10.1007/S10295-019-02166-6
Khan, M. D., Khan, N., Sultana, S., Joshi, R., Ahmed, S., Yu, E., Scott, K., Ahmad, A., & Khan, M. Z. (2017). Bioelectrochemical conversion of waste to energy using microbial fuel cell technology. *Process Biochemistry, 57*, 141–158. https://doi.org/10.1016/J.PROCBIO.2017.04.001
Khater, D. Z., El-Khatib, K. M., & Hassan, H. M. (2017). Microbial diversity structure in acetate single chamber microbial fuel cell for electricity generation. *Journal of Genetic Engineering and Biotechnology, 15*(1), 127–137. https://doi.org/10.1016/j.jgeb.2017.01.008
Kong, F., Ren, H.-Y., Pavlostathis, S. G., Nan, J., Ren, N.-Q., & Wang, A. (2020). Overview of value-added products bioelectro-synthesized from waste materials in microbial electrosynthesis systems. *Renewable and Sustainable Energy Reviews, 125.* https://doi.org/10.1016/j.rser.2020.109816
Krishna Chaitanya, N., Nair, P. S., Rajpurohit, A., & Chatterjee, P. (2024). Impact of cell voltage on synthesis of caproic acid from carbon dioxide and ethanol in direct current powered microbial electrosynthesis cell. *Bioresource Technology, 412.* https://doi.org/10.1016/j.biortech.2024.131383
Langsdorf, A., Schütz, J. P., Ulber, R., Stöckl, M., & Holtmann, D. (2024). Production of polyhydroxybutyrate from industrial flue gas by microbial electrosynthesis. *Journal of CO_2 Utilization, 83.* https://doi.org/10.1016/j.jcou.2024.102800

Lee, Y. R., Lee, S. Y., Lee, J., Kim, H. S., Lee, J.-S., Lee, W.-H., & Lee, S. (2022). Modulation of antioxidant activity enhances photoautotrophic cell growth of Rhodobacter sphaeroides in microbial electrosynthesis. *Energies, 15*(3). https://doi.org/10.3390/en15030935

Lekshmi, G. S., Bazaka, K., Ramakrishna, S., & Kumaravel, V. (2023). Microbial electrosynthesis: Carbonaceous electrode materials for CO_2 conversion. *Materials Horizons, 10*(2), 292–312. https://doi.org/10.1039/D2MH01178F

Lewis, A. J., & Borole, A. P. (2016). Understanding the impact of flow rate and recycle on the conversion of a complex biorefinery stream using a flow-through microbial electrolysis cell. *Biochemical Engineering Journal, 116*, 95–104. https://doi.org/10.1016/j.bej.2016.06.008

Liu, H., & Logan, B. E. (2004). Electricity generation using an air-cathode single chamber microbial fuel cell in the presence and absence of a proton exchange membrane. *Environmental Science & Technology, 38*(14), 4040–4046. https://doi.org/10.1021/es0499344

Liu, X., Jing, X., Ye, Y., Zhan, J., Ye, J., & Zhou, S. (2020). Bacterial vesicles mediate extracellular electron transfer. *Environmental Science and Technology Letters, 7*(1), 27–34. https://doi.org/10.1021/acs.estlett.9b00707

Lobato, J., Cañizares, P., Fernández, F. J., & Rodrigo, M. A. (2012). An evaluation of aerobic and anaerobic sludges as start-up material for microbial fuel cell systems. *New Biotechnology, 29*(3), 415–420. https://doi.org/10.1016/j.nbt.2011.09.004

Lovley, D. R. (2006). Microbial fuel cells: Novel microbial physiologies and engineering approaches. *Current Opinion in Biotechnology, 17*(3), 327–332. https://doi.org/10.1016/J.COPBIO.2006.04.006

Madigan, M. T., Martinko, J. M., & Parker, J. (1997). *Brock biology of microorganisms* (Vol. 11). Prentice Hall.

Marsili, E., Baron, D. B., Shikhare, I. D., Coursolle, D., Gralnick, J. A., & Bond, D. R. (2008). Shewanella secretes flavins that mediate extracellular electron transfer. *PANS, 105*(10), 6–11.

Mateo, S., Zamorano-López, N., Borras, L., Fernandez-Morales, F. J., Cañizares, P., Seco, A., & Rodrigo, M. (2018). Effect of sludge age on microbial consortia developed in MFCs. *Journal of Chemical Technology and Biotechnology, 93*(5), 1290–1299. https://doi.org/10.1002/jctb.5488

Molognoni, D., Vega, M., de Vrije, T., Contreras, A. L., Bosch, M., Bosch-Jimenez, P., & Borràs, E. (2023). Enhancing butanol production by Clostridium beijerinckii through cathodic electrofermentation approach. *Journal of Chemical Technology & Biotechnology, 98*(4). https://doi.org/10.1002/jctb.7324

Nath, D., & Ghangrekar, M. M. (2020). Plant secondary metabolites induced electron flux in microbial fuel cell: Investigation from laboratory-to-field scale. *Scientific Reports, 10*(1), 17185. https://doi.org/10.1038/s41598-020-74092-y

Nevin, K. P., Hensley, S. A., Franks, A. E., Summers, Z. M., Ou, J., Woodard, T. L., Snoeyenbos-West, O. L., & Lovley, D. R. (2011). Electrosynthesis of organic compounds from carbon dioxide is catalyzed by a diversity of acetogenic microorganisms. *Applied and Environmental Microbiology, 77*(9), 2882–2886. https://doi.org/10.1128/AEM.02642-10

Nguyen, H. T. T., Jadhav, D. A., Eisa, T., Nguyen, H. Y., Le, G. T. H., Le, T. T. Q., Jae, M. R., Khoo, K. S., Yang, E., & Chae, K. J. (2024). Sustainable conversion of carbon dioxide to high-value antioxidant astaxanthin through microbial electrosynthesis-assisted microalgae cultivation. *Process Safety and Environmental Protection, 190*. https://doi.org/10.1016/j.psep.2024.07.030

Nwanebu, E., Jezernik, M., Lawson, C., Bruant, G., & Tartakovsky, B. (2024). Impact of cathodic pH and bioaugmentation on acetate and CH_4 production in a microbial electrosynthesis cell. *RSC Advances, 14*(32). https://doi.org/10.1039/D4RA03906H

Pant, D., Van Bogaert, G., Diels, L., & Vanbroekhoven, K. (2010). A review of the substrates used in microbial fuel cells (MFCs) for sustainable energy production. *Bioresource Technology, 101*(6), 1533–1543. https://doi.org/10.1016/j.biortech.2009.10.017

Prévoteau, A., Carvajal-Arroyo, J. M., Ganigué, R., & Rabaey, K. (2020). Microbial electrosynthesis from CO_2: Forever a promise? *Current Opinion in Biotechnology, 62*, 48–57. https://doi.org/10.1016/J.COPBIO.2019.08.014

Qian, F., & Morse, D. E. (2011). Miniaturizing microbial fuel cells. *Trends in Biotechnology, 29*(2), 62–69. https://doi.org/10.1016/j.tibtech.2010.10.003

Rabaey, K., & Rozendal, R. A. (2010). Microbial electrosynthesis—Revisiting the electrical route for microbial production. *Nature Reviews Microbiology, 8*(10). https://doi.org/10.1038/nrmicro2422

Reguera, G., McCarthy, K. D., Mehta, T., Nicoll, J. S., Tuominen, M. T., & Lovley, D. R. (2005). Extracellular electron transfer via microbial nanowires. *Nature, 435*(7045), 1098–1101. https://doi.org/10.1038/nature03661

Rengasamy, K., Ranaivoarisoa, T., Bai, W., & Bose, A. (2020). Magnetite nanoparticle anchored graphene cathode enhances microbial electrosynthesis of polyhydroxybutyrate by Rhodopseudomonas palustris TIE-1. *Nanotechnology, 32*(3), 035103. https://doi.org/10.1088/1361-6528/abbe58

Rezaei, F., Richard, T. L., Brennan, R. A., & Logan, B. E. (2007). Substrate-enhanced microbial fuel cells for improved remote power generation from sediment-based systems. *Environmental Science & Technology, 41*(11), 4053–4058. https://doi.org/10.1021/es070426e

Romans-Casas, M., Perona-Vico, E., Dessì, P., Bañeras, L., Balaguer, M. D., & Puig, S. (2023). Boosting ethanol production rates from carbon dioxide in MES cells under optimal solventogenic conditions. *Science of the Total Environment, 856*. https://doi.org/10.1016/j.scitotenv.2022.159124

Rominger, L., Hackbarth, M., Jung, T., Scherzinger, M., Horn, H., Kaltschmitt, M., & Gescher, J. (2023). Oxic microbial electrosynthesis can be more energy efficient for biomass production than knallgas or photosynthesis based processes. *BioRxiv*, 2023-05. https://doi.org/10.1101/2023.05.24.542149

Roshanravan, B., Younesi, H., Abdollahi, M., Rahimnejad, M., & Pyo, S. H. (2021). Application of proton-conducting sulfonated polysulfone incorporated MIL-100 (Fe) composite materials for polymer-electrolyte membrane microbial fuel cells. *Journal of Cleaner Production, 300*, 126963. https://doi.org/10.1016/j.jclepro.2021.126963

Roveto, P. M., Gupta, A., & Schuler, A. J. (2021). Effects of surface skewness on local shear stresses, biofilm activity, and microbial communities for wastewater treatment. *Bioresource Technology, 320*, 124251. https://doi.org/10.1016/j.biortech.2020.124251

Salar-García, M. J., Montilla, F., Quijada, C., Morallon, E., & Ieropoulos, I. (2020). Improving the power performance of urine-fed microbial fuel cells using PEDOT-PSS modified anodes. *Applied Energy, 278*, 115528. https://doi.org/10.1016/j.apenergy.2020.115528

Santoro, C., Arbizzani, C., Erable, B., & Ieropoulos, I. (2017). Microbial fuel cells: From fundamentals to applications. A review. *Journal of Power Sources, 356*, 225–244. https://doi.org/10.1016/j.jpowsour.2017.03.109

Santoro, C., Soavi, F., Serov, A., Arbizzani, C., & Atanassov, P. (2016). Self-powered supercapacitive microbial fuel cell: The ultimate way of boosting and harvesting power. *Biosensors and Bioelectronics, 78*, 229–235. https://doi.org/10.1016/j.bios.2015.11.026

Shabani, M., Pontié, M., Younesi, H., Nacef, M., Rahimpour, A., Rahimnejad, M., & Bouchenak Khelladi, R. M. (2021). Biodegradation of acetaminophen and its main by-product 4-aminophenol by Trichoderma harzianum versus mixed biofilm of Trichoderma harzianum/Pseudomonas fluorescens in a fungal microbial fuel cell. *Journal of Applied Electrochemistry, 51*(4), 581–596. https://doi.org/10.1007/S10800-020-01518-W/FIGURES/10

Shi, X., Duan, Z., Wang, J., Zhou, W., Jiang, M., Li, T., Ma, H., & Zhu, X. (2023). Simultaneous removal of multiple heavy metals using single chamber microbial electrolysis cells with biocathode in the micro-aerobic environment. *Chemosphere, 318*. https://doi.org/10.1016/j.chemosphere.2023.137982

Singh, S., Noori, M. T., & Verma, N. (2020). Efficient bio-electroreduction of CO_2 to formate on a iron phthalocyanine-dispersed CDC in microbial electrolysis system. *Electrochimica Acta, 338*. https://doi.org/10.1016/j.electacta.2020.135887

Song, T., Xu, Y., Ye, Y., Chen, Y., & Shen, S. (2009). Electricity generation from terephthalic acid using a microbial fuel cell. *Journal of Chemical Technology and Biotechnology, 84*(3), 356–360. https://doi.org/10.1002/JCTB.2047

Surti, P., Arkatkar, A., Kailasa, S. K., & Mungray, A. K. (2022). Role of electroactive biofilms in governing the performance of microbial electrochemical system. In *Scaling up of microbial electrochemical systems: From reality to scalability* (pp. 49–70). https://doi.org/10.1016/B978-0-323-90765-1.00004-6

Surti, P., Kailasa, S. K., & Mungray, A. K. (2021). Genetic engineering strategies for performance enhancement of bioelectrochemical systems: A review. *Sustainable Energy Technologies and Assessments, 47*, 101332. https://doi.org/10.1016/J.SETA.2021.101332

Tian, S., Yao, X., Song, T.-S., Chu, Z., Xie, J., & Jin, W. (2020). Artificial electron mediator with nanocubic architecture highly promotes microbial electrosynthesis from carbon dioxide. *ACS Sustainable Chemistry & Engineering, 8*(17). https://doi.org/10.1021/acssuschemeng.0c01276

Tucci, M., Viggi, C. C., Nunez, A. E., Schievano, A., Rabaey, K., & Aulenta, F. (2021). Empowering electroactive microorganisms for soil remediation: Challenges in the bioelectrochemical removal of petroleum hydrocarbons. *Chemical Engineering Journal, 419*, 130008. https://doi.org/10.1016/j.cej.2021.130008

Xin, S., Shen, J., Liu, G., Chen, Q., Xiao, Z., Zhang, G., & Xin, Y. (2020). Electricity generation and microbial community of single-chamber microbial fuel cells in response to Cu_2O nanoparticles/reduced graphene oxide as cathode catalyst. *Chemical Engineering Journal, 380*, 122446. https://doi.org/10.1016/J.CEJ.2019.122446

Xing, F., Xi, H., Yu, Y., & Zhou, Y. (2021). Anode biofilm influence on the toxic response of microbial fuel cells under different operating conditions. *Science of the Total Environment, 775*, 145048. https://doi.org/10.1016/j.scitotenv.2021.145048

Yang, H.-Y., Hou, N.-N., Wang, Y.-X., Liu, J., He, C.-S., Wang, Y. R., Li, W. H., & Mu, Y. (2021). Mixed-culture biocathodes for acetate production from CO_2 reduction in the microbial electrosynthesis: Impact of temperature. *Science of the Total Environment, 790*. https://doi.org/10.1016/j.scitotenv.2021.148128

Yao, H., Rinta-Kanto, J. M., Vassilev, I., & Kokko, M. (2024). Methanol as a co-substrate with CO_2 enhances butyrate production in microbial electrosynthesis. *Applied Microbiology and Biotechnology, 108*(1). https://doi.org/10.1007/s00253-024-13218-y

Zhang, L., Ong, J., Liu, J., & Li, S. F. Y. (2017a). Enzymatic electrosynthesis of formate from CO_2 reduction in a hybrid biofuel cell system. *Renewable Energy, 108*. https://doi.org/10.1016/j.renene.2017.03.009

Zhang, T., & Tremblay, P. L. (2019). Possible industrial applications for microbial electrosynthesis from carbon dioxide. In *Microbial electrochemical technology* (pp. 825–842). Elsevier. https://doi.org/10.1016/B978-0-444-64052-9.00034-0

Zhang, K., Zhou, Y., Song, T., & Xie, J. (2021). Bioplastic production from the microbial electrosynthesis of acetate through CO_2 reduction. *Energy & Fuels, 35*(19). https://doi.org/10.1021/acs.energyfuels.1c02594

Zhang, S. H., Qiu, C. H., Fang, C. F., Ge, Q. L., Hui, Y. X., Han, B., & Pang, S. (2017b). Characterization of bacterial communities in anode microbial fuel cells fed with glucose, propyl alcohol and methanol. *Applied Biochemistry and Microbiology, 53*, 250–257. https://doi.org/10.1134/S0003683817020193

Zhao, X., Li, X., Li, Y., Zhang, X., Zhai, F., Ren, T., & Li, Y. (2021). Metagenomic analysis reveals functional genes in soil microbial electrochemical removal of tetracycline. *Journal of Hazardous Materials, 408*, 124880. https://doi.org/10.1016/j.jhazmat.2020.124880

Zou, R., Hasanzadeh, A., Khataee, A., Yang, X., Xu, M., Angelidaki, I., & Zhang, Y. (2021). Scaling-up of microbial electrosynthesis with multiple electrodes for in situ production of hydrogen peroxide. *iScience, 24*(2). https://doi.org/10.1016/j.isci.2021.102094

Chapter 13
Application of Bio-electrochemical System for Sustainable Treatment of Waste Materials

Mohd Suhail Chaudhary, Waris, and Mohammad Zain Khan

1 Introduction

The increasing global energy demand and the environmental challenges of fossil fuel dependency underscore the urgent need for alternative energy sources. This growing demand is paralleled by extensive resource use in various sectors, including municipal, industrial, and agricultural activities, all contributing to environmental degradation and exacerbating issues like global warming (Ivase et al., 2020). As these environmental threats continue to intensify, developing and adopting renewable energy sources have become essential for achieving a sustainable future. Transitioning to renewables not only helps to reduce reliance on imported energy but also promotes energy diversification, which is critical for addressing both environmental and energy security concerns (Chae et al., 2009; Hassan et al., 2024). Over the past decade, bio-electrochemical systems (BESs) have gained significant attention as a promising sustainable technology, particularly for their dual role in generating electricity and treating wastewater. This innovative approach harnesses the metabolic activities of microorganisms to convert chemical energy from organic matter into electrical energy, offering a sustainable solution to both energy generation and environmental remediation. Moreover, BESs present unique opportunities for the sustainable production of biofuels and valuable chemicals, which are increasingly

M. S. Chaudhary · Waris
Environmental Research Laboratory, Department of Chemistry, Aligarh Muslim University, Aligarh, India
e-mail: itssuhail26@gmail.com

Waris
e-mail: mwaris3020@gmail.com

M. S. Chaudhary · Waris · M. Z. Khan (✉)
Electrochemical Research Laboratory, Department of Industrial Chemistry, Aligarh Muslim University, Aligarh, India
e-mail: zn.khan1@gmail.com

© The Author(s), under exclusive license to Springer Nature Singapore Pte Ltd. 2025

S. Sevda (ed.), *Electro-Fermentation for Synthesis of Chemicals, Biochemicals and Biofuels*, https://doi.org/10.1007/978-981-96-6358-3_13

in demand as society seeks greener alternatives to traditional industrial processes. Given the multidisciplinary nature of BESs, a vast amount of research has been conducted globally across a wide range of scientific fields. This includes microbiology, which explores the behavior and capabilities of microorganisms involved in these systems, electrochemistry and bioelectrochemistry, which study the transfer of electrons between microorganisms and electrodes; and biotechnology, which seeks to enhance the efficiency and scalability of BES applications. Additionally, research in environmental science focuses on the system's potential for pollution control, while materials science contributes by improving electrode materials and enhancing the overall performance of BESs. This collaborative effort underscores the complexity and potential of BESs as a key technology in the transition toward sustainable energy and environmental management (ElMekawy et al., 2015; Resch et al., 2008; Venkata Mohan et al., 2010).

In electrochemical systems, the generation of electric current is driven by the potential difference between the oxidation reaction occurring at the anode and the reduction reaction at the cathode. This potential difference facilitates the spontaneous movement of electrons from the anode, where the potential is lower, to the cathode, where the potential is higher. The flow of electrons through an external circuit, resulting from this process, is what we measure as electric current. When microorganisms or enzymes catalyze the oxidation at the anode, the reduction at the cathode, or both, the system is classified as a microbial electrochemical system (MES). More broadly, such systems fall under BESs. In cases where microorganisms facilitate the anodic oxidation process, the electrode is referred to as a bioanode. Conversely, when microorganisms catalyze the cathodic reduction reactions, the electrode is known as a biocathode. These bioelectrodes enhance the system's efficiency and enable the development of sustainable energy solutions by utilizing biological processes for electrochemical reactions (Harnisch & Schröder, 2010; Patil et al., 2015).

Microbial fuel cells (MFCs) represent a specialized form of microbial electrochemical systems (MES) that produce electricity by decomposing organic materials in the anode compartment. In these setups, microorganisms on the anode side oxidize organic substances, creating a low redox potential. Concurrently, at the cathode, the oxygen reduction reaction (ORR) takes place, which generally exhibits a high redox potential. This difference in redox potentials between the anode and cathode creates a driving force that facilitates electron flow through an external circuit, leading to the generation of bioelectricity. In contrast, microbial electrolysis cells (MECs) function similarly but require the application of an external voltage to overcome the energy barrier at the cathode. This applied voltage drives electrochemical reactions, enabling the production of valuable chemicals such as hydrogen or methane. Unlike MFCs, which generate electricity, MECs focus on chemical production by utilizing the electron flow and the applied external voltage to convert organic substrates into useful compounds. Both MFCs and MECs are key technologies in the broader field of BESs, with potential applications in renewable energy generation and sustainable chemical production (Singh et al., 2015).

In BESs, when carbon dioxide (CO_2) or organic molecules are reduced at the cathode to produce higher-value organic compounds, the process is referred to as microbial electrosynthesis. Microbial electrosynthesis leverages the metabolic capabilities of microorganisms to drive the electrochemical conversion of simple molecules into more complex, valuable products, offering a sustainable pathway for chemical synthesis. Beyond microbial electrosynthesis, BESs have found numerous other applications. One notable example is microbial desalination cells (MDCs), which are designed to desalinate water while simultaneously generating electricity. Additionally, sediment or plant microbial fuel cells (PMFCs) utilize microbes living in sediment or associated with plant roots to produce bioelectricity.

These systems highlight the versatility of BESs in combining energy generation with environmental applications. BESs have shown considerable promise in addressing environmental challenges, especially in the remediation of persistent pollutants and highly toxic wastewater. This method, referred to as bio-electrochemical treatment (BET) or microbial electroremediation, leverages the electrochemical activity of microorganisms to degrade complex contaminants into simpler, less toxic compounds. This innovative approach offers an effective solution for wastewater treatment and pollution management (Cao et al., 2009; Chiranjeevi et al., 2012; ElMekawy et al., 2014b; Mohanakrishna et al., 2010; Rabaey & Rozendal, 2010).

2 Principles of Microbial Electrochemical Systems

MESs are integrated platforms that utilize microorganisms to simultaneously facilitate wastewater treatment and energy generation. These systems operate through complex microbial electrochemical reactions that convert chemical energy into bioelectrical energy (Luo et al., 2016). According to Mohanakrishna et al., bio-electrochemical systems (BESs), a subset of MESs, employ microorganisms or enzymes as biocatalysts to drive these conversions through various electrochemical processes (Ivase et al., 2020; Mohanakrishna et al., 2015b). Figure 1 provides an overview of the various electrochemical devices, processes, and techniques that constitute BES.

A BES consists of two main compartments: the anodic fuel cell compartment, which is considered the more stable component of the BES (Pham et al., 2009). This compartment generally holds a higher concentration of biomass compared to the cathodic compartment, along with the biocatalyst (either microorganisms or enzymes) and the anode (positive electrode). In contrast, the cathodic fuel cell functions primarily as a chemical fuel cell, incorporating a cathode (negative electrode) and relying on oxygen as an electron acceptor (Risgaard-Petersen & Rotaru, 2023). Separating these compartments is a proton exchange membrane (PEM), which prevents undesired charge transfer while minimizing oxygen diffusion into the anode (Kracke et al., 2015a; Oliot et al., 2016). The substrate (or fuel) is stored in a separate container connected to an electrical circuit, as depicted in Fig. 2 (Clauwaert et al., 2008a).

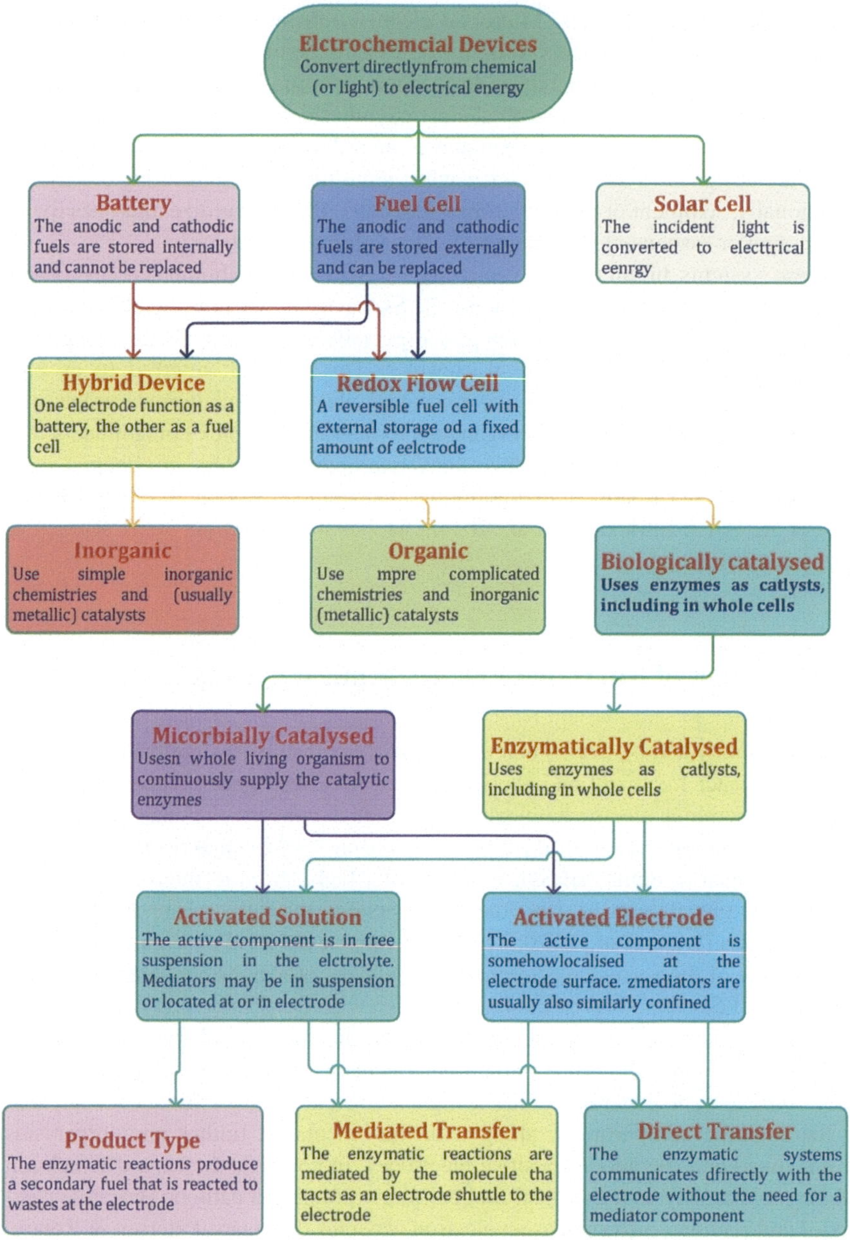

Fig. 1 Classification of electrochemical devices

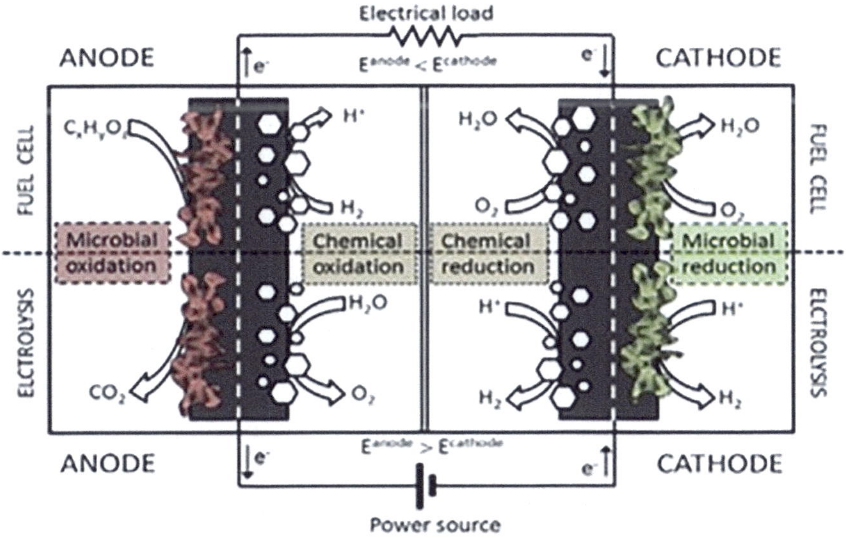

Fig. 2 Schematic of a bio-electrochemical fuel cell (Clauwaert et al., 2008a)

The microbial electron transport chain (METC) is fundamental to microbial metabolism, as it facilitates the oxidation of organic substrates, which serve as primary energy sources for microbial cells (Hassan et al., 2021). Assessing microbial respiration efficiency and electron transport chain activity provides essential insights into cellular functionality. These processes are particularly significant for the growth and reproduction of aerobic microorganisms (Santoro et al., 2017). Early studies focused on measuring dissolved oxygen consumption as an indicator of microbial viability. In microbial electrochemical systems (MESs), the transfer of electrons from microorganisms to electrodes is utilized in applications such as microbial fuel cells (Ng et al., 2017) and rapid microbial activity assessment tools (Cui et al., 2020; Hassan & Wollenberger, 2019).

Consequently, various MES-based methodologies have been developed and optimized for biological applications. The electrical current generated in these systems is directly linked to microbial metabolic activity, as viable cells actively participate in electrochemical reactions, whereas non-viable or non-culturable cells do not contribute to electrochemical signals. The bio-electrochemical responses observed in MESs provide valuable information about anodic respiration, intracellular redox activities, and fundamental microbial interactions. Given their direct connection to metabolic processes, there is increasing interest in advancing high-performance MESs for applications in environmental remediation, sustainable energy generation, and biomedical diagnostics. These developments position MES technology as a promising field for future research and innovation.

2.1 Extracellular Electron Transfer

The core functionality of microbial electrochemical systems (MESs) is driven by extracellular electron transfer (EET), which occurs either through direct electron transfer (DET) from microbial cells to an external electron acceptor or indirectly via electron shuttles in mediated electron transfer (MET). Within MESs, bio-electrochemical signals originate from microbial metabolic processes, including the degradation of organic substrates. For example, glycolysis breaks down organic compounds into pyruvate, a key metabolic intermediate that fuels bioenergy production. This electron transfer process is fundamental to MES operation, enabling efficient energy conversion and expanding the potential applications of MESs in sustainable energy technologies and biotechnological advancements (Fig. 3) (Hassan et al., 2021).

In the absence of oxygen as the terminal electron acceptor, anodic respiration plays a crucial role in facilitating electron transfer from microbial cells to the electrode surface. Under aerobic conditions, electron transport typically begins with the enzymatic oxidation of NADH or NADPH by NADH dehydrogenase, the first complex in the electron transport chain (ETC). This process releases two electrons, which are subsequently transferred to coenzyme Q (CoQ), converting it to ubisemiquinone and then to ubiquinol, a key electron carrier. In aerobic respiration, these electrons continue along the ETC, passing through complex III (bc1 complex), cytochrome c, and complex IV (cytochrome c oxidase) before reaching oxygen, the final electron acceptor, leading to water formation. The energy generated during this electron flow is harnessed to pump protons across the membrane via complexes I, III, and IV, establishing a proton gradient that drives ATP synthesis. In contrast, MESs utilize alternative electron acceptors, redirecting electron flow toward the electrode

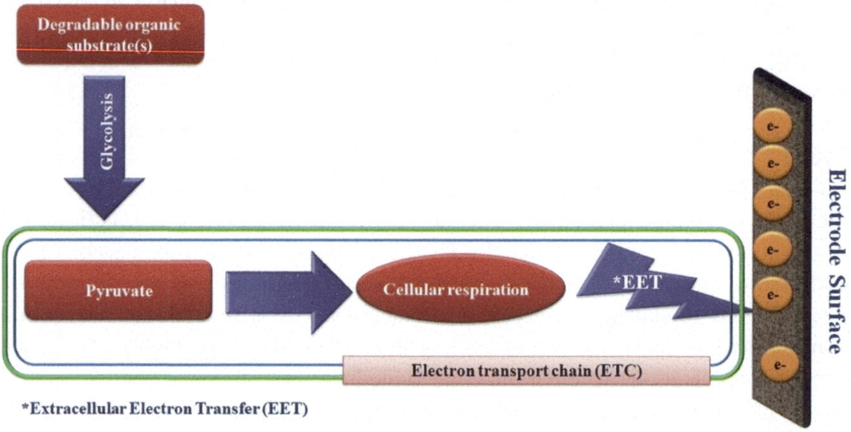

Fig. 3 Fundamental concept of microbe-electrode interactions in typical MES involves the transfer of biologically generated electrons from living microbes, which adhere to the electrode surface, producing bio-electrochemical signals (Hassan et al., 2021)

surface, thereby enabling bioelectricity generation and sustainable energy applications (Matsumae et al., 2014). Since the electron transport chain is enclosed within cytoplasmic membranes and further shielded by structural barriers such as cell walls, peptidoglycans, or outer membranes, the direct transfer of biologically generated electrons to the electrode surface becomes a crucial challenge in microbial electrochemical systems (Zhao et al., 2021). Electron transfer mechanisms from microbial intracellular compartments to the electrode surface have been extensively studied in bacterial systems, positioning bacteria as the primary candidates for microbial fuel cell (MFC) technology. As illustrated in Fig. 4, microbial cells can transfer electrons to the electrode via two distinct pathways: direct electron transfer (DET) and mediated electron transfer (MET) (Hassan et al., 2021). The following section provides a comprehensive discussion of these mechanisms, highlighting their roles in facilitating efficient electron transport within microbial electrochemical systems.

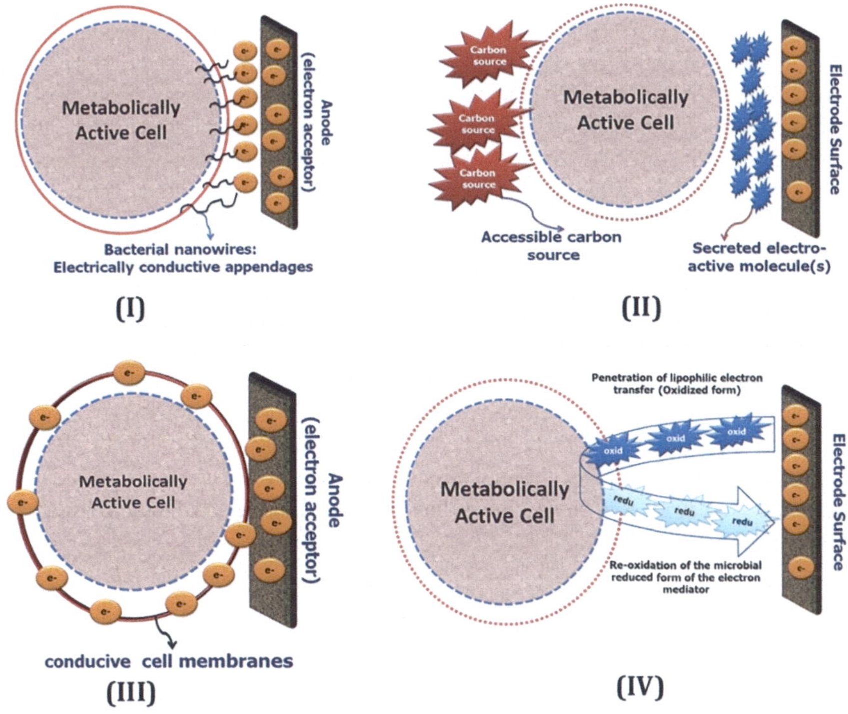

Fig. 4 Illustration of extracellular electron transfer (EET) in microbial electrochemical systems. Direct electron transfer (DET) occurs either through the extension of conductive bacterial nanowires (I) or the development of a conductive biofilm on the microbial cell surface (III). In contrast, mediated electron transfer (MET) involves electron transport via naturally secreted redox mediators (II) or externally supplied artificial electron shuttles (IV) (Hassan et al., 2021)

2.1.1 Direct Electron Transfer Via Electroactive Microbes

Certain microbial species, known as exoelectrogens or electroactive microorganisms, play a crucial role in anodic respiration by facilitating extracellular electron transfer (EET) and directly interacting with electrode surfaces in microbial electrochemical systems (Logan et al., 2019). These microbes can transfer electrons to conductive materials without requiring external mediators, making the electron transfer process highly efficient. They are also referred to as anode-respiring bacteria, electrochemically active bacteria, or electricigens, emphasizing their significance in microbial electrochemical applications (Kracke et al., 2015b). DET occurs when there is direct contact between the microbial redox centers and the electrode surface (Shi et al., 2016). This interaction is enabled by electroactive components such as cell-wall cytochromes, which allow electron flow from the microorganism to the electrode (Xu et al., 2018). Alternatively, a conductive layer can form in the outer membranes to facilitate electron conduction (Myers & Myers, 1992). For instance, Shewanella oneidensis MR-1 produces bacterial nanowires (pili) in its outer membrane, which act as conductive pathways, facilitating extracellular electron transport (Pirbadian et al., 2014). Another electron transfer mechanism involves the secretion of electroactive metabolites into the extracellular matrix, allowing microbes to mediate their own electron transport (Hu et al., 2015; Shi et al., 2016). For example, *Shewanella putrefaciens* releases water-soluble natural electron mediators, which function as quorum-sensing molecules while enhancing electron transfer efficiency (Hassan et al., 2021). Additionally, phenazines produced by *Pseudomonas aeruginosa* serve as quorum-sensing molecules that modulate current production during anodic respiration. DET is not exclusive to bacteria; some eukaryotic microorganisms, such as *Candida albicans*, have also been identified as potential candidates for this mechanism (Venkataraman et al., 2010). In these scenarios, the efficiency of the mitochondrial respiratory chain plays a crucial role in determining electron transfer capacity. Disrupting the respiratory chain, either through chemical inhibition using specific enzyme inhibitors or genetic modifications that knock out key genes in respiratory complexes, leads to a significant reduction in the bio-electrochemical response (Venkataraman et al., 2010).

2.1.2 Mediated Electron Transfer

Unlike exoelectrogens, many electrochemically inactive microorganisms cannot transfer electrons to the surface of the electrode without the assistance of soluble redox mediators (Hernandez & Osma, 2020). To overcome this limitation, artificial redox mediators, also known as exogenous electron shuttles, are introduced to facilitate electron transfer in a process known as mediated electron transfer (MET) (Schmitz & Rosenbaum, 2018). MET is widely applied in microbial electrochemical systems (MESs) due to its well-characterized redox reaction mechanisms, making it valuable for applications such as rapid cell viability and cytotoxicity assessment (Kwon et al., 2018). For example, 2,6-dichlorophenolindophenol (DCIP) has been utilized as a redox mediator to monitor cell viability in *Candida* and *Saccharomyces* species (Hassan & Bilitewski, 2011). DCIP accepts electrons from complex I of the respiratory chain (mitochondrial NADH dehydrogenases), a process absent in the respiratory chain of *Saccharomyces cerevisiae*. This property allows for the selective evaluation of complex I activity in *Candida albicans*. Additionally, a dual-mediator system combining 2,3,5,6-tetramethyl-*p*-phenylenediamine (TMPD) or menadione with hexacyanoferrate(III) (ferricyanide, FCN) has been used to investigate intracellular redox activity and metabolic pathways in *Saccharomyces cerevisiae* and Chinese hamster ovary (CHO) cells (Rawson et al., 2014). This approach enhances electron transfer efficiency and sensitivity compared to single-mediator systems. A similar strategy employing DCIP and FCN has been applied to study the intracellular redox processes of *Staphylococcus aureus*, particularly in response to organic substrates such as glucose or acetate as electron donors. When only the hydrophilic mediator FCN was used, no bio-electrochemical responses were generated, indicating a lack of communication between living bacterial cells and the electrode surface. However, using the lipophilic mediator DCIP alongside FCN produced strong bio-electrochemical signals, demonstrating the potential for connecting living cells to solid conductive surfaces. This dual-mediator system significantly enhanced the electric current response by facilitating electron transfer from intracellular NAD(P)H-dependent oxidoreductases (NQO), thereby improving the overall efficiency and sensitivity of the assay (Pirbadian et al., 2014). A schematic representation of direct and mediated electron transfer mechanisms between living microorganisms and the electrode surface is provided in Fig. 4.

3 Types of Bio-electrochemical Systems (BESs)

BESs can be classified into several categories based on their specific goals, each designed to address distinct challenges in energy production, chemical synthesis, desalination, and environmental remediation. Electrogenesis systems, such as MFCs, focus on generating electricity by harnessing the microbial degradation of organic matter at the anode. Electrohydrogenesis systems, like microbial electrolysis cells (MECs), utilize the application of an external voltage to drive the production of hydrogen or other valuable chemicals through microbial catalysis. Microbial desalination systems (MDSs), such as microbial desalination cells (MDCs), aim to desalinate water while simultaneously generating electricity, utilizing microbial metabolism alongside ion exchange. Microbial electrosynthesis systems leverage the capabilities of microorganisms to reduce CO_2 or organic molecules at the cathode, facilitating the production of high-value chemicals in a carbon-neutral manner. Lastly, bio-electrochemical treatment systems (BETs) are focused on environmental applications, where microbial electroremediation is used to degrade persistent pollutants and treat toxic wastewater. Each of these BES types plays a vital role in advancing sustainable solutions for energy generation, resource recovery, and environmental protection.

3.1 Microbial Fuel Cells (MFCs)

MFCs produce electrical energy by harnessing microbial metabolism to oxidize organic compounds, transferring electrons to a solid electrode that acts as the final electron acceptor (Rabaey et al., 2011). The anode in MFCs provides a surface for microbial attachment and facilitates the oxidation of organic compounds, resulting in the release of electrons. These electrons are then transferred through an external circuit, which includes an external load, to the cathode. To maintain electroneutrality in the system, ions are transported through an ion-permeable medium or membrane between the anode and cathode compartments. The process of electron flow through the external circuit produces electricity. The first MFC system, demonstrated by Potter in 1911, generated a voltage ranging between 0.3 and 0.5 V, marking the early beginnings of microbial electrochemical technologies (Potter, 1911). In BESs, electroactive bacterial biofilms that form on the anode act as the primary electrocatalytic units, playing a crucial role in electricity generation. These biofilms consist of microorganisms, specifically *exoelectrogens*, which have the unique capability to transfer electrons from reduced organic substrates to the anode surface. The production of bioelectricity in MFCs depends heavily on the efficiency and abundance of these *exoelectrogenic* microbes. As these bacteria oxidize organic compounds, particularly in complex environments such as wastewater, they release electrons that are captured by the anode and directed through an external circuit to produce electric current (Kiely et al., 2011; Logan, 2009). The degradation of complex organic matter, particularly in wastewater, relies on a consortium of microbial species that collaboratively break down different compounds through metabolic interactions. Different bacteria contribute to different stages of organic matter degradation, ensuring that a broad spectrum of substances can be processed. Research has shown that numerous bacterial species, often in complex microbial communities, possess the ability to generate electric current. This diversity is essential for optimizing the performance of MFCs, as it enables the effective degradation of a wide variety of organic pollutants while maximizing electron transfer for electricity generation. The understanding and utilization of these microbial communities open up possibilities for the development of more efficient and sustainable energy systems.

The *Shewanella* and *Geobacter* species were among the earliest microorganisms identified for their ability to transfer electrons extracellularly. They have been widely studied for their crucial roles in microbial fuel cells (MFCs) due to their efficient electron transport mechanisms. While these pure cultures demonstrate significant electron transfer capabilities, it has been observed that MFCs employing mixed microbial cultures at the anode tend to achieve more stable and higher current outputs. The presence of *Geobacteraceae* in the anode community is particularly associated with elevated power densities in MFCs. For instance, MFCs utilizing an anode enriched with *Geobacter* species have exhibited remarkable performance. The highest recorded current density, reaching 390 A m^{-2}, was achieved using a mixed microbial culture on a layered-corrugated carbon anode designed to maximize surface area. Several factors influence microbial activity and electron transfer in MFCs, including the nature of the substrate, the potential at the anode, and the chemistry of the electrolyte. These parameters are critical in shaping the efficiency of the microbial electrochemical processes. Notably, higher electrolyte conductivity enhances the performance of MFCs, as it facilitates better ion transport, which is essential for maintaining electroneutrality and maximizing current production. However, the electrolyte conductivity must be carefully controlled to avoid exceeding the tolerance limits of the bacterial communities, as overly conductive environments can inhibit microbial growth and reduce overall system efficiency (Chen et al., 2012; Fan et al., 2008; Pant et al., 2010; Potter, 1911).

Halophilic bacteria have shown significant potential in BESs by generating high current densities under saline conditions. For example, anodes inoculated with microbial communities from salt marsh environments and operated in electrolytes with salinity levels 1.5 times that of seawater have demonstrated current densities reaching up to 85 A m^{-2}. These findings suggest that halophilic bacteria can enhance BES performance in high-salinity environments, making them promising candidates for energy generation in marine and hypersaline conditions. The efficiency of microbial fuel cells (MFCs) is largely dependent on the electrode material and its structural properties. Research has indicated that roughened graphite electrodes facilitate greater microbial attachment and electron transfer, leading to improved power output compared to smooth graphite surfaces. Furthermore, employing high surface area materials such as graphite brushes, carbon felt, and carbon nanotubes as anode substrates has significantly enhanced current density. The increased surface area of these materials promotes bacterial colonization and strengthens the interaction between electroactive microbes and the electrode, ultimately improving electron transfer efficiency and boosting overall system performance (Chen et al., 2011; Feng et al., 2010; Mohanakrishna et al., 2012; Rousseau et al., 2013; ter Heijne et al., 2008).

Beyond carbon-based materials, microbial electron transfer has also been observed with metal electrodes, expanding the range of materials suitable for bioelectrochemical applications. Dimensionally stable anodes (DSAs), composed of titanium substrates coated with iridium or tantalum oxide, have been explored for their durability and conductivity. Among these, platinum has exhibited the highest current densities, though porous carbon anodes typically outperform most metal electrodes (Dumas et al., 2008; Pocaznoi et al., 2012). The shift from planar to three-dimensional electrodes with optimized microstructures has also been associated with substantial improvements in bioelectrode performance. Anodes with multilayered corrugated carbon configurations have shown significant increases in current density, particularly when mass transport limitations are addressed through flow-through electrodes. This reduction in internal resistance has resulted in enhanced current and power densities. Cathode materials play a critical role in determining the overall power output of MFCs, as demonstrated by studies that compare systems with similar anode materials. The materials used in the construction, design, and configuration of MFCs directly impact internal resistance, which in turn governs the overall performance of the system. By optimizing both anode and cathode materials and addressing structural and resistance-related challenges, significant improvements in

MFC efficiency can be achieved (Chen et al., 2012; Sharma et al., 2014; Sleutels et al., 2012; Zhang et al., 2011).

3.2 Microbial Electrolysis Cells (MECs)

MECs leverage bacterial metabolism to generate electrical energy while simultaneously driving water electrolysis (Venkatramanan et al., 2021). In MECs, an external voltage is applied to facilitate electron transfer from the anode to the cathode, enabling hydrogen evolution at the cathode (Jeremiasse et al., 2010). Unlike MFCs, the cathodic chamber in MECs is maintained under anaerobic conditions, which is essential for efficient hydrogen production. However, the presence of CO_2 and methanogenic microbes in this environment can lead to methane formation, reducing hydrogen yield. Various approaches have been suggested to mitigate methane production, including intermittent aeration of the cathode chamber, pH adjustments, reduced retention times, heat shock treatment of the microbial inoculum, and the use of chemical inhibitors to suppress methanogen activity (Logan et al., 2008). Hydrogen generation in MECs is influenced by system design and the type of substrates utilized. Compared to MFCs, MECs generally produce higher current densities due to the applied external voltage, which alleviates cathodic limitations (Sleutels et al., 2013; Wang et al., 2011). Notably, MECs can be powered by MFCs, where multiple MFCs are connected to provide the necessary voltage for hydrogen evolution. A significant advantage of MECs is the reduced reliance on expensive metal catalysts such as platinum. Instead, microbial communities can be enriched on carbon-based cathodes, leading to biocathodes that shorten system startup times and generate current densities comparable to anodes (Jeremiasse et al., 2010). Beyond hydrogen production, MECs hold the potential for generating valuable chemicals through bio-electrochemical processes (Rabaey & Rozendal, 2010). Cathodic reactions can include proton reduction to hydrogen, oxygen reduction to hydrogen peroxide, and CO_2 conversion into methane or acetate. These advancements highlight the role of bio-electrochemical systems in sustainable biofuel production and the synthesis of high-value chemicals, such as organic acids and alcohols, from cost-effective feedstocks or carbon dioxide (He et al., 2002; Logan et al., 2008; Mohanakrishna et al., 2015a; Nevin et al., 2010; Rozendal et al., 2009; Sharma et al., 2013).

3.3 Microbial Electrosynthesis (MES)

Microbial electrosynthesis (MES), also known as bioelectrosynthesis, represents an innovative application of BESs. MES leverages the reducing power generated from anodic oxidation to drive the formation of value-added products at the cathode. In this process, cathodic biocatalysts, typically in the form of attached biofilms, play a crucial role in reducing terminal electron acceptors to synthesize these products

(Marshall et al., 2012; Mohanakrishna et al., 2015a; Rabaey & Rozendal, 2010). The efficiency and selectivity of the bioelectrosynthesis process are influenced by multiple factors, including the type of biocatalyst facilitating the redox reaction, the choice of terminal electron acceptor, and the availability of electrochemically active redox mediators or reducing equivalents. Biocathodes play a crucial role in microbial electrosynthesis (MES), as they are composed of microbial communities that drive cathodic reduction reactions, leading to the formation of various reduced products. Some of the well-known products formed through microbial electrosynthesis include acetate, ethanol, and butyrate. These reduced compounds are synthesized by utilizing the electron transfer capabilities of biocathodes, which are optimized for specific reactions depending on the type of microorganism used and the conditions provided (Bajracharya et al., 2015).

MES primarily focuses on the conversion of CO_2 to multi-carbon organic compounds via electricity-driven microbial catalysis at the cathode. This aspect of MES has garnered significant attention as it offers a promising alternative for capturing electrical energy in the form of chemical bonds in organic products, essentially transforming electrical energy into chemical energy. The ability of MES to drive the reduction of CO_2 into useful organic compounds, such as acetate and other multi-carbon molecules, has demonstrated its potential as a carbon capture and utilization technology. Additionally, MES can be used for the reduction or oxidation of other organic feedstocks, expanding its scope beyond CO_2 reduction. The concept of MES was demonstrated in early studies, which showed that microbial catalysis could effectively reduce CO_2 to form organic compounds when supplied with electrical current at the cathode. This proof-of-concept opened the door to further research into MES as a novel strategy for the sustainable production of chemical compounds using renewable electricity (Nevin et al., 2010; Rabaey & Rozendal, 2010). Beyond its environmental benefits, MES is also being explored for its economic potential, as it provides a pathway to produce value-added chemicals in a manner that integrates renewable energy sources with microbial catalysis (Desloover et al., 2012; Rabaey & Rozendal, 2010; Rabaey et al., 2011).

3.4 Enzymatic Fuel Cells (EFCs)

Enzymatic fuel cells (EFCs) utilize specific enzymes to catalyze fuel oxidation, enabling chemical reactions for diverse applications. These enzymes can either be immobilized on the electrode surface or dispersed within the electrolyte. Early research by Davis and Yarbrough demonstrated the feasibility of employing microbes and enzymes in biofuel cells, particularly emphasizing glucose oxidase as a key biocatalyst (Davis & Yarbrough, 1962). Due to its exceptional thermostability and high specificity for glucose electro-oxidation, glucose oxidase has been widely integrated into biomedical devices such as pacemakers, indicator lights, micro-actuators, and micropumps. EFCs function by leveraging selective enzymatic activity at both the anode and cathode, eliminating the necessity for a membrane between the compartments. However, a major limitation of enzymatic catalysts is their partial oxidation of fuel, which can lead to unwanted heat generation from side reactions, ultimately affecting enzyme stability (Minteer et al., 2007; Palmore & Whitesides, 1994). To improve enzyme durability, researchers have explored methods such as encapsulating enzymes within micelle polymers, which provide pH buffering and a hydrophobic environment to mitigate degradation (Minteer et al., 2007). Significant advancements in EFC technology have emerged through the integration of flavin adenine dinucleotide-dependent glucose dehydrogenase (FADGDH) with osmium-polymer-based electrodes. This combination has demonstrated high glucose sensitivity and impressive current densities, making it a promising candidate for bioanodes in glucose-based EFCs (Zafar et al., 2012). Additionally, pyrroquinoline quinone-dependent alcohol dehydrogenase (PQQ-ADH), an enzyme from acetic acid bacteria, has been explored for use in EFCs and biosensors due to its quasi-reversible electrochemistry and ability to interact with various redox mediators, facilitating efficient electron transfer (Gellett et al., 2010; Yakushi & Matsushita, 2010).

Although EFCs exhibit high energy densities, their relatively low power output makes them particularly well-suited for low-power applications such as wireless sensor systems, video monitoring devices, and portable power sources. Researchers have investigated the use of fructose dehydrogenase from *Gluconobacter* as an anodic catalyst and *laccase*, a copper-containing enzyme derived from mushrooms, as a cathodic catalyst. These enzymes offer high turnover rates, stability, and selectivity, further enhancing EFC performance (Calabrese Barton et al., 2004; Gellett et al., 2010; Kamitaka et al., 2007; Zafar et al., 2012). To enhance EFC efficiency, multi-enzyme systems are generally preferred over single-enzyme configurations, as they enable greater substrate conversion and improved overall system performance. The integration of enzyme cascades in EFCs facilitates complex biochemical reactions, while three-dimensional electrode structures increase the available surface area for enzymatic interactions. Despite these advancements, the high production cost of purified enzymes remains a major challenge hindering the large-scale commercialization of EFCs (Minteer et al., 2007).

3.5 Microbial Solar Cells (MSCs)

Microbial solar cells (MSCs) integrate photoautotrophic microorganisms or higher plants with electroactive bacteria to harness solar energy for electricity generation or the synthesis of valuable compounds such as hydrogen, methane, ethanol, and hydrogen peroxide. The process initiates with photosynthesis, leading to the production of organic substrates that serve as electron donors. These compounds undergo microbial oxidation at the anode, where electroactive microorganisms facilitate electron transfer to the electrode. The electrons subsequently migrate to the cathode, where oxygen reduction reactions take place, forming water (ElMekawy et al., 2014a). Phototrophic biofilms composed of organisms like *Synechocystis*, *Chlorophyta*, and other electroactive species are commonly found on MSC anodes.

A notable application of MSCs involved their use in addressing algal blooms in freshwater systems, where *Microcystis aeruginosa* and *Chlorella vulgaris* biomass, combined with lake water, served as a substrate in microbial fuel cells (MFCs). This system enabled concurrent electricity generation and wastewater treatment. Another study employed algal biomass harvested from a photobioreactor, which was subsequently processed in an anaerobic digester to serve as an MFC feedstock.

A specific MSC design utilizing a photobioreactor with *Chlorella* at the anode exhibited a light-to-electricity conversion efficiency of 0.04%, producing an average power density of 14 mW m^2. Furthermore, *Synechocystis* PCC6803 achieved a current density of 539 mA m^2 under high light intensity (10,000 lx) while capturing 625 mmol CO_2 m^3 without requiring a mediator or external carbon source in the anolyte. Enhancing MSC efficiency can be achieved by refining photobioreactor configurations, optimizing electron transfer between algae and the bioanode, increasing electrode surface area, and promoting electroactive biofilm growth. The dual benefit of CO_2 sequestration and direct solar energy conversion into electricity underscores the potential of MSCs as a sustainable energy technology (De Schamphelaire & Verstraete, 2009; Madiraju et al., 2012; Strik et al., 2008b, 2011; Wang et al., 2012a).

3.6 Plant Microbial Fuel Cells (PMFCs)

PMFCs are a unique and innovative approach to sustainable energy production, combining biological and electrochemical processes. In PMFCs, the root system of a living plant is integrated into the anode of the MFC, enabling the organic compounds released by the plant roots, known as rhizodeposits, to serve as an energy source for electroactive microorganisms (Strik et al., 2008a). These microorganisms oxidize the rhizodeposits, generating electrons that can be harvested for electricity production. Thus, PMFCs offer a novel method of harnessing solar energy indirectly by converting the organic matter produced during the plant's natural metabolic processes into a clean and renewable energy form. Rhizodeposits, which form the basis for electricity generation in PMFCs, consist of various substances released by the plant roots into the surrounding soil. These include root exudates such as sugars, amino acids, and organic acids, which provide nutrients for microbial growth. Additionally, root secretions, composed of complex polymers like carbohydrates and enzymes, further enhance microbial activity. Lysates, which result from the breakdown of dead root cells and various gases produced by the plant, contribute to the diversity of substrates available for microbial consumption. The overall amount of rhizodeposits released by the plant plays a crucial role in determining the efficiency of electricity generation, as these compounds provide the necessary fuel for the electroactive microbes (Strik et al., 2008a). The choice of plant species for PMFCs is of paramount importance, as it directly influences the quantity and composition of rhizodeposits available to the microbial community. Plants that release more exudates and other organic compounds into the soil generally facilitate higher bioelectricity production. Several studies have explored different plant species for PMFC applications, with notable examples including 'Reed manna grass' ('*Glyceria maxima*'), rice ('*Oryza sativa*'), and '*Spartina anglica*'. Among these, PMFCs utilizing 'Pennisetum setaceum', also known as fountain grass, have demonstrated the highest sustainability in terms of power generation. In these systems, the maximum power output achieved was 163 mW/m^2, making 'Pennisetum setaceum' a promising candidate for future PMFC applications. In another study, PMFCs based on 'Spartina anglica' were enhanced by the incorporation of oxygen-reducing biocathodes, significantly boosting long-term performance. These systems achieved a peak power output of 240 mW m^{-2} over a sustained period of two weeks, highlighting the potential of using integrated biocathodes to improve PMFC efficiency. Such advancements underscore the importance of not only selecting optimal plant species but also integrating complementary technologies to enhance power output (De Schamphelaire et al., 2010; Helder et al., 2012; Strik et al., 2008a; Wang et al., 2012a; Wetser et al., 2015). PMFCs have also been combined with MSCs to form hybrid energy systems. In these systems, phototrophic biofilms containing light-absorbing microorganisms are used alongside the plant root systems, enabling dual energy capture from both solar radiation and organic substrate oxidation. This combined approach is seen as a promising direction for future energy harvesting technologies. Interestingly, studies have revealed that the total amount of rhizodeposits available for microbial oxidation at the anode is a more critical factor for bioelectricity production than the photosynthesis process itself. While photosynthesis is essential for plant growth and metabolism, the energy captured and stored in rhizodeposits ultimately drives the microbial activity responsible for electricity generation. This insight points to the potential of engineering plant root systems and microbial communities for enhanced PMFC performance, independent of photosynthetic efficiency (Timmers et al., 2012).

3.7 Microbial Desalination Cells (MDCs)

Microbial desalination cells (MDCs) are an advanced technology that harnesses the electrochemical potential created between the anode and cathode in MFCs to enable the direct removal of salts from water. This system was first introduced by Cao et al. as the initial demonstration of water desalination combined with MFC technology (Cao et al., 2009). The distinguishing feature of MDCs is the incorporation of a third, or middle, compartment between the anode and cathode, specifically designed for desalination purposes. This middle compartment is separated from the anode by an anion exchange membrane (AEM) and from the cathode by a cation exchange membrane (CEM), allowing selective ion transport during the process. The basic principle of MDC operation is driven by the metabolic activity of bacteria at the anode. These bacteria oxidize biodegradable organic substrates, releasing electrons and protons. The electrons are then transferred to the cathode via an external circuit. Simultaneously, anions such as sulfate (SO_4^{2-}) and chloride (Cl^-) present in the saline water of the desalination compartment migrate through the AEM toward the anode, while cations like potassium (K^+) and sodium ions (Na^+) move across the CEM toward the cathode. This ion migration facilitates the removal of salts from the water, thereby achieving desalination of the solution in the middle chamber. One of the key benefits of MDC technology is the simultaneous production of electrical energy and desalination (Cao et al., 2009; Luo et al., 2011, 2012). The migration of ions from the saline water to the anode and cathode enhances the conductivity of the anolyte and catholyte, respectively, which in turn increases the efficiency of power generation in the MDC. This improvement in conductivity leads to better mass transfer, making MDCs more efficient than conventional MFCs. However, this ion migration also results in increased salinity in the anolyte and catholyte, which can negatively impact the bacterial communities at both the anode and cathode, potentially affecting the overall system performance and effluent water quality (Luo et al., 2012).

MDCs offer the flexibility to be used for simultaneous organic matter degradation, salt removal, and electricity generation. Alternatively, they can be integrated into conventional desalination processes such as reverse osmosis (RO). When used as a pretreatment step, MDCs can significantly lower the salinity of feed water, reducing the energy consumption of subsequent RO processes. It has been estimated that MDCs can generate up to 58% of the energy required for the downstream RO system, making this a highly energy-efficient solution. In a comparative analysis, Jacobson et al. found that conventional RO systems typically consume around 2.2 kWh of energy to desalinate 1 m^3 of seawater. In contrast, MDCs not only desalinate water but also produce approximately 1.8 kWh of electricity during the treatment of the same volume of seawater. This results in a net energy benefit of around 4 kWh per cubic meter of water treated, highlighting the potential of MDCs to reduce the overall energy costs associated with seawater desalination. Furthermore, MDCs can be operated in electrolysis mode to produce hydrogen gas at the cathode, offering additional value beyond desalination and power generation. When operated in this mode, MDCs have been reported to recover between 180 and 231% of the input electrical energy in the form of hydrogen (H_2), providing a highly efficient method of energy recovery with the added benefit of desalination. This dual functionality makes MDCs a promising technology for sustainable water treatment and energy production, potentially offering solutions to both the global water and energy crises (Jacobson et al., 2011; Luo et al., 2011; Mehanna et al., 2010).

4 Parameters in Evaluating BES Performance

4.1 Electrode Potential and Power Output

All BES processes operate on the same principle: exoelectrogenic bacteria (EAB) in the anode chamber oxidize biodegradable substrates, generating an electron flow (current) that reduces electron acceptors in the cathode chamber. A direct way to evaluate the performance of a BES is by measuring the circuit voltage (V). The electric current (I) flowing through an external resistor (R_{ext}) can be determined using Ohm's law:

$$I = \frac{V}{R_{ext}}$$

Here, V is the direct output of the BES reactor. The voltage in a BES can also be expressed as:

$$V = \mathrm{OCV} = IR_{int}$$

Here, OCV represents the open circuit voltage, and R_{int} refers to the internal resistance of the system, which comprises contributions from the ohmic resistance of the system, activation losses due to electrode surface reactions, and charge transfer resistance associated with mass transfer limitations (Logan et al., 2006). Since internal resistance directly affects system efficiency, it is often modeled as being in series with the external resistance (R_{ext}).

The power density (P) of a MFC follows an inverse relation to the square of total resistance, as given by:

$$P = \frac{V^2 R_{ext}}{(R_{int} + R_{ext})^2}$$

This equation highlights that optimizing power output requires a balance between internal and external resistance. Notably, maximum power is achieved when the external resistance matches the internal resistance ($R_{ext} = R_{int}$). Laboratory-scale MFCs typically determine power density by maintaining a constant external resistance or applied potential and subsequently generating polarization data by varying R_{ext} at defined intervals (Logan et al., 2006; Lyon et al., 2010; Pinto et al., 2011; Ren et al., 2011; Ren et al., 2011).

Compared to conventional alternative energy technologies, BES and MFCs generally produce lower power outputs, largely due to thermodynamic constraints. The anode potential in MFCs is typically around −0.3 V versus the normal hydrogen electrode (NHE), as dictated by bacterial respiratory enzymes and the metabolic pathways of electron donors. For instance, when acetate (5 mM, pH = 7) is used as a substrate, the oxidation reaction occurring at the anode is:

$$\text{Anode: } 2HCO_3^- + 9H^+ + 8e^- \rightarrow CH_3COO^- + 4H_2O$$

$$\mathrm{OCP}_{anode} = -0.296\ \mathrm{V}$$

This inherently low anode potential, coupled with system losses, imposes constraints on the overall power output of MFCs. Despite these challenges, improvements in electrode design and system architecture have led to enhanced BES efficiency in various applications.

In contrast, the cathode potential in MFCs can reach approximately +0.8 V when oxygen acts as the final electron acceptor (Logan et al., 2006). The cathodic reaction in oxygen-based MFCs is represented as:

$$\text{Cathode: } O_2 + 4H^+ + 4e^- \rightarrow 2H_2O$$

$$OCP_{\text{cathode}} = 0.80\ \text{V}$$

However, due to overpotentials and system inefficiencies, the voltage output of air-cathode MFCs typically remains below 0.8 V, and the current is often limited to a few milliamps. While chemical oxidants such as ferricyanide or permanganate can yield higher cathode potentials, oxygen-based cathodes remain the most sustainable option because of their cost-effectiveness and widespread availability (Logan et al., 2008; Ren et al., 2007). Enhancements in power generation can be realized through advancements in electronic energy harvesting strategies and the development of large-scale MFC stack configurations. Such innovations have the potential to improve the scalability and energy output of MFCs, making them more viable for real-world applications (Aelterman et al., 2006; Dewan et al., 2008; Wang et al., 2012b). Continued research in this field aims to address the existing limitations and maximize the efficiency of BES technologies for various bio-electrochemical applications.

Power density is a critical parameter used to evaluate the performance of MFC reactors. Traditionally, power density is normalized to the projected electrode area, termed surface power density, allowing for comparative analysis across different systems. However, this method does not always provide a fair comparison when different systems utilize varying electrode configurations and materials. For instance, replacing a carbon cloth anode (~100 $m^2\ m^{-3}$) with a high surface area brush anode (~9,600 $m^2\ m^{-3}$) significantly increases the normalized cathode power density from 0.6 to 2.4 Wm^{-2}, while the normalized anode power density remains relatively unchanged (Logan et al., 2007). This variation in results highlights the limitations of surface power density in accurately reflecting system performance. To address this, recent studies are increasingly using volumetric power density, which normalizes power output by reactor volume. This metric offers a more comprehensive evaluation of overall power generation, accounting for the size of the reactor and its capacity for wastewater treatment. Higher electrode density generally results in better volumetric power density due to increased available surface area for electrochemical reactions. However, in real-world applications, systems with extremely high surface areas may encounter practical challenges, such as clogging, especially when processing actual wastewater as the substrate. The shift toward volumetric power density reflects an effort to better represent the full operational potential of BES like MFCs, particularly when scaling up for environmental applications. By considering reactor size and other operational factors, volumetric power density offers a more practical and equitable way to evaluate system performance, especially for larger, more complex configurations designed for wastewater treatment and other real-world applications.

4.2 Chemical Density and Chemical Production Route

In BES designed for chemical production and desalination, current density is prioritized overpower density. This is because current density serves as a direct indicator of the substrate oxidation rate and correlates closely with the efficiency of chemical synthesis processes occurring at the cathode. Higher current densities indicate faster electron flow from the substrate to the electrode, which, in turn, results in increased production rates for target chemicals. For instance, in microbial electrolysis cells (MECs), hydrogen gas production can be directly linked to the volumetric current density. The maximum hydrogen production rate, Q_{max} ($m^3\ H_2/m^3$ day), is proportional to the current density. This relationship is critical for systems aimed at chemical production, as increasing the current density enhances the rate at which hydrogen or other chemicals are synthesized. In these BES applications, optimizing current density is essential for maximizing productivity, as it reflects both the electron transfer efficiency and the substrate utilization rate. This is particularly important for industrial-scale applications, where high current densities translate to higher production rates, making the system more efficient and economically viable (Logan et al., 2008).

$$Q_{max} = \frac{I_v\left(\frac{A}{m^3}\right) r_{cat}\left[\left(\frac{1\,C}{s}\right) A\right]\left(0.5\ \frac{mol\ H_2}{mol}\right)\left(86400\ \frac{s}{day}\right)}{F\left(9.65 \times 10^4\ \frac{C}{mol}\right) c_g\left(\frac{mol\ H_2}{L}\right)\left(10^3\ \frac{L}{m^3}\right)} = \frac{43.2 I_v r_{cat}}{F c_g(T)}$$

where:

- I_v (A/m^3) is the current density averaged over a specified time period.
- r_{cat} is the catalytic efficiency or reaction rate constant at the cathode.
- F is Faraday's constant, 9.65×10^4 C/mol.
- $c_g(T)$ is the molar density of gas at a given temperature T (mol/L), typically calculated for standard conditions of 298.15 K and 1 bar.
- 43.2 is a conversion factor that results from the units provided.

The equation demonstrates that the volumetric hydrogen production rate Q_{max} is directly proportional to the current density I_v, meaning that increasing the current density directly leads to an increase in hydrogen production, assuming the efficiency of gas capture is maximized. This highlights the importance of enhancing current density to achieve higher chemical production in BES. As reported by Logan et al. (2008), hydrogen production rates in MECs have reached up to 3.12 m^3 H_2/m^3 day with an applied voltage of 0.8 V, and a maximum current density of $I_v = 186$ A/m^3 was achieved at $V = 0.6$. These findings suggest that optimizing the applied voltage and current density is crucial for maximizing hydrogen output in microbial electrolysis systems.

4.3 CE and EE

Coulombic efficiency (CE) represents the efficiency with which electrons from an electron donor (substrate) are recovered as electrical current through the external circuit in MFCs or BESs. It quantifies how much of the substrate's energy is converted into electrical output, as opposed to being lost through side reactions or microbial metabolism (Logan et al., 2006).

For MFCs operated in **batch mode**, CE can be calculated using the following equation:

$$\mathrm{CE} = \frac{\int_0^t I\mathrm{d}t}{Fb\upsilon\Delta C}$$

where:

- $\int_0^t I\mathrm{d}t$ is the total recovered coulombs (C_p) obtained by integrating the current (I) over time (t).
- F is Faraday's constant (96,485 C/mol).
- b is the number of electrons produced per mole of substrate (e.g., $b = 24$ for glucose, $b = 8$ for acetate).
- υ is the volume of the anolyte (L).
- ΔC is the concentration difference of the substrate (mol/L) from time 0 to t.

For MFCs operated in **continuous flow mode** under steady-state conditions, where the current output is assumed to be constant, the CE equation simplifies to:

$$\mathrm{CE} = \frac{I}{FbQ\Delta C}$$

where:

- I is the constant current (A).
- F is Faraday's constant.
- b is the number of electrons produced per mole of substrate.
- Q is the anolyte flow rate (L/s).
- ΔC is the concentration difference between the influent and effluent (mol/L).

In both batch and continuous modes, CE provides a measure of the fraction of electrons from the substrate that is captured as electrical current, offering insight into the reactor's performance and efficiency.

Energy efficiency (EE) is a crucial metric for evaluating the performance of BESs, particularly MFCs and MECs. It measures the amount of energy recovered from these systems relative to the theoretical energy available from the substrate being consumed.

For Microbial Fuel Cells (MFCs)

In the context of MFCs, EE is defined as the ratio between the electrical energy produced and the combustion energy of the substrate consumed. The equation can be expressed as follows:

$$\mathrm{EE} = \frac{\int_0^t VI\mathrm{d}t}{\Delta H \Delta C Q}$$

where:

- $\int_0^t VI\mathrm{d}t$ represents the total electrical energy produced by integrating the product of voltage (V) and current (I) over time.
- ΔH is the combustion energy of the substrate (J/mol).
- ΔC is the concentration difference of the substrate (mol/L).
- Q is the flow rate of the substrate (L/s).

For Microbial Electrolysis Cells (MECs)

In MECs, where external voltage is applied to drive the electrochemical reactions, EE is calculated based on the energy content of the product (e.g., hydrogen gas) produced, relative to the energy input from both the substrate and the external power source. This can be expressed as:

$$\mathrm{EE} = \frac{E_{\mathrm{product}}}{E_{\mathrm{substrate}} + E_{\mathrm{external}}}$$

where:

- E_{product} is the energy content of the produced hydrogen or other chemicals.
- $E_{\mathrm{substrate}}$ is the energy derived from the substrate consumed.
- E_{external} is the energy input from the external power supply used to facilitate the reaction.

EE provides insight into the effectiveness of BESs in converting the chemical energy of substrates into useful electrical energy or other chemical products. A higher EE indicates a more efficient conversion process, which is vital for the sustainability and viability of bio-electrochemical technologies.

5 Application of BESs

5.1 Wastewater Treatment

BESs have emerged as an innovative alternative to conventional wastewater treatment methods, offering cost-effective solutions by reducing aeration and sludge disposal expenses while simultaneously enabling resource recovery. This dual advantage makes BES a promising technology for both wastewater treatment and environmental sustainability. While anaerobic digestion (AD) has been widely recognized for its efficiency in treating high-strength wastewater, BESs excel in processing low-strength wastewater and operate effectively under lower temperature conditions (Pham et al., 2006; Rozendal et al., 2008). The integration of AD with BES can enhance energy recovery by utilizing AD as the primary treatment stage for handling high-strength wastewater, followed by BES for secondary treatment to further purify effluent and optimize resource recovery. This combination not only improves energy production but also enhances the overall efficiency of wastewater treatment. Beyond conventional wastewater treatment, BESs have been explored for processing a variety of organic substrates. For instance, microbial fuel cells (MFCs) utilizing waste-activated sludge in their anodic compartments have demonstrated significant reductions in organic matter content while simultaneously generating electrical energy (Ghadge et al., 2015). Beyond conventional wastewater treatment, BESs have been explored for processing a variety of organic substrates. For instance, microbial fuel cells (MFCs) utilizing waste-activated sludge in their anodic compartments have demonstrated significant reductions in organic matter content while simultaneously generating electrical energy (Rodríguez Arredondo et al., 2015).

Dual-chamber MFCs, constructed using clayware cylinders, were employed by Jadhav and Ghangrekar for the simultaneous removal of ammonia and organic matter from wastewater Jadhav & Ghangrekar, (2015). The use of dual-chamber MFCs has shown promising results in wastewater treatment, particularly in the simultaneous removal of ammonia and organic pollutants. These systems have exhibited substantial ammonium-nitrogen and chemical oxygen demand (COD) removal efficiencies, highlighting their capability to manage wastewater with high organic loads and ammonia concentrations. Additionally, the successful incorporation of anaerobic ammonium oxidation (anammox) processes in MFCs has demonstrated the potential for ammonium oxidation while concurrently generating electricity. For enhanced removal of organic matter in wastewater, a two-stage process involving fermentation by *Aspergillus awamori* followed by MFC treatment has been investigated (Ghosh Ray & Ghangrekar, 2015). his approach has achieved high COD removal efficiencies and significant reductions in suspended solids, making it a viable method for industrial wastewater treatment. Additionally, sediment microbial fuel cells (SMFCs) have been employed for in situ remediation of aquaculture pond sediments, showcasing their ability to improve water quality while harnessing bio-electrochemical energy (Sajana et al., 2014). These studies suggest that BES can replace conventional activated sludge processes after anaerobic digestion, achieving effluent quality that meets discharge standards without requiring the energy-intensive aeration typically needed in activated sludge processes.

5.2 Recovery of Metals

Recent advancements in BESs present an innovative approach to metal recovery, providing a versatile platform capable of facilitating both oxidation and reduction reactions within a single system (Wang & Ren, 2014). Traditionally, bio-electrochemical metal recovery involves utilizing metals as electron acceptors in the cathodic chamber, while organic waste serves as the electron donor in the anodic chamber (Mathuriya & Yakhmi, 2014). However, BES offers a unique advantage in that the treatment of heavy metal-contaminated wastewater can occur in both the anode and cathode chambers. For instance, metals such as vanadium and arsenic have been effectively removed in the anodic chamber of MFCs (Li et al., 2009; Mathuriya & Yakhmi, 2014).

There are four primary mechanisms by which metal recovery can occur at the cathode:

1. **Direct metal recovery using abiotic cathodes**: In this method, metals with high redox potentials, such as gold, silver, vanadium, chromium, copper, and iron, act as electron acceptors at the cathode. These metals undergo reduction and precipitate onto the abiotic cathode without the need for external energy input (Mathuriya & Yakhmi, 2014)
2. **Metal recovery using abiotic cathodes with external power sources**: For metals with lower redox potentials than the potential of the BES anode, such as nickel, direct electron transfer from the cathode is not energetically favorable. In such cases, an external power source is applied to drive electrons from the higher potential anode to the lower potential cathode, facilitating the reduction of these metals (Wang & Ren, 2014)
3. **Metal recovery using biocathodes**: In this process, certain metal-reducing bacteria facilitate the reduction of metals by using them as external electron acceptors. These microorganisms, through dissimilatory metal reduction, convert more toxic metal forms into less toxic ones. For example, *Trichococcus pasteurii* and *Pseudomonas aeruginosa* have reduced hexavalent chromium Cr(VI) to the less toxic trivalent form Cr(III) (Holden, 2003; Tandukar et al., 2009).
4. **Metal recovery using biocathodes with external power sources**: For metals with low redox potentials, the recovery process can be enhanced by applying external power to biocathodes. In such cases, direct electrochemical reduction may not occur; however, metals can be removed from the solution and captured by biofilms on the electrodes. Microorganisms within these biofilms facilitate metal ion reduction through microbial respiration, enabling efficient metal recovery (Gregory & Lovley, 2005).

These mechanisms enable the recovery of valuable metals from wastewater and contaminated environments, offering a sustainable and energy-efficient alternative to conventional methods. By leveraging BES's flexibility, metal recovery processes can be tailored to accommodate various types of metals and wastewater conditions, contributing to both environmental remediation and resource recovery.

5.3 Synthesis of Value-Added Product

BESs involve the microbial catalysis of chemical reactions within an electrochemical cell (Rabaey & Rozendal, 2010). These systems utilize two key pathways for microbial synthesis: one that relies on CO_2 as a substrate and another that originates from organic compounds. Methane is a frequent by-product of anaerobic fermentation. In BESs, electrons and CO_2 generated through microbial oxidation of organic matter at the anode can be directed to the cathode, where methane is produced via the following reaction:

$$CO_2 + 8H^+ + 8e^- \rightarrow CH_4 + H_2O$$

Initially, methane production was considered an unwanted side reaction during hydrogen generation in microbial electrolysis cells (MECs). However, methane and hydrogen differ in their energy yields. The energy content of hydrogen and methane, based on Gibbs free energy changes, is 4.12 and 3.52 kWh kg^{-1} of COD equivalents, respectively. While hydrogen is preferred for energy recovery due to its higher energy content, its conversion to methane through methanogenesis leads to an approximate thermodynamic energy loss of 15%.

Methane formation is an unavoidable aspect of biological hydrogen production, posing challenges in achieving high-purity hydrogen. The purification process is energy-intensive, leading to higher production costs and limiting its direct use as a fuel. Conversely, methane generation through bio-electrochemical systems (BES) presents advantages over conventional biogas production. A significant benefit is the decoupling of organic matter oxidation from methane synthesis, eliminating the need to shield methanogenic microbes from inhibitory substances present in waste streams. This separation facilitates the efficient storage and transportation of methane, enhancing the overall sustainability and viability of the process for energy recovery (Cheng et al., 2009; Rabaey & Rozendal, 2010).

Under standard conditions, the theoretical voltage required for the reduction of CO_2 to CH_4 is approximately −0.224 V. However, in practical applications, more negative potentials are required due to significant overpotentials associated with the process. The ability of microorganisms to catalyze the reduction using an electrode as a direct electron donor was reported by Cheng et al. (2009). In this study, CH_4 was generated at cathode potentials more negative than −0.7 V (vs. Ag/AgCl), with a CH_4 production efficiency reaching 96% at a potential of −1.0 V (vs. Ag/AgCl), corresponding to a production rate of 200 mmol CH_4 day^{-1} m^{-2}. In another study by Jiang et al., methane production was reported at a rate of 17.3 mL h^{-1} L^{-1}, with a current capture efficiency of 94%.

A major challenge in utilizing CO_2 as an electron acceptor is its high electron demand for synthesizing organic compounds. Although the theoretical reduction potentials of butyrate to butanol (−0.37 V vs. SHE) and CO_2 to butanol (−0.30 V vs. SHE) are comparable, their electron requirements differ significantly. The transformation of butyrate into butanol requires only four electrons, whereas converting CO_2 to butanol demands 24 electrons, leading to a sixfold increase in current demand and a substantially higher energy requirement. Moreover, CO_2 conversion to butanol likely involves multiple intermediate steps, each contributing to efficiency losses. Additionally, CO_2 is typically sourced from atmospheric emissions or industrial waste gases, such as those from coal power plants, which can limit reaction kinetics or create geographical constraints on production. These factors add complexity to the practical application of BESs for CO_2 reduction, making efficient and scalable production of methane or butanol more difficult to achieve (Rabaey & Rozendal, 2010).

Recent progress in BES technology has enabled the production of valuable chemicals, including ethanol from acetate, butanol from butyrate, and succinate from butanol (Park et al., 1999; Steinbusch et al., 2008, 2010). Several studies have reported ethanol synthesis from acetate, along with butanol production from butyrate and succinate generation from butanol. Additionally, BESs show promise for synthesizing high-value compounds. One such example is the transformation of 6-bromo-2-tetralone into 6-bromo-2-tetralol, an intermediate used in the production of MK-0499, a potassium channel blocker under consideration for achiral drug development (Peguin et al., 1994). This process benefits from improved reaction kinetics in BES; however, its economic viability is still under evaluation. For high-value compounds, the costs associated with energy and chemical inputs are relatively minor compared to overall production expenses, and the advantages of redox control or electron supply via a cathode over traditional methods remain under investigation.

Biocatalyzed cathodes in BES can be implemented through three distinct configurations. The first involves both the anode and cathode being biologically active, where organic matter in wastewater powers the anode while microbial electrosynthesis occurs at the cathode. The second and third setups utilize a chemically driven anode, which is either directly connected to a biocathode or indirectly supports biological reactions at the cathode. These configurations offer versatility in optimizing BES systems for diverse bio-electrochemical applications (Clauwaert et al., 2008b; Rabaey & Rozendal, 2010).

6 Current Challenges of BES

Over the past two decades, microbial fuel cells (MFCs) have made significant advancements; however, several obstacles continue to hinder their large-scale commercialization. The primary challenges include limited power output, high costs of electrode materials and membrane separators, scalability issues, and inefficiencies in energy conversion. The operational mechanism of an MFC is comparable to that of a voltage source with inherent internal resistance, which constrains its overall power generation. As electrical current flows through an external circuit, voltage losses occur due to internal resistances, causing the actual output voltage to drop below the theoretical maximum. These losses can be classified into four main types: activation losses (η_{act}), which arise from reaction kinetics; the concentration losses (η_{conc}), resulting from mass transport limitations; ohmic losses (η_{ohm}), due to resistance in the systems; and Coulombic losses, associated with the electron transfer inefficiencies (Rinaldi et al., 2008). In MFCs, activation losses occur because energy is required to drive electrochemical reactions. These losses are mainly due to slow microbial electron transfer processes and sluggish oxygen reduction reactions, leading to high overpotential. Concentration losses take place when reactants, particularly oxygen at the cathode, are consumed faster than they can be replenished, thereby restricting power generation. Ohmic losses arise from resistance to both electron and ion transport within the system, which is influenced by factors such as poor electrode conductivity, external circuit resistance, and ionic resistance in the electrolyte or separator, all contributing to voltage reductions. Coulombic losses occur when electrons are diverted to non-electricity-generating reactions, such as methanogenesis, or when oxidants migrate to the anode, further diminishing efficiency. To overcome these challenges, employing advanced electrode materials, optimizing system design, and enhancing microbial electron transfer mechanisms are essential strategies for improving the performance and commercial viability of MFCs.

BESs hold great potential for sustainable bioelectricity generation; however, they face significant challenges that hinder large-scale implementation. Despite notable advancements, these systems remain primarily confined to laboratory research and small-scale demonstration projects. Key limitations include low efficiency, reduced production rates, and difficulties in optimizing essential components such as fuel substrates, electrode materials, and microbial communities. A wide variety of substrates can be used in BESs, but their effectiveness depends on microbial nutrient availability. The cost and accessibility of these substrates also influence the overall feasibility of the technology. Commonly utilized substrates include simple and complex sugars, cellulose, starch, and chemically derived materials. Additionally, landfill leachate and wastewater have been explored as alternative feedstocks, though they may pose environmental and health risks. The use of food-based substrates remains controversial due to ethical concerns surrounding resource allocation. Addressing these challenges requires innovative strategies for substrate selection, system optimization, and material advancements.

In bio-electrochemical systems (BESs), electrode materials play a crucial role in determining overall performance. Ideal materials should exhibit high electrical conductivity to facilitate efficient electron transfer, along with substantial surface area and porosity to support microbial adhesion and biofilm formation. Additionally, mechanical strength, scalability, biocompatibility, and chemical stability are essential for long-term operation. However, commonly used materials such as graphite, platinum black, and platinum (Pt) offer excellent conductivity but are often costly, limiting their large-scale application and economic viability. The significant challenge in BESs is the use of membranes to separate electrode chambers. While membranes help maintain ion exchange, they also introduce several drawbacks, including reduced proton permeability, substrate loss, and oxygen back diffusion. These issues contribute to increased internal resistance, pH imbalances, and biofouling, all of which hinder system efficiency. The complexity of membrane maintenance further complicates scalability, making the development of alternative strategies a key focus in BES research.

Scaling up bio-electrochemical systems (BESs) remains a significant challenge, as most studies have been conducted at laboratory scale with small anodic chamber volumes, typically ranging from milliliters to liters. The power densities achieved in these small-scale setups are promising, with reports of up to 2.15 kW/m^3 from a 0.335 mL anodic chamber and 1.55 kW/m^3 from a 2.5 mL chamber. However, translating these power outputs to larger systems is complex due to increased internal resistance, mass transfer limitations, and challenges in maintaining efficient electron transfer (Fan et al., 2007; Nevin et al., 2008). To make MFCs viable for large-scale applications such as wastewater treatment plants, achieving high power densities in larger systems is essential. While MFCs hold promise for simultaneous wastewater treatment and energy recovery, limited studies have focused on scaling up these systems. Existing research suggests that large-scale implementation is feasible; however, significant improvements are required, particularly in enhancing power output through cost-effective materials. Developing low-cost electrode catalysts capable of sustaining high current densities (greater than 5 A/m^2) is crucial for improving performance. Additionally, innovative reactor designs are necessary to minimize energy losses and accommodate higher anodic chamber volumes suitable for industrial applications. Microbial electron transfer mechanisms present another critical challenge in bio-electrochemical systems (BESs). These mechanisms vary among bacterial species, making microbial selection a key factor in optimizing performance. While mixed bacterial cultures can enhance substrate utilization and system resilience, they may also reduce power output due to competing metabolic pathways. The chemical complexity of biological waste requires a diverse microbial community for effective organic matter degradation. Moreover, temperature fluctuations significantly impact BES efficiency, as microbial activity declines at lower temperatures, reducing reaction rates and overall system performance. Addressing these challenges requires advancements in microbial selection, system design, and material engineering to improve scalability and operational stability.

7 Future Prospects and Conclusion

MESs offer significant potential for a wide range of sustainable applications, particularly in the production of renewable chemicals and fuels, as well as in environmental remediation. As a recent development in BESs, MES has attracted considerable interest due to its ability to integrate microbial catalysis with electrochemical processes. However, several challenges remain to be addressed before the full potential of MES can be realized, particularly in terms of scalability, electron transfer efficiency, and optimization of the biocatalytic processes. One of the key prospects of MES lies in its scalability, which remains a major challenge. The reproducibility of laboratory-scale performance in large-scale industrial systems is critical for making MES economically viable. Upscaling has proven difficult, especially due to issues related to system design, overpotentials, and the optimization of bio-electrochemical interfaces. Future advancements must focus on improving electrode designs, optimizing the spatial arrangement of cells, and addressing mass and charge transfer limitations. These improvements will help bridge the gap between bench-scale experimentation and real-world applications, making MES more practical and cost-effective. The potential of MES to be integrated into existing infrastructure, such as solid waste and wastewater treatment plants, also holds great promise. The ability to treat waste while simultaneously producing valuable byproducts, such as hydrogen or other chemicals, is a major advantage of this technology. While MFCs have demonstrated electricity generation from wastewater, MECs are emerging as a more feasible solution for simultaneous waste treatment and hydrogen production. MECs could play a vital role in advancing MES applications for industrial and environmental purposes.

Another area of focus for future MES research is the electron transfer mechanisms and the behavior of microbial biofilms, particularly on the cathode. In anodic MFC mechanisms, mixed microbial cultures are effective biocatalysts when using wastewater as a substrate. However, much less is known about the behavior of cathodic biofilms, planktonic cells, and the mechanisms of electron transfer in MES. A deeper understanding of the interface between biofilms and planktonic cells is necessary, as well as the impact of microbial metabolism on electron transfer efficiency. For example, mixed and pure homo-acetogenic cultures have shown potential in CO_2 reduction processes, but the electron transfer pathways and mechanisms in these systems require further study. The selection of the biocatalyst in MES is another critical factor that influences system performance. The choice of biocatalyst depends not only on the microbial species but also on the terminal electron acceptors and the specific metabolic pathways involved. Both intracellular and extracellular factors, such as enzyme activity, membrane transport processes, and the presence of redox mediators, can significantly impact the metabolic rates and, consequently, the overall efficiency of MES. Further research is required to fully understand how these factors interact in bio-electrochemical systems and to develop strategies for optimizing biocatalyst performance.

A particularly exciting development in the field is the combination of BES technology with microbial fermentation, known as electro-fermentation (EF). EF uses electrical energy as a source of reducing power for microbial fermentation, enabling the generation of valuable products through controlled microbial metabolism. This integration of MES with fermentation opens new avenues for industrial biotechnology, offering a versatile approach to the production of chemicals, fuels, and other high-value products. The use of electricity to drive microbial processes in EF can enhance the efficiency of fermentation, reduce reliance on expensive chemical inputs, and allow for more precise control of metabolic pathways. Researchers are actively exploring the potential of EF to overcome the biological, electrochemical, and logistical challenges associated with BES technologies. From a technological perspective, the integration of MES with EF has the potential to overcome several challenges. EF can address the bottlenecks of electron transfer and metabolic control by leveraging electricity as a modulator of microbial activity. This integration could enable more efficient production processes for bio-based chemicals, further enhancing the versatility of MES technologies. In addition, EF may help reduce the overall energy input required for these processes, making them more sustainable and economically viable. Combining BES and EF technologies could also improve the range of products that can be generated, including biofuels, pharmaceuticals, and specialty chemicals.

References

Aelterman, P., Rabaey, K., Pham, H. T., Boon, N., & Verstraete, W. (2006). Continuous electricity generation at high voltages and currents using stacked microbial fuel cells. *Environmental Science and Technology, 40*, 3388–3394. https://doi.org/10.1021/es0525511

Bajracharya, S., ter Heijne, A., Dominguez Benetton, X., Vanbroekhoven, K., Buisman, C. J., Strik, D. P., & Pant, D. (2015). Carbon dioxide reduction by mixed and pure cultures in microbial electrosynthesis using an assembly of graphite felt and stainless steel as a cathode. *Bioresource Technology, 195*, 14–24. https://doi.org/10.1016/j.biortech.2015.05.081

Calabrese Barton, S., Gallaway, J., & Atanassov, P. (2004). Enzymatic biofuel cells for implantable and microscale devices. *Chemical Reviews, 104*, 4867–4886. https://doi.org/10.1021/cr020719k

Cao, X., Huang, X., Liang, P., Xiao, K., Zhou, Y., Zhang, X., & Logan, B. E. (2009). A new method for water desalination using microbial desalination cells. *Environmental Science and Technology, 43*, 7148–7152. https://doi.org/10.1021/es901950j

Chae, K.-J., Choi, M.-J., Kim, K.-Y., Ajayi, F. F., Chang, I. S., & Kim, I. S. (2009). A solar-powered microbial electrolysis cell with a platinum catalyst-free cathode to produce hydrogen. *Environmental Science and Technology, 43*, 9525–9530. https://doi.org/10.1021/es9022317

Chen, S., He, G., Liu, Q., Harnisch, F., Zhou, Y., Chen, Y., Hanif, M., Wang, S., Peng, X., Hou, H., & Schröder, U. (2012). Layered corrugated electrode macrostructures boost microbial bioelectrocatalysis. *Energy & Environmental Science, 5*, 9769. https://doi.org/10.1039/c2ee23344d

Chen, S., Hou, H., Harnisch, F., Patil, S. A., Carmona-Martinez, A. A., Agarwal, S., Zhang, Y., Sinha-Ray, S., Yarin, A. L., Greiner, A., & Schröder, U. (2011). Electrospun and solution blown three-dimensional carbon fiber nonwovens for application as electrodes in microbial fuel cells. *Energy & Environmental Science, 4*, 1417. https://doi.org/10.1039/c0ee00446d

Cheng, S., Xing, D., Call, D. F., & Logan, B. E. (2009). Direct biological conversion of electrical current into methane by electromethanogenesis. *Environmental Science and Technology, 43*, 3953–3958. https://doi.org/10.1021/es803531g

Chiranjeevi, P., Mohanakrishna, G., & Venkata Mohan, S. (2012). Rhizosphere mediated electrogenesis with the function of anode placement for harnessing bioenergy through CO_2 sequestration. *Bioresource Technology, 124*, 364–370. https://doi.org/10.1016/j.biortech.2012.08.020

Clauwaert, P., Aelterman, P., Pham, T. H., De Schamphelaire, L., Carballa, M., Rabaey, K., & Verstraete, W. (2008a). Minimizing losses in bio-electrochemical systems: The road to applications. *Applied Microbiology and Biotechnology, 79*, 901–913. https://doi.org/10.1007/s00253-008-1522-2

Clauwaert, P., Tolêdo, R., van der Ha, D., Crab, R., Verstraete, W., Hu, H., Udert, K. M., & Rabaey, K. (2008b). Combining biocatalyzed electrolysis with anaerobic digestion. *Water Science and Technology, 57*, 575–579. https://doi.org/10.2166/wst.2008.084

Cui, S., Li, M., Hassan, R. Y. A., Heintz-Buschart, A., Wang, J., & Bilitewski, U. (2020). Inhibition of respiration of Candida albicans by small molecules increases phagocytosis efficacy by macrophages. mSphere, *5*. https://doi.org/10.1128/mSphere.00016-20

Davis, J.B., & Yarbrough, H. F. (1962). Preliminary experiments on a microbial fuel cell. *Science (80-), 137*, 615–616. https://doi.org/10.1126/science.137.3530.615

De Schamphelaire, L., Cabezas, A., Marzorati, M., Friedrich, M. W., Boon, N., & Verstraete, W. (2010). Microbial community analysis of anodes from sediment microbial fuel cells powered by rhizodeposits of living rice plants. *Applied and Environment Microbiology, 76*, 2002–2008. https://doi.org/10.1128/AEM.02432-09

De Schamphelaire, L., & Verstraete, W. (2009). Revival of the biological sunlight-to-biogas energy conversion system. *Biotechnology and Bioengineering, 103*, 296–304. https://doi.org/10.1002/bit.22257

Desloover, J., Arends, J. B. A., Hennebel, T., & Rabaey, K. (2012). Operational and technical considerations for microbial electrosynthesis. *Biochemical Society Transactions, 40*, 1233–1238. https://doi.org/10.1042/BST20120111

Dewan, A., Beyenal, H., & Lewandowski, Z. (2008). Scaling up microbial fuel cells. *Environmental Science and Technology, 42*, 7643–7648. https://doi.org/10.1021/es800775d

Dumas, C., Basseguy, R., & Bergel, A. (2008). DSA to grow electrochemically active biofilms of Geobacter sulfurreducens. *Electrochimica Acta, 53*, 3200–3209. https://doi.org/10.1016/j.electacta.2007.10.066

ElMekawy, A., Hegab, H. M., & Pant, D. (2014a). The near-future integration of microbial desalination cells with reverse osmosis technology. *Energy & Environmental Science, 7*, 3921–3933. https://doi.org/10.1039/C4EE02208D

ElMekawy, A., Hegab, H. M., Vanbroekhoven, K., & Pant, D. (2014b). Techno-productive potential of photosynthetic microbial fuel cells through different configurations. *Renewable and Sustainable Energy Reviews, 39*, 617–627. https://doi.org/10.1016/j.rser.2014.07.116

ElMekawy, A., Srikanth, S., Bajracharya, S., Hegab, H. M., Nigam, P. S., Singh, A., Mohan, S. V., & Pant, D. (2015). Food and agricultural wastes as substrates for bioelectrochemical system (BES): The synchronized recovery of sustainable energy and waste treatment. *Food Research International, 73*, 213–225. https://doi.org/10.1016/j.foodres.2014.11.045

Fan, Y., Hu, H., & Liu, H. (2007). Sustainable power generation in microbial fuel cells using bicarbonate buffer and proton transfer mechanisms. *Environmental Science and Technology, 41*, 8154–8158. https://doi.org/10.1021/es071739c

Fan, Y., Sharbrough, E., & Liu, H. (2008). Quantification of the internal resistance distribution of microbial fuel cells. *Environmental Science and Technology, 42*, 8101–8107. https://doi.org/10.1021/es801229j

Feng, Y., Yang, Q., Wang, X., & Logan, B. E. (2010). Treatment of carbon fiber brush anodes for improving power generation in air-cathode microbial fuel cells. *Journal of Power Sources, 195*, 1841–1844. https://doi.org/10.1016/j.jpowsour.2009.10.030

Gellett, W., Kesmez, M., Schumacher, J., Akers, N., & Minteer, S. D. (2010). Biofuel cells for portable power. *Electroanalysis, 22*, 727–731. https://doi.org/10.1002/elan.200980013

Ghadge, A. N., Jadhav, D. A., Pradhan, H., & Ghangrekar, M. M. (2015). Enhancing waste activated sludge digestion and power production using hypochlorite as catholyte in clayware microbial fuel cell. *Bioresource Technology, 182*, 225–231. https://doi.org/10.1016/j.biortech.2015.02.004

Ghosh Ray, S., & Ghangrekar, M. M. (2015). Enhancing organic matter removal, biopolymer recovery and electricity generation from distillery wastewater by combining fungal fermentation and microbial fuel cell. *Bioresource Technology, 176*, 8–14. https://doi.org/10.1016/j.biortech.2014.10.158

Gregory, K. B., & Lovley, D. R. (2005). Remediation and recovery of uranium from contaminated subsurface environments with electrodes. *Environmental Science and Technology, 39*, 8943–8947. https://doi.org/10.1021/es050457e

Harnisch, F., & Schröder, U. (2010). From MFC to MXC: Chemical and biological cathodes and their potential for microbial bioelectrochemical systems. *Chemical Society Reviews, 39*, 4433. https://doi.org/10.1039/c003068f

Hassan, Q., Viktor, P., Al-Musawi, T. J., Ali, B. M., Algburi, S., Alzoubi, H. M., Al-Jiboory, A. K., Sameen, A. Z., Salman, H. M., & Jaszczur, M. (2024). The renewable energy role in the global energy Transformations. *Renewable Energy Focus, 48*, 100545. https://doi.org/10.1016/j.ref.2024.100545

Hassan, R. Y. A., & Bilitewski, U. (2011). A viability assay for Candida albicans based on the electron transfer mediator 2,6-dichlorophenolindophenol. *Analytical Biochemistry, 419*, 26–32. https://doi.org/10.1016/j.ab.2011.07.025

Hassan, R. Y. A., Febbraio, F., & Andreescu, S. (2021). Microbial electrochemical systems: principles, construction and biosensing applications. *Sensors, 21*, 1279. https://doi.org/10.3390/s21041279

Hassan, R. Y. A., & Wollenberger, U. (2019). Direct determination of bacterial cell viability using carbon nanotubes modified screen-printed electrodes. *Electroanalysis, 31*, 1112–1117. https://doi.org/10.1002/elan.201900047

He, J., Sung, Y., Dollhopf, M. E., Fathepure, B. Z., Tiedje, J. M., & Löffler, F. E. (2002). Acetate versus hydrogen as direct electron donors to stimulate the microbial reductive dechlorination process at chloroethene-contaminated sites. *Environmental Science and Technology, 36*, 3945–3952. https://doi.org/10.1021/es025528d

Helder, M., Strik, D. P., Hamelers, H. V., & Buisman, C. J. (2012). The flat-plate plant-microbial fuel cell: The effect of a new design on internal resistances. *Biotechnology for Biofuels, 5*, 70. https://doi.org/10.1186/1754-6834-5-70

Hernandez, C. A., & Osma, J. F. (2020). Microbial electrochemical systems: deriving future trends from historical perspectives and characterization strategies. *Frontiers in Environmental Science, 8*. https://doi.org/10.3389/fenvs.2020.00044

Holden, J. (2003). Microbe–metal interactions in marine hydrothermal environments. *Current Opinion in Chemical Biology, 7*, 160–165. https://doi.org/10.1016/S1367-5931(03)00026-7

Hu, Y., Yang, Y., Katz, E., & Song, H. (2015). Programming the quorum sensing-based AND gate in Shewanella oneidensis for logic gated-microbial fuel cells. *Chemical Communications, 51*, 4184–4187. https://doi.org/10.1039/C5CC00026B

Ivase, T. J.-P., Nyakuma, B. B., Oladokun, O., Abu, P. T., & Hassan, M. N. (2020). Review of the principal mechanisms, prospects, and challenges of bioelectrochemical systems. *Environmental Progress & Sustainable Energy, 39*. https://doi.org/10.1002/ep.13298

Jacobson, K. S., Drew, D. M., & He, Z. (2011). Use of a liter-scale microbial desalination cell as a platform to study bioelectrochemical desalination with salt solution or artificial seawater. *Environmental Science and Technology, 45*, 4652–4657. https://doi.org/10.1021/es200127p

Jadhav, D. A., & Ghangrekar, M. M. (2015). Effective ammonium removal by anaerobic oxidation in microbial fuel cells. *Environmental Technology, 36*, 767–775. https://doi.org/10.1080/09593330.2014.960481

Jeremiasse, A. W., Hamelers, H. V. M., & Buisman, C. J. N. (2010). Microbial electrolysis cell with a microbial biocathode. *Bioelectrochemistry, 78*, 39–43. https://doi.org/10.1016/j.bioelechem.2009.05.005

Kamitaka, Y., Tsujimura, S., Setoyama, N., Kajino, T., & Kano, K. (2007). Fructose/dioxygen biofuel cell based on direct electron transfer-type bioelectrocatalysis. *Physical Chemistry Chemical Physics: PCCP, 9*, 1793. https://doi.org/10.1039/b617650j

Kiely, P. D., Regan, J. M., & Logan, B. E. (2011). The electric picnic: Synergistic requirements for exoelectrogenic microbial communities. *Current Opinion in Biotechnology, 22*, 378–385. https://doi.org/10.1016/j.copbio.2011.03.003

Kracke, F., Vassilev, I., KrÃmer, & J. O. (2015a). Microbial electron transport and energy conservation—The foundation for optimizing bioelectrochemical systems. *Frontiers in Microbiology, 6*. https://doi.org/10.3389/fmicb.2015.00575

Kracke, F., Vassilev, I., & KrÃmer, J. O. (2015b). Microbial electron transport and energy conservation—The foundation for optimizing bioelectrochemical systems. *Frontiers in Microbiology, 6*, 1707–1714. https://doi.org/10.3389/fmicb.2015.00575

Kwon, J., Cho, E.-M., Nandhakumar, P., Yang, S. I., & Yang, H. (2018). Rapid and sensitive detection of Aspergillus niger using a single-mediator system combined with redox cycling. *Analytical Chemistry, 90*, 13491–13497. https://doi.org/10.1021/acs.analchem.8b03417

Li, H., Feng, Y., Zou, X., & Luo, X. (2009). Study on microbial reduction of vanadium matallurgical waste water. *Hydrometallurgy, 99*, 13–17. https://doi.org/10.1016/j.hydromet.2009.05.019

Logan, B. E. (2009). Exoelectrogenic bacteria that power microbial fuel cells. *Nature Reviews Microbiology, 7*, 375–381. https://doi.org/10.1038/nrmicro2113

Logan, B. E., Call, D., Cheng, S., Hamelers, H. V., Sleutels, T. H., Jeremiasse, A. W., & Rozendal, R. A. (2008). Microbial electrolysis cells for high yield hydrogen gas production from organic matter. *Environmental Science and Technology, 42*, 8630–8640. https://doi.org/10.1021/es801553z

Logan, B., Cheng, S., Watson, V., & Estadt, G. (2007). Graphite fiber brush anodes for increased power production in air-cathode microbial fuel cells. *Environmental Science and Technology, 41*, 3341–3346. https://doi.org/10.1021/es062644y

Logan, B. E., Hamelers, B., Rozendal, R., Schröder, U., Keller, J., Freguia, S., Aelterman, P., Verstraete, W., & Rabaey, K. (2006). Microbial fuel cells: Methodology and technology. *Environmental Science and Technology, 40*, 5181–5192. https://doi.org/10.1021/es0605016

Logan, B. E., Rossi, R., Ragab, A., & Saikaly, P. E. (2019). Electroactive microorganisms in bioelectrochemical systems. *Nature Reviews Microbiology, 17*, 307–319. https://doi.org/10.1038/s41579-019-0173-x

Luo, H., Jenkins, P. E., & Ren, Z. (2011). Concurrent desalination and hydrogen generation using microbial electrolysis and desalination cells. *Environmental Science and Technology, 45*, 340–344. https://doi.org/10.1021/es1022202

Luo, H., Xu, P., Roane, T. M., Jenkins, P. E., & Ren, Z. (2012). Microbial desalination cells for improved performance in wastewater treatment, electricity production, and desalination. *Bioresource Technology, 105*, 60–66. https://doi.org/10.1016/j.biortech.2011.11.098

Luo, S., Sun, H., Ping, Q., Jin, R., & He, Z. (2016). A review of modeling bioelectrochemical systems: Engineering and statistical aspects. *Energies, 9*, 111. https://doi.org/10.3390/en9020111

Lyon, D. Y., Buret, F., Vogel, T. M., & Monier, J.-M. (2010). Is resistance futile? Changing external resistance does not improve microbial fuel cell performance. *Bioelectrochemistry, 78*, 2–7. https://doi.org/10.1016/j.bioelechem.2009.09.001

Madiraju, K. S., Lyew, D., Kok, R., & Raghavan, V. (2012). Carbon neutral electricity production by Synechocystis sp. PCC6803 in a microbial fuel cell. *Bioresource Technology, 110*, 214–218. https://doi.org/10.1016/j.biortech.2012.01.065

Marshall, C. W., Ross, D. E., Fichot, E. B., Norman, R. S., & May, H. D. (2012). Electrosynthesis of commodity chemicals by an autotrophic microbial community. *Applied and Environment Microbiology, 78*, 8412–8420. https://doi.org/10.1128/AEM.02401-12

Mathuriya, A. S., & Yakhmi, J. V. (2014). Microbial fuel cells to recover heavy metals. *Environmental Chemistry Letters, 12*, 483–494. https://doi.org/10.1007/s10311-014-0474-2

Matsumae, Y., Takahashi, Y., Ino, K., Shiku, H., & Matsue, T. (2014). Electrochemical monitoring of intracellular enzyme activity of single living mammalian cells by using a double-mediator system. *Analytica Chimica Acta, 842*, 20–26. https://doi.org/10.1016/j.aca.2014.06.047

Mehanna, M., Kiely, P. D., Call, D. F., & Logan, B. E. (2010). Microbial electrodialysis cell for simultaneous water desalination and hydrogen gas production. *Environmental Science and Technology, 44*, 9578–9583. https://doi.org/10.1021/es1025646

Minteer, S. D., Liaw, B. Y., & Cooney, M. J. (2007). Enzyme-based biofuel cells. *Current Opinion in Biotechnology, 18*, 228–234. https://doi.org/10.1016/j.copbio.2007.03.007

Mohanakrishna, G., Seelam, J. S., Vanbroekhoven, K., & Pant, D. (2015a). An enriched electroactive homoacetogenic biocathode for the microbial electrosynthesis of acetate through carbon dioxide reduction. *Faraday Discussions, 183*, 445–462. https://doi.org/10.1039/C5FD00041F

Mohanakrishna, G., Srikanth, S., & Pant, D. (2015b). Bioelectrochemical systems (BES) for microbial electroremediation: An advanced wastewater treatment technology. In *Applied environmental biotechnology: Present scenario and future trends* (pp. 145–167). Springer India.

Mohanakrishna, G., Krishna Mohan, S., & Venkata Mohan, S. (2012). Carbon based nanotubes and nanopowder as impregnated electrode structures for enhanced power generation: Evaluation with real field wastewater. *Applied Energy, 95*, 31–37. https://doi.org/10.1016/j.apenergy.2012.01.058

Mohanakrishna, G., Venkata Mohan, S., & Sarma, P. N. (2010). Bio-electrochemical treatment of distillery wastewater in microbial fuel cell facilitating decolorization and desalination along with power generation. *Journal of Hazardous Materials, 177*, 487–494. https://doi.org/10.1016/j.jhazmat.2009.12.059

Myers, C. R., & Myers, J. M. (1992). Localization of cytochromes to the outer membrane of anaerobically grown Shewanella putrefaciens MR-1. *Journal of Bacteriology, 174*, 3429–3438. https://doi.org/10.1128/jb.174.11.3429-3438.1992

Nevin, K. P., Woodard, T. L., Franks, A. E., Summers, Z. M., & Lovley, D. R. (2010). Microbial electrosynthesis: Feeding microbes electricity to convert carbon dioxide and water to multicarbon extracellular organic compounds. *MBio*, 1. https://doi.org/10.1128/mBio.00103-10

Nevin, K. P., Richter, H., Covalla, S. F., Johnson, J. P., Woodard, T. L., Orloff, A. L., Jia, H., Zhang, M., & Lovley, D. R. (2008). Power output and columbic efficiencies from biofilms of Geobacter sulfurreducens comparable to mixed community microbial fuel cells. *Environmental Microbiology, 10*, 2505–2514. https://doi.org/10.1111/j.1462-2920.2008.01675.x

Ng, I.-S., Hsueh, C.-C., & Chen, B.-Y. (2017). Electron transport phenomena of electroactive bacteria in microbial fuel cells: A review of Proteus hauseri. *Bioresour Bioprocess, 4*, 53. https://doi.org/10.1186/s40643-017-0183-3

Oliot, M., Galier, S., Roux de Balmann, H., & Bergel, A. (2016). Ion transport in microbial fuel cells: Key roles, theory and critical review. *Applied Energy, 183*, 1682–1704. https://doi.org/10.1016/j.apenergy.2016.09.043

Palmore, G. T. R., & Whitesides, G. M. (1994). *Microbial and enzymatic biofuel cells* (pp. 271–290).

Pant, D., Van Bogaert, G., Diels, L., & Vanbroekhoven, K. (2010). A review of the substrates used in microbial fuel cells (MFCs) for sustainable energy production. *Bioresource Technology, 101*, 1533–1543. https://doi.org/10.1016/j.biortech.2009.10.017

Park, D. H., Laivenieks, M., Guettler, M. V., Jain, M. K., & Zeikus, J. (1999). Microbial utilization of electrically reduced neutral red as the sole electron donor for growth and metabolite production. *Applied and Environment Microbiology, 65*, 2912–2917. https://doi.org/10.1128/AEM.65.7.2912-2917.1999

Patil, S. A., Gildemyn, S., Pant, D., Zengler, K., Logan, B. E., & Rabaey, K. (2015). A logical data representation framework for electricity-driven bioproduction processes. *Biotechnology Advances, 33*, 736–744. https://doi.org/10.1016/j.biotechadv.2015.03.002

Peguin, S., Delorme, P., Goma, G., & Soucaille, P. (1994). Enhanced alcohol yields in batch cultures of Clostridium acetobutylicum using a three-electrode potentiometric system with methyl viologen as electron carrier. *Biotechnology Letters, 16*, 269–274. https://doi.org/10.1007/BF00134624

Pham, T. H., Aelterman, P., & Verstraete, W. (2009). Bioanode performance in bioelectrochemical systems: Recent improvements and prospects. *Trends in Biotechnology, 27*, 168–178. https://doi.org/10.1016/j.tibtech.2008.11.005

Pham, T. H., Rabaey, K., Aelterman, P., Clauwaert, P., De Schamphelaire, L., Boon, N., & Verstraete, W. (2006). Microbial fuel cells in relation to conventional anaerobic digestion technology. *Engineering in Life Sciences, 6*, 285–292. https://doi.org/10.1002/elsc.200620121

Pinto, R. P., Srinivasan, B., Guiot, S. R., & Tartakovsky, B. (2011). The effect of real-time external resistance optimization on microbial fuel cell performance. *Water Research, 45*, 1571–1578. https://doi.org/10.1016/j.watres.2010.11.033

Pirbadian, S., Barchinger, S. E., Leung, K. M., Byun, H. S., Jangir, Y., Bouhenni, R. A., Reed, S. B., Romine, M. F., Saffarini, D. A., Shi, L., & Gorby, Y. A. (2014). Shewanella oneidensis MR-1 nanowires are outer membrane and periplasmic extensions of the extracellular electron transport components. *Proceedings of the National Academy of Sciences, 111*, 12883–12888. https://doi.org/10.1073/pnas.1410551111

Pocaznoi, D., Erable, B., Delia, M.-L., & Bergel, A. (2012). Ultra microelectrodes increase the current density provided by electroactive biofilms by improving their electron transport ability. *Energy & Environmental Science, 5*, 5287–5296. https://doi.org/10.1039/C1EE01469B

Potter, M. C. (1911). Electrical effects accompanying the decomposition of organic compounds. *Royal Society of London. Series B, Containing Papers of a Biological Character, 84*, 260–276. https://doi.org/10.1098/rspb.1911.0073

Rabaey, K., Girguis, P., & Nielsen, L. K. (2011). Metabolic and practical considerations on microbial electrosynthesis. *Current Opinion in Biotechnology, 22*, 371–377. https://doi.org/10.1016/j.copbio.2011.01.010

Rabaey, K., & Rozendal, R. A. (2010). Microbial electrosynthesis—Revisiting the electrical route for microbial production. *Nature Reviews Microbiology, 8*, 706–716. https://doi.org/10.1038/nrmicro2422

Rawson, F. J., Downard, A. J., & Baronian, K. H. (2014). Electrochemical detection of intracellular and cell membrane redox systems in Saccharomyces cerevisiae. *Science and Reports, 4*, 5216. https://doi.org/10.1038/srep05216

Ren, Z., Ward, T. E., & Regan, J. M. (2007). Electricity production from cellulose in a microbial fuel cell using a defined binary culture. *Environmental Science and Technology, 41*, 4781–4786. https://doi.org/10.1021/es070577h

Ren, Z., Yan, H., Wang, W., Mench, M. M., & Regan, J. M. (2011). Characterization of microbial fuel cells at microbially and electrochemically meaningful time scales. *Environmental Science and Technology, 45*, 2435–2441. https://doi.org/10.1021/es103115a

Resch, G., Held, A., Faber, T., Panzer, C., Toro, F., & Haas, R. (2008). Potentials and prospects for renewable energies at global scale. *Energy Policy, 36*, 4048–4056. https://doi.org/10.1016/j.enpol.2008.06.029

Rinaldi, A., Mecheri, B., Garavaglia, V., Licoccia, S., Di Nardo, P., & Traversa, E. (2008). Engineering materials and biology to boost performance of microbial fuel cells: A critical review. *Energy & Environmental Science, 1*, 417. https://doi.org/10.1039/b806498a

Risgaard-Petersen, N., & Rotaru, A.-E. (2023). Editorial: Electromicrobiology—From electrons to ecosystems, volume II. *Frontiers in Microbiology, 14*. https://doi.org/10.3389/fmicb.2023.1253550

Rodríguez Arredondo, M., Kuntke, P., Jeremiasse, A. W., Sleutels, T. H., Buisman, C. J., & Ter Heijne, A. J. (2015). Bioelectrochemical systems for nitrogen removal and recovery from wastewater. *Environmental Science: Water Research & Technology, 1*, 22–33. https://doi.org/10.1039/C4EW00066H

Rousseau, R., Dominguez-Benetton, X., Délia, M.-L., & Bergel, A. (2013). Microbial bioanodes with high salinity tolerance for microbial fuel cells and microbial electrolysis cells. *Electrochemistry Communications, 33*, 1–4. https://doi.org/10.1016/j.elecom.2013.04.002

Rozendal, R. A., Hamelers, H. V. M., Rabaey, K., Keller, J., & Buisman, C. J. (2008). Towards practical implementation of bioelectrochemical wastewater treatment. *Trends in Biotechnology, 26*, 450–459. https://doi.org/10.1016/j.tibtech.2008.04.008

Rozendal, R. A., Leone, E., Keller, J., & Rabaey, K. (2009). Efficient hydrogen peroxide generation from organic matter in a bioelectrochemical system. *Electrochemistry Communications, 11*, 1752–1755. https://doi.org/10.1016/j.elecom.2009.07.008

Sajana, T. K., Ghangrekar, M. M., & Mitra, A. (2014). Effect of presence of cellulose in the freshwater sediment on the performance of sediment microbial fuel cell. *Bioresource Technology, 155*, 84–90. https://doi.org/10.1016/j.biortech.2013.12.094

Santoro, C., Arbizzani, C., Erable, B., & Ieropoulos, I. (2017). Microbial fuel cells: From fundamentals to applications. A review. *Journal of Power Sources, 356*, 225–244. https://doi.org/10.1016/j.jpowsour.2017.03.109

Schmitz, S., & Rosenbaum, M. A. (2018). Boosting mediated electron transfer in bioelectrochemical systems with tailored defined microbial cocultures. *Biotechnology and Bioengineering, 115*, 2183–2193. https://doi.org/10.1002/bit.26732

Sharma, M., Aryal, N., Sarma, P. M., Vanbroekhoven, K., Lal, B., Benetton, X. D., & Pant, D. (2013). Bioelectrocatalyzed reduction of acetic and butyric acids via direct electron transfer using a mixed culture of sulfate-reducers drives electrosynthesis of alcohols and acetone. *Chemical Communications, 49*, 6495. https://doi.org/10.1039/c3cc42570c

Sharma, M., Bajracharya, S., Gildemyn, S., Patil, S. A., Alvarez-Gallego, Y., Pant, D., Rabaey, K., & Dominguez-Benetton, X. (2014). A critical revisit of the key parameters used to describe microbial electrochemical systems. *Electrochimica Acta, 140*, 191–208. https://doi.org/10.1016/j.electacta.2014.02.111

Shi, L., Dong, H., Reguera, G., Beyenal, H., Lu, A., Liu, J., Yu, H. Q., & Fredrickson, J. K. (2016). Extracellular electron transfer mechanisms between microorganisms and minerals. *Nature Reviews Microbiology, 14*, 651–662. https://doi.org/10.1038/nrmicro.2016.93

Singh, A., Sevda, S., Abu Reesh, I., Vanbroekhoven, K., Rathore, D., & Pant, D. (2015). Biohydrogen production from lignocellulosic biomass: Technology and sustainability. *Energies, 8*, 13062–13080. https://doi.org/10.3390/en81112357

Sleutels, T. H. J. A., Ter, H. A., Buisman, C. J. N., & Hamelers, H. V. M. (2013). Steady-state performance and chemical efficiency of microbial electrolysis cells. *International Journal of Hydrogen Energy, 38*, 7201–7208. https://doi.org/10.1016/j.ijhydene.2013.04.067

Sleutels, T. H. J. A., Ter Heijne, A., Buisman, C. J. N., & Hamelers, H. V. M. (2012). Bioelectrochemical systems: An outlook for practical applications. *Chemsuschem, 5*, 1012–1019. https://doi.org/10.1002/cssc.201100732

Steinbusch, K. J. J., Hamelers, H. V. M., & Buisman, C. J. N. (2008). Alcohol production through volatile fatty acids reduction with hydrogen as electron donor by mixed cultures. *Water Research, 42*, 4059–4066. https://doi.org/10.1016/j.watres.2008.05.032

Steinbusch, K. J. J., Hamelers, H. V. M., Schaap, J. D., Kampman, C., & Buisman, C. J. (2010). Bioelectrochemical ethanol production through mediated acetate reduction by mixed cultures. *Environmental Science and Technology, 44*, 513–517. https://doi.org/10.1021/es902371e

Strik, D. P. B. T. B., Hamelers (Bert), H. V. M., Snel, J. F. H., & Buisman, C. J. N. (2008a). Green electricity production with living plants and bacteria in a fuel cell. *International Journal of Energy Research, 32*, 870–876. https://doi.org/10.1002/er.1397

Strik, D. P. B. T. B., Terlouw, H., Hamelers, H. V. M., & Buisman, C. J. N. (2008b). Renewable sustainable biocatalyzed electricity production in a photosynthetic algal microbial fuel cell (PAMFC). *Applied Microbiology and Biotechnology, 81*, 659–668. https://doi.org/10.1007/s00253-008-1679-8

Strik, D. P. B. T. B., Timmers, R. A., Helder, M., Steinbusch, K. J., Hamelers, H. V., & Buisman, C. J. (2011). Microbial solar cells: Applying photosynthetic and electrochemically active organisms. *Trends in Biotechnology, 29*, 41–49. https://doi.org/10.1016/j.tibtech.2010.10.001

Tandukar, M., Huber, S. J., Onodera, T., & Pavlostathis, S. G. (2009). Biological chromium(VI) reduction in the cathode of a microbial fuel cell. *Environmental Science and Technology, 43*, 8159–8165. https://doi.org/10.1021/es9014184

ter Heijne, A., Hamelers, H. V. M., Saakes, M., & Buisman, C. J. N. (2008). Performance of non-porous graphite and titanium-based anodes in microbial fuel cells. *Electrochimica Acta, 53*, 5697–5703. https://doi.org/10.1016/j.electacta.2008.03.032

Timmers, R. A., Strik, D. P. B. T. B., Arampatzoglou, C., Buisman, C. J., & Hamelers, H. V. (2012). Rhizosphere anode model explains high oxygen levels during operation of a Glyceria maxima PMFC. *Bioresource Technology, 108*, 60–67. https://doi.org/10.1016/j.biortech.2011.10.088

Venkata Mohan, S., Mohanakrishna, G., Velvizhi, G., Babu, V. L., & Sarma, P. N. (2010). Bio-catalyzed electrochemical treatment of real field dairy wastewater with simultaneous power generation. *Biochemical Engineering Journal, 51*, 32–39. https://doi.org/10.1016/j.bej.2010.04.012

Venkataraman, A., Rosenbaum, M., Arends, J. B. A., Halitschke, R., & Angenent, L. T. (2010). Quorum sensing regulates electric current generation of Pseudomonas aeruginosa PA14 in bioelectrochemical systems. *Electrochemistry Communications, 12*, 459–462. https://doi.org/10.1016/j.elecom.2010.01.019

Venkatramanan, V., Shah, S., & Prasad, R. (2021). A critical review on microbial fuel cells technology: Perspectives on wastewater treatment. *Open Biotechnology Journal, 15*, 131–141. https://doi.org/10.2174/1874070702115010131

Wang, A., Sun, D., Cao, G., Wang, H., Ren, N., Wu, W. M., & Logan, B. E. (2011). Integrated hydrogen production process from cellulose by combining dark fermentation, microbial fuel cells, and a microbial electrolysis cell. *Bioresource Technology, 102*, 4137–4143. https://doi.org/10.1016/j.biortech.2010.10.137

Wang, H., Liu, D., Lu, L., Zhao, Z., Xu, Y., & Cui, F. (2012a). Degradation of algal organic matter using microbial fuel cells and its association with trihalomethane precursor removal. *Bioresource Technology, 116*, 80–85. https://doi.org/10.1016/j.biortech.2012.04.021

Wang, H., Park, J.-D., & Ren, Z. (2012b). Active energy harvesting from microbial fuel cells at the maximum power point without using resistors. *Environmental Science and Technology, 46*, 5247–5252. https://doi.org/10.1021/es300313d

Wang, H., & Ren, Z. J. (2014). Bioelectrochemical metal recovery from wastewater: A review. *Water Research, 66*, 219–232. https://doi.org/10.1016/j.watres.2014.08.013

Wetser, K., Sudirjo, E., Buisman, C. J. N., & Strik, D. P. B. T. B. (2015). Electricity generation by a plant microbial fuel cell with an integrated oxygen reducing biocathode. *Applied Energy, 137*, 151–157. https://doi.org/10.1016/j.apenergy.2014.10.006

Xu, S., Barrozo, A., Tender, L. M., Krylov, A. I., & El-Naggar, M. Y. (2018). Multiheme cytochrome mediated redox conduction through Shewanella oneidensis MR-1 cells. *Journal of the American Chemical Society, 140*, 10085–10089. https://doi.org/10.1021/jacs.8b05104

Yakushi, T., & Matsushita, K. (2010). Alcohol dehydrogenase of acetic acid bacteria: Structure, mode of action, and applications in biotechnology. *Applied Microbiology and Biotechnology, 86*, 1257–1265. https://doi.org/10.1007/s00253-010-2529-z

Zafar, M. N., Beden, N., Leech, D., Sygmund, C., Ludwig, R., & Gorton, L. (2012). Characterization of different FAD-dependent glucose dehydrogenases for possible use in glucose-based biosensors and biofuel cells. *Analytical and Bioanalytical Chemistry, 402*, 2069–2077. https://doi.org/10.1007/s00216-011-5650-7

Zhang, F., Pant, D., & Logan, B. E. (2011). Long-term performance of activated carbon air cathodes with different diffusion layer porosities in microbial fuel cells. *Biosensors & Bioelectronics, 30*, 49–55. https://doi.org/10.1016/j.bios.2011.08.025

Zhao, J., Li, F., Cao, Y., Zhang, X., Chen, T., Song, H., & Wang, Z. (2021). Microbial extracellular electron transfer and strategies for engineering electroactive microorganisms. *Biotechnology Advances, 53*, 107682. https://doi.org/10.1016/j.biotechadv.2020.107682

Chapter 14
Production of Biofuels and Valuable Chemicals from Syngas by Electro-Fermentation

Arindam Sinharoy and Kannan Pakshirajan

1 Introduction

The ever-growing knowledge and awareness about climate change, fossil fuel depletion, and environmental deterioration necessitates a transition from a fossil fuel-based economy to a renewable energy-based one (Holechek et al., 2022; Pan et al., 2024). Fossil fuels are a major source of energy for the current scenario, yet they also contribute majorly toward pollution and greenhouse gas emissions (Kalair et al., 2021; Sinharoy et al., 2020a). The negative effects of pollutant emission and climate change have necessitated the need for mitigating the challenges by switching to alternative renewable energy sources (Manirethan et al., 2023). Renewable energy is a sustainable substitute for limited fossil fuels since it originates from natural processes that are continually renewed (Perera, 2018). There are major advantages of switching toward renewable energy, which include environmental, economic, and social benefits. Environmental imperatives of using renewable energy sources are reduction of air and water pollution, mitigating climate change, and conserving natural resources (Zafar et al., 2020; Zhang et al., 2022). Renewable energy can provide economic benefits as well which include job creation and stimulating economic growth of a country. According to the "International Renewable Energy Agency", the renewable energy industry employed 11.5 million people globally in 2019 (Trypolska et al., 2021). There is significant number of jobs in manufacturing, installation, maintenance, and research and development in renewable energy industry sector. It also helps in energy independence and security by reducing the over dependence on

A. Sinharoy
Department of Environmental Science and Biotechnology, Jeonju University, Jeonju, Republic of Korea

K. Pakshirajan (✉)
Department of Biosciences and Bioengineering, Indian Institute of Technology Guwahati, Guwahati, Assam, India
e-mail: pakshi@iitg.ac.in

© The Author(s), under exclusive license to Springer Nature Singapore Pte Ltd. 2025
S. Sevda (ed.), *Electro-Fermentation for Synthesis of Chemicals, Biochemicals and Biofuels*, https://doi.org/10.1007/978-981-96-6358-3_14

imported fossil fuel heavily impacted with geopolitical risks and market volatility, thereby stabilizing their energy supply and prices (Pan et al., 2024). Additional social benefits in the form of health improvements, energy access, poverty reduction, and community empowerment through community-based renewable energy initiatives can be achieved (Ucal & Xydis, 2020). This can lead to greater social cohesion and economic benefits for the entire society, particularly toward rural impoverished communities in developing or underdeveloped countries (Zafar et al., 2020).

Among the different renewable energy sources, bioenergy, also known as biomass-based renewable energy, can support a diverse and sustainable energy portfolio and has a lot of promise (Destek et al., 2021; Sinharoy et al., 2019a). Despite many obstacles, the full potential of bioenergy may be realized via technological developments, sustainable practices, and supportive regulations (Negi & Sinharoy, 2024). Biofuels can be broadly categorized into three generations based on their feedstock and production processes. First-generation biofuels, such as bioethanol and biodiesel, are produced from food crops such as corn, sugarcane, and soybeans (Cavelius et al., 2023). Second-generation biofuels are derived from non-food biomass sources, including agricultural residues, forestry waste, and dedicated energy crops like switchgrass and Miscanthus (Sinharoy & Pakshirajan, 2024a). Second-generation biofuels tackle the food-versus-fuel debate and overcome the challenges associated with first-generation biofuels (Pramanik et al., 2023). Third-generation biofuels are primarily sourced from algae and other microorganisms (Sinharoy & Pakshirajan, 2024b). Microalgae can be cultivated using wastewater and saline water, making them highly suitable for production of not only biofuels, but also carotenoids, antioxidants, and other high value products (Sinharoy & Pakshirajan, 2024c).

Various processes which broadly falls under biological/biochemical, chemical, and thermal/thermochemical methods can be utilized for the production of biofuels. Different methods for biofuel production include fermentation, anaerobic digestion, transesterification, gasification, etc. (Sinharoy & Pakshirajan, 2021). Among these processes, gasification involves conversion of biomass into a mixture of gases, i.e., carbon monoxide (CO), hydrogen (H_2), and carbon dioxide (CO_2), known as synthesis gas or syngas, by heating the biomass under an oxygen-limiting condition. The syngas can then be converted into liquid fuels like Fischer–Tropsch diesel through catalytic synthesis (Mazurova et al., 2023). This is more advantageous over traditional biological/biochemical conversion process as biomass feedstock used for such processes often requires one or more pretreatment steps to breakdown their complex structure to simple syngas for fermentation using microorganisms (Martinelli et al., 2020). In contrast, thermochemical conversion of second-generation biomass feedstock to syngas by biological (e.g., fermentation) and other catalytic processes (e.g., Fischer–Tropsch synthesis) is energetically favorable and more economically feasible (Paul & Sinharoy, 2024). Compared with Fischer–Tropsch synthesis for conversion of gaseous feedstock to industrial products, microbe-based syngas fermentation is advantageous in terms of ambient temperature–pressure operating conditions, better conversion efficiency and product yield, higher product selectivity, higher tolerance to catalyst poisoning to gas impurities, and capability to utilize syngas with wide ranging CO_2 to H_2 ratios (Sinharoy & Pakshirajan, 2021; Sinharoy et al., 2022). Hence, this process of gasification–fermentation has become more attractive over conventional fermentation and thermochemical gas conversion methods such as FT synthesis.

Despite its many advantages, certain challenges still exist in the case of syngas fermentation which are related to poor gas solubility and mass transfer, toxicity to microorganisms, and metabolic energy limitation (Dhakal & Acharya, 2021; Puiman et al., 2022). However, limitations due to poor gas mass transfer or toxicity can be overcome with modified bioreactors design, genetic engineering or through optimization approaches (Sinharoy et al., 2022). In the case of metabolic energetic limitations, an external energy supply can help in overcoming the energy barrier of syngas fermentation by commonly used acetogenic bacteria. In this regard, microbial electrosynthesis (MES) has gained interest to increase the process efficiency of syngas fermentation in the recent times (Sinharoy & Pakshirajan, 2024c). The term "MES" describes electricity-driven process to convert CO_2 into chemicals and biofuel by either in situ H_2 evolution or direct electron input to the microbial cell (Wood et al., 2021). This system can also be used for microbial conversion of industrial waste gases with high efficiency which can help in pollution mitigation. Syngas electrofermentation operates at the nexus of microbiology, biotechnology, and renewable energy (Bajracharya et al., 2022). Syngas electrofermentation is a sustainable method of converting gaseous carbon sources into useful products like biofuels (hydrogen, methane, ethanol, and butanol), biochemicals (acetate, butyrate, and caproate), and bioplastics by combining microbial metabolism with electrochemical processes (Barbosa et al., 2021; Sivalingam et al., 2022). Microbial electrochemistry may significantly increase the bioproduction of useful products from syngas, CO, and CO_2 when coupled to fermentation processes, making the integrated system more economically viable (Kumar et al., 2023). This review covers the principles, applications, challenges, and future prospects of syngas electrofermentation, examining the state-of-the-art presently available.

2 Syngas—Production, Composition, and Uses

Scientists and engineers are constantly investigating substitutes that reduce our dependency on conventional fossil fuels while minimizing environmental effects in their pursuit of sustainable energy solutions (Maitlo et al., 2022). In this regard, the synthesis gas, commonly known as syngas, can be a viable option as it has the advantages of adaptability, high efficiency, and environmental friendly (Sinharoy et al., 2022). Syngas is one of the energy sources that can be used effectively for the shift toward renewable energy. Syngas is a gas mixture of mainly carbon monoxide (CO) and hydrogen (H_2) (Maitlo et al., 2022). It also contains a variable quantity of carbon dioxide (CO_2), methane (CH_4), and other minor gases. The ratio of hydrogen to carbon monoxide (H_2/CO ratio) in syngas is an important attribute that determines its conversion to desirable products and can be altered appropriately for synthesis of a target product from syngas (Sinharoy et al., 2015). The production of syngas involves different methods such as steam reforming of natural gas (methane), partial oxidation, as well as gasification of coal or biomass (Maitlo et al., 2022). All these methods have their own merits and demerits, and based on these, a suitable method can be selected and applied for a desired application, using the available feedstock.

In the steam reforming process, syngas is produced by the reaction of steam and natural gas (methane) in the presence of a catalyst (e.g., nickel, platinum, rhodium, etc.) at very high temperature (800–900 °C) and pressure (2–3 MPa) conditions (Chen et al., 2020). It is widely applied in industries for the production of syngas as well as hydrogen. Partial oxidation process is another syngas production route, and as the name suggests, it involves the oxidation of fuel under oxygen (or air) limiting condition. This process is extremely exothermic method and frequently employed in small-scale settings or in situations where oxygen supply is constrained (Caballero et al., 2022). Coal gasification converts coal into syngas by its reaction with steam and oxygen at high temperature condition (dos Santos & Alencar, 2020). This alternative process to conventional coal combustion offers higher efficiency and lower emissions. Similar to the coal gasification, gasification of biomass such as wood chips, agricultural leftovers, or organics produces syngas (Maitlo et al., 2022). By heating the biomass in a low-oxygen environment, this thermochemical process yields syngas along with char and tar as by-products. Gasifier used for coal gasification is being developed since early twentieth century. Examples of some of the commonly used gasifier include updraft or counter-current gasifier, downdraft or co-current gasifier, cross-draft gasifier, and fluidized bed gasifier (Speight, 2020). Other types of uncommon gasifier that are currently under developing stage are double fired, entrained bed, and molten bath type gasifier; they are essentially partial modification of conventional coal gasification technology with one or more additional parts incorporating purification of the gas produced (Gao et al., 2023).

Syngas is used for wide ranging applications across various industries involving power generation to hydrogen, chemicals, and liquid fuels production (Caballero et al., 2022). Both gas turbines and internal combustion engines can operate on syngas with regards to the generation of power (Fiore et al., 2020). Due to its flexibility with production from various feedstock such as coal, natural gas, and biomass, it is beneficial for different sectors. Hydrogen and carbon monoxide which are the main constituent of syngas are raw materials for synthesizing different useful chemicals through various chemical reactions (Gao et al., 2023). It is often used to produce valuable compounds such as methanol, ammonia, and synthetic fuels. Hydrogen as a by-product of the syngas production could be further refined and applied as a clean fuel for industries, vehicles, and fuel cells (Ugwu et al., 2021). Other potential conversion products of syngas include liquid hydrocarbons such as jet fuel, diesel, and gasoline through the Fischer–Tropsch synthesis route (Radenahmad et al., 2020). This technique overcomes the reliance on petroleum because it allows the creation of synthetic fuels from abundant feedstock such as coal or biomass.

3 Syngas Fermentation

Syngas fermentation is a biotechnological process that can be used to convert synthesis gas (syngas) into useful chemicals and biofuels (Sinharoy et al., 2022). Thus, the biological conversion of syngas to fermentation products such as biofuels and chemicals as detailed above is a sustainable method as compared to chemical processes. Syngas fermentation can produce acetate, butanol, ethanol, and other chemicals using the metabolic pathways of some specific microorganisms (Kennes-Veiga et al., 2023). For this purpose, different anaerobic bacteria and archaea are known to metabolize the syngas components including CO, CO_2, and H_2. These group of microorganisms contain specific enzymes (i.e., hydrogenase and carbon monoxide dehydrogenase) and metabolic pathways for utilizing syngas components as sole carbon and energy source (Sinharoy et al., 2019a). The main metabolic pathway involved in syngas fermentation process is Wood–Ljungdahl pathway which produces acetyl-CoA as the main intermediate (Gunes, 2021). Acetyl-CoA is then converted to various useful products such as acetate, ethanol, and butanol via different conversion routes. For example, production of acetate from acetyl-CoA is a straightforward process involving the generation of additional ATP. The production of ethanol from acetyl-CoA is a reductive process with acetaldehyde as an intermediatory compound (Ahuja et al., 2023). In the case of butanol, acetyl-CoA undergoes chain elongation to form butyryl-CoA, which is then reduced to butanol in a subsequent step.

Based on the desired product, the pathways can be manipulated to either enhance the product yield and conversion efficiency or suppress certain undesired products, which, however, requires proper understanding of pathways involved and the constituent enzyme systems in a producer microorganism (Kennes-Veiga et al., 2023). The most commonly reported microorganisms capable of performing syngas fermentation belong to *Clostridium* genera (Martin et al., 2016). This is a group of chemolithoautotrophic bacteria, well known for their capability to produce acetone–butanol–ethanol (ABE) through syngas fermentation. For example, *Clostridium ljungdahlii* is reported for producing ethanol and acetate via the Wood–Ljungdahl pathway (Philipps et al., 2019). *Clostridium carboxidivorans* can produce butanol, ethanol, and acetone from syngas (Shen et al., 2017). Microbes belonging to other genera such as *Moorella thermoacetica* are reported to mainly produce acetate, and functions operate at an elevated temperature (thermophilic) (Redl et al., 2017). *Butyribacterium methylotrophicum* is another strain known for producing butyrate and butanol from syngas (Pacheco et al., 2021). From an environmental perspective, syngas fermentation can be linked to wastewater treatment, particularly for biological reductive processes such as sulfate or selenite reduction where syngas or products of syngas fermentation (e.g., acetate, hydrogen, etc.) can serve as electron donor (Sinharoy & Lens, 2022; Sinharoy et al., 2015, 2019b). This type of treatment process offers much cleaner effluent as it avoids the use of soluble organic carbon source as electron donor, residual of which may remain in the effluent leading to increased organic load in the treated water, requiring additional treatment (Sinharoy et al., 2020b).

The efficiency, yield, and economic feasibility of syngas fermentation to produce useful chemicals and biofuels are all determined by certain important parameters. For a successful scale up and applications of this technology at an industrial scale, it is essential to understand and optimize these parameters. One of the most important factors influencing this process is the composition of syngas. Microbial growth and product formation are strongly influenced by the composition of syngas, specifically the ratios of CO, H_2, and CO_2. For example, CO levels can be toxic to microorganisms, and therefore, CO concentration needs to be optimum. H^+ ions act as an electron donor, requiring optimum H_2 levels to reduce CO_2 and CO to organic compounds. CO_2 in syngas serves as the carbon source for autotrophic microbes due to which an optimum CO_2 concentration is required for their metabolism. Temperature and pH are the key environmental factors that affect syngas fermentation due to their direct influence on the microbial activity and stability. Hence, optimum pH and temperature conditions need to be maintained in order to obtain maximum product yield. Microorganisms are strongly dependent on specific pH and temperature conditions for their optimal growth and product formation. For most of the syngas fermenting microbes, a near neutral pH (6–7) is reported to be optimum. However, many of the strains belonging to *Clostridium* genera require slightly acidic pH (4.5–5.0) to perform solventogenesis, i.e., production of ethanol. In the case of temperature effect, mesophiles require a temperature in the range of 30–37 °C, whereas thermophilic strains require temperature >55 °C. Headspace gas pressure is another important parameter that affect the gas–liquid mass transfer; both gas solubility and mass transfer increase with an increase in the headspace gas pressure. The reactor design, agitation speed, and sparging technique are some other factors that help in enhancing the mass transfer. Adequate supply of macro and micro nutrients such as nitrogen, phosphorus, sulfur, trace metals, and vitamins is also crucial for microbial metabolism in syngas fermentation. Hence, optimization of medium compositions to meet the specific needs of the selected microbial strain can improve fermentation efficiency and product yields. Overcoming product inhibition on microbial growth and metabolism is another critical issue in syngas fermentation as accumulation of products can inhibit microbial activity. The remedial measure in such case involves efficient in situ product separation and recovery from the fermentation broth.

Proper choice of reactor design is important to guarantee microbial stability, scalability, and efficient gas–liquid mass transfer in syngas fermentation (Sinharoy et al., 2022). Different types of bioreactors, namely continuous stirred tank reactor, bubble column reactor, trickle bed reactor, air/gas lift reactor, and membrane reactor are commonly used for syngas fermentation, and each of these reactors have their own advantages and drawbacks. For example, continuous stirred tank reactor provides thorough mixing, but may offer poor gas–liquid mass transfer. Bubble column reactors can overcome the gas–liquid mass transfer limitations, but the biomass retention is low (Asimakopoulos et al., 2018). Air lift reactor or gas lift reactors offers better mixing with lower shear stress; however, this reactor type is difficult to scale up and involve high cost (Sinharoy et al., 2022). Trickle bed reactors are advantageous in terms of good biomass retention due to higher surface area available for biomass growth, whereas problems arise due to excessive biomass growth, leading to reactor clogging and operational difficulties. Membrane reactors provide enhanced mass transfer and are highly suited for selective product removal, improving overall productivity of the system, but could be expensive in comparison with other reactor configuration (Sinharoy et al., 2022). Other innovative bioreactor configurations are also explored for syngas fermentation (Gunes, 2021; Sinharoy et al., 2022).

As mentioned earlier, syngas fermentation is widely reported, but scale-up of the process largely relies on overcoming obstacles including gas–liquid mass transfer limitations, microbial inhibition at higher concentrations of syngas constituents, and challenges associated with effective separation and purification of fermentation products (Ahuja et al., 2023; Munasinghe & Khanal, 2010). In order to overcome such challenges, the following is suggested (a) advanced bioreactor designs for improved mass transfer and microbial stability, (b) metabolic engineering to improve tolerance to syngas components (particularly toward CO) and increase product yield, and (c) developing integrated systems for syngas production and fermentation or combining chemical catalysis and biological fermentation (Asimakopoulos et al., 2018; Sinharoy et al., 2022). Besides, more studies are warranted to reach the full potential of this technology. For syngas fermentation to be successful in the future, interdisciplinary approaches combining microbiology, engineering, and process integration will be essential.

4 Syngas Electrofermentation

4.1 Mechanisms Involved

Among the many challenges with syngas fermentation technology, energy limitation in terms of ATP production is a major concern (Bajracharya et al., 2022). This constraint has negative effect on cell growth, product yield, and formation rate. Controlling the redox potential by adding electrodes to the fermentation medium is a recent method for enhancing the gas fermentation efficiency. Additional electric supply to the system can facilitate the reducing capability of the microorganisms for an efficient conversion of CO/CO_2 to desire products (Sadhukhan et al., 2016). In this regard, syngas electrofermentation represents a promising avenue for integrating microbial metabolism with electrochemical processes to convert gaseous carbon sources into valuable products, including biofuels, biochemicals, and bioplastics (Sivalingam et al., 2022).

The fermentation takes place in an electrochemical cell, which consists of an anode and cathode, wherein, respectively, oxidation and reduction reactions take place (Fig. 1). The electrochemical cell contains an electrolyte which facilitates ionic conductivity between the anode and cathode (Wood et al., 2021). The anode

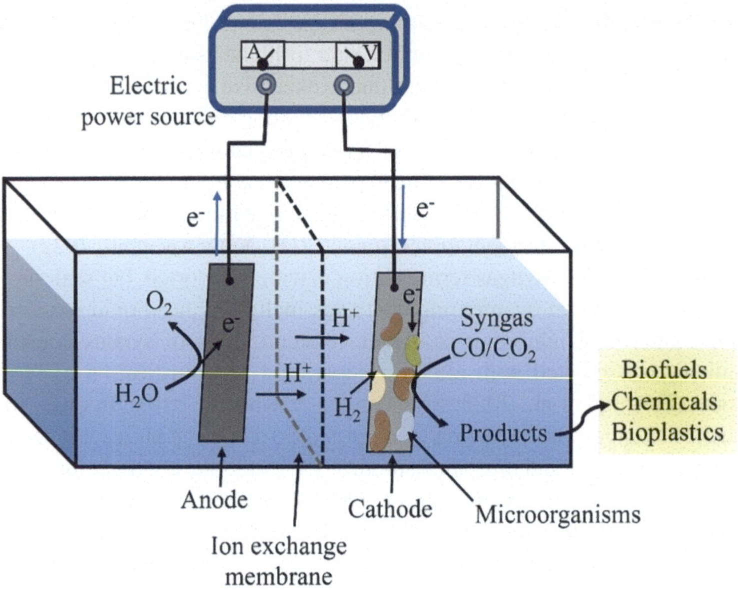

Fig. 1 Bioelectrochemical systems for syngas electrofermentation to produce biofuels and chemicals

and cathode chambers are separated by an ion-exchange membrane through which ion gets transported between them to maintain electroneutrality (Kumar et al., 2023). The electrodes are connected to an external power source that provides the necessary electrical energy.

Electrical energy supplied to the cathode is used to drive the reduction reactions, hydrogen ions (H^+) are reduced to hydrogen gas (H_2), and CO_2 can be reduced to formate ($HCOO^-$) or other intermediates (Bajracharya et al., 2022). The use of a cathodic potential amplifies these reactions and makes this energy demanding reaction feasible. The electrochemical reaction at cathode is as follows (Eqs. (1) and (2)):

$$2H^+ + 2e^- \rightarrow H_2 \quad (1)$$

$$CO_2 + 2H^+ + 2e^- \rightarrow HCOOH \quad (2)$$

At the anode, oxidation reactions occur, often involving water splitting to produce oxygen (O_2), electron (e^-), and protons (H^+). The reaction scheme for this oxidation reaction is provided below (Eq. (3)):

$$H_2O \rightarrow {}^1\!/_2\, O_2 + 2H^+ + 2e^- \quad (3)$$

The electrofermentation process further depends on the specific microorganisms present on the cathode surface which act as a biocatalyst in the system (Bhagchandanii et al., 2020; Kumar et al., 2023). Some microorganisms can directly accept electrons from the cathode. In this process, microbes can reduce CO_2 to organic compounds by using cathodic electrons as per the reaction scheme given below (Eq. (4)):

$$CO_2 + 4H^+ + 4e^- \rightarrow CH_4 + 2H_2O \quad (4)$$

Many of the archaeal community are capable of performing this reaction to produce methane. In the case of indirect electron transfer, microbes use intermediates produced at the cathode (e.g., H_2 or formate) as electron donors for carrying out the reduction reaction (Wood et al., 2021). For example, acetogens can ferment CO, CO_2, and H_2 through the Wood–Ljungdahl pathway to produce acetic acid or ethanol (Eq. (5)) (Sinharoy et al., 2022).

$$4H_2 + 2CO_2 \rightarrow CH_3COOH + 2H_2O \quad (5)$$

4.2 *Factors Influencing Syngas Electrofermentation*

Syngas electrofermentation is a complex system and is subject to several factors that impact its viability, productivity, and efficiency (Fig. 2). Gaining an understanding of these crucial elements is necessary to optimize the process in order to obtain the desired results. Many of the important factors that affect syngas electrofermentation are similar to syngas fermentation. For example, selecting a suitable microbial strain is essential for desired product formation from syngas as the ability of various bacteria to use syngas components and generate certain products varies. Acetogenic bacteria, such as *Clostridium ljungdahlii*, and methanogens, such as *Methanosarcina barkeri*, are commonly used. Syngas composition, i.e., the relative ratios of CO, CO_2, and H_2 in the syngas influences the microbial activity and product formation. Presence of certain impurities such as sulfur compounds, tar, and particulates can negatively influence the process due to their inhibitory effect on microbial growth and metabolic activity. In order to avoid the toxic effects, efficient removal of the impurities is essential. Operation and environmental parameters such as gas pressure, pH, and temperature significantly influence the process efficiency by affecting solubility and bioavailability of gaseous substrate, microbial metabolism, and enzyme activity.

Some of the most important parameters that are unique to electrochemical systems are electrode materials and its configuration, and the electrochemical conditions followed (Bajracharya et al., 2022). The cathode material has an impact on the electron transfer rates. Metals, metal oxides, and electrodes made of carbon are examples of commonly used cathode materials (Chandrasekhar et al., 2021). All these materials have their own merits and demerits in terms of their cost, biocompatibility, and conductivity. The material properties such as surface area and porosity also have

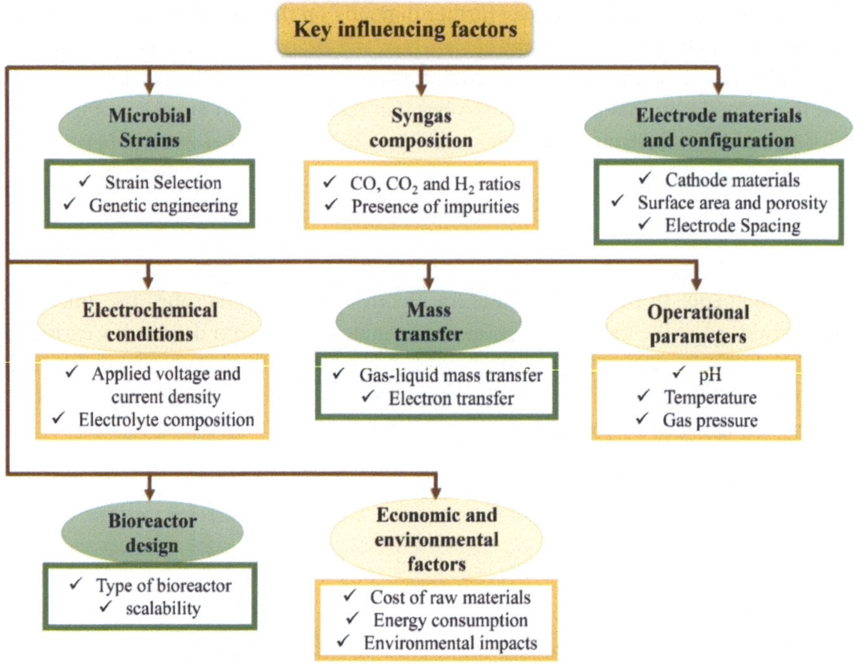

Fig. 2 Key factors influencing syngas electrofermentation process

influence on the process efficiency as a high value of these parameters enhances electron transfer and microbial attachment which in turn increases the electrode efficiency (Bhagchandanii et al., 2020). The spacing between the two electrodes is also important as they are deciding factors for the resistance and effectiveness of the electrochemical cell. The rate of electron transfer and the overall energy efficiency of the process are determined by the voltage and current density applied to the electrochemical cell (Barbosa et al., 2021). The choice of electrolyte (e.g., buffer and salts) affects ionic conductivity and microbial compatibility. The choice of bioreactor is another key factor influencing process efficiency and scalability (Chandrasekhar et al., 2021; Sadhukhan et al., 2016). There are several bioreactor designs that offer good gas–liquid mass transfer and control of operating parameters. However, the reactor constructions need to be economical while still being scalable to industrial levels.

4.3 Applications of Syngas Electrofermentation

There is great potential for several applications of syngas electrofermentation in a variety of sectors (industries). A schematic showing various product formation routes with important products is provided in Fig. 3. Acetogenic bacteria is primarily used as inoculum for the production of biofuels such as ethanol and butanol from syngas, offering a renewable alternative to fossil fuels (Munasinghe & Khanal, 2010). However, there are limited studies on syngas electrofermentation for the production of useful products as most of them have focused on CO_2 as the gaseous substrate in the feed. Tharak and Mohan (2022) explored a single-chambered microbial electrosynthesis system with carbon cloth/graphite as a working/counter electrode for ethanol production from syngas. The study reported a maximum ethanol concentration of 0.2 g/L along with 3.8 g/L of acetic acid at an applied potential of −0.5 V. Characterization of the enriched microbial community revealed the presence of *Actinobacteria*, *Firmicutes*/*Clostridia*, and *Bacilli* which were involved in syngas fermentation. Zhu et al. (2022) used an innovative strategy for syngas electrofermentation by incorporating highly selective cobalt phthalocyanine catalysts into new planar 2D and porous 3D MES cathodes to produce syngas in situ to enhance the overall bioconversion rates of CO_2 into acetate and ethanol as the fermentation products. The reason behind this strategy was to overcome the low organic production rates of typical MESs by direct electron transfer mechanisms. The electrochemically produced H_2 is often used to provide reducing equivalents for improved carbon utilization in the Wood–Ljungdahl pathway; however, CO is a more favorable electron donor than H_2 and can potentially improve CO_2 conversion to useful products. The results showed that the 3D porous electrodes are better compared to 2D planar cathodes as higher maximum ethanol (0.9 g/L) and acetate (3.8 g/L) were produced by *Clostridium ljungdahlii*. The improved product formation using 3D porous cathodes was due to more stable evolution of CO (0.08–0.06 mL/min) and H_2 (0.16–0.06 mL/min) in the system. Another drawback of the 2D planar cathodes is that it requires a higher potential to maintain similar current densities as in the case of 3D porous cathodes, leading to higher energy requirement to perform similar task.

Vassilev et al. (2018) applied a pH modulation strategy to shift the metabolic pathway of the microbial community from acetogenesis (acetate production) to solventogenesis (ethanol production) by microbial electrosynthesis and using CO_2 as the substrate. This simultaneous production of acetate and ethanol further broadened the product spectrum by formation of C4 and C6 carboxylic acids (e.g., particular butyric, isobutyric, and caproic acids) via the reverse β-oxidation chain elongation pathway at an optimum pH of 5. A maximum ethanol titre of 56.1 mM-C (mM of carbon) during the initial phase of continuous operation which was later consumed and a lower titre of 18.6 mM-C (mM of carbon) was obtained during the later phase. In the case of other alcohols, during initial phases their concentration was low (<10 mM-C), which, however, increased in the subsequent phases to reach their maximum values of 44.0 mM-C butanol, 12.8 mM-C hexanol, and 11.3 mM-C isobutanol. These results confirm the sequential production process comprising acetogenesis, solventogenesis, and chain elongation. Microbial community analysis showed the dominance of species belonging to the *Clostridium* genus in both the biofilm and the planktonic microbial communities in the MES system.

Acetate from syngas is used as a precursor for producing useful compounds including long-chain hydrocarbons and methane. Previous studies have evaluated

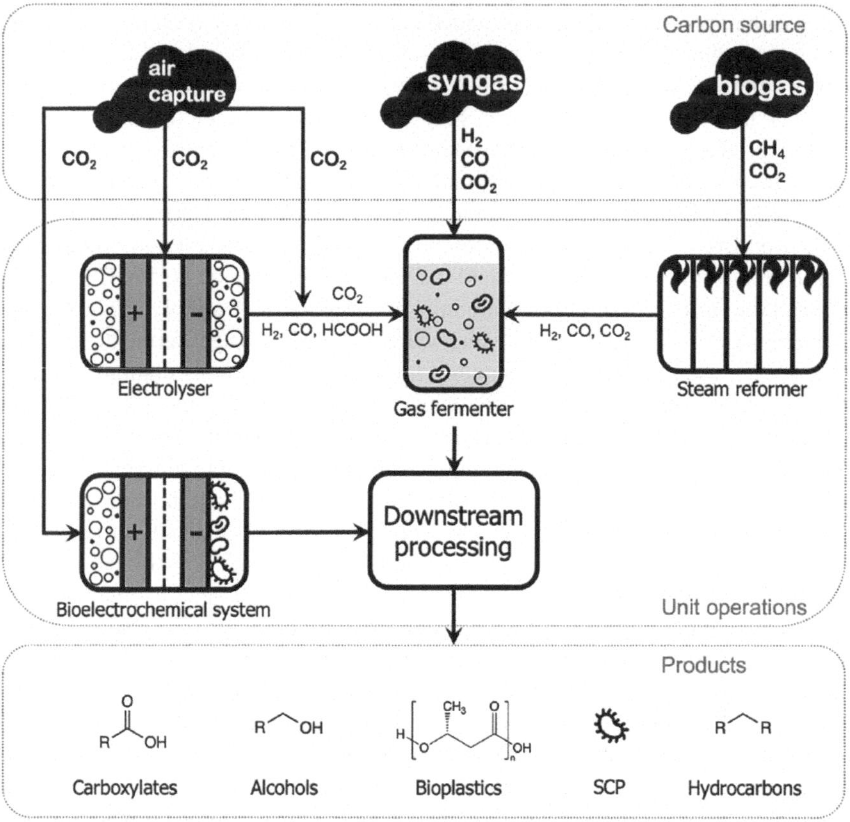

Fig. 3 Schematic showing syngas electrofermentation for the production of various products including biofuels. Reused with permission from Wood et al. (2021), copyright Bioresource Technology

effect of different factors including applied power, electrode composition and configuration, microbial community, and gas composition for enhancing the product yield. Sivalingam et al. (2022) examined the effect of electrochemical reducing power (– 50 to –400 mV) and CO concentration on acetic acid production from syngas using a microbial electrosynthesis system. The results showed that –150 mV was the optimum potential for maximum acetic acid synthesis rate (0.263 mmol/L h). There was no significant inhibition by CO and 60% of the gas was consumed under optimum condition. In another study a different strategy was applied where rather than using syngas, off-gases (CO_2 and H_2) produced during electrofermentation was recycled back to the cathodic chamber for VFA production (Zhou et al., 2019). A total VFAs production of 0.57 g-VFAs/g-VS was achieved through this reuse of acidogenic off-gases strategy in the system which corresponds to a 48.7% increase in VFA production compared to the control system without off-gases reuse. Through this method the overall conversion efficiency of carbon to VFAs could also be improved significantly (13.92%).

Tharak et al. (2023) examined a two-stage strategy involving gas and gas-electro fermentation systems for acetic acid production from CO_2 by an enriched microbial consortium. Selective enrichment was performed by heat-shock, acid-shock, and by using methanogenic inhibitor (2-Bromoethano sulphonate (BESA)) which achieved one-fold increment in acetic acid production. A maximum of 0.84 g/L acetic acid production was obtained from CO_2 in the first stage (gas fermentation) with heat pretreated inoculum. For the second stage, a single-chambered microbial electrochemical system (MES; Gas-electro fermentation) under different applied potentials (−0.4 and −0.6 V vs. Ag/AgCl [3.5 M KCl]) was used. The maximum acetic acid titre of 1.02 g/L was attained during electrofermentation with −0.6 V applied potential. The microbial community shifted toward CO_2 utilizing species as shown by the dominance of *Proteobacteria*, *Firmicutes*, and Actinobacteria phylum in the inoculum.

Pu et al. (2024) studied the production of medium-chain fatty acids from CO_2 through combined acidic electrolysis and syngas fermentation process. For the production of caproate, CO may act as the only energy source or, in the case of cofeeding with acetate, as the electron donor. Additionally, studies on gas diffusion electrodes, their structure and engineering reveal that constant, tunable syngas production from acidic CO_2 electrolysis (at pH 1) is possible by the use of carbon black, either alone or in conjunction with graphite. The CO selectivity increased from 0 to 43.5% (0.05 M K^+) and then to 92.4% (0.5 M K^+) with the addition of carbon black layer. Rather than producing a confined alkaline environment for CO_2-to-CO conversion, this improvement in performance was attributable to the enhancement in K^+ accumulation, stabilizing catalytically active sites.

The primary product of microbial electrosynthesis (MES) from carbon dioxide (CO_2) till date is acetate. However, some of the previous studies have reported butyrate production from CO_2 using MESs. Ganigué et al. (2015) demonstrated for the first time the bioelectrochemical synthesis of butyrate from CO_2 as the only carbon source. The maximum butyrate production rate of 1.82 mMC/day was attained in this study which corresponds to a maximum butyrate concentration of 20.2 mMC. The electrochemical characterization revealed that hydrogen is mainly involved in the reduction of CO_2 to butyrate. Butanol and ethanol were also produced in the system, suggesting biofuel production through this process.

Batlle-Vilanova et al. (2017) focused on enhancing the production of butyrate over that of other organic products through the use of a tubular bioelectrochemical MES at a constant cathode potential of −0.8 V. Initially, butyrate production could be initiated by limiting the CO_2 supply and allowing hydrogen partial pressure to increase. This resulted in butyrate being the main end product, with a concentration of 59.7 mMC compared to 20.3 mMC of acetate. In order to further increase the butyrate production, CO_2 supply was reduced which resulted in maximum 34.7 mMC of acetate and 87.5 mMC of butyrate production. The consumption of ethanol and the presence of other products (such as caproate) in the biocathode indicated that the butyrate production is driven by chain elongation due to certain microbes present in this system. The dominant microbial community was *Megasphaera sueciensis* with >39% relative abundance in the cathode chamber of the MES. Following extraction and concentration of butyrate using a liquid membrane, the butyrate/acetate ratio increased to 16.4, and a concentration phase containing 252.4 mMC of butyrate was obtained.

The main biopolymers with characteristics akin to that of petroleum-derived plastics is polyhydroxyalkanoates (PHAs) can be produced from syngas. In bacterium, PHAs serve as a reserve carbon source in the form of lipid granules under nutritional imbalance conditions. Such nutritional imbalance is often created as a part of bioreactor feeding strategy to initiate microorganisms for PHA production. This polymer has the added benefit of being biodegradable and biocompatible in nature, making its thermomechanical properties quite similar to those of synthetic plastics. Syngas may be converted by microorganisms into biopolymers such as PHA and other sustainable substitutes for plastics made from petroleum (Dhakal & Acharya, 2021). Examples of microorganisms capable of producing PHA from syngas are *Rhodospirillum rubrum*, *Cupriavidus necator*, etc. (Dhakal & Acharya, 2021; Revelles et al., 2016). However, no studies on PHA production from syngas using electrofermentation process have been reported thus far. Previous studies on PHA production by electrofermentation have mostly used soluble substrates such as glucose and other hexose sugars, acetate, and glycerol (Kumar et al., 2018). Hence, electrofermentation of syngas provides unique opportunities for future research including two-stage processes where soluble products formed during the electrofermentation of syngas can be utilized as substrate for PHA production in the next stage.

5 Challenges and Future Research Directions

Despite its potential, syngas electrofermentation faces several challenges that hinder its commercialization (Fig. 4). First and most important challenge is related with the microbial efficiency (Bhagchandanii et al., 2020). The low metabolic yields and slow growth rates of acetogenic bacteria restrict the productivity of syngas fermentation. Acetogenic bacteria can be genetically and metabolically engineered to increase growth rates, substrate utilization, and product selectivity (Awate et al., 2017). For long-term microbial activity to be supported in electrochemical systems, stable electrodes and effective electron transfer interfaces are necessary. It is still very difficult to regulate acetogen metabolic pathways in order to maximize the formation of desired products and minimize by-products. Another challenge is the scalability and economic viability of syngas production, purification, and fermentation processes (Chandrasekhar et al., 2021). Improving electron transfer kinetics and electrode stability by creating new electrode materials, architectures, and surface treatments are key to maximizing product yield and quality as well as energy efficiency in electrofermentation systems (Liu et al., 2023). Integration of syngas electrofermentation system with existing industrial framework will facilitate sustainable and economical production of biofuel and valuable chemicals.

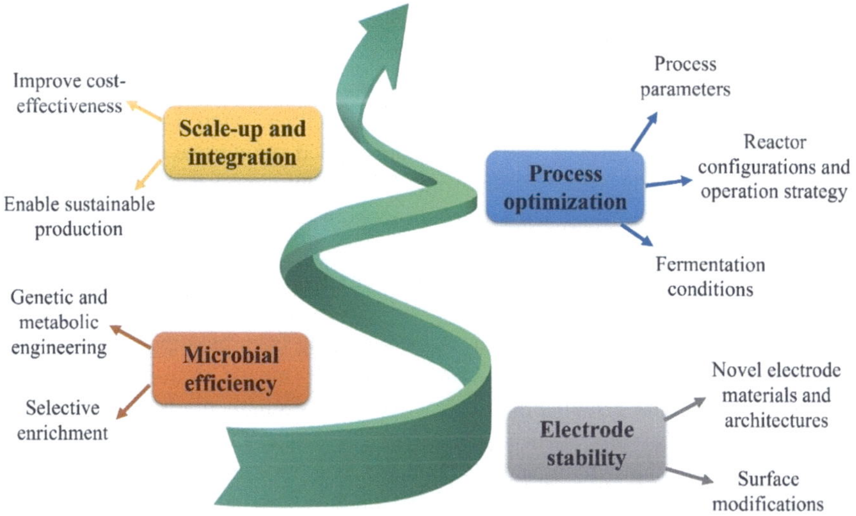

Fig. 4 Key challenges in syngas electrofermentation and future research avenues to overcome those

In order to build a more circular and sustainable system, the syngas electrofermentation process can be implemented concomitantly with other systems such as gasification, anaerobic digestion, or wastewater treatment. This approach will help to improve the process economics, and a better utilization of available resources could be achieved. The process can be made more sustainable by utilizing renewable sources of electrical power for electrochemical cell, such as the solar or wind energy.

A major cost involved in this process is related to the production and purification of syngas, and concentrated efforts toward optimizing this step will improve economic viability of the entire process (Sinharoy et al., 2022). The other vital economic considerations include the energy input to the electrochemical cell and overall efficiency of the hybrid process. Furthermore, environmental effects of this system should be assessed employing available sustainability analysis tools and compared with other similar conversion processes. For this purpose, life cycle analysis could be performed which helps in assessing the sustainability of a product or prototype by analyzing longevity, renewability, and efficiency (Sinharoy et al., 2022). These elements make sure that researchers are aware of community values that support beneficial effects on the environment, society, and economy. In conclusion, syngas electrofermentation presents a promising biotechnological approach for converting gaseous carbon sources into valuable products. Steady research and improvements in technology could subsequently bring syngas electrofermentation to its commercialization potential to push through a circular and efficient bioeconomy.

6 Conclusions

Syngas is expected to play a significant and important role in the world energy scenario due to increasing awareness of climate change, environmental pollution and energy security. However, more efforts are needed to improve the process efficiency and optimize the variables involved. Additionally, market incentives, policy support, and ongoing developments in clean energy technologies are necessary for the widespread adoption of syngas-based technologies. With regards to syngas conversion technologies, electrofermentation process can be regarded as the most prominent among all, which combines microbiology with electrochemistry to develop a highly effective biochemical process. Various aspects ranging from biological to chemical parameters involved in this hybrid process need to be optimized for improving its efficiency, scalability, and sustainability. This will require strengthening the research and development on materials science, process integration, and microbial engineering. Future technology deployment and commercialization in this field will be fueled by sustained innovation and encouraged by investments and policies, opening the door to a cleaner, more sustainable energy source.

References

Ahuja, V., Bhatt, A. K., Ravindran, B., Yang, Y. H., & Bhatia, S. K. (2023). A mini-review on syngas fermentation to bio-alcohols: Current status and challenges. *Sustainability, 15*(4), 3765.

Asimakopoulos, K., Gavala, H. N., & Skiadas, I. V. (2018). Reactor systems for syngas fermentation processes: A review. *Chemical Engineering Journal, 348*, 732–744.

Awate, B., Steidl, R. J., Hamlischer, T., & Reguera, G. (2017). Stimulation of electro-fermentation in single-chamber microbial electrolysis cells driven by genetically engineered anode biofilms. *Journal of Power Sources, 356*, 510–518.

Bajracharya, S., Krige, A., Matsakas, L., Rova, U., & Christakopoulos, P. (2022). Advances in cathode designs and reactor configurations of microbial electrosynthesis systems to facilitate gas electro-fermentation. *Bioresource Technology, 354*, 127178.

Barbosa, S. G., Peixoto, L., Alves, J. I., & Alves, M. M. (2021). Bioelectrochemical systems (BESs) towards conversion of carbon monoxide/syngas: A mini-review. *Renewable and Sustainable Energy Reviews, 135*, 110358.

Batlle-Vilanova, P., Ganigué, R., Ramió-Pujol, S., Bañeras, L., Jiménez, G., Hidalgo, M., Balaguer, M. D., Colprim, J., & Puig, S. (2017). Microbial electrosynthesis of butyrate from carbon dioxide: Production and extraction. *Bioelectrochemistry, 117*, 57–64.

Bhagchandanii, D. D., Babu, R. P., Sonawane, J. M., Khanna, N., Pandit, S., Jadhav, D. A., Khilari, S., & Prasad, R. (2020). A comprehensive understanding of electro-fermentation. *Fermentation, 6*(3), 92.

Caballero, J. J. B., Zaini, I. N., & Yang, W. (2022). Reforming processes for syngas production: A mini-review on the current status, challenges, and prospects for biomass conversion to fuels. *Applications in Energy and Combustion Science, 10*, 100064.

Cavelius, P., Engelhart-Straub, S., Mehlmer, N., Lercher, J., Awad, D., & Brück, T. (2023). The potential of biofuels from first to fourth generation. *PLoS Biology, 21*(3), e3002063.

Chandrasekhar, K., Kumar, A. N., Kumar, G., Kim, D. H., Song, Y. C., & Kim, S. H. (2021). Electro-fermentation for biofuels and biochemicals production: Current status and future directions. *Bioresource Technology, 323*, 124598.

Chen, G., Jamro, I. A., Samo, S. R., Wenga, T., Baloch, H. A., Yan, B., & Ma, W. (2020). Hydrogen-rich syngas production from municipal solid waste gasification through the application of central composite design: An optimization study. *International Journal of Hydrogen Energy, 45*(58), 33260–33273.

Destek, M. A., Sarkodie, S. A., & Asamoah, E. F. (2021). Does biomass energy drive environmental sustainability? An SDG perspective for top five biomass consuming countries. *Biomass and Bioenergy, 149*, 106076.

Dhakal, N., & Acharya, B. (2021). Syngas fermentation for the production of bio-based polymers: A review. *Polymers, 13*(22), 3917.

dos Santos, R. G., & Alencar, A. C. (2020). Biomass-derived syngas production via gasification process and its catalytic conversion into fuels by Fischer Tropsch synthesis: A review. *International Journal of Hydrogen Energy, 45*(36), 18114–18132.

Fiore, M., Magi, V., & Viggiano, A. (2020). Internal combustion engines powered by syngas: A review. *Applied Energy, 276*, 115415.

Ganigué, R., Puig, S., Batlle-Vilanova, P., Balaguer, M. D., & Colprim, J. (2015). Microbial electrosynthesis of butyrate from carbon dioxide. *Chemical Communications, 51*(15), 3235–3238.

Gao, Y., Wang, M., Raheem, A., Wang, F., Wei, J., Xu, D., Song, X., Bao, W., Huang, A., Zhang, S., & Zhang, H. (2023). Syngas production from biomass gasification: Influences of feedstock properties, reactor type, and reaction parameters. *ACS Omega, 8*(35), 31620–31631.

Gunes, B. (2021). A critical review on biofilm-based reactor systems for enhanced syngas fermentation processes. *Renewable and Sustainable Energy Reviews, 143*, 110950.

Holechek, J. L., Geli, H. M., Sawalhah, M. N., & Valdez, R. (2022). A global assessment: Can renewable energy replace fossil fuels by 2050? *Sustainability, 14*(8), 4792.

Kalair, A., Abas, N., Saleem, M. S., Kalair, A. R., & Khan, N. (2021). Role of energy storage systems in energy transition from fossil fuels to renewables. *Energy Storage, 3*(1), e135.

Kennes-Veiga, D. M., Fernández-Blanco, C., Veiga, M. C., & Kennes, C. (2023). Ethanol production from syngas. In *Advances in synthesis gas: Methods, technologies and applications* (pp. 147–171). Elsevier.

Kumar, M., Sinharoy, A., Priyanka, U., & Singh, K. S. (2023). Industrial wastewater treatment using bioelectrochemical systems and the potential for energy recovery. In *Development in wastewater treatment research and processes* (pp. 219–238). Elsevier.

Kumar, P., Chandrasekhar, K., Kumari, A., Sathiyamoorthi, E., & Kim, B. S. (2018). Electro-fermentation in aid of bioenergy and biopolymers. *Energies, 11*(2), 343.

Liu, Z., Xue, X., Cai, W., Cui, K., Patil, S. A., & Guo, K. (2023). Recent progress on microbial electrosynthesis reactor designs and strategies to enhance the reactor performance. *Biochemical Engineering Journal, 190*, 108745.

Maitlo, G., Ali, I., Mangi, K. H., Ali, S., Maitlo, H. A., Unar, I. N., & Pirzada, A. M. (2022). Thermochemical conversion of biomass for syngas production: Current status and future trends. *Sustainability, 14*(5), 2596.

Manirethan, V., Shajan, N., Sajan, A. S., Priyanka, U., & Sinharoy, A. (2023). Development of bioelectrochemical systems integrated nanocomposite membranes for wastewater management. In *Development in wastewater treatment research and processes* (pp. 191–217). Elsevier.

Martin, M. E., Richter, H., Saha, S., & Angenent, L. T. (2016). Traits of selected *Clostridium* strains for syngas fermentation to ethanol. *Biotechnology and Bioengineering, 113*(3), 531–539.

Martinelli, M., Gnanamani, M. K., LeViness, S., Jacobs, G., & Shafer, W. D. (2020). An overview of Fischer-Tropsch synthesis: XtL processes, catalysts and reactors. *Applied Catalysis A: General, 608*, 117740.

Mazurova, K., Miyassarova, A., Eliseev, O., Stytsenko, V., Glotov, A., & Stavitskaya, A. (2023). Fischer-Tropsch synthesis catalysts for selective production of diesel fraction. *Catalysts, 13*(8), 1215.

Munasinghe, P. C., & Khanal, S. K. (2010). Biomass-derived syngas fermentation into biofuels: Opportunities and challenges. *Bioresource Technology, 101*(13), 5013–5022.

Negi, B. B., & Sinharoy, A. (2024). A detailed overview of anaerobic digestion of sewage sludge with different process intensification strategies. In *Application of sewage sludge in industrial wastewater treatment* (pp. 39–63).

Pacheco, M., Pinto, F., Ortigueira, J., Silva, C., Gírio, F., & Moura, P. (2021). Lignin syngas bioconversion by *Butyribacterium methylotrophicum*: Advancing towards an integrated biorefinery. *Energies, 14*(21), 7124.

Pan, A., Xu, S., & Zaidi, S. A. H. (2024). Environmental impact of energy imports: Natural resources income and natural gas production profitability in the Asia-Pacific Economic Cooperation countries. *Geoscience Frontiers, 15*(2), 101756.

Paul, T., & Sinharoy, A. (2024). Emerging issues and their solutions related to the use of sewage sludge in waste treatment. In *Application of sewage sludge in industrial wastewater treatment* (pp. 21–37).

Perera, F. (2018). Pollution from fossil-fuel combustion is the leading environmental threat to global pediatric health and equity: Solutions exist. *International Journal of Environmental Research and Public Health, 15*(1), 16.

Philipps, G., de Vries, S., & Jennewein, S. (2019). Development of a metabolic pathway transfer and genomic integration system for the syngas-fermenting bacterium *Clostridium ljungdahlii. Biotechnology for Biofuels, 12*, 1–14.

Pramanik, A., Sinha, A., Chaubey, K. K., Hariharan, S., Dayal, D., Bachheti, R. K., Bachheti, A., & Chandel, A. K. (2023). Second-generation bio-fuels: Strategies for employing degraded land for climate change mitigation meeting United Nation-sustainable development goals. *Sustainability, 15*(9), 7578.

Pu, Y., Wang, Y., Wu, G., Wu, X., Lu, Y., Yu, Y., Chu, N., He, X., Li, D., Zeng, R. J., & Jiang, Y. (2024). Tandem acidic CO_2 electrolysis coupled with syngas fermentation: A two-stage process for producing medium-chain fatty acids. *Environmental Science & Technology, 58*(17), 7445–7456.

Puiman, L., Abrahamson, B., van der Lans, R. G., Haringa, C., Noorman, H. J., & Picioreanu, C. (2022). Alleviating mass transfer limitations in industrial external-loop syngas-to-ethanol fermentation. *Chemical Engineering Science, 259*, 117770.

Radenahmad, N., Azad, A. T., Saghir, M., Taweekun, J., Bakar, M. S. A., Reza, M. S., & Azad, A. K. (2020). A review on biomass derived syngas for SOFC based combined heat and power application. *Renewable and Sustainable Energy Reviews, 119*, 109560.

Redl, S., Sukumara, S., Ploeger, T., Wu, L., Ølshøj Jensen, T., Nielsen, A. T., & Noorman, H. (2017). Thermodynamics and economic feasibility of acetone production from syngas using the thermophilic production host *Moorella thermoacetica. Biotechnology for Biofuels, 10*, 1–17.

Revelles, O., Tarazona, N., García, J. L., & Prieto, M. A. (2016). Carbon roadmap from syngas to polyhydroxyalkanoates in *Rhodospirillum rubrum. Environmental Microbiology, 18*(2), 708–720.

Sadhukhan, J., Lloyd, J. R., Scott, K., Premier, G. C., Eileen, H. Y., Curtis, T., & Head, I. M. (2016). A critical review of integration analysis of microbial electrosynthesis (MES) systems with waste biorefineries for the production of biofuel and chemical from reuse of CO_2. *Renewable and Sustainable Energy Reviews, 56*, 116–132.

Shen, Y., Brown, R. C., & Wen, Z. (2017). Syngas fermentation by *Clostridium carboxidivorans* P7 in a horizontal rotating packed bed biofilm reactor with enhanced ethanol production. *Applied Energy, 187*, 585–594.

Sinharoy, A., Baskaran, D., & Pakshirajan, K. (2019a). Sustainable biohydrogen production by dark fermentation using carbon monoxide as the sole carbon and energy source. *International Journal of Hydrogen Energy, 44*(26), 13114–13125.

Sinharoy, A., Baskaran, D., & Pakshirajan, K. (2019b). A novel carbon monoxide fed moving bed biofilm reactor for sulfate rich wastewater treatment. *Journal of Environmental Management, 249*, Article 109402.

Sinharoy, A., Kumar, M., & Pakshirajan, K. (2020a). An overview of bioreactor configurations and operational strategies for dark fermentative biohydrogen production. *Bioreactors*, 249–288. https://doi.org/10.1016/B978-0-12-821264-6.00014-0

Sinharoy, A., & Lens, P. N. L. (2022). Biological selenate and selenite reduction by waste activated sludge using hydrogen as electron donor. *Journal of Environmental Management, 319*, 115745.

Sinharoy, A., Manikandan, N. A., & Pakshirajan, K. (2015). A novel biological sulfate reduction method using hydrogenogenic carboxydotrophic mesophilic bacteria. *Bioresource Technology, 192*, 494–500.

Sinharoy, A., & Pakshirajan, K. (2024a). Biohydrogen production from agroindustrial wastes: Pretreatment, process engineering and techno-economic analysis. In: *Biohydrogen—Advances and processes* (pp. 287–309). Springer Nature Switzerland.

Sinharoy, A., & Pakshirajan, K. (2024b). Biohydrogen and the circular economy: Insertion of biohydrogen production in biorefineries aiming at zero-waste processes. In: *Biohydrogen—Advances and processes* (pp. 565–599). Springer Nature Switzerland.

Sinharoy, A., & Pakshirajan, K. (2024c). Bioprocessing of domestic wastewater and sewage sludge for biohydrogen production: Different routes and pretreatment strategies. In: *Biohydrogen—Advances and processes* (pp. 311–336). Springer Nature Switzerland.

Sinharoy, A., & Pakshirajan, K. (2021). Methane free biohydrogen production from carbon monoxide using a continuously operated moving bed biofilm reactor. *International Journal of Hydrogen Energy, 46*(1), 306–313.

Sinharoy, A., Pakshirajan, K., & Lens, P. N. L. (2022). Syngas fermentation for bioenergy production: Advances in bioreactor systems. In *Renewable energy technologies for energy efficient sustainable development* (pp. 325–358).

Sinharoy, A., Pakshirajan, K., & Lens, P. N. L. (2020b). Biological sulfate reduction using gaseous substrates to treat acid mine drainage. *Current Pollution Reports, 6*, 328–344.

Sivalingam, V., Winkler, D., Haugen, T., Wentzel, A., & Dinamarca, C. (2022). Syngas fermentation and microbial electrosynthesis integration as a single process unit. *Bioresource Technology, 356*, 127314.

Speight, J. G. (2020). *Handbook of gasification technology: Science, processes, and applications*. John Wiley & Sons.

Tharak, A., Katakojwala, R., Kajla, S., & Mohan, S. V. (2023). Chemolithoautotrophic reduction of CO_2 to acetic acid in gas and gas-electro fermentation systems: Enrichment, microbial dynamics, and sustainability assessment. *Chemical Engineering Journal, 454*, 140200.

Tharak, A., & Mohan, S. V. (2022). Syngas fermentation to acetate and ethanol with adaptative electroactive carboxydotrophs in single chambered microbial electrochemical system. *Micromachines, 13*(7), 980.

Trypolska, G., Kryvda, O., Kurbatova, T., Andrushchenko, O., Suleymanov, C., & Brydun, Y. (2021). Impact of new renewable electricity generating capacities on employment in Ukraine in 2021–2030. *International Journal of Energy Economics and Policy, 11*(6), 98–105.

Ucal, M., & Xydis, G. (2020). Multidirectional relationship between energy resources, climate changes and sustainable development: Technoeconomic analysis. *Sustainable Cities and Society, 60*, 102210.

Ugwu, A., Zaabout, A., Donat, F., van Diest, G., Albertsen, K., Muller, C., & Amini, S. (2021). Combined syngas and hydrogen production using gas switching technology. *Industrial & Engineering Chemistry Research, 60*(9), 3516–3531.

Vassilev, I., Hernandez, P. A., Batlle-Vilanova, P., Freguia, S., Krömer, J. O., Keller, J., Ledezma, P., & Virdis, B. (2018). Microbial electrosynthesis of isobutyric, butyric, caproic acids, and corresponding alcohols from carbon dioxide. *ACS Sustainable Chemistry & Engineering, 6*(7), 8485–8493.

Wood, J. C., Grové, J., Marcellin, E., Heffernan, J. K., Hu, S., Yuan, Z., & Virdis, B. (2021). Strategies to improve viability of a circular carbon bioeconomy—A techno-economic review of microbial electrosynthesis and gas fermentation. *Water Research, 201*, 117306.

Zafar, M. W., Shahbaz, M., Sinha, A., Sengupta, T., & Qin, Q. (2020). How renewable energy consumption contribute to environmental quality? The role of education in OECD countries. *Journal of Cleaner Production, 268*, 122149.

Zhang, Y., Khan, I., & Zafar, M. W. (2022). Assessing environmental quality through natural resources, energy resources, and tax revenues. *Environmental Science and Pollution Research, 29*(59), 89029–89044.

Zhou, M., Yan, B., Lang, Q., & Zhang, Y. (2019). Elevated volatile fatty acids production through reuse of acidogenic off-gases during electro-fermentation. *Science of the Total Environment, 668*, 295–302.

Zhu, X., Jack, J., Leininger, A., Yang, M., Bian, Y., Lo, J., Xiong, W., Tsesmetzis, N., & Ren, Z. J. (2022). Syngas mediated microbial electrosynthesis for CO_2 to acetate conversion using *Clostridium ljungdahlii*. *Resources, Conservation and Recycling, 184*, 106395.

Chapter 15
Shewanella as a Model System: Deciphering Exo-electrogenic Pathways for Advanced Electro-Fermentation

Vashisht Vishakha and Meenu Chhabra

1 Introduction

Electro-fermentation (EF) consolidates electrochemistry and microbial metabolism to perform all electrochemical redox processes (Fernandez et al., 2016). It is possible to circumvent the thermochemical limitations of conventional anaerobic fermentation technology by combining electrochemical and biological operations because of higher efficiency in substrate conversion and better electron transfer which can handle a wider range of material, including complex organic wastes (Bundhoo, 2017). This system is collectively called a bioelectrochemical system (BES). BES can produce electricity and valuable products while treating wastewater, reducing unwanted by-products. They also operate flexibly under various conditions, including anaerobic environments, high temperatures, and alkaline/acidic pH, which lowers the operational costs. BES captures and utilizes carbon dioxide (CO_2), supporting sustainable processes. By integrating with renewable energy systems, BES can store excess energy, making them a superior choice for bioenergy production and waste treatment. BES demonstrates the spawning of electrical power or the accomplishment of a redox reaction by facilitating the transfer of electrons between an electron acceptor and an electron donor while maintaining a certain potential (Fernandez et al., 2016). A multitude of microorganisms can transfer electrons to electrodes located outside their cells. Under anoxic conditions, the gram-negative bacteria *Geobacter* can utilize electrodes as the terminal electron acceptors for anaerobic respiration which results in electricity generation (Bond & Lovley, 2003). Similar to this, the gram-negative bacteria *Shewanella* has the potential to couple the process of oxidizing organic materials intracellularly with extracellular electrode reduction (Baron et al., 2009). Gram-negative bacteria *Acidithiobacillus ferrooxidans* and *Rhodopseudomonas palustris*

V. Vishakha · M. Chhabra (✉)
Department of Bioscience and Bioengineering, Indian Institute of Technology Jodhpur, Jodhpur, Rajasthan, India
e-mail: meenuchhabra@iitj.ac.in

© The Author(s), under exclusive license to Springer Nature Singapore Pte Ltd. 2025

S. Sevda (ed.), *Electro-Fermentation for Synthesis of Chemicals, Biochemicals and Biofuels*, https://doi.org/10.1007/978-981-96-6358-3_15

TIE-1 instantly obtain electrons from electrodes and utilize them to fix atmospheric CO_2 (Bose et al., 2014; Ishii et al., 2015). Not only gram-negative bacteria but also gram-positive bacteria and archaea are capable of transferring electrons to electrodes (Lohner et al., 2014). Microorganisms that can transfer electrons extracellularly by oxidizing organic substrates are commonly known as exoelectrogens (Lovley, 2017), the systems that assist these redox reactions are called microbial fuel cells (MFCs) and microbial electrolytic cells (MECs) (Fig. 1a, b). The operational progress of MFCs and MECs is sustained by microbial catalysts utilizing numerous feeding schemes, including pure nutrients or nutrient-rich waste from various wastewater sources (Wang & Sangeetha, 2019). MFCs have effectively utilized BES with various microbial species including *Shewanella oneidensis* MR-1 (Bretschger et al., 2007), *Shewanella putrefaciens* (Yang et al., 2021), and *Escherichia coli* DH5α (Li et al., 2018a, 2018b) which produced maximum power density ranging from 3800 to 4400 mW/m^2 (Li et al., 2018a, 2018b). MFCs are anticipated to be utilized for power generation from biomass waste, while MECs are specifically developed to produce hydrogen gas, a clean energy source free of CO_2 emissions, with high levels of efficiency. Microbial electrosynthesis is appealing due to its ability to produce organic molecules from CO_2 sustainably. Researchers concentrate their efforts on comprehending the physiology and genetics of electrochemically active bacteria (EAB), as these bacteria play crucial functions in these BESs. This understanding is essential for maximizing the utilization of EAB in BESs (Dessì et al., 2021; Munoz-Cupa et al., 2021; Yang et al., 2021).

2 *Shewanella*: The Model Organisms

The *Shewanella* genus consists of more than 70 species. It encompasses bacterial strains obtained from various environments, such as deep-sea sediments, oil brine, fish bodies, freshwater sediment, and ruined foods (Nealson & Scott, 2006). These bacteria are gram-negative, facultatively anaerobic, and heterotrophic which belongs to γ-proteobacterium (Hau & Gralnick, 2007; Nealson & Scott, 2006). Some of them are known to cause diseases in fish and animals. The majority of these bacteria may be readily cultivated in nutrient-rich media (Lemaire et al., 2020). Currently, over 200 strains of the genus *Shewanella* have had their genomes sequenced. Through comparative genomic analysis, it has been indicated that these bacteria possess similar genetic characteristics (Yang et al., 2017). The ability to respire a wide range of inorganic and organic compounds, including oxidized metals such as Mn(III, IV) and Fe(III), metal oxides such as hematite and magnetite, elemental sulfur, nitrate, thiosulfate, sulfite, fumarate, dimethyl sulfoxide, and trimethylamine-N-oxide is a shared characteristic of genus *Shewanella* (Hau & Gralnick, 2007; Nealson & Scott, 2006). Their ability to live in redox-stratified environments, where the availability of electron acceptors frequently changes in response to environmental perturbations, is because of terminal reductases and overlapping electron-transport routes

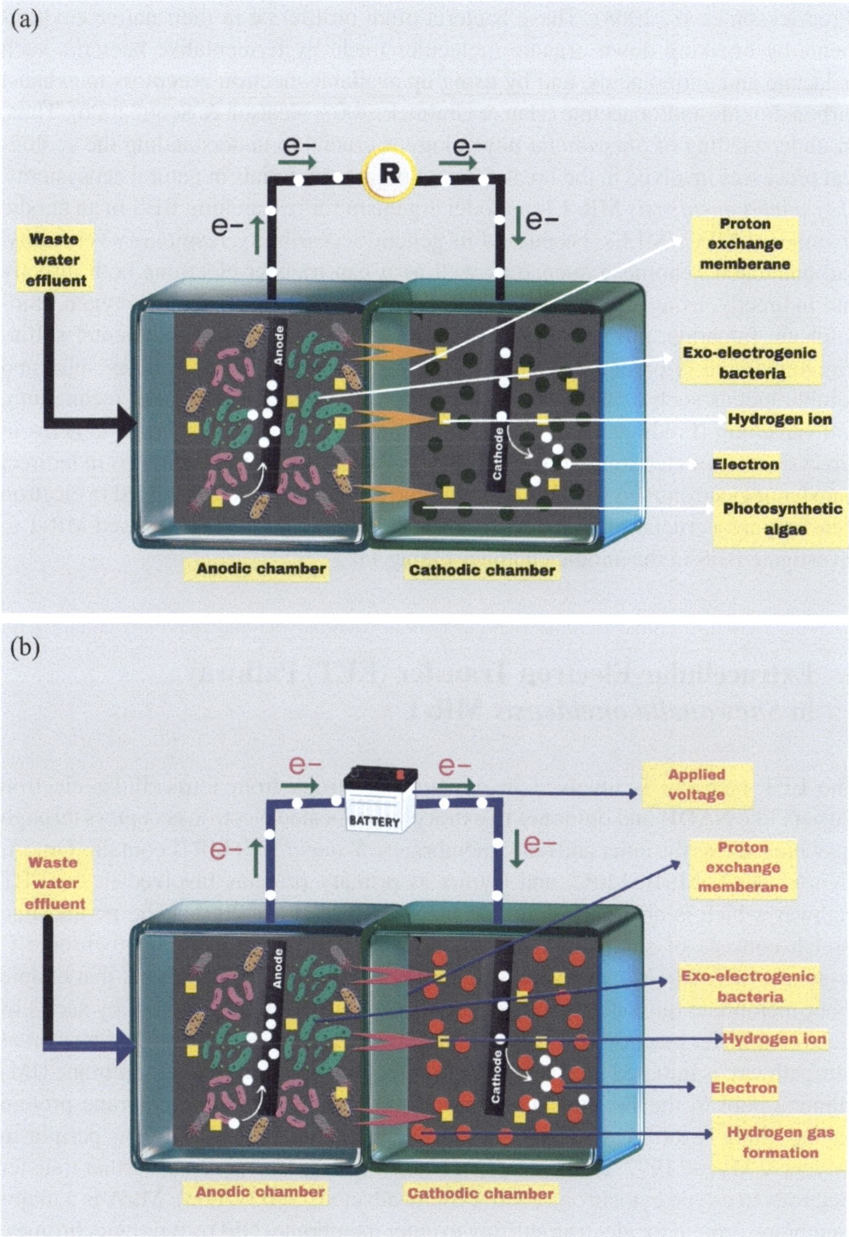

Fig. 1 **a** Microbial fuel cell, **b** Microbial electrolytic cell

(Fredrickson et al., 2008). These bacteria often proliferate in their native environments by breaking down organic molecules made by fermentative bacteria, such as lactate and amino acids, and by using up available electron acceptors to exhaust carbon dioxide and/or acetate (Hau & Gralnick, 2007; Nealson & Scott, 2006). Thus, an understanding of *Shewanella* physiology is crucial to understanding the ecological processes involved in the breakdown of organic materials in natural ecosystems. *Shewanella oneidensis* MR-1 is a model organism for researching BES in an anodic chamber of MFCs/MECs, because of its genetic accessibility, respiratory versatility, and annotated genome sequence as well as it can transfer electrons both directly and indirectly, using many organic and inorganic substrates, such as oxygen, thiosulphate, fumarate, nitrate, nitrite, dimethyl sulfoxide (DMSO), elemental sulfur, trimethylamine N-oxide, and anthraquinone-2,6-disulfonate, as well as solid and soluble metals such as cobalt, uranium, manganese, iron, chromium, technetium, and vanadium (Fredrickson et al., 2008; Kim et al., 1999). Cytochrome C helps in direct transfer of electrons from cell to electrode whereas riboflavin helps in indirect transfer of electrons (Fig. 2). The extracellular pathways that are involved in electron transfer play a crucial role in electron BES. Various researchers have used MR-1 to investigate BES in the anodic chamber (Table 1).

3 Extracellular Electron Transfer (EET) Pathway in *Shewanella oneidensis* MR-1

The EET pathway is involved in directing electrons from intracellular electron carriers like NADH and quinones to extracellular located electron acceptors through spawning across the inner and outer membranes. *S. oneidensis* MR-1 contains OmcA, CymA, MtrA, MtrB, MtrC, and CymA as primary proteins involved in the EET pathway which is also known as Mtr pathway (Shi et al., 2007). The periplasmic stretch consists of small tetraheme cytochromes (STCs) and flavocytochrome C protein involved in Mtr pathway, which act as major electron channels that couple inner membrane quinone pool to extracellular solid electron receptors by series of electron transfer reactions (Fonseca et al., 2013; Sturm et al., 2015). The first step of Mtr pathway is initiated by transferring of electrons from the Inner membrane (IM) quinone pool to the IM-attached cymA. CymA is an intrinsic membrane protein having short N-terminal outside to IM and C-region poking out into periplasm (Myers & Myers, 1997, 2000). The C-terminal of CymA protein will further transfer electrons to c-type cytochrome, MtrA (Schwalb et al., 2002, 2003). MtrA is a major membrane protein for electron shifting to outer membrane (OM) c-type cytochromes, OmcA, and MtrC. MtrA is inevitable because of its periplasmic localization and biochemical attachment to MtrB and MtrC (Hartshorne et al., 2009; Schuetz et al., 2009). MtrB is a β-barrel transmembrane protein located on OM which further transfers electrons to metal and electrode and is helpful in perfect localization and insertion

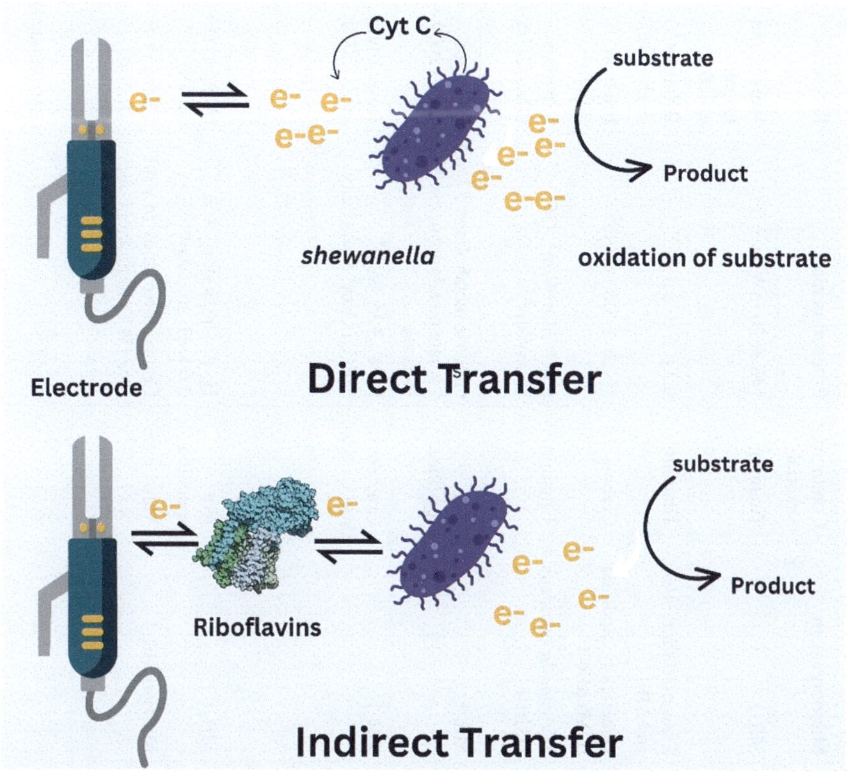

Fig. 2 Direct and indirect transfer of electrons from cell to electrode and vice versa

of OmcA and MtrC into the OM (Myers & Myers, 2002). The MtrC and OmcA act as the terminal reductases for extracellular electron acceptors.

Metabolically *Shewanella oneidensis* MR-1 has more tendency to utilize low molecular weight substrates such as amino acid, lactate, and pyruvate rather than glucose (Yang et al., 2006). However, in marine environments, MR-1 utilizes N-acetylglucosamine (NAG) sugar for metabolism which is present in profuse amounts in seawater (Yang et al., 2006). MR-1 also uses d-lactate as substrate in the absence of electron acceptor to produce acetate, the catabolic end product (Kasai et al., 2019a, 2019b). According to genomic analysis, defects in the phosphorylation stage during glycolysis may be the cause of the impairment in utilizing glucose (Kouzuma et al., 2015). Global regulatory systems govern the expression of numerous genes involved in catabolic and electron-transport pathways in MR-1; two of these systems are thought to be crucial: the cyclic AMP (cAMP)/cAMP-receptor protein (CRP) system and the aerobic respiration control (Arc) system (Hirose et al., 2019) (Fig. 3a). cAMP/CRP system is responsible for the expression of genes involved in Mtr pathway

Table 1 Comparative analysis of MR-1 investigated by various researchers

Sl. No.	Electro-fermentation (EF)	Electrolytic cell	Electrode material	Microorganism	Conducting material	Production rate	Reference
1	Anodic EF	MFC	Pt catalyst air cathode constructed from carbon cloth	MR-1	Titanium	148 ± 20 mW m ± 2	Watson and Logan (2010)
2	Anodic EF	MFC	Titanium sheet coated with carbon	*Shewanella japonica* and MR-1	Titanium	$0.3\ \mu A \pm 2\ \mu A$	Biffinger et al. (2011)
3	Anodic EF	MFC	Carbon cloth	Genetically modified (GM) MR-1	Carbon	$167.6 \pm 3.6\ mW/m^2$	Liu et al. (2015)
4	Anodic EF	MFC	Reticulated vitreous carbon (RVC) electrode + graphite felt (GF)	*Shewanella* oneidensis DSP10	Titanium	RVC: 24 mW/m^2, GF: 10 mW/m^2	Ringeisen et al. (2006)
5	Anodic EF	MFC voltage base screening assay (VBSA)	Single-sided carbon-coated titanium flags	MR-1	Not specified	Glucose metabolism: 17 mA Lactate metabolism: twofold increase	Biffinger et al. (2009)
6	Anodic EF	Current production in MFC Hydrogen production in MEC	Carbon-fiber-fabric	MR-1	Graphite rod	MFC: 3.56 W/m^3 MEC: 0.3 $m^3\ H_2/m^3$ day	Rosenbaum et al. (2010)
7	Anodic EF	MEC	Carbon cloth	MR-1	Not specified	H_2 production: 61.8 ± 2.0 L-H_2 $m^{-3}day^{-1}$	Wang et al. (2019)
8	Anodic EF	MFC, single-chamber three-electrode MEC	Carbon felt	Mutated MR-1	Metal wire	Comparative study of wild type (WT) to mutated one	Cheng et al. (2020)

(continued)

Table 1 (continued)

Sl. No.	Electro-fermentation (EF)	Electrolytic cell	Electrode material	Microorganism	Conducting material	Production rate	Reference
9	Anodic EF	MFC	AXF-5Q graphite	GM MR-1	Platinum wire	Comparative study of WT to GM	Ross et al. (2011)
10	Anodic EF	MEC	5X-AQ carbon electrode	MR-1	Platinum wire	110 μA ± 5.1, n = 3	Baron et al. (2009)

including MtrA, B, C, and omcA (Kasai et al., 2015). With the help of genetic engineering and metabolomics, these metabolic pathways can be modified for glucose oxidation which is the most abundant source of energy.

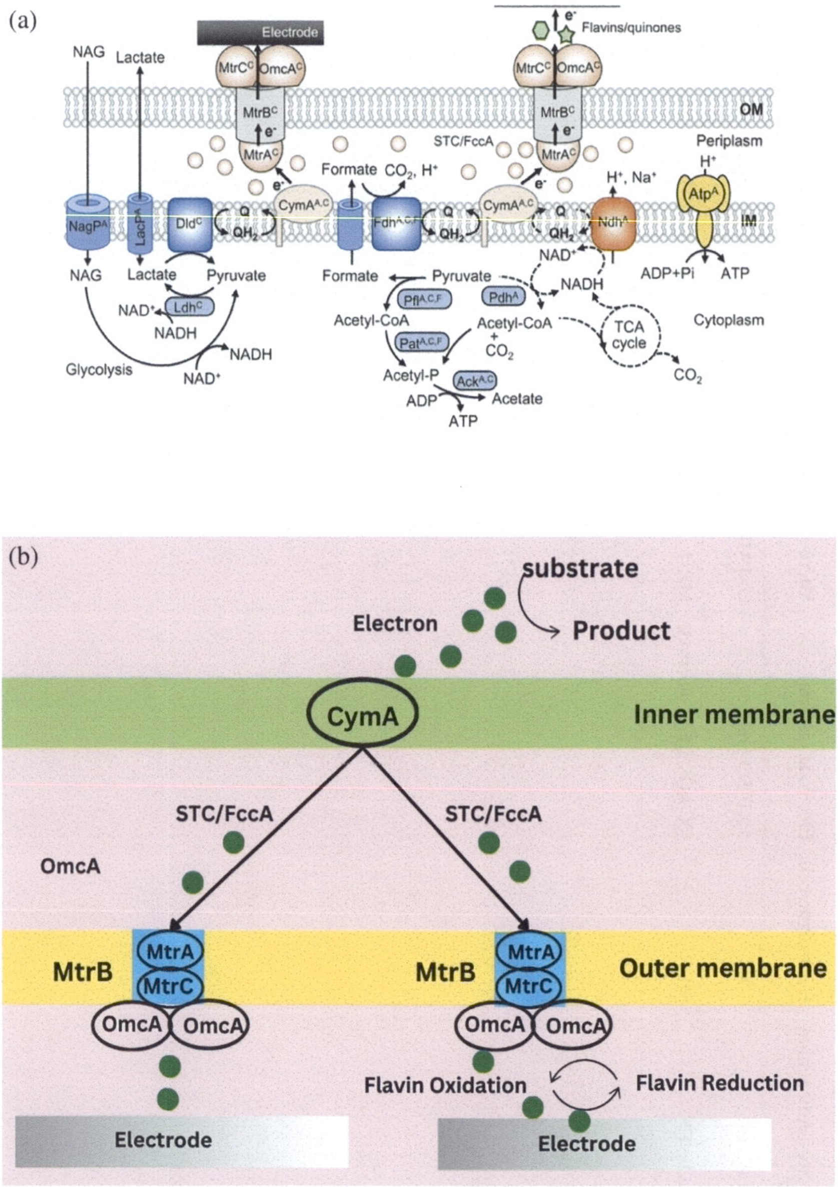

Fig. 3 **a** The Mtr pathway of *Shewanella oneidensis* MR-1 (Ikeda et al., 2021). **b** Flow diagram of transferring of electron from cell to electrode

4 Biofilm Formation by *Shewanella oneidensis* MR-1

Biofilms have been generally cognate to nosocomial infections or material biodeterioration but they play an important role in the electron exchange process. Upon adhering to a surface, bacteria frequently build intricate biofilm structures. According to Davey and O'toole (2000), this includes molecular processes such as aggregation, adhesion, and community growth mechanisms. Some auto aggregation elements such as type IV pili, curli protein, fimbriae, and surface adhesions also help in biofilm formation (Béchet & Blondeau, 2003; Bieber et al., 1998; Schembri et al., 2003). These elements also enable bacteria to auto-aggregate, a behavior that may be seen as flocculation from liquid suspensions and easily recognized under a microscope as a clumping of cells (Schembri et al., 2003). Immediate association of MR-1 with, or close vicinity to solid surface expedites the accumulation of electrons (Das & Caccavo Frank, 2000). MR-1 species has versatility for respiration which allows it to use different organic and inorganic compounds as terminal electron acceptors (Myers & Nealson, 1988). In aerobic hydrodynamic conditions, it forms biofilm by first attaching a single cell to a solid surface, then covering the surface, and then forming distinct three-dimensional (3D) structures (Fig. 4) (Thormann et al., 2004). This 3D structure is constructed of extracellular polysaccharides, such as extracellular DNA (eDNA), proteins, carbohydrates, and lipids but its exact architect constituent material is still unknown (Thormann et al., 2004). In anaerobic conditions, it shows commendable reducing ability by utilizing Fe(III), Mn(IV), and some ribonucleotide as terminal electron acceptors (Myers & Nealson, 1988), but it forms less dense and thick biofilm as compared to other species (e.g., *Geobacter*) (Malvankar NS). Cyclic di-GMP (c-di-GMP) plays an important role in biofilm formation, virulence, motility, and cell cycle progression (Chao et al., 2013). The elevated level of c-di-GMP facilitates the productivity of adhesive matrix components which increases the biofilm formation whereas a lower level is advantageous for the planktonic cells (Fig. 4) (Hengge, 2009). Diguanylate cyclases (DGC) having GGDEF amino acid sequence motif is responsible for the synthesis of c-di-GMP. Many GGDEF (protein no = 50) and EAL-domain proteins (protein no = 31) are anticipated to be encoded by the *S. oneidensis* MR-1 genome, with a huge portion predicted to be multidomain (e.g., GGDEF-EAL-domain) containing enzymes (Hengge, 2009). Additional Per-Arnt-Sim (PAS) sensory domains are present in ten of those multidomain proteins which are responsible for controlling external environment stimuli including light, oxygen, and redox status in a multitude of animals. However, c-di-GMP pathway of targeting biofilm formation in *Shewanella oneidensis* MR-1 is not clear.

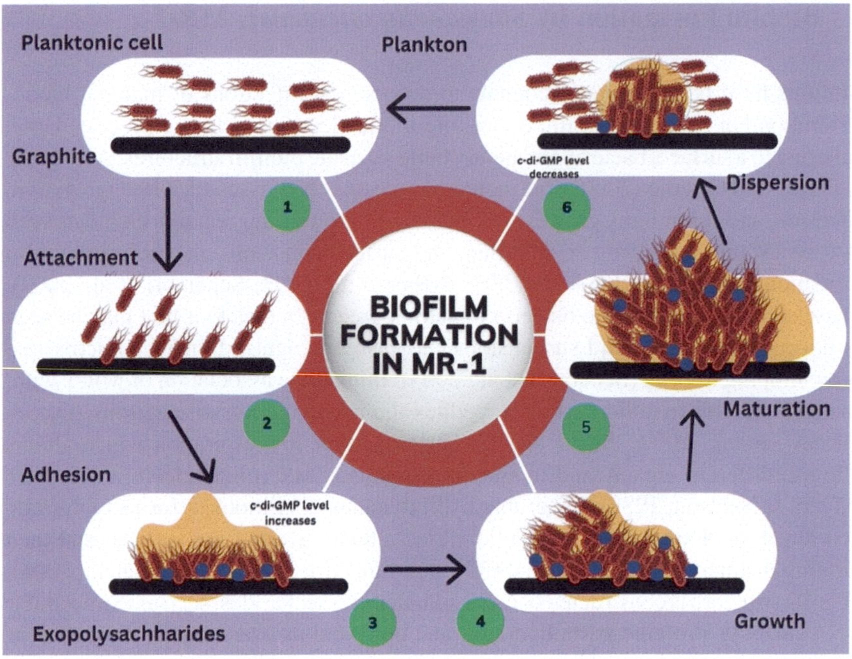

Fig. 4 Biofilm formation in MR-1

5 Genetic Engineering Insights of *Shewanella oneidensis* MR-1 for Enhanced Electron Flow

Modern bioelectric systems have huge potential and give promising results but are not widely used and have a narrow range of industrial applications due to their low power output. To improve the electrochemical system's capacity and increase bioelectricity productivity, scientists have genetically altered the Shewanella bacterium (Table 2), including multiple gene modifications. Genetic engineering escalates the EET rate of microbial catalysts in BES by modifying genes to expand the spectrum of substrate utilization, increase the intracellular pool of releasable electrons, change the membrane's permeability to electron shuttle molecules, and overexpress electron shuttles or cytochrome pathways for electron transport, genetic engineering enables the enhancement of the EET rate of microbial catalyst in BES (Surti et al., 2021). Dozens of genetic methods are available to modify genetic material of any organism, such as adding new genes, removing undesirable genes, or replacing current genes with new ones. Novel genes can be integrated into chromosomes using integrating viral vectors expressed from extrachromosomal genetic material known as plasmid. For heterologous (plasmid-based) or chromosomal expression of genes, the desired gene for a novel function can be amplified from the source by cutting with a site-specific restriction endonuclease, joined into the desired plasmid or viral vector, and then injected into the host organism using polymerase chain reaction (PCR). Other methods include site-specific or transposon-mediated mutagenesis disruptions of desired genes. The regulatory components that manage the expression of genes downstream are known as gene promoters. Modification in the promoter region can also allow overexpression or downregulation of any gene. Compared to traditional genetic alteration methods, genome-scale editing approaches such as clustered regularly spaced short palindromic repeat (CRISPR)-Cas9 are more reliable and less time-consuming. CRISPR technique does not rely on plasmid stability, antibiotic resistance selection markers, and metabolic or genetic load on the host (Surti et al., 2021). The availability of the Genome sequence of *Shewanella* and advanced genetic tools have opened up new paths in the field of genetic engineering. As presented in Table 2, many researchers have targeted direct and indirect pathways of electron transfer (Fig. 3) to be stepped up EET, which will be further detailed below.

5.1 Targeting EET Pathway

Numerous studies have been conducted on MR-1's bidirectional electron transport across cell membranes. In Mtr pathway, outward current flows from the inside of cells to the outer membrane (OM) and extracellular anodes. Electrons from NADH, flow through the menaquinone pool, CymA in IM, tetraheme c-type cytochromes [c-Cyts], MtrA (periplasmic decaheme c-Cyts), MtrB (β-barrel trans-OM protein), and finally MtrC and OmcA (two OM decaheme c-Cyts) (Fig. 1). There is still debate on the mechanism of interfacial electron exchange between OM c-Cyts and electrodes, despite the membrane channel of OmcA-MtrCAB being well established. A concept of interfacial EET comprises two methods: shuttle-mediated EET through naturally secreted soluble flavins, such as flavin mono-nucleotide and riboflavin, and direct contact-based EET through OM c-Cyts or nanowires (Bretschger et al., 2007; Fonseca et al., 2013; Shi et al., 2007; Sturm et al., 2015). Flavin acts as a cofactor that binds to OmcA or MtrC and helps in EET which is considered to be rate rate-limiting step (Okamoto et al., 2013, 2014a, 2014b, 2014c). Genetic modification in flavin protein will directly affect the electron bidirectional flow and enhance the EF process. Yang et al. (2015) heterologously modified the flavin biosynthetic pathway from *Bacillus subtilis* into MR-1, resulting in a 25.7-fold increase in secreted flavin concentration (Yang et al., 2015).

5.2 Genetic Modification in MtrB Gene

MtrB gene presents downstream of OmcB which encodes protein responsible for the reduction of Fe(III) and Mn(IV) (Beliaev et al., 2001). Researchers identify that MtrA, MtrB, and OmcB are transcribed as polycistronic mRNA in nature (Beliaev et al., 2001). To better understand MtrB function in metal reduction, research using

Table 2 A survey of GM *Shewanella* species for EF

Sl. No.	Electro-fermentation (EF)	Electrode material	Conducting material	Microorganism	Targeting gene/ pathway	Plasmid designed	WT production rate	GM production rate	Refs.
1	Anodic EF in MFC	Graphite felt	Platinum (Pt) wire	MR-1 + *Escherichia coli*	CymA	ccm (pEC86), mtrA (I5024), and mtrCAB (I5023)	Current production: ~19 μA Iron reduction capability = 62 ± 19 μM Fe(II)	Current production: ~90 μA Iron reduction: twofold increase	Jensen et al. (2016)
2	Single chambered MFC	Pt-loaded carbon cloth	Pt wire	*Pseudomonas aeruginosa*	phzM (methyltransferase encoding gene)	pBBR1MCS-5 vector	Current density: 142.31 μA/cm^2	Current density: 487.97 μA/cm^2	Yong et al. (2014)
3	Anodic EF in MFC	Carbon cloth	Metal wire	MR-1 + *Bacillus subtilis*	Flavin biosynthesis pathway	pYYDT plasmid	Current density: 15.5 μA/cm^2	Current density 255.3 μA/cm^2	Yang et al. (2015)
4	Anodic EF in MFC	Carbon cloth	Metal wire	*S. oneidensis*	Ribosome binding site engineering for increased flavin biosynthetic pathway	Graphene oxide used for formation engineered *Shewanella*-reduced GO	Current density: 0.18 A/m^2	Current density: 18.78 A/m^2	Lin et al. (2018)
5	Anodic EF in MFC	Carbon cloth	Metal wire	*S. oneidensis*	Redirect metabolic flux toward NAD$^+$ by de novo (nadA, B, C gene), salvage (pncB gene), and universal biosynthesis pathway (nadE, M, D gene)	Overexpression of genes by vector pYYDT	Power density: 30.3 ± 3.4 mW m^{-2}	Power density: 162.8 ± 5.6 mW m^{-2}	Li et al. (2018a, 2018b)

(continued)

Table 2 (continued)

Sl. No.	Electro-fermentation (EF)	Electrode material	Conducting material	Microorganism	Targeting gene/ pathway	Plasmid designed	WT production rate	GM production rate	Refs.
6	Anodic EF in MFC	Graphite felt	Metal wire	*S. oneidensis*	Gene mtrF containing a C-terminal Strep-tag® introduced before gene mtrA and B	pMQ150	–	43.47 ± 5.9 μA cm^2/mV	Golitsch et al. (2013)
7	Anodic EF in MEC	Carbon felt	Metal wire	MR-1	Insertion of an ISSod2 element into the flrA (encoding the master regulator for flagella synthesis and assembly) of MR-1 resulted in the non-motile and biofilm-deficient phenotypes in nonmotile mutant cells	pRE112, E. coli WM3064	86 mA/m^2	144 mA/m^2	Cheng et al. (2020)

an MtrB gene replacement mutant showed that MtrB is required to correctly include OmcA and OmcB cytochromes into the OM of MR-1. This significantly impacts the metal reduction deficit linked to MtrB loss (Myers & Myers, 2002). Myres and Myers (2002) identified the role of MtrB protein in *S. putrefaciens* MR-1 by using a gene replacement mutant MTRB1 which lacks MtrB gene. Only 36% cytochrome content was present in the outer membrane of MTRB1. However, the absence of MtrB impacted the localization of OmcA and OmcB in the CM even though they were found in the OM and IM fractions of MTRB1 which results in OmcA dragging in the soluble portion of MTRB1. MR-1 was impervious to triton X-100 and was not dissolved OmcA and OmcB in the presence of Mg(II). However, MTRB1 was sensitive to triton X-100 and gladly solubilized OmcA and OmcB. Myres et al. findings point to the necessity of MtrB for the accurate localization and insertion of OmcA and OmcB into the OM of MR-1 (Myers & Myers, 2002). With this finding, overexpression in the MtrB gene will indirectly impact the bidirectional flow of the electrons through OmcA and OmcB.

5.3 Genetic Engineering of Adenylate Cyclase

Cyclic AMP receptor protein (CRP) stimulates the transcription of the gene cluster that codes for Mtr complex (mentioned above). The cyclic adenosine mono phosphate (cAMP) finding raises the prospect that altering the cAMP/CRP-dependent regulation system could alter MR-1's electrochemical capacity to produce current in BES. Kasai et al. genetically overexpressed cyaC gene which encodes for a membrane-bound adenylate cyclase class III. Kasai et al. (2019a, 2019b) constructed plasmids pBBR1MCS-2, pBBR-cyaC, and pBBR-cyaC-crp to introduce desired gene into MR-1 by filter mating with E. coli WM6026. The MR-1 strain containing pBBR-cyaC was designated cyaC-OE. More than five times cAMP is found intracellularly in cyaC-OE than in wild-type MR-1. Additionally, cya-OE produces around twice as much current in BES as the wild-type strain (Kasai et al., 2019a, 2019b).

5.4 Alternate Approaches to Multiple Gene Modifications

In fermentation, recycling intracellular electron carriers necessitates converting substrate into undesirable products, whereas respiration requires a continuous supply of electron acceptors (mentioned above). However, these limitations might be solved by balancing biotransformation with electrode-based electron acceptors. Some researchers have focused on product-based gene modification of desired strains to balance the EF process. Flynn et al. engineered *S. oneidensis* for stoichiometrically producing ethyl alcohol from glycerol. This biotransformation will not proceed until two electrons are eliminated via foreign reaction, such as electrode reduction. Flynn et al. (2010) constructed a single plasmid consisting glycerol module of genes *glpK*, *glpF*, *glpA*, and *tpiD* from *E. coli* and an ethyl alcohol module containing *adh* and *pdc* from *Zymomonas mobilis*. These genes were transferred to *S. oneidensis* to enhance ethyl formation. The phosphotransacetylase (pta) gene was knocked out from *S. oneidensis* to divert the metabolic flux from acetate to ethyl alcohol production. The observed ethyl alcohol production increased from 89 to 188 mA/m^2 (Flynn et al., 2010).

6 Future Perspective

By applying synthetic biology techniques, researchers have been able to successfully overexpress the EET pathways of model electroactive bacteria in wild-type electroactive microbes and extend this property to industrially significant non-electroactive microbes. It has been possible to create microbial factories for novel substrate oxidation or product synthesis by extrapolating the metabolic potential of the microbial catalyst in BES. All effective genetic engineering techniques, however, are constrained by our knowledge of the microbial catalyst's EET routes and metabolic controls. Bacterial populations exhibit remarkable sensitivity to environmental fluctuations in terms of genetic variation. Prior research has demonstrated that exposure to antibiotics can alter the pattern of bacterial mutations as well as increase their pace (Maharjan & Ferenci, 2018). While genetically engineered strains provide opportunities for innovations in agriculture, medicine, and industry, the concerns raised necessitate thorough safety evaluations, environmental monitoring, and regulatory oversight to balance benefits with potential risks. In terms of the environment, there's a chance that modified genes will move to populations that aren't genetically engineered, upsetting ecosystems and lowering biodiversity. Another worry is the emergence of superweeds or pests resistant to pesticides. Potential allergenicity and the adoption of marker genes resistant to antibiotics could be harmful to one's health. Precaution is necessary since the long-term impacts of genetically engineered organisms on the environment and human health are yet unclear. GM strains offer chances for innovation, but to weigh the advantages against the risks, they need to be carefully regulated and overseen. The pilot scale study of GM strains in the BES system is not widely recognized or present in the industrial system despite encouraging results from laboratory-scale research. The most popular application of BES has been as an integration into current wastewater treatment and anaerobic codigestion processes. For these kinds of applications, a mixed consortium of microorganisms is more adaptable in terms of organic breakdown than a single bacterium. Therefore, it is important to investigate the syntrophic relationships between genetically engineered microbial catalysts in BES. The goal of future studies can be to apply genetically modified BES to on-site studies and at larger scales production of commercial products.

References

Baron, D., LaBelle, E., Coursolle, D., Gralnick, J. A., & Bond, D. R. (2009). Electrochemical measurement of electron transfer kinetics by Shewanella oneidensis MR-1. *Journal of Biological Chemistry, 284*(42), 28865–28873.

Béchet, M., & Blondeau, R. (2003). Factors associated with the adherence and biofilm formation by *Aeromonas caviae* on glass surfaces. *Journal of Applied Microbiology, 94*(6), 1072–1078. https://doi.org/10.1046/j.1365-2672.2003.01931.x

Beliaev, A. S., Saffarini, D. A., McLaughlin, J. L., & Hunnicutt, D. (2001). MtrC, an outer membrane decahaem *c* cytochrome required for metal reduction in *Shewanella putrefaciens* MR-1. *Molecular Microbiology, 39*(3), 722–730. https://doi.org/10.1046/j.1365-2958.2001.02257.x

Bieber, D., Ramer, S. W., Wu, C.-Y., Murray, W. J., Tobe, T., Fernandez, R., & Schoolnik, G. K. (1998). Type IV Pili, transient bacterial aggregates, and virulence of enteropathogenic *Escherichia coli. Science, 280*(5372), 2114–2118. https://doi.org/10.1126/science.280.5372.2114

Biffinger, J. C., Fitzgerald, L. A., Ray, R., Little, B. J., Lizewski, S. E., Petersen, E. R., Ringeisen, B. R., Sanders, W. C., Sheehan, P. E., Pietron, J. J., Baldwin, J. W., Nadeau, L. J., Johnson, G. R., Ribbens, M., Finkel, S. E., & Nealson, K. H. (2011). The utility of Shewanella japonica for microbial fuel cells. *Bioresource Technology, 102*(1), 290–297. https://doi.org/10.1016/j.biortech.2010.06.078

Biffinger, J. C., Ray, R., Little, B. J., Fitzgerald, L. A., Ribbens, M., Finkel, S. E., & Ringeisen, B. R. (2009). Simultaneous analysis of physiological and electrical output changes in an operating microbial fuel cell with *Shewanella oneidensis. Biotechnology and Bioengineering, 103*(3), 524–531. https://doi.org/10.1002/bit.22266

Bond, D. R., & Lovley, D. R. (2003). Electricity production by *Geobacter sulfurreducens* attached to electrodes. *Applied and Environmental Microbiology, 69*(3), 1548–1555. https://doi.org/10.1128/AEM.69.3.1548-1555.2003

Bose, A., Gardel, E. J., Vidoudez, C., Parra, E. A., & Girguis, P. R. (2014). Electron uptake by iron-oxidizing phototrophic bacteria. *Nature Communications, 5*(1), 3391. https://doi.org/10.1038/ncomms4391

Bretschger, O., Obraztsova, A., Sturm, C. A., Chang, I. S., Gorby, Y. A., Reed, S. B., Culley, D. E., Reardon, C. L., Barua, S., Romine, M. F., Zhou, J., Beliaev, A. S., Bouhenni, R., Saffarini, D., Mansfeld, F., Kim, B.-H., Fredrickson, J. K., & Nealson, K. H. (2007). Current production and metal oxide reduction by *Shewanella oneidensis* MR-1 wild type and mutants. *Applied and Environmental Microbiology, 73*(21), 7003–7012. https://doi.org/10.1128/AEM.01087-07

Bundhoo, Z. M. A. (2017). Coupling dark fermentation with biochemical or bioelectrochemical systems for enhanced bio-energy production: A review. *International Journal of Hydrogen Energy, 42*(43), 26667–26686. https://doi.org/10.1016/j.ijhydene.2017.09.050

Chao, L., Rakshe, S., Leff, M., & Spormann, A. M. (2013). PdeB, a cyclic di-GMP-specific phosphodiesterase that regulates Shewanella oneidensis MR-1 motility and biofilm formation. *Journal of Bacteriology, 195*(17), 3827–3833.

Cheng, L., Min, D., Liu, D.-F., Zhu, T.-T., Wang, K.-L., & Yu, H.-Q. (2020). Deteriorated biofilm-forming capacity and electroactivity of Shewanella oneidensis MR-1 induced by insertion sequence (IS) elements. *Biosensors and Bioelectronics, 156*, 112136. https://doi.org/10.1016/j.bios.2020.112136

Das, A., & Caccavo Frank, J. (2000). Dissimilatory Fe (III) oxide reduction by Shewanella alga BrY requires adhesion. *Current Microbiology, 40*, 344–347.

Davey, M. E., & O'toole, G. A. (2000). Microbial biofilms: From ecology to molecular genetics. *Microbiology and Molecular Biology Reviews, 64*(4), 847–867. https://doi.org/10.1128/MMBR.64.4.847-867.2000

de los Ángeles Fernandez, M., de los Ángeles Sanromán, M., Marks, S., Makinia, J., Gonzalez del Campo, A., Rodrigo, M., & Fernandez, F. J. (2016). A grey box model of glucose fermentation and syntrophic oxidation in microbial fuel cells. *Bioresource Technology, 200*, 396–404. https://doi.org/10.1016/j.biortech.2015.10.010

Dessì, P., Rovira-Alsina, L., Sánchez, C., Dinesh, G. K., Tong, W., Chatterjee, P., Tedesco, M., Farràs, P., Hamelers, H. M. V., & Puig, S. (2021). Microbial electrosynthesis: Towards sustainable biorefineries for production of green chemicals from CO_2 emissions. *Biotechnology Advances, 46*, 107675. https://doi.org/10.1016/j.biotechadv.2020.107675

Flynn, J. M., Ross, D. E., Hunt, K. A., Bond, D. R., & Gralnick, J. A. (2010). Enabling unbalanced fermentations by using engineered electrode-interfaced bacteria. *MBio, 1*(5). https://doi.org/10.1128/mBio.00190-10

Fonseca, B. M., Paquete, C. M., Neto, S. E., Pacheco, I., Soares, C. M., & Louro, R. O. (2013). Mind the gap: Cytochrome interactions reveal electron pathways across the periplasm of *Shewanella oneidensis* MR-1. *Biochemical Journal, 449*(1), 101–108. https://doi.org/10.1042/BJ20121467

Fredrickson, J. K., Romine, M. F., Beliaev, A. S., Auchtung, J. M., Driscoll, M. E., Gardner, T. S., Nealson, K. H., Osterman, A. L., Pinchuk, G., Reed, J. L., Rodionov, D. A., Rodrigues, J. L. M., Saffarini, D. A., Serres, M. H., Spormann, A. M., Zhulin, I. B., & Tiedje, J. M. (2008). Towards environmental systems biology of Shewanella. *Nature Reviews Microbiology, 6*(8), 592–603. https://doi.org/10.1038/nrmicro1947

Golitsch, F., Bücking, C., & Gescher, J. (2013). Proof of principle for an engineered microbial biosensor based on Shewanella oneidensis outer membrane protein complexes. *Biosensors and Bioelectronics, 47*, 285–291. https://doi.org/10.1016/j.bios.2013.03.010

Hartshorne, R. S., Reardon, C. L., Ross, D., Nuester, J., Clarke, T. A., Gates, A. J., Mills, P. C., Fredrickson, J. K., Zachara, J. M., Shi, L., Beliaev, A. S., Marshall, M. J., Tien, M., Brantley, S., Butt, J. N., & Richardson, D. J. (2009). Characterization of an electron conduit between bacteria and the extracellular environment. *Proceedings of the National Academy of Sciences, 106*(52), 22169–22174. https://doi.org/10.1073/pnas.0900086106

Hau, H. H., & Gralnick, J. A. (2007). Ecology and biotechnology of the genus *Shewanella*. *Annual Review of Microbiology, 61*(1), 237–258. https://doi.org/10.1146/annurev.micro.61.080706.093257

Hengge, R. (2009). Principles of c-di-GMP signalling in bacteria. *Nature Reviews Microbiology, 7*(4), 263–273.

Hirose, A., Kasai, T., Koga, R., Suzuki, Y., Kouzuma, A., & Watanabe, K. (2019). Understanding and engineering electrochemically active bacteria for sustainable biotechnology. *Bioresources and Bioprocessing, 6*, 1–15.

Ikeda, S., Takamatsu, Y., Tsuchiya, M., Suga, K., Tanaka, Y., Kouzuma, A., & Watanabe, K. (2021). *Shewanella oneidensis* MR-1 as a bacterial platform for electro-biotechnology. *Essays in Biochemistry, 65*(2), 355–364. https://doi.org/10.1042/EBC20200178

Ishii, T., Kawaichi, S., Nakagawa, H., Hashimoto, K., & Nakamura, R. (2015). From chemolithoautotrophs to electrolithoautotrophs: CO_2 fixation by Fe(II)-oxidizing bacteria coupled with direct uptake of electrons from solid electron sources. *Frontiers in Microbiology, 6*. https://doi.org/10.3389/fmicb.2015.00994

Jensen, H. M., TerAvest, M. A., Kokish, M. G., & Ajo-Franklin, C. M. (2016). CymA and exogenous flavins improve extracellular electron transfer and couple it to cell growth in Mtr-expressing Escherichia coli. *ACS Synthetic Biology, 5*(7), 679–688.

Kasai, T., Kouzuma, A., Nojiri, H., & Watanabe, K. (2015). Transcriptional mechanisms for differential expression of outer membrane cytochrome genes omcA and mtrC in Shewanella oneidensis MR-1. *BMC Microbiology, 15*, 1–12.

Kasai, T., Suzuki, Y., Kouzuma, A., & Watanabe, K. (2019a). Roles of d-lactate dehydrogenases in the anaerobic growth of *Shewanella oneidensis* MR-1 on sugars. *Applied and Environmental Microbiology, 85*(3). https://doi.org/10.1128/AEM.02668-18

Kasai, T., Tomioka, Y., Kouzuma, A., & Watanabe, K. (2019b). Overexpression of the adenylate cyclase gene cyaC facilitates current generation by Shewanella oneidensis in bioelectrochemical systems. *Bioelectrochemistry, 129*, 100–105. https://doi.org/10.1016/j.bioelechem.2019.05.010

Kim, B.-H., Kim, H.-J., Hyun, M.-S., & Park, D.-H. (1999). Direct electrode reaction of Fe(III)-reducing bacterium, Shewanella putrefaciens. *Journal of Microbiology and Biotechnology, 9*(2), 127–131.

Kouzuma, A., Kasai, T., Hirose, A., & Watanabe, K. (2015). Catabolic and regulatory systems in Shewanella oneidensis MR-1 involved in electricity generation in microbial fuel cells. *Frontiers in Microbiology, 6*. https://doi.org/10.3389/fmicb.2015.00609

Lemaire, O. N., Méjean, V., & Iobbi-Nivol, C. (2020). The *Shewanella* genus: Ubiquitous organisms sustaining and preserving aquatic ecosystems. *FEMS Microbiology Reviews, 44*(2), 155–170. https://doi.org/10.1093/femsre/fuz031

Li, F., Li, Y.-X., Cao, Y.-X., Wang, L., Liu, C.-G., Shi, L., & Song, H. (2018a). Modular engineering to increase intracellular NAD (H/+) promotes rate of extracellular electron transfer of Shewanella oneidensis. *Nature Communications, 9*(1), 3637.

Li, H., Liao, B., Xiong, J., Zhou, X., Zhi, H., Liu, X., Li, X., & Li, W. (2018b). Power output of microbial fuel cell emphasizing interaction of anodic binder with bacteria. *Journal of Power Sources, 379*, 115–122. https://doi.org/10.1016/j.jpowsour.2018.01.040

Lin, T., Ding, W., Sun, L., Wang, L., Liu, C.-G., & Song, H. (2018). Engineered Shewanella oneidensis-reduced graphene oxide biohybrid with enhanced biosynthesis and transport of flavins enabled a highest bioelectricity output in microbial fuel cells. *Nano Energy, 50*, 639–648. https://doi.org/10.1016/j.nanoen.2018.05.072

Liu, T., Yu, Y., Deng, X., Ng, C. K., Cao, B., Wang, J., Rice, S. A., Kjelleberg, S., & Song, H. (2015). Enhanced *Shewanella* biofilm promotes bioelectricity generation. *Biotechnology and Bioengineering, 112*(10), 2051–2059. https://doi.org/10.1002/bit.25624

Lohner, S. T., Deutzmann, J. S., Logan, B. E., Leigh, J., & Spormann, A. M. (2014). Hydrogenase-independent uptake and metabolism of electrons by the archaeon *Methanococcus maripaludis*. *The ISME Journal, 8*(8), 1673–1681. https://doi.org/10.1038/ismej.2014.82

Lovley, D. R. (2017). Happy together: Microbial communities that hook up to swap electrons. *The ISME Journal, 11*(2), 327–336. https://doi.org/10.1038/ismej.2016.136

Maharjan, R. P., & Ferenci, T. (2018). The impact of growth rate and environmental factors on mutation rates and spectra in *Escherichia coli*. *Environmental Microbiology Reports, 10*(6), 626–633. https://doi.org/10.1111/1758-2229.12661

Munoz-Cupa, C., Hu, Y., Xu, C., & Bassi, A. (2021). An overview of microbial fuel cell usage in wastewater treatment, resource recovery and energy production. *Science of the Total Environment, 754*, 142429. https://doi.org/10.1016/j.scitotenv.2020.142429

Myers, C. R., & Myers, J. M. (1997). Cloning and sequence of cymA, a gene encoding a tetraheme cytochrome c required for reduction of iron(III), fumarate, and nitrate by Shewanella putrefaciens MR-1. *Journal of Bacteriology, 179*(4), 1143–1152. https://doi.org/10.1128/jb.179.4.1143-1152.1997

Myers, C. R., & Myers, J. M. (2002). MtrB is required for proper incorporation of the cytochromes OmcA and OmcB into the outer membrane of *Shewanella putrefaciens* MR-1. *Applied and Environmental Microbiology, 68*(11), 5585–5594. https://doi.org/10.1128/AEM.68.11.5585-5594.2002

Myers, C. R., & Nealson, K. H. (1988). Bacterial manganese reduction and growth with manganese oxide as the sole electron acceptor. *Science, 240*(4857), 1319–1321. https://doi.org/10.1126/science.240.4857.1319

Myers, J. M., & Myers, C. R. (2000). Role of the tetraheme cytochrome CymA in anaerobic electron transport in cells of *Shewanella putrefaciens* MR-1 with normal levels of menaquinone. *Journal of Bacteriology, 182*(1), 67–75. https://doi.org/10.1128/JB.182.1.67-75.2000

Nealson, K. H., & Scott, J. (2006). Ecophysiology of the genus Shewanella. *The Prokaryotes, 6*, 1133–1151.

Okamoto, A., Hashimoto, K., & Nealson, K. H. (2014a). Flavin redox bifurcation as a mechanism for controlling the direction of electron flow during extracellular electron transfer. *Angewandte Chemie International Edition, 53*(41), 10988–10991. https://doi.org/10.1002/anie.201407004

Okamoto, A., Hashimoto, K., Nealson, K. H., & Nakamura, R. (2013). Rate enhancement of bacterial extracellular electron transport involves bound flavin semiquinones. *Proceedings of the National Academy of Sciences, 110*(19), 7856–7861. https://doi.org/10.1073/pnas.1220823110

Okamoto, A., Nakamura, R., Nealson, K. H., & Hashimoto, K. (2014b). Bound flavin model suggests similar electron-transfer mechanisms in *Shewanella* and *Geobacter*. *ChemElectroChem, 1*(11), 1808–1812. https://doi.org/10.1002/celc.201402151

Okamoto, A., Saito, K., Inoue, K., Nealson, K. H., Hashimoto, K., & Nakamura, R. (2014c). Uptake of self-secreted flavins as bound cofactors for extracellular electron transfer in Geobacter species. *Energy & Environmental Science, 7*(4), 1357–1361.

Ringeisen, B. R., Henderson, E., Wu, P. K., Pietron, J., Ray, R., Little, B., Biffinger, J. C., & Jones-Meehan, J. M. (2006). High power density from a miniature microbial fuel cell using Shewanella oneidensis DSP10. *Environmental Science & Technology, 40*(8), 2629–2634.

Rosenbaum, M., Cotta, M. A., & Angenent, L. T. (2010). Aerated *Shewanella oneidensis* in continuously fed bioelectrochemical systems for power and hydrogen production. *Biotechnology and Bioengineering, 105*(5), 880–888. https://doi.org/10.1002/bit.22621

Ross, D. E., Flynn, J. M., Baron, D. B., Gralnick, J. A., & Bond, D. R. (2011). Towards electrosynthesis in Shewanella: Energetics of reversing the Mtr pathway for reductive metabolism. *PLoS ONE, 6*(2), e16649. https://doi.org/10.1371/journal.pone.0016649

Schembri, M. A., Kjærgaard, K., & Klemm, P. (2003). Global gene expression in *Escherichia coli* biofilms. *Molecular Microbiology, 48*(1), 253–267. https://doi.org/10.1046/j.1365-2958.2003.03432.x

Schuetz, B., Schicklberger, M., Kuermann, J., Spormann, A. M., & Gescher, J. (2009). Periplasmic electron transfer via the *c*-type cytochromes MtrA and FccA of *Shewanella oneidensis* MR-1. *Applied and Environmental Microbiology, 75*(24), 7789–7796. https://doi.org/10.1128/AEM.01834-09

Schwalb, C., Chapman, S. K., & Reid, G. A. (2002). The membrane-bound tetrahaem c-type cytochrome CymA interacts directly with the soluble fumarate reductase in *Shewanella*. *Biochemical Society Transactions, 30*(4), 658–662. https://doi.org/10.1042/bst0300658

Schwalb, C., Chapman, S. K., & Reid, G. A. (2003). The tetraheme cytochrome CymA is required for anaerobic respiration with dimethyl sulfoxide and nitrite in *Shewanella oneidensis*. *Biochemistry, 42*(31), 9491–9497. https://doi.org/10.1021/bi034456f

Shi, L., Squier, T. C., Zachara, J. M., & Fredrickson, J. K. (2007). Respiration of metal (hydr)oxides by *Shewanella* and *Geobacter* : A key role for multihaem *c*-type cytochromes. *Molecular Microbiology, 65*(1), 12–20. https://doi.org/10.1111/j.1365-2958.2007.05783.x

Sturm, G., Richter, K., Doetsch, A., Heide, H., Louro, R. O., & Gescher, J. (2015). A dynamic periplasmic electron transfer network enables respiratory flexibility beyond a thermodynamic regulatory regime. *The ISME Journal, 9*(8), 1802–1811. https://doi.org/10.1038/ismej.2014.264

Surti, P., Kailasa, S. K., & Mungray, A. K. (2021). Genetic engineering strategies for performance enhancement of bioelectrochemical systems: A review. *Sustainable Energy Technologies and Assessments, 47*, 101332. https://doi.org/10.1016/j.seta.2021.101332

Thormann, K. M., Saville, R. M., Shukla, S., Pelletier, D. A., & Spormann, A. M. (2004). Initial phases of biofilm formation in *Shewanella oneidensis* MR-1. *Journal of Bacteriology, 186*(23), 8096–8104. https://doi.org/10.1128/JB.186.23.8096-8104.2004

Wang, C.-T., & Sangeetha, T. (2019). *Microbial fuel cell: A synergistic flow approach for energy power generation and wastewater treatment* (pp. 309–334). https://doi.org/10.1007/978-981-32-9804-0_14

Wang, W., Zhang, B., & He, Z. (2019). Bioelectrochemical deposition of palladium nanoparticles as catalysts by Shewanella oneidensis MR-1 towards enhanced hydrogen production in microbial electrolysis cells. *Electrochimica Acta, 318*, 794–800. https://doi.org/10.1016/j.electacta.2019.06.038

Watson, V. J., & Logan, B. E. (2010). Power production in MFCs inoculated with *Shewanella oneidensis* MR-1 or mixed cultures. *Biotechnology and Bioengineering, 105*(3), 489–498. https://doi.org/10.1002/bit.22556

Yang, C., Rodionov, D. A., Li, X., Laikova, O. N., Gelfand, M. S., Zagnitko, O. P., Romine, M. F., Obraztsova, A. Y., Nealson, K. H., & Osterman, A. L. (2006). Comparative genomics and experimental characterization of N-acetylglucosamine utilization pathway of Shewanella oneidensis. *Journal of Biological Chemistry, 281*(40), 29872–29885.

Yang, E., Omar Mohamed, H., Park, S.-G., Obaid, M., Al-Qaradawi, S. Y., Castaño, P., Chon, K., & Chae, K.-J. (2021). A review on self-sustainable microbial electrolysis cells for electro-biohydrogen production via coupling with carbon-neutral renewable energy technologies. *Bioresource Technology, 320*, 124363. https://doi.org/10.1016/j.biortech.2020.124363

Yang, L., Deng, W., Zhang, Y., Tan, Y., Ma, M., & Xie, Q. (2017). Boosting current generation in microbial fuel cells by an order of magnitude by coating an ionic liquid polymer on carbon anodes. *Biosensors and Bioelectronics, 91*, 644–649. https://doi.org/10.1016/j.bios.2017.01.028

Yang, Y., Ding, Y., Hu, Y., Cao, B., Rice, S. A., Kjelleberg, S., & Song, H. (2015). Enhancing bidirectional electron transfer of Shewanella oneidensis by a synthetic flavin pathway. *ACS Synthetic Biology, 4*(7), 815–823.

Yong, X.-Y., Shi, D.-Y., Chen, Y.-L., Jiao, F., Lin, X., Zhou, J., Wang, S.-Y., Yong, Y.-C., Sun, Y.-M., OuYang, P.-K., & Zheng, T. (2014). Enhancement of bioelectricity generation by manipulation of the electron shuttles synthesis pathway in microbial fuel cells. *Bioresource Technology, 152*, 220–224. https://doi.org/10.1016/j.biortech.2013.10.086

Chapter 16
Electro-Fermentation for the Circular Carbon Economy

Prasann Kumar, Komal Agarwal, Debjani Choudhury, and Amit Raj

1 Introduction

Electro-fermentation (EF) is an innovative bioelectrochemical process that integrates traditional fermentation with electrochemical systems to enhance the synthesis of biofuels and valuable chemicals. EF is distinct from conventional fermentation because it uses an external electric current to control microbial metabolism rather than relying solely on organic substrates as electron donors. This technique employs electroactive microorganisms that can transfer electrons to electrodes, leading to a more controlled and efficient fermentation (Dahiya et al., 2021; dos Santos et al., 2023). The primary significance of EF lies in its ability to precisely control redox conditions, optimise metabolic pathways, and increase product yields. This manipulation can redirect metabolic fluxes towards the synthesis of specific compounds, reduce the production of undesirable by-products, and improve the overall efficiency of the process. This technological breakthrough shows immense potential for generating biofuels and creating various industrial chemicals. As a result, it has significant relevance and applicability in biochemistry, environmental science, and sustainable development. EF is a critical component of the circular carbon economy, which aims to repurpose and reutilise carbon-based resources to reduce inefficiency

P. Kumar (✉) · A. Raj
Department of Agronomy, School of Agriculture, Lovely Professional University, Phagwara, Punjab, India
e-mail: prasann0659@gmail.com

K. Agarwal (✉)
Department of Microbiology, School of Bioengineering and Biosciences, Lovely Professional University, Phagwara, Punjab, India
e-mail: komalagarwal935@gmail.com

D. Choudhury
Department of Plant Pathology, School of Agriculture, Lovely Professional University, Phagwara, Punjab, India

© The Author(s), under exclusive license to Springer Nature Singapore Pte Ltd. 2025
S. Sevda (ed.), *Electro-Fermentation for Synthesis of Chemicals, Biochemicals and Biofuels*, https://doi.org/10.1007/978-981-96-6358-3_16

and address the environmental impact of carbon emissions. In this economy, carbon atoms undergo a perpetual cycle of conversion, where CO_2 emissions from industrial processes are transformed into biofuels or other valuable chemicals. This approach differs from the traditional linear economy, in which resources are extracted, used, and thrown away as waste. The circular carbon economy aims to establish a highly sustainable and resilient economic framework through efficient management and mitigation of carbon emissions. EF contributes to this framework by converting waste carbon into valuable resources, such as agricultural residues and industrial effluents. This reduces the carbon emissions generated by various industries and offers a sustainable source of chemicals and fuels. Consequently, it aids global efforts to address climate change and transition to sustainable development. This chapter examines the role of electro-fermentation in promoting a circular carbon economy, focusing on its technological advancements, practical applications, and potential benefits (Ahmed et al., 2023; Kumareswaran et al., 2024). This extensive analysis seeks to provide a comprehensive understanding of the principles of EF and its significance in enhancing the efficiency and long-term sustainability of biochemical processes. The paper comprehensively investigates the underlying mechanisms of EF, recent progress in bioelectrochemical reactors, and the integration of renewable energy sources for powering these systems. Furthermore, the paper will investigate EF's various applications in producing biofuels, organic acids, and bioplastics and its role in waste management and extracting valuable resources. The chapter seeks to evaluate EF's economic and environmental consequences, explicitly focusing on the financial implications. This analysis highlights the feasibility and advantages of implementing this technology on a larger scale. In addition, this discussion will address the challenges and potential prospects for the field of EF, including technical and operational barriers, regulatory and policy considerations, and possible directions for future research. This paper aims to emphasise the ability of electro-fermentation to achieve a sustainable and circular carbon economy. It ultimately supports a more robust research, development, and investment focus in this promising field (Table 1).

2 Principles of Electro-fermentation

Electro-fermentation (EF) demonstrates the innovative combination of electrochemical and biological systems. It facilitates the improvement of microbial metabolic processes by using external electrical inputs. Electroactive microorganisms in EF establish a direct interaction with electrodes, enabling a two-way exchange of electrons between the biological and electrochemical realms. This interface allows for precise control of microbial metabolic pathways by manipulating the redox environment in the bioelectrochemical system. Electroactive microorganisms, including specific types of bacteria and archaea, have specialised proteins and complexes that facilitate the transfer of electrons outside their cells, known as extracellular electron transfer (EET). These features consist of cytochromes, conductive pili, and other redox-active molecules that are integrated into their cell membranes and can

Table 1 The primary mechanisms involved in the recycling of carbon within a circular carbon economy

Procedure	Explanation	Benefits	Illustration
Carbon capture and storage (CCS)	Controlling the release of carbon dioxide gas from many sources, including power stations, factories, and the air itself, by the application of various technologies	Reduces atmospheric CO_2 and supplies fuel for future usage, both of which contribute to lessening the effects of climate change	"Direct Air Capture" (DAC) chemical absorbents
Carbon capture and utilisation (CCU)	Creating new products from recycled carbon dioxide include fuels, chemicals, and construction materials	Creates profit from CO_2 waste; reduces reliance on fossil fuels; encourages ecologically conscious company practices	Hydrogenation in metabolism leads to methanol production from CO_2
Biomass conversion	The process of transforming organic materials into valuable compounds by a variety of biochemical and thermochemical reactions (e.g., fermentation and pyrolysis)	Produces biodegradable goods using renewable resources while also reducing waste	Producing bioethanol from organic matter
Electrochemical carbon reduction	Electrochemical cells can convert carbon dioxide (CO_2) into valuable compounds and fuels using renewable energy and modern catalysts	Reduces emissions of greenhouse gases while producing valuable chemicals and fuels using renewable energy	The electrochemical process involving a copper catalyst transforms carbon dioxide gas into ethylene
Anaerobic digestion	Anaerobic biological processes decompose organic waste, producing digestate and biogas (a mixture of methane and carbon dioxide)	Creates renewable energy, reduces waste, and recovers plant fertilizer	Anaerobic digestion of municipal solid waste is used to produce biogas
Gasification	Carbon, hydrogen, and carbon dioxide are produced from carbonaceous materials by high-temperature processing in an oxygen-poor environment	Maximises energy use while recycling trash into helpful chemical and energy production syngas	Fuel produced by the gasification of agricultural waste

(continued)

Table 1 (continued)

Procedure	Explanation	Benefits	Illustration
Plastic and polymer recycling	Chemical or mechanical processes that transform carbon-containing polymers into new ones	Reduces waste for landfills and helps save resources; it's a win–win for a circular economy	Technologies for chemically regenerating PET polymers
Carbon sequestration	Biological processes, such as afforestation or different geological formations, can store carbon dioxide for a long time	The re-entry of carbon dioxide into the atmosphere is prevented, reducing climate change's impact	Carbon dioxide is being stored in shale gas and oil deposits through geological processes
Synthetic biology	Genetic engineering can be used on plants or microbes to enhance CO_2 fixation and create valuable bio-based goods	Enhanced CO_2 use and environmentally friendly chemical and material production are conceivable	Biofuels are produced by genetically modified algae using carbon dioxide
Biochar production	The pyrolysis of biomass produces biochar, which has dual purposes: improving soil health and sequestering carbon	Soil fertility is improved, greenhouse gas emissions are decreased, and carbon is sequestered	Biochar, produced by the pyrolysis of agricultural waste, is used as a soil fertilizer

directly interact with the electrode surface. Researchers can optimise the production of desired compounds and reduce the formation of unwanted by-products by adjusting the electrical potential of the electrodes to create favourable conditions for specific metabolic reactions. The principles of bioelectrochemistry are the fundamental factors that govern the mechanisms of electron transfer in EF. During this process, microorganisms engage in either anodic reactions, donating electrons to an electrode, or cathodic reactions, accepting electrons from an electrode. Anodic reactions generally entail the oxidation of organic substrates, during which microorganisms oxidise these compounds to obtain energy and transfer the resulting electrons to the anode (Awasthi et al., 2020; Chatterjee & Mazumder, 2024; Zhao et al., 2024). The flow of electrons can be utilised to catalyse additional chemical reactions, such as reducing protons to generate hydrogen gas. Microorganisms on the cathode side can utilise electrons to facilitate reductive biosynthesis processes. For example, when converting carbon dioxide into methane or other hydrocarbons, microorganisms like methanogens utilise the electrons from the cathode to decrease the CO_2, incorporating them into their metabolic processes to generate methane. This electron transfer occurs through either direct or mediated mechanisms. During direct electron transfer (DET), microorganisms make direct physical contact with the surface of the electrode, enabling the direct exchange of electrons. This is achieved through conductive structures, such as nanowires or pili, which connect the microbial cells to the electrode. On the other hand, mediated electron transfer (MET) refers to the process in which redox mediators, which are small and diffusible molecules, transport electrons between microbial cells and the electrode. These mediators can consist of naturally occurring compounds such as flavins or artificial electron shuttles introduced into the system to increase the speed of electron transfer. The selection between DET (Direct Electron Transfer) and MET (Mediated Electron Transfer) is contingent upon the specific criteria of the EF (Electron Transfer) process, which encompasses the particular microorganisms employed and the targeted level of electron transfer efficiency. Several factors, such as the electrode materials, bioelectrochemical reactor configuration, and operational conditions (such as pH, temperature, and substrate concentration), influence the efficiency and effectiveness of electron transfer in EF. Utilising nanomaterials to modify carbon-based electrodes can significantly increase their surface area and conductivity. This enhancement promotes improved microbial adhesion and facilitates faster electron transfer rates. Reactor configurations, such as bioelectrochemical cells (BECs) or microbial electrolysis cells (MECs), enhance the interaction between microorganisms and electrodes, thereby maximising electron transfer efficiency. Furthermore, the redox potential applied to the electrodes plays a crucial role in directing the metabolic pathways of the microorganisms (Bains et al., 2024; Knežević et al., 2022; Leininger et al., 2023; Wainaina et al., 2020). Adjusting the electrode potential makes it feasible to selectively stimulate or suppress particular enzymatic activities, thus directing the microbial metabolism towards synthesising desired substances. The precise regulation of metabolic processes is a significant benefit of EF, ensuring the customised production of biofuels and biochemicals from renewable and waste materials. The electro-fermentation interaction is a complex interplay between electrochemical and biological systems that utilise microbial metabolism to optimise production processes based on bioelectrochemical principles. Electron transfer mechanisms, whether direct or mediated, are fundamental to the ability of EF to regulate and enhance microbial activities, thus making it a potent tool for advancing sustainable biotechnological applications (Table 2).

2.1 Microbial Electrochemical Systems

Microbial electrochemical systems (MES) combine microbiology and electrochemistry to take advantage of the distinctive abilities of electroactive microorganisms in promoting bioelectrochemical processes. These systems are leading the field of bioengineering by utilising the metabolic activities of certain microorganisms capable of transferring electrons to and from electrodes. This allows for the integration of biological and electrochemical domains. Electroactive microorganisms, such as Geobacter, Shewanella, and certain methanogens, have specific proteins and structures that transfer electrons outside their cells (EET). These microorganisms can convert organic substrates into oxidised forms and transfer the electrons to an anode.

Table 2 Principles of electro-fermentation

Principle	Description	Mechanism	Key microorganisms involved	Electron transfer mode	Applications	Advantages/ challenges
Electron transfer mechanisms	It involves the transfer of electrons between microbes and electrodes to drive fermentation	Direct electron transfer (DET) and mediated electron transfer (MET) using redox mediators	*Geobacter*, *Shewanella*, *Clostridium*	Direct (cytochromes, nanowires) or mediated (flavins, quinones)	Biofuel production, organic acid synthesis	It enhances efficiency but requires optimised electron mediators
Redox potential control	Adjusting the redox environment to regulate microbial metabolism and improve fermentation yields	Electrodes serve as electron donors/ acceptors, influencing metabolic pathways	*Escherichia coli*, *Saccharomyces cerevisiae*	Direct or mediated	Industrial bioprocess optimisation	Improves product selectivity but requires precise voltage control
Bioelectrochemical interactions	Interaction between microbes and electrodes influences metabolic activity and fermentation products	Electron flow modulates metabolic fluxes and gene expression	*Geobacter sulfurreducens*, *Shewanella oneidensis*	Direct electron uptake or release	Waste valorisation, bioremediation	It can be complex and requires specialised reactor design

(continued)

Table 2 (continued)

Principle	Description	Mechanism	Key microorganisms involved	Electron transfer mode	Applications	Advantages/ challenges
Microbial consortia involvement	Mixed microbial communities enhance fermentation efficiency through synergistic interactions	Co-cultures improve electron transfer and metabolic cooperation	*Clostridium acetobutylicum*, *Methanobacterium*	Mixed-mode transfer (direct and mediated)	Anaerobic digestion, biogas upgrading	Increases robustness but requires community stability management
Electroactive biofilms	Biofilms on electrodes facilitate electron transfer and stabilise microbial communities	The formation of conductive biofilms enables efficient electron flow	*Geobacter*, *Shewanella*	Primarily direct transfer via biofilm matrix	Energy storage, microbial fuel cells	It enhances stability but may require long biofilm maturation times
Electrode materials	The choice of electrode materials affects electron transfer efficiency and microbial adhesion	Conductive materials like carbon, graphite, and metal oxides enhance performance	N/A	N/A	Bioelectrochemical systems, electro-fermentation reactors	High-cost materials may limit scalability

(continued)

Table 2 (continued)

Principle	Description	Mechanism	Key microorganisms involved	Electron transfer mode	Applications	Advantages/ challenges
External circuit configurations	Configurations determine electron flow, affecting fermentation rates and efficiency	Open or closed circuit setups influence microbial electron exchange	N/A	N/A	Bioreactors, microbial electrolysis cells	Optimized design can enhance performance but requires engineering expertise
Real-world applications	Electro-fermentation is used in diverse industries for sustainable production	Controlled electron flow enables selective biosynthesis	*Clostridium*, *Saccharomyces*, *Escherichia*	Mixed electron transfer modes	Biofuel production, pharmaceuticals, CO_2 reduction	Reduces reliance on chemical additives but requires scale-up strategies

Alternatively, they can accept electrons from a cathode to facilitate reduction reactions. Critical components such as cytochromes, multiheme complexes, and conductive pili or nanowires facilitate electron transfer, establishing physical connections between microbial cells and the electrodes. The exceptional metabolic adaptability of these microorganisms renders them indispensable for producing electricity, restoring environmental quality, and creating biofuels and valuable chemical compounds. Bioelectrochemical reactors are purposely designed environments where electroactive microorganisms and electrodes interact to enhance microbial processes. The reactors are available in different arrangements, such as microbial fuel cells (MFCs), microbial electrolysis cells (MECs), and microbial electrosynthesis systems (MESs). Microbial fuel cells involve the oxidation of organic substrates by microorganisms, transferring electrons to the anode. These electrons then pass through an external circuit, producing electricity. In addition, this process generates protons, which then react with electrons at the cathode to convert oxygen into water. This action completes the circuit and establishes a continuous stream of electricity derived from organic material (Amulya et al., 2024; Gautam et al., 2024; Naresh Kumar et al., 2022; Virdis et al., 2022). Microbial electrolysis cells, however, specifically prioritise the generation of hydrogen. In this process, an external voltage is utilised to facilitate the reduction of protons at the cathode, generating hydrogen gas. This method increases the amount of hydrogen produced from organic substances beyond what can be achieved solely through biological fermentation. Microbial electrosynthesis is an advanced application in which electroactive microorganisms use electrons from a cathode to convert CO_2 or other simple molecules into complex organic compounds like acetate, methane, or biofuels. This process is essential for carbon capture and utilisation, as it converts greenhouse gases into valuable products and helps to establish a circular carbon economy. The efficiency and effectiveness of bioelectrochemical reactors heavily rely on their design and operation. The rates of microbial electron transfer are significantly influenced by factors such as the material of the electrode, its surface area, and conductivity. Contemporary reactors frequently utilise sophisticated materials such as carbon nanofibers or graphene to improve these characteristics. Furthermore, the reactor's configuration must guarantee the most efficient interaction between microorganisms and electrodes. This is accomplished by considering factors such as the positioning of electrodes, flow patterns, and the duration for which the microbial biomass is retained. Precise control of operational parameters such as pH, temperature, and substrate concentration is essential to maintain microbial activity and stability in these reactors (Battista et al., 2022; Dahiya et al., 2023; Deng et al., 2023; Kumar et al., 2023). Furthermore, the redox potential of the electrodes can be accurately manipulated to direct microbial metabolism towards specific pathways, thereby optimising the synthesis of desired compounds while minimising the formation of unwanted by-products. By incorporating renewable energy sources like solar or wind power into these bioelectrochemical systems, their sustainability is further improved, offering an environmentally friendly alternative to processes reliant on fossil fuels. Microbial electrochemical systems utilise electroactive microorganisms in advanced bioelectrochemical reactors, providing a promising and adaptable platform for sustainable bioenergy production and environmental remediation. These systems enhance the effectiveness of microbial processes and create new opportunities for transforming waste and renewable resources into valuable products, in line with the objectives of a circular carbon economy. MES can significantly impact the shift towards more environmentally friendly industrial practices by progressing in reactor design, materials science, and microbial engineering.

2.2 Electrochemical Control of Metabolic Pathways

Electro-fermentation (EF) utilises electrochemical control of metabolic pathways to precisely manipulate the biochemical processes in microorganisms. Scientists can control microbial metabolism to create particular substances, enhance the efficiency of processes, and reduce the generation of undesired by-products. This can be achieved by managing the redox potential and carefully selecting electron donors and acceptors. The ability to exert this level of control is facilitated by the intricate interplay between electrochemical principles and microbial physiology, enabling a tailored approach to biotechnological applications. Redox potential, or oxidation–reduction potential (ORP), measures the tendency of a chemical species to acquire electrons and undergo reduction. Controlling the redox potential of the system is a powerful technique in electrochemical fluorination (EF) for modulating microbial metabolic pathways (Liu et al., 2023; Ngoc-Dan Cao et al., 2022; Soo et al., 2024; Vassilev et al., 2021). Microorganisms possess enzymes sensitive to their environment's redox state, leading to a consequential influence on the flow of electrons within their metabolic networks. Finely tuning the electrode potential can create conditions that encourage specific metabolic reactions. Establishing a heightened oxidising atmosphere at the anode makes enhancing microorganisms' proficient oxidation of organic substrates feasible. This process entails the transfer of the electrons generated to the electrode. This has the potential to enhance electricity production in microbial fuel cells or expedite the decomposition of waste compounds in bioremediation processes. Alternatively, increasing the reducing potential of the cathode facilitates the reduction of protons or carbon dioxide, thereby enhancing the production of hydrogen gas or reduced carbon compounds like methane and acetate. Controlling important metabolic enzymes and pathways becomes possible by modulating the redox potential. Redox alterations can impact enzymes involved in energy production, such as those in the electron transport chain. By accurately manipulating the electrode potential, it is possible to enhance the performance of these enzymes, thereby increasing the overall efficiency of microbial metabolism. Accurate control of this mechanism can lead to higher production of desired products and improved process stability, making EF a versatile and resilient technique for various biotechnological uses. Electro-fermentation requires the meticulous choice of electron donors and acceptors to facilitate the desired metabolic processes efficiently. Electron donors are substances that provide electrons to microbial cells, whereas electron acceptors are compounds that receive electrons from the cells. The choice of these electron carriers is pivotal in determining the direction and efficiency of the metabolic pathways. Common electron donors in EF include organic substrates like glucose, acetate, and other compounds derived from biomass. Microorganisms undergo oxidation of these substrates, leading to the liberation of electrons, which are subsequently transferred to the anode. The characteristics of the electron donor significantly impact the activated metabolic pathways, thus influencing the types of products generated. Glucose oxidation can produce ethanol, acetate, and various other fermentation products. The particular result is determined by the specific type of microorganism involved and the existing redox conditions. Electron acceptors play a critical role in facilitating reductive processes at the cathode (Cardeña et al., 2024; Jiang et al., 2019; Roy Chowdhury et al., 2024; Shehata et al., 2023). Oxygen is frequently employed as an electron acceptor in aerobic processes. Nevertheless, in anaerobic reactions, alternative acceptors such as protons (H^+) or carbon dioxide (CO_2) are commonly used to assist the reactions in EF. For instance, protons can convert into hydrogen gas, whereas carbon dioxide can be transformed into methane or organic acids through microbial electrosynthesis. The efficacy and specificity of these reactions depend on the microbial species involved and the electrochemical conditions established by the reactor. Specific mediators can enhance electron transfer efficiency between the microorganisms and the electrodes. Redox mediators are chemical substances that aid in the movement of electrons between microbial cells and electrode surfaces, thereby increasing the electron transfer rate. Mediators can be composed of naturally occurring substances like flavins and quinones or artificially created chemicals designed to improve electron transfer efficiency. Electro-fermentation utilises manipulating redox potential and carefully selecting electron donors and acceptors to exert control over metabolic pathways. By carefully manipulating these variables, scientists can direct microbial metabolism towards specific goals, thereby enhancing the production of biofuels, chemicals, and other valuable commodities. This method improves the efficiency of microbial processes and opens up new possibilities for eco-friendly biotechnological applications, thus supporting the advancement of a circular carbon economy (Table 3).

3 Technological Advancements in Electro-fermentation

3.1 Development of Bioelectrochemical Reactors

Bioelectrochemical reactors (BERs) are critical in advancing electro-fermentation (EF) technologies. These technologies utilise electroactive microorganisms to convert organic substrates into valuable chemicals and biofuels. The design and configuration of these reactors and selecting and modifying electrode materials are crucial in maximising the efficiency and scalability of bioelectrochemical processes (Raj et al., 2024; Rovira-Alsina et al., 2021; Serrano-Jiménez et al.,

Table 3 Exploring different approaches to enhance the sustainability of a circular carbon bioeconomy

Approach	Description	Pros	Example
Advanced catalysis engineering	The invention of catalysts that exchange CO_2 into usable biomass with great selectivity and efficiency	Lowers energy consumption and maximises response efficiency while increasing product specificity	Producing nanoparticle catalysts for the methanol-from-carbon dioxide reaction
Renewable energy integration	Solar, wind, and biomass powering bioeconomic processes	Contributes to more environmentally friendly energy systems, mitigates pollutants, and minimises reliance on fossil fuels	Solar-powered electrochemical systems for carbon dioxide (CO_2) conversion
Biomass valorisation	Utilizing advanced biological processes and techniques on biomass converts raw resources into final commodities	The process generates biodegradable products with minimum waste by utilizing renewable raw materials	Producing bioplastics and bioethanol from waste products in the agricultural sector
Process intensification	Optimisation of thermochemical and biochemical processes for higher output with lower energy consumption	Reduces operating expenses and increases overall economic viability to maximise process efficiency	The gasification of biomass produces synthetic gas under high pressure
Carbon capture and utilisation (CCU)	Reusing carbon dioxide gas by making usable products using cutting-edge biological and chemical procedures	Reduces emissions of greenhouse gases and promotes industrial sustainability while producing trading CO_2 products	Using modern facilities catalytic techniques to convert carbon dioxide into polymers
Circular design principles	When creating goods and services, consider the reusability, recycling, and reducing environmental impact	Advocates for sustainable consumption through waste reduction and resource conservation	Creating packaging that naturally decomposes using eco-friendly materials
Policy and economic incentives	Supporting the growth of a circular bioeconomy by creating favourable policies, subsidies, and incentives	Encourages industry-wide adoption, sparks innovation, and reduces financial risk	Government subsidies are available for renewable energy initiatives and bio-based industries

(continued)

Table 3 (continued)

Approach	Description	Pros	Example
Collaborative research and development initiatives	To promote innovation in bioeconomy technology through multidisciplinary collaborations between academic institutions, companies, and governments	Promotes rapid technological advancement, facilitates knowledge sharing, and supports broad adoption	Biorefineries can be constructed through collaboration between the public and private sectors
Education and awareness	Publicising and disseminating information on the carbon bioeconomy's techniques for regeneration and their benefits	Educates the public, encourages the purchase of environmentally friendly products, and pushes for constructive behavioural changes	Courses cover topics including bioeconomy benefits and agricultural sustainability
Supply chain optimisation	Supply chain optimisation and integration to improve material and product flow in the bioeconomy	It saves money, works better, and makes better use of what is available	Incorporating agricultural and industrial techniques into new biorefineries

2023a; Tharak et al., 2023). The bioelectrochemical reactor design aims to optimise the interface between microorganisms and electrodes to facilitate efficient electron transfer and enhance metabolic activity. Various reactor configurations have been created, each designed for specific applications and operational conditions. Common configurations typically consist of microbial fuel cells (MFCs), microbial electrolysis cells (MECs), and microbial electrosynthesis systems (MESs). Microbial fuel cells (MFCs) are specifically engineered to produce electrical energy by oxidising organic substrates using electroactive microorganisms. In a conventional MFC configuration, the reactor comprises an anode and a cathode, divided by a proton exchange membrane. Microorganisms inhabit the anode, breaking down organic substances and releasing electrons. These electrons then pass through an external circuit, producing electrical energy. Protons are generated and transported across the membrane to the cathode during oxidation. At the cathode, they react with electrons and an electron acceptor, usually oxygen, to produce water. The performance of microbial fuel cells (MFCs) relies on key factors, including the electrode's surface area, the materials' conductivity, and the proximity of microorganisms to the electrode surface. Microbial electrolysis cells (MECs) are primarily designed to produce hydrogen. MECs, like MFCs, comprise an anode and a cathode. However, in contrast to microbial fuel cells (MFCs), an external voltage surpasses the thermodynamic obstacle of hydrogen production. Within Microbial Electrochemical

Cells (MECs), electroactive microorganisms located at the anode undergo oxidation of organic substrates, resulting in the liberation of electrons and protons. The applied voltage propels the electrons towards the cathode, where they chemically convert protons into hydrogen gas. Microbial electrolysis cells (MECs) have the potential to achieve greater hydrogen yields compared to conventional fermentation methods, thus making them an up-and-coming technology for the sustainable production of hydrogen. Microbial electrosynthesis systems (MESs) are designed to harness cathodic electrons' reducing power to transform CO_2 or other simple molecules into intricate organic compounds (Arshad et al., 2022; Blank et al., 2020; Kabir et al., 2024; Lozano & García-Verdugo, 2023). Within these systems, the cathode is the source of electrons, which the microorganisms utilise to power reductive biosynthesis processes. Metallic electrocatalytic systems (MESs) play a crucial role in carbon capture and utilisation by enabling the conversion of greenhouse gases into valuable chemicals and biofuels. This process contributes to the establishment of a circular carbon economy. The selection of electrode materials and their alterations are crucial to the efficiency of bioelectrochemical reactors. An optimal electrode material should possess exceptional electrical conductivity, a substantial surface area, biocompatibility, and chemical stability. Several materials, such as carbon-based electrodes, metals, and conductive polymers, have been investigated to fulfil these criteria. Carbon-based electrodes, such as graphite, carbon cloth, and carbon paper, are extensively utilised in BERs because of their exceptional conductivity, large surface area, and relatively low cost. Nevertheless, unprocessed carbon materials frequently necessitate alterations to optimise their functionality. A prevalent method involves utilising nanostructured carbon materials, such as carbon nanotubes (CNTs) and graphene. These materials exhibit a substantially larger surface area and enhanced electrical conductivity than conventional carbon materials. Integrating carbon nanotubes (CNTs) or graphene into electrode structures can improve the attachment of microorganisms and increase the rates of electron transfer, resulting in improved efficiency of reactors. Besides carbon materials, metal-based electrodes, including stainless steel, platinum, and gold, have been employed in BERs. Metals generally exhibit superior conductivity and stability, although they can be costly and may not always offer the optimal surface for microbial colonisation. To overcome these restrictions, applying biocompatible coatings or surface treatments to metal electrodes is common practice, enhancing their interaction with microorganisms. Coating metal electrodes with a thin layer of conductive polymers or incorporating metal nanoparticles can improve their performance by augmenting surface roughness and facilitating microbial attachment. Conductive polymers, such as polyaniline (PANI) and polypyrrole (PPy), are used to modify electrodes. These polymers can be applied onto electrode surfaces to form a conductive and biocompatible interface. Conductive polymers can adjust their chemical and physical characteristics, allowing for customisation to suit particular uses. Modifying the polymer's composition can enhance the electrode's electrochemical properties, improve microbial adhesion, and boost reactor performance. Surface modification techniques, such as chemical vapour deposition, electrodeposition, and plasma treatment, are frequently used to improve the characteristics of electrode materials (Annie Modestra et al.,

2022; Aresta & Dibenedetto, 2024; Awasthi et al., 2022; Lu et al., 2024). Chemical vapour deposition is a technique that can produce thin and even nanomaterial coatings, enhancing the conductivity and surface area of the electrode. Electrodeposition enables accurate manipulation of coating thickness and composition, facilitating the production of customised electrode surfaces with improved characteristics. Plasma treatment can alter the chemical composition of electrodes' surfaces, creating functional groups that enhance microbial adhesion and activity. Another crucial factor in electrode design is the integration of three-dimensional (3D) structures. Conventional two-dimensional (2D) electrodes, like flat carbon plates, may restrict the available surface area for microbial colonisation and electron transfer. In contrast, 3D electrodes, such as carbon foam or metal meshes, offer a substantially larger surface area and enhanced mass transfer properties. These structures enable higher microbial loading and more effective electron transfer, thereby improving the overall performance of bioelectrochemical reactors. There is a growing interest in advancing hybrid electrodes, which involve combining various materials and structures. Combining carbon-based materials with metal nanoparticles or conductive polymers can achieve synergistic effects, enhancing electrode performance. Hybrid electrodes can benefit from different materials, such as the excellent conductivity of metals, the extensive surface area of nanomaterials, and the biocompatibility of conductive polymers. The development of bioelectrochemical reactors requires meticulous attention to design and configuration, along with the careful selection and modification of electrode materials. Optimising these factors is crucial for improving the effectiveness and expandability of bioelectrochemical processes. Researchers can utilise cutting-edge materials and inventive designs to develop reactors, optimising microbial activity and electron transfer. This advancement opens up possibilities for the sustainable production of biofuels, chemicals, and environmental remediation. To fully harness the potential of electro-fermentation, conducting continuous research and development in reactor design and electrode materials is imperative.

3.2 Integration with Renewable Energy Sources

Integrating bioelectrochemical reactors (BERs) with renewable energy sources, such as solar and wind energy, significantly advances sustainable and efficient bioelectrochemical processes. This integration takes advantage of the intermittent nature of renewable energy and improves the overall sustainability of bioelectrochemical systems. Integrating solar and wind energy with bioenergy recovery systems (BERs) guarantees a consistent and dependable energy supply, enhancing microbial processes' stability and efficiency for producing biofuels and chemicals. Integrating solar energy with building energy requirements (BERs) entails using photovoltaic (PV) cells to directly convert sunlight into electrical energy. This electricity can fuel bioelectrochemical processes, offering a sustainable and environmentally friendly energy source that does not contribute to carbon emissions. An important benefit of solar energy is its plentiful availability, particularly in areas with high solar insolation. Photovoltaic cells can be implemented significantly, offering a consistent electricity supply throughout daylight. This sustainable energy source can power different bioelectrochemical processes, such as the breakdown of organic substances at the anode or the conversion of CO_2 into valuable products at the cathode (Acosta-Santoyo et al., 2024; Ilmasari et al., 2022; Leong & Chang, 2023; Serrano-Jiménez et al., 2023b; Zheng et al., 2023). Wind energy provides an additional environmentally friendly and long-lasting source of electricity for Building Energy Renovations (BERs). Wind turbines harness the motion energy of wind and transform it into electrical energy, which can be directly supplied to bioelectrochemical systems. Wind energy integration is especially beneficial in areas with reliable and strong wind patterns. Wind energy can supplement solar energy by generating electricity during periods when solar power is inaccessible, such as at night or during overcast conditions. The simultaneous incorporation of solar and wind energy can guarantee a consistent and dependable power provision for BERs, alleviating the problem of irregularity commonly associated with separate renewable sources. Advanced grid technologies and energy management systems can enhance the efficiency of combining solar and wind energy with building energy resources (BERs). These technologies can effectively distribute the energy input from both sources, guaranteeing a consistent and ample electricity supply to the bioelectrochemical processes. In addition, integrating solar and wind energy with energy storage solutions in hybrid systems can improve the dependability and effectiveness of BERs. Energy storage is essential for efficiently integrating renewable energy sources with building energy resources (BERs). Efficient storage solutions are required to ensure a continuous supply of electricity for bioelectrochemical processes due to the intermittent nature of solar and wind energy. Different energy storage technologies, including batteries, supercapacitors, and hydrogen storage, can store surplus electricity produced during periods of high production (Bellini et al., 2022; Hussein et al., 2024; Javanmard et al., 2024; Sana et al., 2024; Sriram et al., 2022; Wood et al., 2021). Renewable energy is often stored using batteries, such as lithium-ion or flow batteries. These batteries can store and discharge substantial electricity as required, ensuring a steady power source for BERs. The accumulated energy can sustain microbial activity and facilitate electron transfer processes during low renewable energy generation periods, such as at night or during calm weather. Supercapacitors, known for their high-power density and quick charge–discharge cycles, can temporarily store energy. This guarantees a consistent supply of electricity to the BERs, even when there are variations in renewable energy inputs. Hydrogen storage is a novel method of connecting renewable energy with BERs. Surplus electricity generated by solar and wind energy can be utilised to produce hydrogen via water electrolysis. This hydrogen can be stored and used as an energy source for bioelectrochemical processes or other applications. Hydrogen can be introduced into microbial electrolysis cells (MECs) to augment hydrogen production or utilised directly in microbial fuel cells (MFCs) as a substrate for electricity generation. Advanced control systems and innovative grid technologies can optimise the utilisation of stored electrical energy in BERs. These systems can monitor the energy requirements of bioelectrochemical processes and efficiently distribute stored energy to meet those requirements. By incorporating up-to-the-minute information on the production of renewable energy, energy storage levels, and bioelectrochemical reactors' performance, these control systems can guarantee the effective and environmentally friendly functioning of BERs. We can achieve a sustainable and highly efficient approach by combining bioelectrochemical processes (BERs) with renewable energy sources like solar and wind energy and effective energy storage solutions. This integration guarantees an uninterrupted and dependable energy provision while improving bioelectrochemical systems' long-term viability and expandability. The ongoing progress of renewable energy technologies presents a significant opportunity for the future of sustainable biofuel and chemical production when combined with BERs.

3.3 Innovations in Microbial Engineering

Microbial engineering innovations are revolutionising bioelectrochemical reactors (BERs), significantly improving their performance and efficiency. At the core of these advancements are genetic modifications that aim to enhance the abilities of microorganisms and use synthetic biology to create new microbial pathways and systems. These approaches are crucial in expanding the limits of what bioelectrochemical processes can accomplish, resulting in microbial communities that are more efficient, resilient, and adaptable for specific purposes. Genetic modifications entail modifying the genomes of microbes to improve specific characteristics, such as their metabolic efficiency, ability to tolerate stress, and capacity for electron transfer. Researchers can enhance the overall performance of electroactive microorganisms by using tools like CRISPR-Cas9 to introduce accurate genetic modifications. One main objective of genetic modification is to improve electron transfer pathways. Electroactive bacteria, such as Geobacter and Shewanella, have been modified at the genetic level to increase the production of proteins that transfer electrons outside the cell. These proteins include cytochromes and conductive pili. These alterations enhance the efficacy of these microorganisms in transferring electrons to and from electrodes, thereby enhancing the performance of bioelectrochemical systems (Battista et al., 2022; Deng et al., 2023; Jiang et al., 2019; Liu et al., 2023; Roy Chowdhury et al., 2024; Shehata et al., 2023; Soo et al., 2024). Another area of emphasis in genetic engineering is enhancing the metabolic pathways responsible for utilising substrates and producing desired products. Scientists can manipulate microbial metabolism to generate specific substances by altering important enzymes and regulatory networks. For example, researchers are changing the metabolic pathways of Clostridium species to increase the production of biofuels like butanol. They are also modifying the genetic composition of Escherichia coli to achieve higher quantities of valuable chemicals such as succinic acid or 1,3-propanediol. These genetic alterations augment desired goods' output and improve bioelectrochemical processes' overall effectiveness and durability. Synthetic biology surpasses genetic modifications by generating completely novel biological components, apparatuses, and systems that are not naturally present.

This field utilises engineering and design principles to create artificial pathways and organisms with customised functionalities. Synthetic biology can be employed within the framework of BERs to engineer microbial consortia that are capable of executing intricate tasks using cooperative interactions. Consortia can be designed to execute multi-step biochemical processes more efficiently than cultures of a single species. For instance, a synthetic consortium can be engineered to break down intricate organic substances into basic, intermediate compounds, which are subsequently transformed into biofuels or chemicals by another modified microbial strain. An intriguing use of synthetic biology in microbial engineering involves the creation of artificial electron transfer chains. Efficiency and control of electron transfer processes can be enhanced by developing and implementing synthetic redox-active proteins and pathways, as demonstrated by researchers. This innovation can greatly improve bioelectrochemical reactor (BER) performance by optimising the movement of electrons between microbial cells and electrodes. As a result, it can increase the efficiency of energy conversion and the synthesis of products. Synthetic biology facilitates the creation of microbial cell factories with improved biosynthetic capacities. Scientists can develop new types of biofuels, pharmaceuticals, and valuable chemicals using renewable resources by incorporating artificial pathways into the genetic makeup of microorganisms. These genetically modified microorganisms can be customised to efficiently use various substances, including waste materials, thereby promoting the sustainability and circularity of bioelectrochemical processes. For example, genetically modified microorganisms have the potential to transform agricultural leftovers or industrial waste into biofuels, thereby minimising waste and generating sustainable energy. Synthetic biology is utilised to create biosensors and actuators for bioelectrochemical systems. Engineered microorganisms can be programmed to react to particular environmental stimuli or chemical signals, allowing for the immediate supervision and regulation of bioelectrochemical procedures. These biosensors can identify variations in the levels of substances being processed, the oxidation–reduction conditions, or the existence of particular pollutants. This information is crucial for enhancing the reactor's efficiency and maintaining the overall process's stability. The advancements in bioelectrochemical reactors are being propelled by innovations in microbial engineering, which have been achieved through genetic modifications and synthetic biology. Genetic modifications improve the characteristics of microorganisms, such as their ability to transfer electrons and their metabolic performance (Ahmed et al., 2023; Awasthi et al., 2020; Bains et al., 2024; Chatterjee & Mazumder, 2024; Knežević et al., 2022; Leininger et al., 2023; Wainaina et al., 2020; Zhao et al., 2024). On the other hand, synthetic biology applications develop new pathways and systems to make bioelectrochemical processes more efficient and adaptable. These advancements enhance the effectiveness and capacity of bioelectrochemical systems and broaden their possible uses, thereby aiding in the advancement of sustainable biotechnological solutions for energy generation, waste management, and the production of valuable chemicals. Integrating advanced microbial engineering techniques will be essential in fully harnessing the potential of bioelectrochemical technologies as the field continues to develop (Table 4).

Table 4 The technological challenges to achieving a carbon-neutral economy by electrochemical carbon conversion

Classification	Challenges	Description
Efficiency and selectivity of catalysts	Catalyst efficiency	Formulating catalysts with a high turnover rate and a low overpotential need
	Product selectivity	Managing many conflicting reaction pathways while being extremely precise with the target components
Catalyst durability and stability	Catalyst deactivation	Keeping catalysts safe against surface modification, poisoning, and sintering degradation caused by reactions
	Operational stability	Ensuring catalyst stability under long-term, continuous operation circumstances
Electrode engineering	Material conductivity	Producing highly conductive electrode materials that are resistant to corrosion
	Surface area optimisation	Optimizing reaction kinetics without sacrificing structural integrity by maximizing active surface area
Electrolyte development	Ionic conductivity	Improve electrolytes' resistive loss-reducing capabilities by making them more ionically conducting
	Electrolyte compatibility	Verifying that the electrolyte is sustainable and appropriate for use with the electrodes and catalyst
	pH control	Managing the electrolyte pH allows for optimal product distribution and reaction pathway control
Reactor design and configuration	Mass transfer efficiency	Reactor design optimisation aims to increase mass transfer rates while decreasing concentration polarization
	Scale-up challenges	Maximizing reactor efficiency during the transition from laboratory towards industrial scale
	Floe dynamic management	Managing the flow of reactants and products with precision to maintain constant reaction conditions

(continued)

Table 4 (continued)

Classification	Challenges	Description
Energy efficiency	Overpotential reduction	Reducing the overpotential required to initiate electrochemical reactions is the primary objective
	Renewable energy integration	Creating a sustainable future by integrating electrochemical processes with renewable energy
Product separation and purification	Product recovery	Development of efficient methods for obtaining pure final products from reaction mixtures
	Cost-effective purification	Creating products with commercial potential by employing cost-effective purification techniques
Economic viability	Cost reduction	Electrolytes, catalysts, and electrodes may be made more affordable to make it more practical
	Process integration	Adding electrochemical processes to existing industrial infrastructure to make them more cost-effective
Environmental and safety concerns	By-product management	They efficiently manage trash and by-products to lessen their environmental impact
	Operational safety	They protect workers from harm when exposed to potentially harmful substances, high voltage, or current
Regulatory and policy challenges	Compliance with standards	Paying attention to regulations on product safety, environmental impact, and quality
	Policy support	Applying for and receiving governmental funding to promote electrochemical uses

4 Applications of Electro-fermentation

4.1 Biofuel Production

Electro-fermentation (EF) is an advanced biotechnological method that combines microbial fermentation with electrochemical processes to improve the production of biofuels and other valuable chemicals. EF can enhance microbial metabolic pathways by manipulating the redox potential and supplying external electron donors or acceptors, increasing yields and efficiencies. One notable application of EF is the production of biofuels, including ethanol, butanol, methane, and hydrogen. This application has the potential to make a significant contribution to sustainable energy solutions. Ethanol and butanol are extensively researched biofuels in the field of EF applications. Ethanol is the main output of traditional fermentation methods that utilise yeast or bacteria. However, EF can boost its production by enhancing the effectiveness of microbial metabolic pathways. Electroactive microorganisms, such as Saccharomyces cerevisiae or genetically modified Escherichia coli strains, are commonly employed in an EF system for ethanol production. Applying a regulated redox potential using electrodes makes it feasible to enhance the fermentation conditions, thereby minimising the production of by-products such as acetic acid and maximising ethanol yield (Annie Modestra et al., 2022; Arshad et al., 2022; Blank et al., 2020; Kabir et al., 2024; Lozano & García-Verdugo, 2023; Lu et al., 2024; Raj et al., 2024; Tharak et al., 2023). This approach can additionally reduce the energy input necessary for distillation and purification, thereby enhancing the economic feasibility of the process. However, butanol is a more effective biofuel than ethanol because it has a more significant energy content and is more compatible with current fuel infrastructure. Clostridium species are the dominant microorganisms in butanol production through acetone–butanol–ethanol (ABE) fermentation. Electrochemical fermentation (EF) can significantly increase butanol production by supplying a constant stream of electrons, which can be utilised to catalyse the reduction reactions in the microbial metabolic pathways. By precisely adjusting the redox potential and electron supply, EF can alter the metabolic balance in favour of higher butanol production while inhibiting the formation of undesirable by-products such as acetone and ethanol. This method enhances butanol production's overall efficiency and strengthens the microbial cultures' resilience, resulting in a more stable and scalable process. Methane and hydrogen are essential biofuels with substantial promise in renewable energy and sustainable waste management. Electrokinetic fermentation (EF) can increase the generation of these gases by utilising the abilities of electroactive microorganisms and optimising the redox conditions in bioelectrochemical reactors. The process of methane production through anaerobic digestion is a widely recognised and established method that involves the decomposition of organic matter by methanogenic archaea. Electrochemical fermentation (EF) can enhance methane production by supplying extra electrons that methanogens can directly utilise to convert carbon dioxide into methane. Microbial electrosynthesis, which enhances methane production rates and yields, surpasses traditional anaerobic digestion. By manipulating the electrode potential, one can optimise the metabolic pathways of methanogens, improving their efficiency and stability. This method enhances methane production and enables the utilisation of waste streams with low organic content, thereby expanding the variety of feedstocks that can be processed. Hydrogen production in electrolytic cells is commonly accomplished using microbial electrolysis cells (MECs). In MECs, electroactive bacteria oxidise organic substances and transfer electrons to the cathode, where protons are converted into hydrogen gas. The utilisation of an applied voltage in this process surpasses the hydrogen yields obtained through conventional fermentation techniques by surmounting the thermodynamic obstacles linked to hydrogen production. EF can enhance hydrogen production by strategically choosing and modifying electroactive microorganisms with superior hydrogen-producing abilities, thereby optimising the conditions for achieving the highest possible hydrogen output (Bellini et al., 2022; Hussein et al., 2024; Javanmard et al., 2024; Sriram et al., 2022; Wood et al., 2021). Moreover, incorporating sustainable energy sources, such as solar or wind power, to power the electrolysis process further improves hydrogen production's sustainability and economic feasibility through EF. Electro-fermentation provides a flexible and effective method for generating a range of biofuels, such as ethanol, butanol, methane, and hydrogen. By utilising the principles of electrochemistry and microbial metabolism, EF can enhance the efficiency of biofuel production processes, resulting in increased yields, improved effectiveness, and enhanced sustainability. With the continuous advancement of technology, EF (energy finance) can significantly impact the shift towards renewable energy and the establishment of a circular carbon economy. This will effectively tackle energy and environmental issues.

4.2 Production of Value-Added Chemicals

Electro-fermentation (EF) shows great potential as a sustainable method for producing valuable chemicals. It utilises the combination of microbial metabolism and electrochemical processes to improve productivity and effectiveness. EF produces a variety of chemicals, including organic acids like acetate and succinate, as well as bioplastics and biopolymers. These substances are essential in industry and can contribute to a circular carbon economy. Organic acids are essential foundational components for various industrial uses, such as producing chemicals, pharmaceuticals, and food additives. Acetate and succinate are crucial precursors for a wide range of value-added products. In electro-fermentation (EF), producing these organic acids can be improved by optimising the redox conditions and supplying external electron donors or acceptors to stimulate microbial metabolism. Acetate production via EF generally entails using acetogenic bacteria, such as Acetobacterium woodii or Clostridium species, which transform carbon dioxide and hydrogen into acetic acid. Applying a regulated voltage to the bioelectrochemical reactor enables the direct supply of electrons to the microorganisms, thereby improving their capacity to decrease carbon dioxide and enhance acetate production. This method enhances productivity by utilising carbon dioxide, a notable greenhouse gas, which aids in carbon recycling and reduction initiatives. EF's capacity to effectively convert carbon dioxide and renewable electricity into acetate establishes it as a crucial technology for sustainable chemical synthesis. Succinate is a valuable organic acid precursor that produces various products such as solvents, plastics, and pharmaceuticals. Engineered strains of bacteria, such as Escherichia coli or Corynebacterium glutamicum, can produce succinate through EF by optimising metabolic flux towards succinate formation (Cardeña et al., 2024; Jiang et al., 2019; Naresh Kumar et al., 2022; Roy Chowdhury et al., 2024; Soo et al., 2024; Vassilev et al., 2021; Virdis et al., 2022). Adjusting the redox potential in the bioelectrochemical reactor makes establishing optimal conditions for reducing intermediates feasible, resulting in increased succinate production. This technique enhances the efficacy of succinate synthesis while minimising the generation of undesirable secondary products, simplifying the purification procedure and reducing production expenses. EF's utilisation in bioplastics and biopolymers production signifies noteworthy progress in pursuing sustainable and biodegradable substitutes for traditional plastics. Bioplastics, including polyhydroxyalkanoates (PHAs) and polylactic acid (PLA), are becoming increasingly popular because of their ability to break down naturally and their potential to decrease plastic pollution. Electrofusion (EF) can optimise the production of these biopolymers by delivering a consistent and regulated source of reducing power to the microbial cultures engaged in their synthesis. PHAs are synthesised by different bacteria, such as Cupriavidus necator and Pseudomonas species, which store these biopolymers inside their cells to store nutrients when resources are scarce. Within the context of electrochemical fermentation (EF), it is possible to manipulate the redox potential to optimise the metabolic pathways involved in synthesising polyhydroxyalkanoates (PHA). This manipulation increases both the yield and productivity of these biopolymers. In addition, the utilisation of electroactive microorganisms can enable the direct transmission of electrons from the electrodes to the cells, thereby improving the efficiency of the biosynthetic process. By incorporating electrolysis (EF) with renewable electricity sources, it becomes feasible to generate polyhydroxyalkanoates (PHAs) in an environmentally sustainable manner, further decreasing the ecological impact of bioplastic manufacturing. Polylactic acid (PLA) is a significant biopolymer that can be synthesised using EF. The monomer of PLA, lactic acid, is conventionally generated through fermentation by lactic acid bacteria. Electrofusion (EF) can increase lactic acid production by optimising the redox environment and supplying electrons to facilitate the reduction reactions involved in its synthesis. This method enhances lactic acid production and enables using various raw materials, such as agricultural residues and industrial waste streams. The subsequent polymerisation of lactic acid into PLA can be seamlessly incorporated into a sustainable production process, thereby aiding in advancing biodegradable and renewable plastics. Electro-fermentation provides a flexible and effective method for producing valuable chemicals, such as organic acids and biopolymers. EF can improve the effectiveness and sustainability of chemical production processes by utilising the principles of electrochemistry and microbial metabolism, thereby making substantial contributions to a circular carbon economy. The capacity to generate valuable chemicals from renewable feedstocks and electricity effectively tackles environmental concerns and offers a feasible approach to the sustainable production of crucial industrial materials. Continuing technological advancements allow EF to integrate with advanced microbial engineering and renewable energy sources, which offers new possibilities for sustainable chemical synthesis and material production.

4.3 Waste Treatment and Resource Recovery

Waste treatment and resource recovery are significant issues that contemporary society must address. Electro-fermentation (EF) provides innovative solutions for both cleaning up the environment and efficiently sustainably using waste materials. EF can efficiently tackle pollution problems by utilising the combined effects of microbial metabolism and electrochemical processes, all while extracting valuable resources from different waste streams. Industrial effluents frequently contain an intricate combination of pollutants, such as organic compounds, heavy metals, and other contaminants. If these effluents are released without treatment, they can pose substantial environmental hazards. Using electroactive microorganisms in EF offers an up-and-coming method for treating industrial effluents. These microorganisms can metabolise and break down pollutants into less harmful or inactive substances. Electroactive organisms, such as Geobacter and Shewanella species, can use contaminants as sources of electrons or acceptors of electrons in EF-based bioremediation systems. This enables the organisms to remove or transform the contaminants (Amulya et al., 2024; Bains et al., 2024; Chatterjee & Mazumder, 2024; dos Santos et al., 2023; Knežević et al., 2022; Leininger et al., 2023; Naresh Kumar et al., 2022; Wainaina et al., 2020). Applying a regulated voltage to the bioelectrochemical reactor enables the direct supply of electrons to the microorganisms, thereby increasing their metabolic activity and facilitating the breakdown of stubborn pollutants. Utilising electrochemical methods in bioremediation can expedite the elimination of pollutants from industrial effluents, resulting in enhanced treatment efficiency and reduced costs. In addition, EF can provide benefits compared to traditional bioremediation methods by addressing constraints such as the availability of substrates, microbial activity, and treatment effectiveness. Integrating electrochemical stimulation with microbial metabolism allows for the utilisation of a wider variety of electron donors and acceptors, including those that are not easily broken down without oxygen. In addition, electroactive microorganisms can promote the adhesion of microbes to electrode surfaces, leading to biomass accumulation and facilitating the breakdown of pollutants within biofilms. EF is highly effective in treating complex industrial effluents that contain a wide range of persistent contaminants. Organic waste, such as agricultural residues, food waste, and municipal solid waste, poses a substantial environmental and economic problem because of its large quantity and capacity to cause pollution. EF provides a sustainable method for extracting value from organic waste by transforming it into valuable commodities such as biofuels, chemicals, and bioplastics. This process also helps minimise the negative environmental effects of waste disposal. Electroactive microorganisms in EF-based waste valorisation processes utilise organic waste as a substrate for metabolic activities, resulting in the production of target products through fermentation or electrochemical reactions. Introducing an external voltage to the bioelectrochemical reactor makes it possible to supply electrons to microbial cells directly. This process improves their metabolic efficiency and leads to an increase in the desired product yield. Electrochemically assisted fermentation can greatly enhance the efficiency of converting organic waste into valuable chemicals, leading to optimal resource recovery and minimal waste production. An instance of waste valorisation through the process of enzymatic fermentation (EF) involves the conversion of lignocellulosic biomass into biofuels, specifically ethanol and hydrogen. Electroactive microorganisms can convert sugars obtained from breaking down biomass into biofuels. Their metabolic performance and production yields can be improved by electrochemical stimulation. This comprehensive approach allows using various raw materials, such as agricultural waste and energy crops, to produce biofuels (Acosta-Santoyo et al., 2024; Annie Modestra et al., 2022; Aresta & Dibenedetto, 2024; Awasthi et al., 2022; Ilmasari et al., 2022; Kabir et al., 2024; Lozano & García-Verdugo, 2023; Lu et al., 2024; Serrano-Jiménez et al., 2023b; Zheng et al., 2023). This helps decrease dependence on fossil fuels and reduces greenhouse gas emissions. Furthermore, EF can manufacture bioplastics and biopolymers using organic waste streams. Electroactive microorganisms can convert organic substrates into precursor molecules, which can then be polymerised into biodegradable polymers like polyhydroxyalkanoates (PHAs) and polylactic acid (PLA). Through optimising redox conditions and providing electron donors or acceptors, EF can increase the efficiency of biopolymer production, transforming organic waste into valuable materials for diverse industrial applications. Electro-fermentation presents a potential solution for waste management and extracting valuable resources by integrating microbial metabolism with electrochemical processes. EF facilitates the sustainable utilisation of various raw materials to produce valuable products while reducing the environmental consequences of waste disposal. This includes activities such as bioremediation of industrial effluents and the conversion of organic waste into useful resources. Electric vehicles (EVs) can effectively tackle worldwide issues such as pollution, resource scarcity, and climate change as technological advancements progress. This will aid in the shift towards a more environmentally friendly and efficient economy.

5 Electro-fermentation in the Circular Carbon Economy

5.1 Role in Carbon Capture and Utilisation

Electro-fermentation (EF) is crucial for advancing the circular carbon economy as it allows for the capture and use of carbon dioxide (CO_2), a significant greenhouse gas, to create valuable fuels and chemicals. EF provides cutting-edge solutions for carbon capture and utilisation (CCU) by combining electrochemical processes with microbial metabolism. This helps to decrease greenhouse gas emissions and facilitate the shift towards a more sustainable energy economy. A crucial use of EF in the circular carbon economy involves the transformation of CO_2 into biofuels using microbial electrosynthesis. Electroactive microorganisms, such as acetogenic bacteria and methanogenic archaea, can use CO_2 as a carbon source and obtain electrons from electrodes. This enables them to produce methane, ethanol, and butanol biofuels. Applying a regulated voltage to the bioelectrochemical reactor makes it possible to directly provide electrons to the microorganisms, thereby increasing their metabolic activity and converting CO_2 into biofuels. Methanogenic archaea can convert CO_2 into methane using a sequence of enzymatic reactions, where electrodes act as the source of electrons for the reduction process. The process, called microbial electrosynthesis of methane (MEM), provides a sustainable method for producing methane from CO_2 and renewable electricity. Acetogenic bacteria can transform carbon dioxide (CO_2) into acetate. This acetate can then be converted into biofuels such as ethanol or butanol through microbial fermentation. Electrochemically driven microbial processes facilitate the generation of biofuels from CO_2, thereby closing the carbon loop and diminishing the dependence on fossil fuels. Electrochemical fuel (EF) can be combined with carbon capture technologies to improve the effectiveness of carbon dioxide (CO_2) utilisation and reduce emissions from industrial sources (Ahmed et al., 2023; Amulya et al., 2024; Awasthi et al., 2020; Bains et al., 2024; Gautam et al., 2024; Knežević et al., 2022; Leininger et al., 2023; Naresh Kumar et al., 2022; Virdis et al., 2022; Wainaina et al., 2020). EF can be combined with post-combustion or pre-combustion carbon capture systems to capture and convert CO_2 emissions from power plants, cement kilns, and other industrial sources into valuable products using microbial electrosynthesis. During post-combustion carbon capture, the flue gas, which contains CO_2, is captured using sorbents or membranes. This captured CO_2 is then sent to a bioelectrochemical reactor, where electroactive microorganisms are used to convert the CO_2 into biofuels. Integrating EF (electrochemical fuel) into the current carbon capture infrastructure makes it possible to reduce CO_2 emissions while generating sustainable fuels and chemicals efficiently. This comprehensive strategy provides an economical and enduring solution for eliminating carbon emissions from industrial operations and mitigating the impact of greenhouse gases. Pre-combustion carbon capture technologies, such as steam reforming or gasifying fossil fuels, can also be combined with electrochemical fermentation (EF) to use syngas streams rich in CO_2 for microbial electrosynthesis. EF can mitigate CO_2 emissions from fossil fuel-based industries by converting CO_2 in syngas into biofuels and chemicals. This process helps reduce the environmental impact of these industries and supports the concept of a circular carbon economy. In addition, EF can be integrated with direct air capture (DAC) technologies to extract CO_2 from the atmosphere and transform it into valuable commodities. EF provides a sustainable method for using atmospheric CO_2 to create biofuels and chemicals. This uses renewable electricity sources, like solar or wind power, to fuel the electrochemical conversion process. This method effectively extracts carbon dioxide (CO_2) from the atmosphere while offering a sustainable substitute for fossil fuels, aiding worldwide endeavours to alleviate climate change. Electro-fermentation is essential in the circular carbon economy as it allows for the capture and utilisation of CO_2 to generate biofuels and chemicals. EF provides a sustainable method for transforming CO_2 emissions from industrial sources and the atmosphere into valuable products using microbial electrosynthesis. This process helps to close the carbon cycle and decrease greenhouse gas emissions. Combining EF with carbon capture technologies makes it possible to efficiently reduce CO_2 emissions while simultaneously generating renewable fuels and chemicals, thereby promoting the shift towards a more environmentally friendly and circular economy.

5.2 Enhancing Sustainability and Reducing Carbon Footprint

Electro-fermentation (EF) aims to improve sustainability and decrease the carbon footprint by providing novel approaches to producing fuels and chemicals with a more minor environmental impact than traditional fermentation methods. Two primary strategies for attaining these objectives entail performing life cycle assessments (LCA) of EF processes and contrasting them with conventional fermentation methods. Life cycle assessment (LCA) is a thorough approach to assessing the environmental effects of a product or process from the beginning to the end of its life cycle, encompassing everything from the extraction of raw materials to the disposal of the product at the end of its useful life. Within the environmental footprint analysis (EF) framework, the process of conducting life cycle assessments (LCAs) offers valuable insights into the environmental efficiency of bioelectrochemical systems. It aids in identifying opportunities to decrease resource consumption, minimise emissions, and enhance overall sustainability. EF processes' life cycle assessments (LCAs) typically consider multiple factors, such as energy consumption, greenhouse gas emissions, water usage, and resource depletion. Through quantification, researchers and stakeholders can evaluate EF's environmental advantages and disadvantages compared to traditional fermentation techniques. LCAs have demonstrated that EF can effectively decrease the carbon footprint of biofuel production by utilising renewable electricity sources and converting CO_2 emissions into valuable products. This helps to reduce greenhouse gas emissions and contributes to carbon neutrality. In addition, LCAs can assist in identifying opportunities for process optimisation and technological advancements to enhance the sustainability of EF processes further. By assessing the environmental efficacy of various reactor designs, electrode materials, and operational parameters, scientists can pinpoint possibilities for minimising energy consumption, decreasing resource utilisation, and mitigating environmental repercussions across the entire lifespan of EF systems. Using this iterative method for sustainability assessment allows for ongoing enhancement and optimisation of EF processes, resulting in the development of biotechnological solutions that are both environmentally friendly and economically feasible. Comparative evaluations of EF and conventional fermentation methods offer valuable insights into each approach's environmental advantages and disadvantages. Traditional fermentation methods, such as batch or continuous culture fermentation, have been extensively employed to produce fuels and chemicals using renewable raw materials. Nevertheless, these approaches frequently depend on fossil fuels as their primary energy source and generate substantial waste and emissions, contributing to environmental deterioration and climate change. On the other hand, EF provides numerous environmental benefits compared to traditional fermentation methods. Integrating renewable electricity sources, such as solar or wind power, can significantly reduce greenhouse gas emissions and decrease reliance on fossil fuels. In addition, electrochemical synthesis (EF) allows using waste streams, such as CO_2 emissions and organic waste, as raw materials for producing biofuels and chemicals. This helps to recover resources and reduce waste (Annie Modestra et al., 2022; Arshad et al., 2022; Blank et al., 2020; Cardeña et al., 2024; Jiang et al., 2019; Raj et al., 2024; Rovira-Alsina et al., 2021; Roy Chowdhury et al., 2024; Serrano-Jiménez et al., 2023a; Shehata et al., 2023; Soo et al., 2024; Tharak et al., 2023). In addition, EF can potentially enhance resource utilisation and increase product output compared to traditional fermentation methods. EF improves metabolic pathways and enhances biochemical reactions by directly supplying electrons to microbial cells. This results in increased product yields and decreased resource consumption. This improved efficacy leads to a decrease in production expenses and a reduction in the environmental consequences linked to the extraction and processing of resources. Electro-fermentation is a method that aims to improve sustainability and decrease the carbon footprint by providing innovative solutions for producing fuels and chemicals. This approach offers lower environmental impact compared to traditional fermentation methods. Life cycle assessments (LCA) of EF processes provide valuable insights into their environmental performance and aid in identifying opportunities for optimisation and enhancement. Comparative evaluations of EF and traditional fermentation techniques emphasise the environmental benefits of EF, such as decreased greenhouse gas emissions, improved resource efficiency, and waste reduction. The Environmental Fund (EF) can be pivotal in accelerating the shift towards a more sustainable and circular economy by harnessing these environmental advantages.

5.3 Economic and Environmental Benefits

Electro-fermentation (EF) offers a range of advantages in terms of both the economy and the environment. These benefits include cost analysis, economic viability assessment, evaluation of environmental impact, and the development of strategies to mitigate adverse effects. Comprehending these factors is essential for evaluating EF processes' long-term durability and feasibility. Cost analyses are crucial for assessing the economic viability of EF processes and determining their competitiveness compared to traditional fermentation methods and alternative chemical synthesis pathways. Cost considerations in the field of EF encompass factors such as initial investment, ongoing operational expenses, costs of raw materials, energy usage, the amount of product obtained, and the prevailing market prices for the final goods. Although installing bioelectrochemical reactors and electrode infrastructure for EF may result in higher initial capital costs, it has the potential to generate cost savings in the long run. EF can decrease operational costs by utilising renewable electricity sources, such as solar or wind power, which can be more cost-effective and environmentally friendly than traditional energy sources. In addition, EF allows for the use of waste streams as inputs, which helps decrease the expenses associated with raw materials and improve resource utilisation efficiency. Furthermore, EF can generate income from selling biofuels, chemicals, and other products that have increased value. EF can enhance the economic feasibility of bio-based industries and generate fresh market prospects by extracting valuable compounds from inexpensive raw materials. The economic viability of EF relies on several factors, such as operational efficiency, product output, market need, regulatory incentives, and technological progress. Evaluating the ecological consequences of EF processes is crucial for recognising possible hazards and implementing measures to reduce negative impacts on ecosystems, air quality, water resources, and biodiversity. Environmental impact assessments of EF typically examine factors such as greenhouse gas emissions, water consumption, land utilisation, energy usage, waste production, and ecosystem services. EF provides numerous environmental advantages compared to traditional fermentation methods and chemical synthesis routes. EF can effectively decrease greenhouse gas emissions and address climate change by harnessing renewable electricity sources and capturing CO_2 emissions. In addition, EF allows for the use of waste streams as raw materials, which helps to decrease pollution and encourage the recovery of resources. Furthermore, EF has the potential to enhance resource efficiency and mitigate environmental pollution by minimising the utilisation of fossil fuels, water, and chemical inputs. Several strategies can be employed to reduce the environmental effects of EF processes. These include improving the design of reactors, using better electrode materials, optimising operational parameters, and implementing efficient methods for recovering products. In addition, implementing environmental management systems, such as obtaining ISO 14001 certification, can assist in guaranteeing adherence to environmental regulations and fostering ongoing enhancement in environmental performance. Moreover, it is imperative to involve stakeholders, such as local communities, regulatory agencies, non-governmental organisations, and industry partners, to effectively tackle environmental issues, establish trust, and promote sustainable development. EF projects can improve social acceptance and positively contribute to social and environmental outcomes by promoting transparency, accountability, and stakeholder engagement. Electro-fermentation provides notable economic and environmental advantages, such as cost reduction, revenue generation, decrease in greenhouse gas emissions, waste minimisation, and enhancement of resource efficiency. Stakeholders can assess EF processes' economic viability and environmental sustainability by conducting cost analyses and environmental impact assessments. They can then implement strategies to minimise any adverse effects. Furthermore, involving key individuals and organisations and advocating for openness and clarity is crucial for establishing confidence, nurturing long-lasting progress, and fully harnessing the capabilities of electro-fermentation in tackling worldwide issues about energy, the environment, and the economy (Acosta-Santoyo et al., 2024; Bellini et al., 2022; Javanmard et al., 2024; Leong & Chang, 2023; Serrano-Jiménez et al., 2023b; Sriram et al., 2022; Zheng et al., 2023).

6 Challenges and Future Perspectives

6.1 Technical and Operational Challenges

Addressing technical and operational challenges is essential for realising the full potential of electro-fermentation (EF) and ensuring its commercial viability at scale. Two critical challenges in this regard are scale-up and reactor design optimisation, which require innovative solutions and systematic approaches to overcome. Scaling up EF processes from lab-scale prototypes to commercial production facilities presents significant technical and economic challenges. While lab-scale experiments demonstrate the feasibility of EF for producing biofuels and chemicals, transitioning to large-scale operations requires careful consideration of factors such as reactor design, electrode materials, microbial performance, energy consumption, and cost-effectiveness. One key challenge in scale-up is maintaining the efficiency and stability of electrochemical and microbial processes at larger scales. Mass transfer limitations, electrode fouling, microbial competition, and reactor heterogeneity can affect performance and productivity. Therefore, optimising reactor design and operation parameters is crucial for maximising process efficiency and ensuring consistent product quality. Moreover, the commercial viability of EF depends on its ability to compete with conventional fermentation methods and chemical synthesis routes in terms of cost, product yield, and market demand. While EF offers potential advantages such as renewable feedstocks, reduced greenhouse gas emissions, and value-added products, it also entails higher capital and operational costs, which may limit its competitiveness in specific applications. To address these challenges, collaborative efforts among researchers, engineers, industry partners, and policymakers are needed to develop scalable and cost-effective EF technologies. This includes investing in research and development to improve electrode materials, reactor designs, microbial strains, and process optimisation strategies. Additionally, establishing partnerships with key stakeholders, securing funding support, and navigating regulatory requirements are essential for accelerating the commercialisation of EF technologies. Optimising reactor design is critical for enhancing EF processes' performance, efficiency, and scalability. The design of bioelectrochemical reactors influences factors such as mass transfer, electrode surface area, microbial attachment, and electrochemical activity, all of which affect process performance and product yields. One key consideration in reactor design optimisation is maximizing the surface area and accessibility of electrodes to microbial cells. This can be achieved through various approaches, including using porous electrodes, membrane electrode assemblies, three-dimensional electrode architectures, and biofilm-promoting materials. By increasing the surface area-to-volume ratio and promoting microbial attachment, these design features enhance electron transfer efficiency and biomass retention, leading to improved process performance. Furthermore, optimising reactor configuration and operating conditions is crucial for achieving uniform flow distribution, minimising dead zones, and maximising mass transfer rates (Battista et al., 2022; Dahiya et al., 2023; Deng et al., 2023; Kumar et al., 2023; Liu et al., 2023; Ngoc-Dan Cao et al., 2022; Virdis et al., 2022). This may involve adopting continuous-flow or fed-batch operation modes, optimising flow rates, pH control, temperature management, and substrate-feeding strategies. Process efficiency, substrate-feeding activity, and product quality can be maximised by optimising reactor design and operation while minimising energy consumption and environmental impact. Integrating monitoring and control systems into bioelectrochemical reactors enables real-time feedback and optimisation. This may include online sensors for monitoring key parameters such as pH, temperature, dissolved oxygen, and microbial activity and automated control algorithms for adjusting operating conditions in response to changing process dynamics. By implementing advanced monitoring and control systems, operators can optimise reactor performance, maximise resource utilisation, and minimise operational costs. Addressing technical and operational challenges is essential for realising the commercial potential of electro-fermentation. This includes scaling up EF processes to commercial production levels, optimising reactor design for improved performance and efficiency, and integrating monitoring and control systems for process optimisation. Collaborative efforts among researchers, industry partners, and policymakers are needed to overcome these challenges and accelerate the adoption of EF technologies for sustainable biofuel and chemical production.

6.2 *Regulatory and Policy Considerations*

Regulatory and policy factors significantly influence the development, deployment, and adoption of electro-fermentation (EF) technologies. As the implementation of EF transitions from research laboratories to commercial applications, it becomes increasingly crucial for public and private stakeholders to ensure compliance with standards and regulations and access incentives and funding opportunities. Establishing and enforcing standards and regulations for bioelectrochemical systems is essential to guarantee safety, safeguard the environment, and gain public approval. Although EF is a recent technology, it is subject to existing regulations that govern biotechnology, waste management, water quality, and energy production. For instance, bioelectrochemical reactors utilised in EF applications must adhere to safety standards, electrical codes, and environmental regulations that dictate their design, construction, operation, and maintenance. In addition, EF processes may necessitate permits or authorisations from regulatory agencies responsible for environmental conservation, waste administration, water purity, and biosecurity. In addition, using genetically modified organisms (GMOs) in EF could result in implementing regulations and supervision by government agencies accountable for biotechnology, biosafety, and public health. It is crucial to adhere to rules that govern GMOs, which include evaluating risks, controlling the spread of GMOs, providing clear information on labels, and monitoring compliance. This is necessary to reduce potential hazards and gain public approval of EF technologies. As EF progresses and becomes more commercialised, stakeholders must collaborate with regulatory agencies, standard-setting organisations, industry associations, and other relevant parties to establish agreed-upon standards and regulations that prioritise safety, environmental protection, and innovation while reducing regulatory obstacles and uncertainties. Accessing incentives and funding opportunities can substantially affect EF technologies' development, deployment, and adoption by offering financial assistance, technical support, and market incentives. Various entities such as governments, international organisations, private foundations, and industry consortia may provide different incentives and funding mechanisms to facilitate the research, development, demonstration, and implementation of EF technologies. Government incentives, such as grants, tax credits, loans, and subsidies, can assist in reducing the initial expenses of implementing EF systems, encouraging innovation, and boosting market demand. Public–private partnerships, technology incubators, and industry consortia can offer technical support, networking opportunities, and access to expertise and resources to facilitate the development and commercialisation of EF (Acosta-Santoyo et al., 2024; Aresta & Dibenedetto, 2024; Awasthi et al., 2022; Bellini et al., 2022; Ilmasari et al., 2022; Leong & Chang, 2023; Sana et al., 2024; Serrano-Jiménez et al., 2023b; Sriram et al., 2022; Zheng et al., 2023). In addition, financial support from research funding agencies, venture capitalists, angel investors, and corporate partners can be crucial for funding EF research, pilot projects, and commercial ventures. By utilising both public and private funding sources, stakeholders can expedite the advancement and implementation of EF technologies, surmount obstacles related to technology and finance, and take advantage of potential market prospects. Furthermore, policy tools such as requirements for renewable energy, mechanisms for pricing carbon, systems for trading emissions, and targets for blending biofuels can generate market demand and motivations for environmentally friendly technologies by incorporating the costs of environmental damage, encouraging the use of renewable energy, and decreasing the release of greenhouse gases. Regulatory and policy factors influence electro-fermentation technologies' development, deployment, and adoption. Standards and regulations that govern bioelectrochemical systems, genetically modified organisms, and waste management are crucial for ensuring safety, safeguarding the environment, and gaining public acceptance. Obtaining incentives and funding from government agencies, international organisations, private foundations, and industry consortia can offer crucial financial support and market incentives for developing, developing, demonstrating, and deploying EF technologies. Stakeholders can effectively address regulatory and policy challenges, secure financial support, and expedite the implementation of EF technologies for sustainable biofuel and chemical production by collaborating with regulators, policymakers, industry stakeholders, and funding agencies.

6.3 Future Research Directions

Future electro-fermentation (EF) research aims to stimulate innovation and fully exploit the potential of microbial electrosynthesis for sustainable biofuel and chemical production. Primary areas of emphasis encompass progress in microbial electrosynthesis and the investigation of novel microbial strains and consortia. Advancements in microbial electrosynthesis hold the potential to enhance the efficiency, productivity, and versatility of electrochemical fermentation processes. Microbial electrosynthesis depends on electroactive microorganisms that can effectively capture electrons from electrodes and direct them into metabolic pathways to generate desired products. An area of research concentration involves the manipulation of electroactive microorganisms to augment their ability to transfer electrons and improve their metabolic pathways. Genetic engineering and synthetic biology methods can alter microbial strains to enhance their ability to take in electrons, utilise substrates, and produce desired products. Researchers can enhance the performance and productivity of microbial cells for specific EF applications by optimising the expression of crucial enzymes and transporters involved in electron transfer and metabolic pathways. Furthermore, electrode materials and reactor design improvements can significantly enhance microbial electrosynthesis. Advancements in electrode materials, such as nanostructured surfaces, conductive polymers, and catalyst coatings, can enhance the rate at which electrons are transferred, increase the available surface area of the electrode, and improve the compatibility of the electrode with microbial cells. Similarly, innovative reactor designs, such as flow-through systems, membrane-integrated reactors, and immobilised cell reactors, can optimise mass transfer, reduce electrode fouling, and enhance process efficiency. Moreover, comprehending the basic mechanisms of microbial electrosynthesis is crucial for developing and enhancing electrochemical fermentation processes. Research endeavours centred on understanding the kinetics, thermodynamics, and regulatory mechanisms of electron transfer and metabolic pathways can offer valuable insights into manipulating and regulating microbial metabolism to enhance product yields and selectivity. Investigating novel microbial strains and consortia shows potential for broadening the products generated through EF and enhancing their production in various environmental conditions. Although there has been significant research on electroactive microorganisms like acetogens and methanogens, known for their ability to produce biofuels and chemicals, many unexplored microbial species still have the potential for applications in electrochemical fermentation. One area of research involves investigating new electroactive microorganisms found in various environmental habitats, including deep-sea sediments, hot springs, and extreme environments. Metagenomic and metatranscriptomic methods can identify and describe microbial communities with distinct metabolic abilities, such as new ways of transferring electrons, utilising different substances, and synthesising various products. Moreover, manipulating microbial communities shows potential for improving the efficiency and resilience of EF processes. Researchers can create synthetic ecosystems by combining microbial consortia with complementary metabolic activities. These ecosystems are capable of efficiently converting various feedstocks into valuable products. Synthetic ecology methods, such as community modelling, consortia design, and engineering interspecies interactions, can enhance the performance of microbial consortia for targeted EF applications and specific environmental conditions. Furthermore, investigating the collaborative relationships between electroactive microorganisms and non-electroactive counterparts, such as syntrophic bacteria, archaea, and fungi, has the potential to broaden the variety of products generated through electrochemical fermentation (EF). By leveraging the metabolic diversity and cooperative interactions between various microbial species, researchers can create cutting-edge biotechnological solutions for the sustainable production of biofuels and chemicals. Future research in electro-fermentation aims to enhance microbial electrosynthesis and investigate novel microbial strains and consortia for the sustainable production of biofuels and chemicals. The progress in microbial electrosynthesis, which encompasses genetic engineering, electrode materials, and reactor designs, can potentially improve efficiency, productivity, and adaptability. Similarly, the investigation of new electroactive microorganisms and microbial consortia shows potential for broadening the variety of products generated through electrochemical fermentation (EF) and enhancing their production in different environmental conditions. By tackling these research obstacles, scientists and engineers can fully exploit the capabilities of electro-fermentation and significantly contribute to the shift towards a more environmentally friendly and cyclical economy.

7 Conclusion

Upon reviewing the main aspects, electro-fermentation emerges as an up-and-coming technology potentially transforming the circular carbon economy. Electro-fermentation is a process that combines microbial fermentation with electrochemistry, providing a new and innovative method for bioconversion processes. This collaboration brings various benefits, such as increased product output, decreased energy usage, and reduced greenhouse gas emissions. Electro-fermentation utilises electricity to drive fermentation reactions, offering a sustainable approach to biomanufacturing. Furthermore, this technology showcases its adaptability, with a wide range of applications from biofuel production to pharmaceuticals. Therefore, it signifies a crucial advancement in pursuing sustainable resolutions to worldwide difficulties. The transformative impact of electro-fermentation on the circular carbon economy is significant. This approach optimises resource utilisation and reduces environmental impact by combining biological processes with electrochemical principles. One of the main benefits of this technology is its ability to use renewable sources of electricity, such as solar or wind power, to power fermentation reactions. This process transforms carbon dioxide into valuable commodities, effectively completing the carbon cycle and decreasing dependence on fossil fuels. Moreover, electro-fermentation provides a means to convert waste streams into valuable commodities, thus increasing their worth. This technology can open up new opportunities for sustainable production in various industries by combining microbial metabolism with electrocatalysts. The potential uses are extensive, ranging from carbon-neutral fuels to biodegradable plastics, indicating a significant change towards a circular economy powered by renewable resources. To fully achieve the potential of electro-fermentation, it is crucial to concentrate on research and development. Investing in interdisciplinary collaboration is essential for understanding the underlying mechanisms and improving process efficiency. Furthermore, it is imperative to conduct scale-up studies to verify the practicality of electro-fermentation on a sizeable industrial level. Overcoming technical challenges and promoting technology adoption necessitates collaboration among academics, industry professionals, and policymakers. Furthermore, it is necessary to modify regulatory frameworks to accommodate the distinctive features of electro-fermentation and encourage its adoption. We can expedite the shift towards a durable circular carbon economy by cultivating a network of inventive ideas and assistance. Let us use electro-fermentation to redefine the future of biomanufacturing and contribute to creating a more environmentally friendly and robust world.

References

Acosta-Santoyo, G., Treviño-Reséndez, J., Robles, I., Godínez, L. A., & García-Espinoza, J. D. (2024). A review on recent environmental electrochemistry approaches for the consolidation of a circular economy model. *Chemosphere, 346*, 140573. https://doi.org/10.1016/j.chemosphere.2023.140573

Ahmed, S. F., Kabir, M., Mehjabin, A., Oishi, F. T. Z., Ahmed, S., Mannan, S., Mofijur, M., Almomani, F., Badruddin, I. A., & Kamangar, S. (2023). Waste biorefinery to produce renewable energy: Bioconversion process and circular bioeconomy. *Energy Reports, 10*, 3073–3091. https://doi.org/10.1016/j.egyr.2023.09.137

Amulya, K., Dahiya, S., & Venkata Mohan, S. (2024). *Chapter 22—Building circular bio-based economy through sustainable waste management* (M. N. V. B. T.-B. and B. (Second E. Prasad (ed.); pp. 639–666). Elsevier. https://doi.org/10.1016/B978-0-443-16120-9.00023-6

Annie Modestra, J., Matsakas, L., Rova, U., & Christakopoulos, P. (2022). Prospects and trends in bioelectrochemical systems: Transitioning from CO_2 towards a low-carbon circular bioeconomy. *Bioresource Technology, 364*, 128040. https://doi.org/10.1016/j.biortech.2022.128040

Aresta, M., & Dibenedetto, A. (2024). Merging the green-H_2 production with carbon recycling for stepping towards the carbon cyclic economy. *Journal of CO_2 Utilization, 80*, 102688. https://doi.org/10.1016/j.jcou.2024.102688

Arshad, R. N., Abdul-Malek, Z., Roobab, U., Ranjha, M. M. A. N., Režek Jambrak, A., Qureshi, M. I., Khan, N., Manuel Lorenzo, J., & Aadil, R. M. (2022). Nonthermal food processing: A step towards a circular economy to meet the sustainable development goals. *Food Chemistry: X, 16*, 100516. https://doi.org/10.1016/j.fochx.2022.100516

Awasthi, M. K., Sarsaiya, S., Patel, A., Juneja, A., Singh, R. P., Yan, B., Awasthi, S. K., Jain, A., Liu, T., Duan, Y., Pandey, A., Zhang, Z., & Taherzadeh, M. J. (2020). Refining biomass residues for sustainable energy and bio-products: An assessment of technology, its importance, and strategic applications in circular bio-economy. *Renewable and Sustainable Energy Reviews, 127*, 109876. https://doi.org/10.1016/j.rser.2020.109876

Awasthi, M. K., Singh, E., Binod, P., Sindhu, R., Sarsaiya, S., Kumar, A., Chen, H., Duan, Y., Pandey, A., Kumar, S., Taherzadeh, M. J., Li, J., & Zhang, Z. (2022). Biotechnological strategies for bio-transforming biosolid into resources toward circular bio-economy: A review. *Renewable and Sustainable Energy Reviews, 156*, 111987. https://doi.org/10.1016/j.rser.2021.111987

Bains, A., Sridhar, K., Dhull, S. B., Chawla, P., Sharma, M., Sarangi, P. K., & Gupta, V. K. (2024). Circular bioeconomy in carbon footprint components of nonthermal processing technologies towards sustainable food system: A review. *Trends in Food Science & Technology, 149*, 104520. https://doi.org/10.1016/j.tifs.2024.104520

Battista, F., Strazzera, G., Valentino, F., Gottardo, M., Villano, M., Matos, M., Silva, F., Reis, M. A., Mata-Alvarez, J., Astals, S., Dosta, J., Jones, R. J., Massanet-Nicolau, J., Guwy, A., Pavan, P., Bolzonella, D., & Majone, M. (2022). New insights in food waste, sewage sludge and green waste anaerobic fermentation for short-chain volatile fatty acids production: A review. *Journal of Environmental Chemical Engineering, 10*(5), 108319. https://doi.org/10.1016/j.jece.2022.108319

Bellini, S., Tommasi, T., & Fino, D. (2022). Poly(3-hydroxybutyrate) biosynthesis by Cupriavidus necator: A review on waste substrates utilization for a circular economy approach. *Bioresource Technology Reports, 17*, 100985. https://doi.org/10.1016/j.biteb.2022.100985

Blank, L. M., Narancic, T., Mampel, J., Tiso, T., & O'Connor, K. (2020). Biotechnological upcycling of plastic waste and other non-conventional feedstocks in a circular economy. *Current Opinion in Biotechnology, 62*, 212–219. https://doi.org/10.1016/j.copbio.2019.11.011

Cardeña, R., Valencia-Ojeda, C., Chazaro-Ruiz, L. F., & Razo-Flores, E. (2024). Regulation of the dark fermentation products by electro-fermentation in reactors without membrane. *International Journal of Hydrogen Energy, 49*, 107–116. https://doi.org/10.1016/j.ijhydene.2023.06.253

Chatterjee, B., & Mazumder, D. (2024). A critical review of the advances in valorizing agro-industrial wastes through mixed culture fermentation. *Journal of Environmental Chemical Engineering, 12*(1), 111838. https://doi.org/10.1016/j.jece.2023.111838

Dahiya, S., Lingam, Y., & Venkata Mohan, S. (2023). Understanding acidogenesis towards green hydrogen and volatile fatty acid production—Critical analysis and circular economy perspective. *Chemical Engineering Journal, 464*, 141550. https://doi.org/10.1016/j.cej.2023.141550

Dahiya, S., Chatterjee, S., Sarkar, O., & Mohan, S. V. (2021). Renewable hydrogen production by dark-fermentation: Current status, challenges and perspectives. *Bioresource Technology, 321*, Article 124354.

Deng, S., Ngo, H. H., Guo, W., You, N., & Peng, S. (2023). *Chapter 15—Biological nutrient recovery from wastewater for circular economy* (G. Mannina, A. Pandey, & R. B. T.-C. D. in B. and B. Sirohi (Eds.), pp. 355–412). Elsevier. https://doi.org/10.1016/B978-0-323-99920-5.00010-X

dos Santos, E. V., Martínez-Huitle, C. A., & Rodrigo, M. A. (2023). The electro-refinery in organics: A new arising concept for valorization of wastes. *Current Opinion in Electrochemistry, 39*, 101267. https://doi.org/10.1016/j.coelec.2023.101267

Gautam, R., Ress, N. V, Wilckens, R. S., & Ghosh, U. K. (2024). Hydrogen production in microbial electrolysis cell and reactor digestate valorization for biochar—A noble attempt towards circular economy. *International Journal of Hydrogen Energy, 52*, 649–668. https://doi.org/10.1016/j.ijhydene.2023.07.190

Hussein, S., Shehata, N., Al-wahbi, A. A., & Abdelkareem, M. A. (2024). *Biosorbent; definitions, types, application and circular economy*. Elsevier. https://doi.org/10.1016/B978-0-443-15738-7.00020-9

Ilmasari, D., Kamyab, H., Yuzir, A., Riyadi, F. A., Khademi, T., Al-Qaim, F. F., Kirpichnikova, I., & Krishnan, S. (2022). A review of the biological treatment of leachate: Available technologies and future requirements for the circular economy implementation. *Biochemical Engineering Journal, 187*, 108605. https://doi.org/10.1016/j.bej.2022.108605

Javanmard, A., Wan Daud, W. M. A., Patah, M. F. A., Zuki, F. M., Ai, S. P., Azman, D. Q., & Chen, W.-H. (2024). Breaking barriers for a green future: A comprehensive study on pre-treatment techniques for empty fruit bunches in the bio-based economy. *Process Safety and Environmental Protection, 182*, 535–558. https://doi.org/10.1016/j.psep.2023.11.053

Jiang, Y., May, H. D., Lu, L., Liang, P., Huang, X., & Ren, Z. J. (2019). Carbon dioxide and organic waste valorization by microbial electrosynthesis and electro-fermentation. *Water Research, 149*, 42–55. https://doi.org/10.1016/j.watres.2018.10.092

Kabir, M. M., Sabur, G. M., Akter, M. M., Nam, S. Y., Im, K. S., Tijing, L., & Shon, H. K. (2024). Electrodialysis desalination, resource and energy recovery from water industries for a circular economy. *Desalination, 569*, 117041. https://doi.org/10.1016/j.desal.2023.117041

Knežević, K., Rastädter, K., Quehenberger, J., Spadiut, O., Krampe, J., & Kreuzinger, N. (2022). Circular production—Evaluation of membrane technologies for nutrient recycling from a microbial fermentation effluent. *Journal of Cleaner Production, 377*, 134436. https://doi.org/10.1016/j.jclepro.2022.134436

Kumar, V., Vangnai, A. S., Sharma, N., Kaur, K., Chakraborty, P., Umesh, M., Singhal, B., Utreja, D., Carrasco, E. U., Andler, R., Awasthi, M. K., & Taherzadeh, M. J. (2023). Bioengineering of biowaste to recover bioproducts and bioenergy: A circular economy approach towards sustainable zero-waste environment. *Chemosphere, 319*, 138005. https://doi.org/10.1016/j.chemosphere.2023.138005

Kumareswaran, K., Ranasinghe, S., Jayasinghe, G. Y., & Dassanayake, K. B. (2024). Systematic review on liquid organic waste (LOW) characteristics, processing technologies, and their potential applications: Towards circular economy and resource efficiency. *Journal of Cleaner Production, 447*, 141286. https://doi.org/10.1016/j.jclepro.2024.141286

Leininger, A., Chen, J., Ramaswami, A., & Ren, Z. J. (2023). Urban circular carbon economy through electrochemically influenced microbiomes. *One Earth, 6*(3), 278–289. https://doi.org/10.1016/j.oneear.2023.02.011

Leong, Y. K., & Chang, J.-S. (2023). Lutein biosynthesis from microalgae—Recent advances and circular economy. *Environmental Technology & Innovation, 30*, 103097. https://doi.org/10.1016/j.eti.2023.103097

Liu, Z., Yin, X., Ni, B., Chen, X., Xie, F., Guo, Z., Li, D., Liu, W., Yue, X., & Zhou, A. (2023). Synchronous vivianite and hydrogen recovery from waste activated sludge fermentation liquid via electro-fermentation mediated by iron anode. *Chemical Engineering Journal, 474*, 145442. https://doi.org/10.1016/j.cej.2023.145442

Lozano, P., & García-Verdugo, E. (2023). From green to circular chemistry paved by biocatalysis. *Green Chemistry, 25*(18), 7041–7057. https://doi.org/10.1039/d3gc01878d

Lu, S., Leininger, A., Jassby, D., Hoek, E. M. V, May, H. D., & Ren, Z. J. (2024). Temperature dependence of acids recovery from wastewater electro-fermentation. *Resources, Conservation and Recycling, 203*, 107407. https://doi.org/10.1016/j.resconrec.2024.107407

Naresh Kumar, A., Sarkar, O., Chandrasekhar, K., Raj, T., Narisetty, V., Mohan, S. V., Pandey, A., Varjani, S., Kumar, S., Sharma, P., Jeon, B.-H., Jang, M., & Kim, S.-H. (2022). Upgrading the value of anaerobic fermentation via renewable chemicals production: A sustainable integration for circular bioeconomy. *Science of The Total Environment, 806*, 150312. https://doi.org/10.1016/j.scitotenv.2021.150312

Ngoc-Dan Cao, T., Mukhtar, H., Yu, C.-P., Bui, X.-T., & Pan, S.-Y. (2022). Agricultural waste-derived biochar in microbial fuel cells towards a carbon-negative circular economy. *Renewable and Sustainable Energy Reviews, 170*, 112965. https://doi.org/10.1016/j.rser.2022.112965

Raj, T., Verma, S., Kumar, N., & Agrawal, R. (2024). Chapter 11—Role of woody biomass in carbon capture, circular bioeconomy, and biomanufacturing. In D. Kumar, S. Kumar, K. Rajendran, & R. C. B. T.-S. B. of W. B. to B. and B. Ray (Eds.), *Applied biotechnology reviews* (pp. 291–318). Woodhead Publishing. https://doi.org/10.1016/B978-0-323-91187-0.00002-3

Rovira-Alsina, L., Balaguer, M. D., & Puig, S. (2021). Thermophilic bio-electro carbon dioxide recycling harnessing renewable energy surplus. *Bioresource Technology, 321*, 124423. https://doi.org/10.1016/j.biortech.2020.124423

Roy Chowdhury, P., Medhi, H., Bhattacharyya, K. G., & Mustansar Hussain, C. (2024). Layered double hydroxides derived from waste for highly efficient electrocatalytic water splitting: Challenges and implications towards circular economy driven green energy. *Coordination Chemistry Reviews, 501*, 215547. https://doi.org/10.1016/j.ccr.2023.215547

Sana, A., Mahmood, A., & Ibrahim, M. (2024). 17—Recovery of resources and circular economy from biomass-derived waste through aerobic and anaerobic digestion-based technique. In K. Rizwan & M. B. T.-N. in B. C. Bilal (Eds.), *Woodhead series in bioenergy* (pp. 411–463). Woodhead Publishing. https://doi.org/10.1016/B978-0-443-13500-2.00014-6

Serrano-Jiménez, J., de Lucas-Consuegra, A., Sánchez, P., Romero, A., & de la Osa, A. R. (2023b). Electrochemical reforming of a fusel oil stream from the winery industry: New insights for a circular economy based on renewable hydrogen. *Fuel, 350*, 128728. https://doi.org/10.1016/j.fuel.2023.128728

Serrano-Jiménez, J., de la Osa, A. R., Rodríguez-Gómez, A., Sánchez, P., Romero, A., & de Lucas-Consuegra, A. (2023a). Electro-reforming of bioethanol produced by sugar fermentation on a Pt-Ni anodic catalyst supported on graphene nanoplatelets. *Journal of Environmental Chemical Engineering, 11*(3), 109703. https://doi.org/10.1016/j.jece.2023.109703

Shehata, N., Egirani, D., Olabi, A. G., Inayat, A., Abdelkareem, M. A., Chae, K.-J., & Sayed, E. T. (2023). Membrane-based water and wastewater treatment technologies: Issues, current trends, challenges, and role in achieving sustainable development goals, and circular economy. *Chemosphere, 320*, 137993. https://doi.org/10.1016/j.chemosphere.2023.137993

Soo, A., Kim, J., & Shon, H. K. (2024). Technologies for the wastewater circular economy—A review. *Desalination and Water Treatment, 317*, 100205. https://doi.org/10.1016/j.dwt.2024.100205

Sriram, S., Wong, J. W. C., & Pradhan, N. (2022). Recent advances in electro-fermentation technology: A novel approach towards balanced fermentation. *Bioresource Technology, 360*, 127637. https://doi.org/10.1016/j.biortech.2022.127637

Tharak, A., Katakojwala, R., Kajla, S., & Venkata Mohan, S. (2023). Chemolithoautotrophic reduction of CO_2 to acetic acid in gas and gas-electro fermentation systems: Enrichment, microbial dynamics, and sustainability assessment. *Chemical Engineering Journal, 454*, 140200. https://doi.org/10.1016/j.cej.2022.140200

Vassilev, I., Averesch, N. J. H., Ledezma, P., & Kokko, M. (2021). Anodic electro-fermentation: Empowering anaerobic production processes via anodic respiration. *Biotechnology Advances, 48*, 107728. https://doi.org/10.1016/j.biotechadv.2021.107728

Virdis, B., Hoelzle, R. D., Marchetti, A., Boto, S. T., Rosenbaum, M. A., Blasco-Gómez, R., Puig, S., Freguia, S., & Villano, M. (2022). Electro-fermentation: Sustainable bioproductions steered by electricity. *Biotechnology Advances, 59*, 107950. https://doi.org/10.1016/j.biotechadv.2022.107950

Wainaina, S., Awasthi, M. K., Sarsaiya, S., Chen, H., Singh, E., Kumar, A., Ravindran, B., Awasthi, S. K., Liu, T., Duan, Y., Kumar, S., Zhang, Z., & Taherzadeh, M. J. (2020). Resource recovery and circular economy from organic solid waste using aerobic and anaerobic digestion technologies. *Bioresource Technology, 301*, 122778. https://doi.org/10.1016/j.biortech.2020.122778

Wood, J. C., Grové, J., Marcellin, E., Heffernan, J. K., Hu, S., Yuan, Z., & Virdis, B. (2021). Strategies to improve viability of a circular carbon bioeconomy—A techno-economic review of microbial electrosynthesis and gas fermentation. *Water Research, 201*, 117306. https://doi.org/10.1016/j.watres.2021.117306

Zhao, Z.-T., Ding, J., Wang, B.-Y., Bao, M.-Y., Liu, B.-F., Pang, J.-W., Ren, N.-Q., & Yang, S.-S. (2024). Advances in the biomass valorization in dark fermentation systems: A sustainable approach for biohydrogen production. *Chemical Engineering Journal, 481*, 148444. https://doi.org/10.1016/j.cej.2023.148444

Zheng, T., Bian, C., Xiao, B., Chen, X., Wang, J., & Li, L. (2023). Performance enhancement of integrating microbial electrolysis cell on two-stage anaerobic digestion of food waste: Electro-methanogenic stage versus electro-two stages. *Bioresource Technology, 386*, 129562. https://doi.org/10.1016/j.biortech.2023.129562

The manufacturer's authorised representative in the EU is Springer Nature Customer Service Centre GmbH, Europaplatz 3, 69115 Heidelberg, Germany. If you have any concerns regarding our products, please contact ProductSafety@springernature.com

Printed and bound by CPI Group (UK) Ltd, Croydon, CR0 4YY
01/10/2026
02225921-0016